SEMI-RIEMANNIAN GEOMETRY

WITH APPLICATIONS TO RELATIVITY

This is a volume in
PURE AND APPLIED MATHEMATICS
A Series of Monographs and Textbooks

SEMI-RIEMANNIAN GEOMETRY

WITH APPLICATIONS TO RELATIVITY

BARRETT O'NEILL

Department of Mathematics
University of California
Los Angeles, California

Academic Press
An Imprint of Elsevier
San Diego New York Boston
London Sydney Tokyo Toronto

Academic Press
An Imprint of Elsevier
525 B Street, Suite 1900, San Diego, California 92101-4495, USA
http://www.academicpress.com

Academic Press
32 Jamestown Road, London NW1 7BY, UK
http://www.academicpress.com

Library of Congress Cataloging-in-Publication

O'Neill, Barrett.
Semi-Riemannian geometry.

(Pure and Applied Mathematics ;)
Bibliography: p.
ISBN-13: 978-0-12-526740-3 ISBN-10: 0-12-526740-1
1. Geometry, Riemannian. 2. Manifolds (Mathematics) 3. Calculus of tensors.
4. Reletivity (Physics) I. Title. II. Series:
Pure and applied mathematics (Academic Press) ;
QA3.P8 [QA649] 510s [516.3'73] 82-13917

ISBN-13: 978-0-12-526740-3
ISBN-10: 0-12-526740-1

Transferred to Digital Printing 2010

CONTENTS

6. SPECIAL RELATIVITY

7. CONSTRUCTIONS

8. SYMMETRY AND CONSTANT CURVATURE

9. ISOMETRIES

13. SCHWARZSCHILD GEOMETRY

14. CAUSALITY IN LORENTZ MANIFOLDS

APPENDIX A. FUNDAMENTAL GROUPS AND COVERING MANIFOLDS 441

APPENDIX B. LIE GROUPS

APPENDIX C. NEWTONIAN GRAVITATION 453

PREFACE

This book is an exposition of *semi-Riemannian geometry* (also called *pseudo-Riemannian geometry*)—the study of a smooth manifold furnished with a metric tensor of arbitrary signature. The principal special cases are Riemannian geometry, where the metric is positive definite, and Lorentz geometry. For many years these two geometries have developed almost independently: Riemannian geometry reformulated in coordinate-free fashion and directed toward global problems, Lorentz geometry in classical tensor notation devoted to general relativity. More recently, this divergence has been reversed as physicists, turning increasingly toward invariant methods, have produced results of compelling mathematical interest.

After establishing the requisite language of manifolds and tensors (Chapters 1 and 2), the plan of the book is to develop the foundations of semi-Riemannian geometry in the simplest way and without regard to signature, allowing the Riemannian and Lorentz cases to appear as needed (Chapters 3–5 and 7). Then in the latter half of the book two threads are followed. One uses the notion of isometry to develop algebraic aspects of semi-Riemannian geometry: manifolds of constant curvature, symmetric spaces, and homogeneous spaces (Chapters 8, 9, and 11); the introductions to these chapters will give a more detailed description of their contents. The other thread applies Lorentz geometry to special and general relativity (Chapters 6,12, and 13). The fact that relativity theory is expressed in terms of Lorentz geometry is lucky for geometers, who can thus penetrate surprisingly quickly into cosmology (redshift, expanding universe, and big bang) and, a topic no less interesting geometrically, the gravitation of a single star (perihelion precession, bending of light, and black holes). The tendency of the spacetimes in Chapters 12 and 13 to have singularities (big bang and black holes) is accounted for in abstract Lorentz terms by two theorems, due respectively to S. W. Hawking and R. Penrose; these are the goals of Chapter 14.

The general approach of the book is coordinate-free; however, coordinates are not neglected. Typically, geometric objects are defined invariantly and then described in terms of coordinates. In particular, the definition of a tensor I have adopted converts almost automatically into the classical coordinate formulation. A number of key proofs are given in classical notation. This attitude is only reasonable in view of the vast literature in each style.

The basic prerequisites for the book are modest: a good working knowledge of multivariable differential calculus, a firm belief in the existence and uniqueness theorems of ordinary differential equations, and an acquaintance with the fundamentals of point set topology and algebra. Later on, a knowledge of fundamental groups, covering spaces, and Lie groups is required; the necessary background in these topics is outlined briefly in Appendixes A and B. A college course in physics (particularly Newtonian mechanics) is required, not to read this book, but to appreciate the transformation and unification of Newtonian concepts effected by Einstein's relativistic geometry and the remarkable way the old and new theories—so different at base—reach approximate agreement on, say, the running of the solar system (Appendix C versus Chapter 13).

In the early chapters (1–5 and 7) the logical ordering is fairly strict. Thereafter the two branches— 8,9,11 and 6,12,13— are almost independent. (Chapters 12 and 13 require only an occasional reference to Chapters 9 and perhaps 8.) Chapter 10 is used in Chapters 11 and 14. Otherwise Chapter 14, though strongly motivated by Chapters 12 and 13, depends logically on only the early chapters.

Following each chapter are a number of exercises; these are meant to be workable without undue strain. In each chapter a single sequence of numbers designates collectively the theorems, lemmas, examples, and so on. For instance, Lemma 5.12 is the twelfth designated item in Chapter 5, not the twelfth lemma. Within a given chapter, the chapter number is omitted. Initials in square brackets, e.g., [SW], direct the reader to the References.

It is a pleasure to express my gratitude to the authors of the following brilliant and very different books: S. W. Hawking and G. F. R. Ellis, *The Large Scale Structure of Space-time*; C. W. Misner, K. S. Thorne, and J. A. Wheeler, *Gravitation*; R. K. Sachs and H. Wu, *General Relativity for Mathematicians*.

NOTATION AND TERMINOLOGY

The following notations are among the most frequently used throughout the book:

M, N	manifolds	p, q	points
f, g, h	real-valued functions	α, β, γ	curves
v, w	vectors	V, W, X, Y	vector fields
ϕ, ψ	mappings	$\mathscr{U}, \mathscr{V}$	open sets

$\xi = (x^1, \ldots, x^n)$ coordinate system

$\boldsymbol{R}$ is the real number field, I denotes an open interval in $\boldsymbol{R}$, and, for example, $[a, b) = \{r \in \boldsymbol{R}: a \leq r < b\}$. The identity map is id; $\phi \circ \psi$ is the composite mapping that sends p to $\phi(\psi p)$. See Appendix B for Lie group notation such as $GL(n, \boldsymbol{R})$.

A mapping $\phi: M \rightarrow N$ is *one-to-one* (injective) if $p \neq q$ implies $\phi p \neq \phi q$. The *image* of ϕ is $\{\phi p : p \in M\} \subset N$, and ϕ is *onto* (surjective) if image $\phi = N$. (Inclusion $B \subset N$ does not exclude equality $B = N$.) If $B \subset N$ then $\phi^{-1}(B) = \{p \in M : \phi p \in B\}$, and when ϕ is one-to-one and onto, ϕ^{-1} also denotes the inverse mapping of ϕ.

If $\pi \circ \tilde{\phi} = \phi$, then $\tilde{\phi}$ is called a *lift* of ϕ through π. A lift of the identity map is called a *cross section* (or merely a *section*).

A *linear isomorphism* of vector spaces is a linear transformation that is one-to-one and onto, hence is *invertible*.

A subset A of a topological space has closure $\bar{A}$, interior int A, and boundary bd A.

1 MANIFOLD THEORY

Generally speaking, a manifold is a topological space that locally resembles Euclidean space. A smooth manifold is a manifold M for which this resemblance is sharp enough to permit the establishment of partial differentiation—in fact, all the essential features of calculus—on M. Smooth manifolds are thus the natural setting for "calculus in the large."

SMOOTH MANIFOLDS

Euclidean n-space $\boldsymbol{R}^n$ is the set of all n-tuples $p = (p_1, \ldots, p_n)$ of real numbers. We assume in particular a familiarity with its structure as a vector space and as a topological space. The natural inner product of $\boldsymbol{R}^n$ is the *dot product* $p \cdot q = \sum p_i q_i$, with *norm* $|p| = \sqrt{p \cdot p}$. The resulting *metric* $d(p, q) = |p - q|$ is compatible with the topology of $\boldsymbol{R}^n$.

A real-valued function f defined on an open set $\mathscr{U}$ of $\boldsymbol{R}^n$ is *smooth* (or equivalently, C^∞) provided all mixed partial derivatives of f—of all orders—exist and are continuous at every point of $\mathscr{U}$.

For $1 \leq i \leq n$, let $u^i: \boldsymbol{R}^n \to \boldsymbol{R}$ be the function that sends each point $p = (p_1, \ldots, p_n)$ to its ith coordinate p_i. Then $u^1, \ldots, u^n$ are the *natural coordinate functions* of $\boldsymbol{R}^n$.

A function ϕ from an open set $\mathscr{U}$ of $\boldsymbol{R}^m$ to $\boldsymbol{R}^n$ is *smooth* provided each real-valued function $u^i \circ \phi$ is smooth $(1 \leq i \leq n)$.

We can now make precise the resemblance to Euclidean space mentioned above. A *coordinate system* (or *chart*) in a topological space S is a homeomorphism ξ of an open set $\mathscr{U}$ of S onto an open set $\xi(\mathscr{U})$ of $\boldsymbol{R}^n$. If we write

$$\xi(p) = (x^1(p), \ldots, x^n(p)) \qquad \text{for each} \quad p \in \mathscr{U},$$

the resulting functions $x^1, \ldots, x^n$ are called the *coordinate functions* of ξ. Thus

$$\xi = (x^1, \ldots, x^n): \mathscr{U} \to \boldsymbol{R}^n.$$

Here we call n the *dimension* of ξ. Note the identity $u^i \circ \xi = x^i$.

Two n-dimensional coordinate systems ξ and η in S *overlap smoothly* provided the functions $\xi \circ \eta^{-1}$ and $\eta \circ \xi^{-1}$ are both smooth. Explicitly, if $\xi: \mathscr{U} \to \boldsymbol{R}^n$ and $\eta: \mathscr{V} \to \boldsymbol{R}^n$, then $\eta \circ \xi^{-1}$ is defined on the open set $\xi(\mathscr{U} \cap \mathscr{V})$ and carries it to $\eta(\mathscr{U} \cap \mathscr{V})$—while its inverse function $\xi \circ \eta^{-1}$ runs in the opposite direction (see Figure 1). These functions are then required to be smooth in the usual Euclidean sense defined above. This condition is considered to hold trivially if $\mathscr{U}$ and $\mathscr{V}$ do not meet.

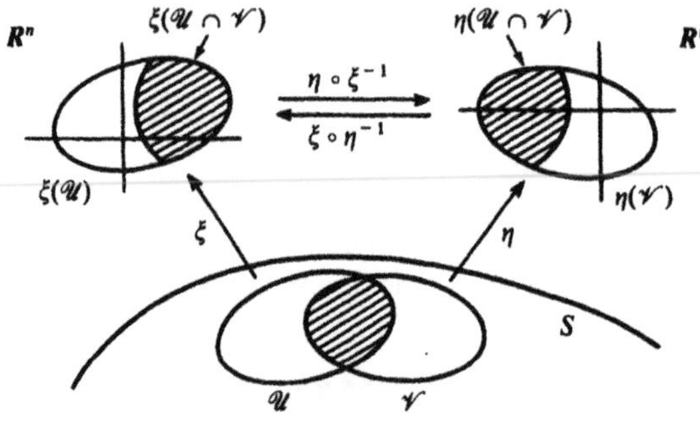

Figure 1.

1. Definition. An *atlas* $\mathscr{A}$ of dimension n on a space S is a collection of n-dimensional coordinate systems in S such that

(A1) each point of S is contained in the domain of some coordinate system in $\mathscr{V}$, and

(A2) any two coordinate systems in $\mathscr{A}$ overlap smoothly.

An atlas on S makes it possible to do calculus consistently on all of S. But different atlases may produce the same calculus, a technical difficulty eliminated as follows. Call an atlas $\mathscr{C}$ on S *complete* if $\mathscr{C}$ contains each coordinate system in S that overlaps smoothly with every coordinate system in $\mathscr{C}$.

2. Lemma. Each atlas $\mathscr{A}$ on S is contained in a unique complete atlas.

Proof. If $\mathscr{A}$ has dimension n, let $\mathscr{A}'$ be the set of all n-dimensional coordinate systems in S that overlap smoothly with every one contained in $\mathscr{A}$.

(a) *$\mathscr{A}'$ is an atlas* (of the same dimension as $\mathscr{A}$).

Since (A1) is obvious, consider (A2). If $\eta_1, \eta_2 \in \mathscr{A}'$, then by symmetry we need only prove that the function $\eta_1 \circ \eta_2^{-1}$ is Euclidean smooth. For any point $p \in \boldsymbol{R}^n$ in its domain, choose a $\xi \in \mathscr{A}$ whose domain contains $\eta_2^{-1}(p)$. As a composition of smooth functions, $(\eta_1 \circ \xi^{-1}) \circ (\xi \circ \eta_2^{-1})$ is smooth. Since this function equals $\eta_1 \circ \eta_2^{-1}$ on a neighborhood of p, the latter is smooth on that neighborhood. Smoothness being a local property, (a) follows.

(b) *$\mathscr{A}'$ is complete.* If a coordinate system ξ in S overlaps smoothly with every element of $\mathscr{A}' \supset \mathscr{A}$, then by definition $\xi \in \mathscr{A}'$.

(c) *$\mathscr{A}'$ is the unique complete atlas containing $\mathscr{A}$.*
If $\mathscr{C}$ is another, then since $\mathscr{C}$ contains $\mathscr{A}$, (A2) guarantees that $\mathscr{C} \subset \mathscr{A}'$. But then (A2) implies $\mathscr{A}' \subset \mathscr{C}$. ∎

3. Definition. A *smooth manifold M* is a Hausdorff space furnished with a complete atlas.

There are many variants of the notion of manifold but for us *manifold* will mean *smooth manifold* as above. Any atlas $\mathscr{A}$ on a Hausdorff space makes it a manifold since we agree always to use the unique complete atlas containing $\mathscr{A}$ to fulfill Definition 3. The *dimension* $n = \dim M$ of a manifold M is the dimension of its atlas, and is often indicated by the notation M^n.

A coordinate system ξ *in a manifold M* is a coordinate system belonging to the complete atlas of M. If the domain $\mathscr{U}$ of ξ contains the point $p \in M$, then ξ is called a coordinate system *at p* and $\mathscr{U}$ a *coordinate neighborhood* of p.

If ξ is a coordinate system in M and $\mathscr{V}$ is an open set contained in the domain of ξ, then by completeness $\xi|\mathscr{V}$ is also a coordinate system in M.

4. Examples of Manifolds. (1) The identity map $(u^1, \ldots, u^n)$ of $\boldsymbol{R}^n$, by itself, is an atlas. From now on, $\boldsymbol{R}^n$ will denote the resulting n-dimensional manifold, called *Euclidean n-space.*

(2) *The sphere S^n.* Let S^n be the subspace $\{a \in \boldsymbol{R}^{n+1} : |a| = 1\}$ of $\boldsymbol{R}^{n+1}$. For each $1 \leq i \leq n + 1$, let $\mathscr{U}_i$ $[\mathscr{U}_i^-]$ be the open hemisphere consisting of all points a with $a_i > 0$ $[a_i < 0]$. The restriction to $\mathscr{U}_i$ or $\mathscr{U}_i^-$ of the coordinate functions $u^1, \ldots, u^{i-1}, u^{i+1}, \ldots, u^{n+1}$ gives a coordinate system in the space S^n. It is easy to check that the $2(n + 1)$ coordinate systems gotten in this way constitute an atlas on S^n making it an n-dimensional manifold.

(3) A two-dimensional manifold is often called a *surface*, and generally speaking, the objects called surfaces in elementary calculus (torus, cylinder, paraboloid, etc.) are two-dimensional manifolds.

We now consider two simple ways to get new manifolds from old.

Let $\mathscr{U}$ be an open set in a manifold M. Let $\mathscr{A}'$ be the set of all coordinates systems ξ in M such that the domain of ξ is contained in $\mathscr{U}$. By the remark preceding Example 4 these domains cover $\mathscr{U}$. Hence $\mathscr{A}'$ is an atlas on $\mathscr{U}$, making it a manifold called an *open submanifold* of M. Open sets of a manifold will always be considered to be open submanifolds.

If M and N are manifolds, let

$$\xi = (x^1, \ldots, x^m) : \mathscr{U} \to \boldsymbol{R}^m \qquad \text{and} \qquad \eta = (y^1, \ldots, y^n) : \mathscr{V} \to \boldsymbol{R}^n$$

be coordinate systems in M and N, respectively. The product function $\xi \times \eta : \mathscr{U} \times \mathscr{V} \to \boldsymbol{R}^{m+n}$ is defined by

$$(\xi \times \eta)(p, q) = (x^1(p), \ldots, x^m(p), y^1(q), \ldots, y^n(q)).$$

Evidently $\xi \times \eta$ is a coordinate system in the Hausdorff space $M \times N$, and it is easy to see that any two such *product coordinate systems* in $M \times N$ overlap smoothly.

5. Lemma. If M and N are manifolds, then the set of all product coordinate systems in $M \times N$ is an atlas on $M \times N$ making it the *product manifold* of M and N.

The dimension of $M \times N$ is $\dim M + \dim N$. This construction extends in an obvious way to the product of any finite number of manifolds. Indeed Euclidean space $\boldsymbol{R}^n$, as in Example 4, is exactly the product manifold $\boldsymbol{R}^1 \times \cdots \times \boldsymbol{R}^1$ (n factors).

SMOOTH MAPPINGS

Consider first the special case of a real-valued function f on a manifold M. If $\xi : \mathscr{U} \to \boldsymbol{R}^n$ is a coordinate system in M, then the composite function $f \circ \xi^{-1} : \xi(\mathscr{U}) \to \boldsymbol{R}^1$ is called the *coordinate expression* for f in terms of ξ. In fact,

$$f = (f \circ \xi^{-1})(x^1, \ldots, x^n) \quad \text{on} \quad \mathscr{U}.$$

(Compare, from elementary calculus, expressing a function $f = f(x, y)$ in terms of polar coordinates.) It is natural then to define a function $f : M \to \boldsymbol{R}$ to be *smooth* provided that for every coordinate system ξ in M the coordinate expression $f \circ \xi^{-1}$ is smooth in the usual Euclidean sense. Let $\mathfrak{F}(M)$ be the set of all smooth real-valued functions on M. If f and g are smooth functions on M so is their sum $f + g$ and product fg. The usual algebraic rules hold for these two operations, making $\mathfrak{F}(M)$ a commutative ring. Multiplicative inverses do not exist in general, but if $f \in \mathfrak{F}(M)$ is never zero, then $1/f \in \mathfrak{F}(M)$.

The notion of smoothness extends from a real-valued function to an arbitrary mapping of manifolds using the same idea: that coordinate expressions must be Euclidean smooth.

6. Definition. Let M^m and N^n be manifolds. A mapping $\phi: M \to N$ is *smooth* provided that for every coordinate system ξ in M and η in N the *coordinate expression* $\eta \circ \phi \circ \xi^{-1}$ is Euclidean smooth (and defined on an open set of $\boldsymbol{R}^m$).

Explicitly, if $\mathscr{U}$ and $\mathscr{V}$ are the domains of ξ and η, then for all $p \in \phi^{-1}(\mathscr{V}) \cap \mathscr{U}$ the coordinates $y^j(\phi p)$, $1 \leq j \leq n$, depend smoothly on the coordinates $x^1(p), \ldots, x^m(p)$.

Comments. (1) It suffices to check the smoothness condition for sufficiently many coordinate systems to cover M and N; smooth overlap then takes care of the rest.

(2) A Euclidean smooth mapping ϕ from an open set $\mathscr{U} \subset \boldsymbol{R}^m$ to $\boldsymbol{R}^n$ is smooth in the above sense, since ϕ is its own coordinate expression relative to the identity coordinate systems on $\mathscr{U}$ and $\boldsymbol{R}^n$. Similarly this definition agrees with the earlier one in the case $f: M \to \boldsymbol{R}$.

(3) The identity map of a manifold is smooth; any composition of smooth mappings is smooth. (Once we have built up a supply of smooth mappings the latter will provide the easiest way to prove smoothness.)

(4) Coordinate systems ξ and coordinate functions x^i are smooth maps on the domain of ξ.

(5) Smoothness is a local property. Explicitly, define $\phi: M \to N$ to be *smooth at* $p \in M$ provided the restriction of ϕ to some neighborhood of p is smooth. Then evidently ϕ is smooth if and only if smooth at every point of M.

(6) Smooth mappings are continuous.

The following consequence of (5) will be used frequently. For each index $\alpha \in A$ let $\mathscr{U}_\alpha$ be an open set in a manifold M and let $\phi_\alpha: \mathscr{U}_\alpha \to N$ be a smooth mapping. If for all $\alpha, \beta \in A$,

$$\phi_\alpha = \phi_\beta \quad \text{on} \quad \mathscr{U}_\alpha \cap \mathscr{U}_\beta,$$

then these mappings combine to give a single mapping $\phi: \bigcup \mathscr{U}_\alpha \to N$ such that $\phi \,|\, \mathscr{U}_\alpha = \phi_\alpha$ for all $\alpha \in A$. Since smoothness is a local property, ϕ is smooth.

7. Definition. A *diffeomorphism* $\phi: M \to N$ is a smooth mapping that has an inverse mapping which is also smooth.

Identity maps of manifolds, compositions of diffeomorphisms, and inverses of diffeomorphisms are all diffeomorphisms. If there exists a diffeomorphism ϕ from M to N, then M and N are said to be *diffeomorphic* under ϕ. For example, any open interval (a, b) in $\boldsymbol{R}^1$ is diffeomorphic to $(-1, 1)$ under a suitable linear map, and $(-1, 1)$ is diffeomorphic to $\boldsymbol{R}^1$ under $\phi(t) = t/(1 - t^2)$. *Manifold theory* can be defined as the study of those objects preserved by diffeomorphisms; thus diffeomorphic manifolds are the same from the viewpoint of manifold theory.

If ϕ is a one-to-one function from a set Σ onto a manifold M, then evidently there is a unique way to make Σ a manifold (that is, a unique topology and complete atlas on Σ) such that ϕ is a diffeomorphism.

Since smooth maps are continuous, a diffeomorphism is in particular a homeomorphism. However a smooth homeomorphism need not be a diffeomorphism, since its inverse may not be smooth. The canonical example is the mapping $t \to t^3$ on the real line.

Every coordinate system ξ is a diffeomorphism from its domain $\mathscr{U}$ to $\xi(\mathscr{U}) \subset \boldsymbol{R}^n$. Conversely, *every diffeomorphism ϕ from an open set $\mathscr{V} \subset M$ to $\phi(\mathscr{V}) \subset \boldsymbol{R}^n$ is a coordinate system of M.* In fact, for any coordinate system ξ the maps $\phi \circ \xi^{-1}$ and $\xi \circ \phi^{-1}$ are smooth, and since the atlas of M is complete, ϕ is in it. Thus following a coordinate system by a diffeomorphism of $\boldsymbol{R}^n$ gives again a coordinate system. In particular, if $p \in M$ there always exist coordinate systems with $\xi(p) = 0 \in \boldsymbol{R}^n$.

The *support* supp f of $f \in \mathfrak{F}(M)$ is the closure of the set $\{p \in M : f(p) \neq 0\}$. Thus $M -$ supp f is the largest open set on which f is identically zero.

8. Lemma. Given any neighborhood $\mathscr{U}$ of a point p in M there is a function $f \in \mathfrak{F}(M)$, called a *bump function* at p, such that

(1) $0 \leq f \leq 1$ on M.
(2) $f = 1$ on some neighborhood of p.
(3) supp $f \subset \mathscr{U}$.

Proof. Start from the celebrated smooth function equal to e^{-1/t^2} if $t > 0$, and zero elsewhere. Using elementary calculus one can then build, for any $\varepsilon > 0$, a smooth function h on $\boldsymbol{R}^1$ such that $h(t) = 1$ for $t \leq \varepsilon$, $h(t) = 0$ for $t \geq 2\varepsilon$, and $0 \leq h \leq 1$.

Now let $\xi: \mathscr{V} \to \boldsymbol{R}^n$ be a coordinate system in M with $\xi(p) = 0$ and $\mathscr{V} \subset \mathscr{U}$. If $\varepsilon > 0$ is sufficiently small, then $\xi(\mathscr{V})$ contains the neighborhood $\{p \in \boldsymbol{R}^n : |p|^2 < 3\varepsilon\}$ of 0 in $\boldsymbol{R}^n$. Let $N = \sum (x^i)^2$ on $\mathscr{V}$. For h as above, define f to be $h \circ N$ on $\mathscr{V}$ and zero elsewhere. Then f is smooth on M and has the required properties. ∎

TANGENT VECTORS

The crucial step in generalizing calculus from $\boldsymbol{R}^n$ to an arbitrary manifold is the following elegant definition, which axiomatizes the *directional derivative* aspect of Euclidean tangent vectors.

9. Definition. Let p be a point of a manifold M. A *tangent vector to M at p* is a real-valued function v: $\mathfrak{F}(M) \to \boldsymbol{R}$ that is

(1) $\boldsymbol{R}$-linear: $v(af + bg) = av(f) + bv(g)$, and
(2) Leibnizian: $v(fg) = v(f)g(p) + f(p)v(g)$ for all $a, b \in \boldsymbol{R}$ and $f, g \in \mathfrak{F}(M)$.

At each point $p \in M$ let $T_p(M)$ be the set of all tangent vectors to M at p. The usual definitions of functional addition and scalar multiplication make $T_p(M)$ a vector space over the real numbers $\boldsymbol{R}$. Explicitly,

$$(v + w)(f) = v(f) + w(f),$$

$$(av)(f) = av(f) \qquad \text{for all } f \in \mathfrak{F}(M), \quad a \in \boldsymbol{R},$$

and $T_p(M)$ is called the *tangent space to M at p.*

To define partial differentiation on a manifold, the scheme is to move the function f back to Euclidean space using a coordinate system, and then take the usual partial derivatives.

10. Definition. Let $\xi = (x^1, \dots, x^n)$ be a coordinate system in M at p. If $f \in \mathfrak{F}(M)$, let

$$\frac{\partial f}{\partial x^i}(p) = \frac{\partial (f \circ \xi^{-1})}{\partial u^i}(\xi p) \qquad (1 \le i \le n),$$

where $u^1, \dots, u^n$ are the natural coordinate functions of $\boldsymbol{R}^n$.

A straightforward computation then shows that the function

$$\partial_i|_p = \left.\frac{\partial}{\partial x^i}\right|_p : \mathfrak{F}(M) \to \boldsymbol{R}$$

sending each $f \in \mathfrak{F}(M)$ to $(\partial f/\partial x^i)(p)$ is a tangent vector to M at p. We can picture $\partial_i|_p$ as an arrow at p tangent to the x^i-coordinate curve through p.

11. Lemma. Let $v \in T_p(M)$. (1) If $f, g \in \mathfrak{F}(M)$ are equal on a neighborhood of p, then $v(f) = v(g)$. (2) If $h \in \mathfrak{F}(M)$ is constant on a neighborhood of p, then $v(h) = 0$.

Proof. (1) By linearity it suffices to show that if $f = 0$ on a neighborhood $\mathscr{U}$ of p, then $v(f) = 0$. Let g be a bump function at p with support in $\mathscr{U}$; then $fg = 0$ on all of M. But $v(0) = v(0 + 0) = v(0) + v(0)$ implies $v(0) = 0$. Thus

$$0 = v(fg) = v(f)g(p) + f(p)v(g) = v(f),$$

since $f(p) = 0$ and $g(p) = 1$.

(2) By (1) we can assume that h has constant value c on all of M. If 1 is the constant function of value 1, then

$$v(1) = v(1 \cdot 1) = v(1)1 + 1v(1) = 2v(1).$$

Hence $v(1) = 0$, and $v(h) = v(c \cdot 1) = cv(1) = 0$. ∎

The preceding lemma is one way to express the fact that tangent vectors are *local* objects. Another version is this: If $\mathscr{U}$ is an open set in M then, as an open submanifold, $\mathscr{U}$ has tangent space $T_p(\mathscr{U})$ at $p \in \mathscr{U}$. If $v \in T_p(\mathscr{U})$ define

$$\tilde{v}(f) = v(f \mid \mathscr{U}) \qquad \text{for all} \quad f \in \mathfrak{F}(M).$$

Evidently $\tilde{v} \in T_p(M)$, and the function $v \to \tilde{v}$ is a linear isomorphism. We ignore this trivial isomorphism henceforth, writing $T_p(\mathscr{U}) = T_p(M)$.

The following result, called *the basis theorem*, is the fundamental link between coordinates and tangent vectors.

12. Theorem. If $\xi = (x^1, \ldots, x^n)$ is a coordinate system in M at p, then its coordinate vectors $\partial_1|_p, \ldots, \partial_n|_p$ form a basis for the tangent space $T_p(M)$; and

$$v = \sum_{i=1}^{n} v(x^i)\, \partial_i|_p \qquad \text{for all} \quad v \in T_p(M).$$

Proof. By the preceding remarks we can work solely on the coordinate neighborhood $\mathscr{U}$ of ξ. Since $v(c) = 0$ there is no loss of generality in assuming $\xi(p) = 0 \in \mathbf{R}^n$. Shrinking $\mathscr{U}$ if necessary gives $\xi(\mathscr{U}) = \{q \in \mathbf{R}^n : |q| < \varepsilon\}$ for some ε.

If g is a smooth function on $\xi(\mathscr{U})$ then for each $1 \leq i \leq n$ define

$$g_i(q) = \int_0^1 \frac{\partial g}{\partial u^i}(tq)\, dt \qquad \text{for all} \quad q \in \xi(\mathscr{U}).$$

It follows using the fundamental theorem of calculus that

$$g = g(0) + \sum g_i u^i \quad \text{on} \quad \xi(\mathscr{U}).$$

Thus if $f \in \mathfrak{F}(M)$, setting $g = f \circ \xi^{-1}$ yields

$$f = f(p) + \sum f_i x^i \quad \text{on} \quad \mathscr{U}.$$

Applying $\partial/\partial x^i$ gives $f_i(p) = (\partial f/\partial x^i)(p)$. Thus applying the tangent vector v to the formula gives

$$v(f) = 0 + \sum v(f_i) x^i(p) + \sum f_i(p) v(x^i) = \sum \frac{\partial f}{\partial x^i}(p) v(x^i).$$

Since this holds for all $f \in \mathfrak{F}(M)$, the tangent vectors v and $\sum v(x^i)\, \partial_i|_p$ are equal.

It remains to show that the coordinate vectors are linearly independent. But if $\sum a_i\, \partial_i|_p = 0$, then application to x^j yields

$$0 = \sum_i a_i \frac{\partial x^j}{\partial x^i}(p) = \sum_i a_i\, \delta_{ij} = a_j. \qquad \blacksquare$$

In particular the (vector space) dimension of $T_p(M)$ is the same as the dimension of M.

DIFFERENTIAL MAPS

The basic idea of differential calculus is to approximate smooth objects by linear objects. In the preceding section a manifold M was approximated near each of its points p by the tangent space $T_p(M)$. Now we approximate a smooth mapping $\phi\colon M \to N$ near each point $p \in M$ by a linear transformation of tangent spaces. Note first that if $v \in T_p(M)$ then the function $v_\phi\colon \mathfrak{F}(N) \to R$ sending each g to $v(g \circ \phi)$ is a tangent vector to N at $\phi(p)$. To prove the Leibnizian property, for example, let $f, g \in \mathfrak{F}(N)$; then

$$\begin{aligned} v_\phi(fg) = v(fg \circ \phi) = v((f \circ \phi)(g \circ \phi)) &= v(f \circ \phi)g(\phi p) + f(\phi p)v(g \circ \phi) \\ &= v_\phi(f)g(\phi p) + f(\phi p)v_\phi(g). \end{aligned}$$

13. Definition. Let $\phi\colon M \to N$ be a smooth mapping. For each $p \in M$ the function

$$d\phi_p\colon T_p(M) \to T_{\phi p}(N)$$

sending v to v_ϕ (as above) is called the *differential map* of ϕ at p (see Figure 2).

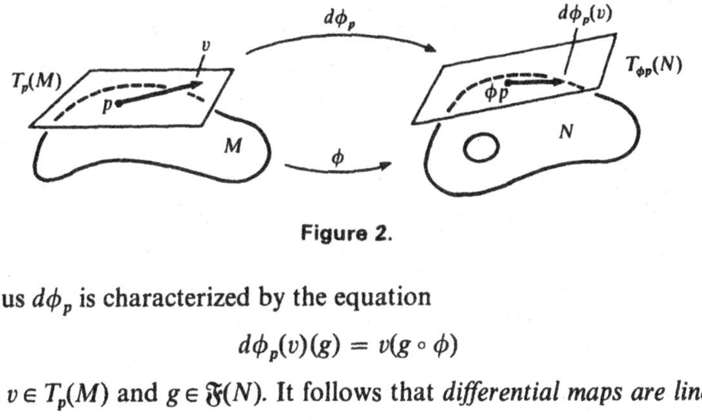

Figure 2.

Thus $d\phi_p$ is characterized by the equation

$$d\phi_p(v)(g) = v(g \circ \phi)$$

for all $v \in T_p(M)$ and $g \in \mathfrak{F}(N)$. It follows that *differential maps are linear.*

14. Lemma. Let $\phi\colon M^m \to N^n$ be a smooth mapping. If ξ is a coordinate system at p in M, and η is a coordinate system at $\phi(p)$ in N, then

$$d\phi_p\left(\frac{\partial}{\partial x^j}\bigg|_p\right) = \sum_{i=1}^{n} \frac{\partial(y^i \circ \phi)}{\partial x^j}(p)\,\frac{\partial}{\partial y^i}\bigg|_{\phi p} \qquad (1 \le j \le m).$$

Proof. Let $w \in T_{\phi p}(N)$ be the left-hand side of this equation. By the basis theorem (12), $w = \sum w(y^i)\, \partial/\partial y^i|_{\phi p}$. But by the definition of a differential map,

$$w(y^i) = d\phi_p\left(\frac{\partial}{\partial x^j}\bigg|_p\right)(y^i) = \frac{\partial(y^i \circ \phi)}{\partial x^j}(p). \qquad \blacksquare$$

Thus the matrix of $d\phi_p$ with respect to these coordinate bases is

$$\left(\frac{\partial(y^i \circ \phi)}{\partial x^j}(p)\right)_{1\leq i\leq n,\, 1\leq j\leq m}$$

called the *Jacobian matrix* of ϕ at p relative to ξ and η. The classical chain-rule formula for the Jacobian matrix of a composite mapping follows immediately from the following expression for it in terms of differential maps.

15. Lemma. If $\phi: M \to N$ and $\psi: N \to P$ are smooth mappings, then for each $p \in M$,

$$d(\psi \circ \phi)_p = d\psi_{\phi p} \circ d\phi_p.$$

Proof. If $v \in T_p(M)$ and $g \in \mathfrak{F}(P)$, then

$$d(\psi \circ \phi)(v)(g) = v(g \circ \psi \circ \phi) = d\phi(v)(g \circ \psi) = (d\psi\, d\phi(v))(g). \quad \blacksquare$$

Henceforth we usually omit the subscript p from $d\phi_p$. In terms of manifold theory the *inverse function theorem* can be stated as follows.

16. Theorem. Let $\phi: M \to N$ be a smooth mapping. The differential map $d\phi_p$ at a point $p \in M$ is a linear isomorphism if and only if there is a neighborhood $\mathscr{V}$ of p in M such that $\phi|\mathscr{V}$ is a diffeomorphism from $\mathscr{V}$ onto a neighborhood $\phi(\mathscr{V})$ of $\phi(p)$ in N.

For a proof, apply the classical theorem to a coordinate expression for ϕ near p.

Because of this result a smooth map $\phi: M \to N$ such that every $d\phi_p$ is a linear isomorphism is called a *local diffeomorphism*. If ϕ is also one-to-one and onto, then it is a diffeomorphism.

CURVES

A *curve* in a manifold M is a smooth mapping $\alpha: I \to M$, where I is an open interval in the real line $\boldsymbol{R}^1$. (We allow I to be half-infinite or all of $\boldsymbol{R}$.) As an open submanifold of $\boldsymbol{R}^1$, I has a coordinate system consisting of the identity map u of I. At each $t \in R$ we can picture the coordinate vector $(d/du)(t) \in T_t(\boldsymbol{R})$ as the unit vector at t in the positive u direction.

17. Definition. Let $\alpha: I \to M$ be a curve. The *velocity vector of* α *at* $t \in I$ is

$$\alpha'(t) = d\alpha\left(\left.\frac{d}{du}\right|_t\right) \in T_{\alpha(t)}(M).$$

Intuitively, $\alpha'(t)$ is the vector rate of change of α at $t \in I$. We list some of its basic properties:

(1) *Directional derivatives.* By the definition of $d\alpha$, the tangent vector $\alpha'(t)$ applied to a function $f \in \mathfrak{F}(M)$ gives

$$\alpha'(t)f = \frac{d(f \circ \alpha)}{du}(t).$$

Thus if α is any curve with say $\alpha'(0) = v$, then $v(f) = (d(f \circ \alpha)/dt)(0)$.

(2) *Coordinate expression.* Let $x^1, \ldots, x^n$ be a coordinate system in M at a point $\alpha(t)$ of α. By the basis theorem and (1):

$$\alpha'(t) = \sum_i \frac{d(x^i \circ \alpha)}{du}(t)\, \partial_i\bigg|_{\alpha(t)}.$$

(3) *Reparametrization.* If $\alpha\colon I \to M$ is a curve and $h\colon J \to I$ is a smooth function on an interval J, then $\beta = \alpha(h)\colon J \to M$ is a curve called a *reparametrization* of α. Then

$$\beta'(s) = (dh/du)(s)\alpha'(h(s)) \qquad \text{for all} \quad s \in J.$$

This follows from the chain rule (Lemma 15), as does the following.

(4) *Effect of a mapping.* If $\alpha\colon I \to M$ is a curve in M, then a mapping $\phi\colon M \to N$ carries α to the curve $\phi \circ \alpha\colon I \to N$ in N. The differential map of ϕ *preserves velocities*, that is,

$$d\phi(\alpha'(t)) = (\phi \circ \alpha)'(t) \qquad \text{for all} \quad t \in I.$$

This is usually a more efficient way to get information about $d\phi$ than are coordinate computations as in Lemma 14.

A curve α is *regular* provided $\alpha'(t) \neq 0$ for all t. If $[a, b]$ is a closed interval in $\boldsymbol{R}$ then a *curve segment* $\alpha\colon [a, b] \to M$ is a function that has a smooth extension to an open interval containing $[a, b]$. Thus α' is well defined even at the endpoints a and b.

A function $\beta\colon [a, b] \to M$ is a *piecewise smooth* curve segment provided there is a partition $a = t_0 < t_1 < \cdots < t_{k+1} = b$ of $[a, b]$ such that each $\beta | [t_i, t_{i+1}]$ is a curve segment. Thus β may well have two velocity vectors at each *break* t_i. For an open interval I, $\beta\colon I \to M$ is piecewise smooth provided that for all $a < b$ in I the restriction $\beta | [a, b]$ is piecewise smooth. (Thus the breaks have no cluster point in I.)

For no class of curves to be considered is the reparametrization $t \to \alpha(t + c)$ important. Thus to simplify notation we shall often assume, without explicit mention, that the domain of a curve contains the number 0.

VECTOR FIELDS

A *vector field* V on a manifold M is a function that assigns to each point $p \in M$ a tangent vector V_p to M at p. Intuitively, V is a collection of arrows, one at each point of M. If V is a vector field on M and $f \in \mathfrak{F}(M)$, then Vf denotes the real-valued function on M given by

$$(Vf)(p) = V_p(f) \qquad \text{for all} \quad p \in M.$$

Then V is *smooth* provided Vf is smooth for all $f \in \mathfrak{F}(M)$.

Vector fields on M are added together, or multiplied by a function $f \in \mathfrak{F}(M)$, in the obvious way:

$$(fV)_p = f(p)V_p,$$
$$(V + W)_p = V_p + W_p \qquad \text{for all} \quad p \in M.$$

If V and W are smooth, then the vector fields $V + W$ and fV are also. These two operations make the set $\mathfrak{X}(M)$ of all smooth vector fields on M a *module* over the ring $\mathfrak{F}(M)$. (The definition of module over a commutative ring with unit is formally the same as that of a vector space over a field.)

If $\xi = (x^1, \ldots, x^n)$ is a coordinate system on $\mathscr{U} \subset M$, then for each $1 \leq i \leq n$ the vector field ∂_i on $\mathscr{U}$ sending each p to $\partial_i|_p$ is called the ith *coordinate vector field* of ξ. These vector fields are smooth, since $\partial_i(f) = \partial f/\partial x^i$. It follows immediately from the basis theorem that for any vector field V

$$V = \sum Vx^i \, \partial_i \quad \text{on} \quad \mathscr{U}.$$

For low-dimensional manifolds index notation can be cumbersome. On the Euclidean plane, for example, we often denote the natural coordinate functions by x, y as in elementary calculus; then the coordinate vector fields are denoted by ∂_x, ∂_y.

A *derivation* on $\mathfrak{F}(M)$ is a function $\mathscr{D}\colon \mathfrak{F}(M) \to \mathfrak{F}(M)$ that is

(1) $\boldsymbol{R}$-linear: $\mathscr{D}(af + bg) = a\mathscr{D}(f) + b\mathscr{D}(g)$, $(a, b \in \boldsymbol{R})$, and
(2) Leibnizian: $\mathscr{D}(fg) = \mathscr{D}(f)g + f\mathscr{D}(g)$.

The definition of tangent vector shows that for a vector field $V \in \mathfrak{X}(M)$ the function $f \to Vf$ is a derivation on $\mathfrak{F}(M)$. Conversely, *every derivation* $\mathscr{D}$ *on* $\mathfrak{F}(M)$ *comes from a vector field.* In fact, for each $p \in M$ define $V_p\colon \mathfrak{F}(M) \to R$ by $V_p(f) = \mathscr{D}(f)(p)$. The derivation properties (1) and (2) above imply that V_p is a tangent vector to M at p; thus V is a well-defined vector field on M. But $Vf = \mathscr{D}(f) \in \mathfrak{F}(M)$ for all $f \in \mathfrak{F}(M)$, so V is smooth and determines the derivation $\mathscr{D}$.

Henceforth whenever convenient *we will consider vector fields to be derivations on* $\mathfrak{F}(M)$. This interpretation leads to a crucial operation on vector fields. If $V, W \in \mathfrak{X}(M)$ let $[V, W] = VW - WV$. This is the function from $\mathfrak{F}(M)$ to $\mathfrak{F}(M)$ sending each f to $V(Wf) - W(Vf)$. An easy computation shows that $[V, W]$ is a derivation of $\mathfrak{F}(M)$, hence a smooth vector field on M. $[V, W]$ is called the *bracket* of V and W.

In terms of the original definition of vector field, $[V, W]$ assigns to each $p \in M$ the tangent vector $[V, W]_p$ such that

$$[V, W]_p(f) = V_p(Wf) - W_p(Vf).$$

18. Lemma. The bracket operation on $\mathfrak{X}(M)$ has the following properties:

(1) $\boldsymbol{R}$-bilinearity: $[aV + bW, X] = a[V, X] + b[W, X]$,
$[X, aV + bW] = a[X, V] + b[X, W]$.

(2) Skew-symmetry: $[W, V] = -[V, W]$.

(3) *Jacobi identity*: $[X, [Y, Z]] + [Y, [Z, X]] + [Z, [X, Y]] = 0$.

Proof. These identities are purely formal, and hold for derivations on any linear algebra: (1) is immediate from the linearity of derivations, and (2) is obvious. For (3), substitute the definition of bracket; then the resulting twelve terms cancel in pairs. ∎

The bracket operation on $\mathfrak{X}(M)$ though $\boldsymbol{R}$-bilinear, is not $\mathfrak{F}(M)$-bilinear. In fact,

$$[fV, gW] = fg[V, W] + f(Vg)W - g(Wf)V.$$

To prove this, check that both sides have the same effect on any function $h \in \mathfrak{F}(M)$.

In two special cases, the bracket is always zero. For any $V \in \mathfrak{X}(M)$, $[V, V] = 0$. In fact, in the presence of $\boldsymbol{R}$-bilinearity this is equivalent to the skew-symmetry (2) above. For any two coordinate vector fields of the same coordinate system, $[\partial_i, \partial_j] = 0$, which is the bracket expression of $\partial^2 f/\partial x^i\, \partial x^j = \partial^2 f/\partial x^j\, \partial x^i$ for smooth functions f.

19. Example. Let x, y be the natural coordinates of $\boldsymbol{R}^2$, and consider the vector fields $V = y\, \partial_y$ and $W = x\, \partial_y$. Then $[V, W] = -W$, since for all $f \in \mathfrak{F}(M)$,

$$\begin{aligned}[V, W]f &= [y\, \partial_y, x\, \partial_y]f = y\, \partial_y(x\, \partial_y f) - x\, \partial_y(y\, \partial_y f) \\ &= yx\frac{\partial^2 f}{\partial y^2} - x\frac{\partial f}{\partial y} - xy\frac{\partial^2 f}{\partial y^2} = -x\, \partial_y f = -Wf.\end{aligned}$$

Such computations can be abbreviated by omitting the function f—and also the second derivatives (since these must cancel.)

The differential map of $\phi\colon M \to N$ moves individual tangent vectors from M to N, but in general provides no way to move vector *fields* from M to N (or the reverse). This difficulty can be overcome to some extent as follows.

20. Definition. Let $\phi\colon M \to N$ be a smooth mapping. Vector fields X on M and Y on N are *ϕ-related*, written $X \underset{\phi}{\sim} Y$ provided

$$d\phi(X_p) = Y_{\phi p} \qquad \text{for all} \quad p \in M.$$

21. Lemma. Vector fields $X \in \mathfrak{X}(M)$ and $Y \in \mathfrak{X}(N)$ are ϕ-related if and only if $X(g \circ \phi) = Yg \circ \phi$ for all $g \in \mathfrak{F}(N)$.

Proof. The following assertions are equivalent:

$$\begin{aligned} X(g \circ \phi) &= Yg \circ \phi, && g \in \mathfrak{F}(N), \\ X(g \circ \phi)(p) &= (Yg \circ \phi)(p), && p \in M, \quad g \in \mathfrak{F}(N), \\ X_p(g \circ \phi) &= (Yg)(\phi p), && p \in M, \quad g \in \mathfrak{F}(N), \\ (d\phi X_p)g &= Y_{\phi p}(g), && p \in M, \quad g \in \mathfrak{F}(N), \\ d\phi(X_p) &= Y_{\phi p}, && p \in M. \quad \blacksquare \end{aligned}$$

Using this criterion repeatedly provides a proof that brackets are preserved by ϕ-relatedness:

22. Lemma. If $X_1 \underset{\phi}{\sim} Y_1$ and $X_2 \underset{\phi}{\sim} Y_2$, then $[X_1, X_2] \underset{\phi}{\sim} [Y_1, Y_2]$.

Let $\phi\colon M \to N$ be a diffeomorphism. For each $X \in \mathfrak{X}(M)$ there is a unique vector field $d\phi(X) \in \mathfrak{X}(N)$ that is ϕ-related to X. In fact, we have no choice but to define $(d\phi X)_q = d\phi(X_p)$ for all $q = \phi(p) \in N$. Then $d\phi(X)$ is smooth, since if $g \in \mathfrak{F}(N)$ the definition leads to the formula $(d\phi X)g = X(g \circ \phi) \circ \phi^{-1} \in \mathfrak{F}(N)$. We call $d\phi(X)$ the *transferred vector field* of X.

ONE-FORMS

The one-forms on a smooth manifold M are the objects dual to vector fields. At a point p of M the dual space $T_p(M)^*$ of the tangent space $T_p(M)$ is called the *cotangent space* of M at p. Elements of $T_p(M)^*$—sometimes called *covectors*—are linear maps of $T_p(M)$ into $\boldsymbol{R}$.

23. Definition. A *one-form* θ on a manifold M is a function that assigns to each point p an element θ_p of the cotangent space $T_p(M)^*$.

Thus θ assigns a number to every tangent vector and is linear on the tangent vectors at each point.

If θ is a one-form on M and X is a vector field on M, denote by θX the real-valued function on M whose value at each point p is the value of θ_p on X_p. A one-form θ is *smooth* provided θX is smooth for all $X \in \mathfrak{X}(M)$.

Let $\mathfrak{X}^*(M)$ be the set of all (smooth) one-forms on M. Two one-forms are added, and a one-form is multiplied by a real-valued function, just as in the dual case of vector fields. Explicitly,

$$(\theta + \omega)_p = \theta_p + \omega_p, \qquad (f\theta)_p = f(p)\theta_p$$

for all $p \in M$. Thus $\mathfrak{X}^*(M)$ becomes a module over $\mathfrak{F}(M)$.

There is a remarkable operation that converts functions into one-forms.

24. Definition. The *differential* of $f \in \mathfrak{F}(M)$ is the one-form df such that $(df)(v) = v(f)$ for every tangent vector v to M.

Clearly df is a one-form since at each point p the function $(df)_p\colon T_p(M) \to \boldsymbol{R}$ is linear, and if $V \in \mathfrak{X}(M)$ the function $(df)(V) = Vf$ is smooth.

If $x^1, \ldots, x^n$ is a coordinate system on $\mathscr{U} \subset M$ we have thus the *coordinate one-forms* $dx^1, \ldots, dx^n$ on $\mathscr{U}$. At each point of $\mathscr{U}$ these provide a dual basis to the coordinate vector fields $\partial_1, \ldots, \partial_n$ since $dx^i(\partial_j) = \partial x^i/\partial x^j = \delta_{ij}$. It follows that for any one-form θ,

$$\theta = \sum \theta(\partial_i)\, dx^i \quad \text{on} \quad \mathscr{U}.$$

This formula corresponds to the basis theorem for vector fields. (To prove it, apply both sides to the coordinate vector fields.) In particular, if $f \in \mathfrak{F}(M)$, then since $df(\partial_i) = \partial f/\partial x^i$,

$$df = \sum \frac{\partial f}{\partial x^i}\, dx^i \quad \text{on} \quad \mathscr{U},$$

25. Lemma. The differential has the following properties:

(1) $d\colon \mathfrak{F}(M) \to \mathfrak{X}^*(M)$ is $\boldsymbol{R}$-linear.
(2) Product rule: If $f, g \in \mathfrak{F}(M)$, then $d(fg) = g\, df + f\, dg$.
(3) If $f \in \mathfrak{F}(M)$ and $h \in \mathfrak{F}(\boldsymbol{R}^1)$, then $d(h(f)) = h'(f)\, df$.

SUBMANIFOLDS

Roughly speaking, a submanifold P of a manifold M is a subset of M that acquires its manifold structure from M. In particular we shall require that P be a topological subspace of M, that is, have the *induced topology*, for which a subset $\mathscr{V}$ of P is open if and only if there is an open set $\tilde{\mathscr{V}}$ of M such that $\tilde{\mathscr{V}} \cap P = \mathscr{V}$.

26. Definition. A manifold P is a *submanifold* of a manifold M provided:

(1) P is a topological subspace of M.
(2) The inclusion map $j: P \subset M$ is smooth and at each point $p \in P$ its differential map dj is one-to-one.

If P is a submanifold of M and $\phi: M \to N$ is a smooth map, then so is the restriction $\phi | P$ of ϕ to P, since $\phi | P$ is just $\phi \circ j$. In particular, if $f \in \mathfrak{F}(M)$, then $f | P \in \mathfrak{F}(P)$. Since j is such an obvious mapping and each $dj: T_p(P) \to T_p(M)$ is one-to-one, it is customary to ignore dj and consider the tangent space $T_p(P)$ as a vector subspace of $T_p(M)$.

Open submanifolds are trivially submanifolds, and the sphere S^n as in Example 4(2) is a submanifold of $\boldsymbol{R}^{n+1}$. Coordinate systems produce submanifolds; for instance the plane $z = 1$ in $\boldsymbol{R}^3$ is a submanifold, diffeomorphic to $\boldsymbol{R}^2$ under $(x, y, 1) \to (x, y)$. More generally, if $\xi: \mathscr{U} \to \boldsymbol{R}^n$ is a coordinate system in a manifold M then holding any $n - m$ of the coordinate functions constant produces an m-dimensional submanifold called a *ξ-coordinate slice* Σ of $\mathscr{U}$. Our goal is to show that every submanifold can be constructed by gluing together such slices.

27. Definition. Let P be a subset of M^n. A coordinate system $\xi: \mathscr{U} \to \boldsymbol{R}^n$ in M is *adapted to* P provided $\mathscr{U} \cap P$ is a ξ-coordinate slice of $\mathscr{U}$.

By permuting the coordinates we can always suppose it is the last $n - m$ that are held constant. Then let $\xi_P: \mathscr{U} \cap P \to \boldsymbol{R}^m$ be the restriction of $(x^1, \ldots, x^m)$ to $\mathscr{U} \cap P$.

In case P is a submanifold it follows using the inverse function theorem that ξ_P is a diffeomorphism onto its (open) image in $\boldsymbol{R}^m$, and hence is a coordinate system in P.

28. Proposition. If P^m is a submanifold of M^n, then at each point of P there is a coordinate system of M adapted to P.

Proof. Let $\xi = (x^1, \ldots, x^n)$ be a coordinate system for M at $p \in M$ and let $y^1, \ldots, y^m$ be a coordinate system for P at p. Since the differential map dj_p is one-to-one its Jacobian matrix

$$\left(\frac{\partial x^i}{\partial y^j}(p)\right)_{1 \leq i \leq n,\, 1 \leq j \leq m}$$

has rank m. Thus (relabeling the x's if necessary) we can suppose its first m rows are linearly independent. Then by Exercise 7 the restrictions to P of the functions $x^1, \ldots, x^m$ form a coordinate system for P on a neighborhood $\mathscr{W}$ of p. If f^k is the coordinate expression for $x^k | P$ relative to these coordinates, then

$$x^k = f^k(x^1, \ldots, x^m) \quad \text{on} \quad \mathscr{W}.$$

It follows that the functions

$$z^k = x^k - f^k(x^1, \ldots, x^m)$$

are well defined on some neighborhood of p in M.

If $\zeta = (x^1, \ldots, x^m, z^{m-1}, \ldots, z^n)$ then at p the Jacobian matrix of ζ relative to ξ has the form

$$\begin{pmatrix} I_m & 0 \\ A & I_{n-m} \end{pmatrix},$$

where I_m is the $m \times m$ identity matrix. Evidently this matrix is invertible, so ζ is a coordinate system on a neighborhood $\mathscr{U}$ of p in M. (1) Because P is a topological subspace of M we can choose $\mathscr{U}$ so that $\mathscr{U} \cap P \subset \mathscr{W}$. (2) Since $\mathscr{O} = (x^1, \ldots, x^m)(\mathscr{U} \cap P)$ is an open set of $\boldsymbol{R}^m$, by shrinking $\mathscr{U}$ if necessary we can also suppose that $(x^1, \ldots, x^m)(\mathscr{U}) \subset \mathscr{O}$.

Now on $\mathscr{U} \cap P$ the functions z^k are all zero, so $\mathscr{U} \cap P$ is contained in the slice

$$\Sigma : z^{m+1} = 0, \ldots, z^n = 0$$

of $\mathscr{U}$. Conversely, one can check that $\Sigma \subset \mathscr{U} \cap P$. Hence $\mathscr{U} \cap P = \Sigma$. ∎

The following seemingly obvious result depends on the fact that submanifolds have the induced topology.

29. Corollary. Let P^m be a submanifold of M^n. If $\phi: N \to M$ is a smooth mapping such that $\phi(N) \subset P$, then the induced mapping $\bar{\phi}: N \to P$ is smooth.

Proof. If $q \in N$ let $x^1, \ldots, x^n$ be a coordinate system on a neighborhood $\mathscr{U} \subset M$ of $\phi(q)$ that is adapted to P. Being smooth, ϕ is continuous, and since P is a topological subspace of M, $\bar{\phi}$ is continuous. Thus there is a neighborhood $\mathscr{V}$ of q in N such that $\phi(\mathscr{V})$ is contained in the neighborhood $\mathscr{U} \cap P$ of $\phi(q)$ in P. Now $x^1 | P, \ldots, x^m | P$ is a coordinate system on $\mathscr{U} \cap P$. If $j: P \subset M$ is the inclusion map, then

$$(x^i | P) \circ \bar{\phi} = x^i \circ j \circ \bar{\phi} = x^i \circ \phi.$$

These functions are smooth since ϕ and the x^is are smooth. It follows by Exercise 1 that $\bar{\phi}$ is smooth. ∎

30. Corollary. A subset P of a smooth manifold M can be made a submanifold of M in at most one way.

Proof. By definition P must be given the induced topology. Suppose two atlases are assigned to the space P, producing submanifolds P_1 and P_2 of M. The inclusion maps of P_1 and P_2 into M are smooth, hence by the preceding corollary the identity maps $P_1 \to P_2$ and $P_2 \to P_1$ are smooth. Thus these identity maps are inverse diffeomorphisms. It follows that the complete atlases of P_1 and P_2 are identical. ■

Thus it makes sense to say that a subset P of a manifold M is (or is not) a submanifold of M. A basic criterion is as follows.

31. Proposition. A subset P of a manifold M is an m-dimensional submanifold if (and only if) at each point p of P there is a coordinate system of M adapted to P by m-dimensional slices.

Proof. Assign P the induced topology. Let $\xi: \mathscr{U} \to \boldsymbol{R}^n$ be a coordinate system of M at $p \in P$ such that $\mathscr{U} \cap P$ is the slice $x^j = x^j(p)$ for $m + 1 \leq j \leq n$. We can suppose $\xi(p) = 0$. Then since ξ is a homeomorphism, the map $\xi_P = (x^1, \ldots, x^m)|P$ is a homeomorphism from $\mathscr{U} \cap P$ to the open set $\xi(\mathscr{U}) \cap \boldsymbol{R}^m$ of $\boldsymbol{R}^m$.

We assert that all such topological coordinate systems ξ_P in P form an atlas. Certainly they cover P, and any two overlap smoothly since for $1 \leq i \leq m$,

$$u^i \circ \xi_P \circ \eta_P^{-1} = u^i \circ (\xi \circ \eta^{-1})|\boldsymbol{R}^m,$$

where $\boldsymbol{R}^m$ is considered as the coordinate m-plane of the first m coordinates of $\boldsymbol{R}^n$.

It remains to show that this atlas makes P a submanifold of M. For a coordinate system ξ as above, the functions $x^j|P = x^i \circ j$ for $1 \leq i \leq m$ are smooth, since they constitute the coordinate system ξ_P. Hence by Exercise 1 the inclusion map $j: P \subset M$ is smooth. Evidently the Jacobian matrix of ξ relative to ξ_P contains an $m \times m$ identity matrix, hence dj is always one-to-one. ■

A vector field X on M is *tangent* to a submanifold P of M provided $X_p \in T_p(P)$ for all $p \in P$. (Recall that for a submanifold, $T_p(P)$ is considered as a subspace of $T_p(M)$.)

32. Proposition. Let P be a submanifold of M. (1) If $X \in \mathfrak{X}(M)$ is tangent to P, then its restriction $X|P$ to P is a smooth vector field on P. (2) Furthermore, if $Y \in \mathfrak{X}(M)$ is also tangent to P, then the bracket $[X, Y]$ is tangent to P and

$$[X, Y]|P = [X|P, Y|P].$$

The proof is straightforward, using an adapted coordinate system for (1) and applying Lemma 22 to (2).

IMMERSIONS AND SUBMERSIONS

This section deals with two special types of smooth mappings defined by hypotheses on differential maps.

33. Lemma. Let $\phi: M^m \to N^n$ be a smooth map, and let p be a point of M. Then the following are equivalent:

(1) The differential map $d\phi_p$ is one-to-one.

(2) The Jacobian matrix of $d\phi_p$ has rank m relative to one (hence every) choice of coordinate systems.

(3) If $y^1, \ldots, y^n$ is a coordinate system in N at $\phi(p)$, then there are integers $1 \leq i_1 \leq \cdots \leq i_m \leq n$ such that the functions $y^{i_1} \circ \phi, \ldots, y^{i_m} \circ \phi$ form a coordinate system on a neighborhood p in M.

The proof is a mild generalization of the first part of the proof of Proposition 28.

34. Definition. An *immersion* $\phi: M \to N$ is a smooth mapping such that $d\phi_p$ is one-to-one for all $p \in M$.

For example, every regular curve (α' never 0) is an immersion.

An *imbedding* of a manifold P into M is a one-to-one immersion $\phi: P \to M$ such that the induced map $P \to \phi(P)$ is a homeomorphism onto the subspace $\phi(P)$ of M. The standard example is the mapping $(a_1, \ldots, a_m) \to (a_1, \ldots, a_m, 0, \ldots, 0)$ of $\boldsymbol{R}^m$ into $\boldsymbol{R}^n$. Lemma 33(3) says that locally every immersion looks like this. Explicitly, the restriction of an immersion to any sufficiently small open set is an imbedding.

Submanifolds and imbeddings are very closely related: If P is a submanifold of M, then the inclusion map $j: P \subset M$ is an imbedding. Conversely, if $\phi: P \to M$ is an imbedding, make its image $\phi(P)$ a manifold so that the induced map $\bar{\phi}: P \to \phi(P)$ is a diffeomorphism. Then $\phi(P)$ is a subspace of M, and the inclusion map $j: \phi(P) \subset M$ is just $\phi \circ \bar{\phi}^{-1}$, which by the chain rule is an immersion. Thus $\phi(P)$ is a submanifold of M.

The term "submanifold" is sometimes applied to a more general object: Let P be a manifold that is merely a subset of a manifold M. If the inclusion map $j: P \subset M$ is an immersion, we call P an *immersed submanifold* of M. Thus submanifolds are immersed submanifolds, but not conversely, since the latter need not have the induced topology (see Exercise 15).

Lemma 33 dualizes as follows.

35. Lemma. Let $\psi: M^m \to N^n$ be a smooth map, and let $p \in M$. The following are equivalent:

(1) The differential map $d\psi_p$ is onto.

(2) The Jacobian matrix of $d\psi_p$ has rank n relative to one (hence every) choice of coordinates at p and $\psi(p)$.

(3) If $y^1, \ldots, y^n$ is a coordinate system for N at $\psi(p)$, there is a coordinate system for M at p of the form $(y^1 \circ \psi, \ldots, y^n \circ \psi, x^{n+1}, \ldots, x^m)$.

Proof. It is easy to see that (2), for one choice of coordinates, implies (1), and that (1) implies (2) for every choice of coordinates. For coordinates as in (3), (2) is certainly true. Assume (2) holds for coordinates $y^1, \ldots, y^n$ at $\psi(p)$ and $x^1, \ldots, x^m$ at p, where necessarily $m \geq n$. At p the $n \times m$ Jacobian matrix $(\partial(y^i \circ \psi)/\partial x^j)$ has, by hypothesis, n linearly independent columns. Relabeling, if necessary, we may assume these are the first n columns. But then by the Jacobian criterion in Exercise 7, $(y^1 \circ \psi, \ldots, y^n \circ \psi, x^{n+1}, \ldots, x^m)$ is a coordinate system at p. ∎

A point $q \in N$ is called a *regular value* of a smooth mapping $\psi: M \to N$ provided that $d\psi_p$ is onto for every $p \in \psi^{-1}(q)$.

36. Corollary. If $q \in \psi(M)$ is a regular value of a smooth mapping $\psi: M \to N$, then $\psi^{-1}(q)$ is a submanifold of M, and $\dim M = \dim N + \dim \psi^{-1}(q)$.

Proof. We use the slice condition from Proposition 31. At $p \in \psi^{-1}(q)$ let the notation be as in the proof above. Reordering the coordinate functions, $(x^{n+1}, \ldots, x^m, y^1 \circ \psi, \ldots, y^n \circ \psi)$ is a coordinate system on a neighborhood $\mathscr{U}$ of p. But the slice of $\mathscr{U}$ through p is just $\mathscr{U} \cap \psi^{-1}(q)$. ∎

A *hypersurface* in a manifold M is a submanifold P whose *codimension* $\dim M - \dim P$ is 1. Applying the previous result with $N = \boldsymbol{R}^1$ gives the following efficient way to get hypersurfaces.

37. Corollary. Let c be a value of the function $f \in \mathfrak{F}(M)$. If at each point of $f^{-1}(c) = \{p \in M : f(p) = c\}$ the differential df_p is nonzero, then $f^{-1}(c)$ is a submanifold of M called a *level hypersurface* of f.

For example, on $\boldsymbol{R}^{n+1}$ let $f = \sum (u^i)^2$. Then $df = 2 \sum u^i\, du^i$, so f and df are each zero only at the origin. Thus the n-sphere $S^n(r) = f^{-1}(r^2)$ of radius $r > 0$ is a hypersurface in $\boldsymbol{R}^{n+1}$.

38. Definition. A *submersion* $\psi: M \to B$ is a smooth mapping onto B such that $d\psi_p$ is onto for all $p \in M$.

Since every value of a submersion is regular, M is partitioned into the submanifolds $\psi^{-1}(q)$ for all $q \in B$. For $m \geq n$ the projection $\boldsymbol{R}^m \to \boldsymbol{R}^n$ sending $(t_1, \ldots, t_m)$ to $(t_1, \ldots, t_n)$ is clearly a submersion. Lemma 35(3) implies that locally every submersion has this form.

TOPOLOGY OF MANIFOLDS

Manifolds are locally Euclidean; that is, each point of a manifold M has a neighborhood homeomorphic (by a coordinate system) to an open set in Euclidean space. Thus manifolds share all the local properties of Euclidean space, notably local connectedness and local compactness. A locally Euclidean space need not be Hausdorff; hence the inclusion of the Hausdorff axiom in Definition 3.

A. Connectedness

As for any topological space, a manifold is *connected* provided it cannot be expressed as the disjoint union of two nonempty open sets. For a manifold its coordinate neighborhoods show that any point can be connected to a sufficiently nearby point by a (smooth) curve segment. It follows that *a manifold is connected if and only if any two of its points can be joined by a piecewise smooth curve segment.*

An arbitrary space S is the disjoint union of its *connected components*, these being the maximal connected subspaces of S. Because a manifold is locally Euclidean its components are open, hence are (connected) manifolds.

B. Second Countability

A topological space S is *second countable* provided its topology has a *countable base* $\mathscr{B}$, that is, a countable collection of open sets such that every open set is the union of some subcollection of $\mathscr{B}$. For example, $\boldsymbol{R}^n$ has a countable basis consisting of all sets $\{p \in \boldsymbol{R}^n : a_i < p_i < b_i\}$ where a_i and b_i are rational numbers. It is easy to verify that (1) any subspace of a second countable space is second countable, (2) a cartesian product of second countable spaces is second countable, and (3) a space that is a countable union of second countable open subspaces is itself second countable.

An *[open] covering* $\mathfrak{C}$ of a topological space S is a collection of [open] subsets of S whose union is S. A second countable space S has the Lindelöf property: *Any open covering $\mathfrak{C}$ of S has a countable subcovering* (subcollection of $\mathfrak{C}$ that still covers S). In particular, a second countable manifold has only countably many connected components. There do exist connected manifolds that are not second countable but they are merely curiosities.

Henceforth we shall assume, whenever convenient, that the manifolds we deal with are second countable.

C. Partitions of Unity

A collection $\mathscr{L}$ of subsets of a space S is *locally finite* provided each point of S has a neighborhood that meets only finitely many elements of $\mathscr{L}$. Let $\{f_\alpha : \alpha \in A\}$ be a collection of smooth functions on a manifold M such that $\{\text{supp } f_\alpha : \alpha \in A\}$ is locally finite. Then the sum $\sum_\alpha f_\alpha$ is a well-defined smooth function on M, since on some neighborhood of each point all but a finite number of f_αs are identically zero.

39. Definition. A *smooth partition of unity* on a manifold M is a collection $\{f_\alpha : \alpha \in A\}$ of functions $f_\alpha \in \mathfrak{F}(M)$ such that

(1) $0 \leq f_\alpha \leq 1$ for all $\alpha \in A$.
(2) $\{\text{supp } f_\alpha : \alpha \in A\}$ is locally finite.
(3) $\sum_\alpha f_\alpha = 1$.

The partition is said to be *subordinate* to an open covering $\mathfrak{C}$ of M provided each set supp f_α is contained in some element of $\mathfrak{C}$.

Partitions of unity are an indispensable tool for assembling locally defined objects into a global object (or decomposing a global object into a sum of local objects). For such purposes partitions of unity with "small" supports are needed, as follows.

40. Proposition. If M is a (second countable) manifold then given any open covering $\mathfrak{C}$ of M there is a smooth partition of unity subordinate to $\mathfrak{C}$.

For a proof of this and related results, see [Mat], [Sp].

The existence of partitions of unity subordinate to arbitrary open coverings is equivalent to the topological property *paracompactness.* Thus by the proposition, a second countable manifold is paracompact. (The converse fails only when the manifold has noncountably many components—a possibility of scant practical importance.)

D. Orientability

This subtle topological property has many different characterizations; for manifolds the following is simplest.

41. Definition. A manifold M is *orientable* provided there exists a collection $\mathcal{O}$ of coordinate systems in M whose domains cover M and such that for each $\xi, \eta \in \mathcal{O}$ the Jacobian determinant function $J(\xi, \eta) = \det(\partial y^i/\partial x^j)$ is positive. ($\mathcal{O}$ is called an *orientation atlas* for M.)

Evidently $\boldsymbol{R}^n$ is orientable, since it can be covered by a single coordinate system. Experiment shows that the sphere and torus are orientable, but Möbius bands and Klein bottles are not.

The standard definition of manifold starts with a topological space, but actually the atlas of a manifold determines its topology, and this approach is often the most practical way to construct manifolds.

42. Proposition. Let Σ be an abstract set, and for each $\alpha \in A$ let ξ_α be a one-to-one function from a subset $\mathcal{U}_\alpha$ of Σ to an open set $\xi_\alpha(\mathcal{U}_\alpha)$ in $\boldsymbol{R}^n$. Suppose the following.

(1) The domains $\{\mathcal{U}_\alpha : \alpha \in A\}$ cover Σ.

(2) For all $\alpha, \beta \in A$ the function $\xi_\beta \circ \xi_\alpha^{-1}$ is Euclidean smooth—its domain $\xi_\alpha(\mathcal{U}_\alpha \cap \mathcal{U}_\beta)$ an open set of $\boldsymbol{R}^n$.

(3) If $p \neq q$ in Σ, then *either* p and q are in a single $\mathcal{U}_\alpha$ *or* there are $\alpha, \beta \in A$ such that $p \in \mathcal{U}_\alpha$, $q \in \mathcal{U}_\beta$, with $\mathcal{U}_\alpha$ and $\mathcal{U}_\beta$ disjoint.

Then there is a unique Hausdorff topology and complete atlas on Σ such that each ξ_α is a coordinate system in the resulting manifold. Furthermore, if countably many $\mathcal{U}_\alpha$ cover Σ the manifold is second countable.

Proof. Since each ξ_α is to be a homeomorphism there is no choice but to define a subset $\mathcal{V}$ of Σ to be *open* if and only if $\xi_\alpha(\mathcal{V} \cap \mathcal{U}_\alpha)$ is open in $\xi_\alpha(\mathcal{U}_\alpha)$—hence in $\boldsymbol{R}^n$—for all $\alpha \in A$. It is easy to check that such open sets constitute a topology on Σ. Note that by (2) the domains $\mathcal{U}_\alpha$ are open.

We assert that each ξ_α is a homeomorphism. If $\mathcal{V}$ is open in $\mathcal{U}_\alpha$ hence in Σ, then by definition $\xi_\alpha(\mathcal{V})$ is open in $\xi_\alpha(\mathcal{U}_\alpha)$. Conversely, if $\mathcal{W}$ is an open set in $\xi_\alpha(\mathcal{U}_\alpha)$ we must show that $\xi_\alpha^{-1}(\mathcal{W})$ is open. If $\beta \in A$ then applying (2) to both α, β and β, α shows that $\xi_\beta \circ \xi_\alpha^{-1}$ is a homeomorphism from $\xi_\alpha(\mathcal{U}_\alpha \cap \mathcal{U}_\beta)$ to $\xi_\beta(\mathcal{U}_\alpha \cap \mathcal{U}_\beta)$. But then the expression on the right-hand side in the formula

$$\xi_\beta(\xi_\alpha^{-1}\mathcal{W} \cap \mathcal{U}_\beta) = (\xi_\beta \xi_\alpha^{-1})(\mathcal{W} \cap \xi_\alpha(\mathcal{U}_\alpha \cap \mathcal{U}_\beta))$$

shows that this set is open. Consequently $\xi_\alpha^{-1}(\mathcal{W})$ is open.

Thus $\{\xi_\alpha : \alpha \in A\}$ is an atlas on the topological space Σ, which by (3) is Hausdorff. The countability assertion is clear, since each $\mathcal{U}_\alpha$, being homeomorphic to an open set in $\boldsymbol{R}^n$, is second countable. ■

SOME SPECIAL MANIFOLDS

A. Product Manifolds

We consider how the calculus on a product manifold $M \times N$ (Lemma 5) derives from that of M and N separately. This decomposition is modeled closely on the way the calculus of the plane $\boldsymbol{R}^2 = \boldsymbol{R}^1 \times \boldsymbol{R}^1$ derives from that of the real line.

Using product coordinate systems on $M \times N$ it is easy to check that

(a) The *projections*

$$\pi: M \times N \to M \quad \text{sending} \quad (p, q) \quad \text{to} \quad p,$$
$$\sigma: M \times N \to N \quad \text{sending} \quad (p, q) \quad \text{to} \quad q$$

are smooth mappings—in fact, submersions.

(b) A mapping $\phi: P \to M \times N$ is smooth if and only if both $\pi \circ \phi$ and $\sigma \circ \phi$ are smooth.

(c) For each $(p, q) \in M \times N$ the subsets

$$M \times q = \{(r, q) \in M \times N : r \in M\},$$
$$p \times N = \{(p, r) \in M \times N : r \in N\}$$

are submanifolds of $M \times N$.

(d) For each (p, q)

$$\pi | M \times q \quad \text{is a diffeomorphism from} \quad M \times q \quad \text{to} \quad M;$$
$$\sigma | p \times N \quad \text{is a diffeomorphism from} \quad p \times N \quad \text{to} \quad N.$$

By (b) the tangent spaces

$$T_{(p,q)}M \equiv T_{(p,q)}(M \times q) \qquad \text{and} \qquad T_{(p,q)}N \equiv T_{(p,q)}(p \times N)$$

are subspaces of the tangent space to $M \times N$ at (p, q).

43. Lemma. $T_{(p,q)}(M \times N)$ is the direct sum of its subspaces $T_{(p,q)}M$ and $T_{(p,q)}N$; that is, each element of $T_{(p,q)}(M \times N)$ has a unique expression as

$$x + v, \qquad \text{where} \quad x \in T_{(p,q)}M \quad \text{and} \quad v \in T_{(p,q)}N.$$

Proof. Since $\pi | p \times N$ is constant, $d\pi$ at (p, q) sends $T_{(p,q)}N$ to 0. But by (c), $d\pi | T_{(p,q)}M$ is an isomorphism. Thus $T_{(p,q)}M \cap T_{(p,q)}N = 0$. The result then follows by linear algebra, since by (c) the sum of the dimensions of these two subspaces is $\dim(M \times N)$. ∎

To relate the calculus of $M \times N$ to that of its factors the crucial notion is that of *lifting*, as follows.

If $f \in \mathfrak{F}(M)$ the *lift* of f to $M \times N$ is $\tilde{f} = f \circ \pi \in \mathfrak{F}(M \times N)$.

If $x \in T_p(M)$ and $q \in N$ then the *lift* $\tilde{x}$ of x to (p, q) is the unique vector in $T_{(p,q)}(M)$ such that $d\pi(\tilde{x}) = x$.

If $X \in \mathfrak{X}(M)$ the *lift* of X to $M \times N$ is the vector field $\tilde{X}$ whose value at each (p, q) is the lift of X_p to (p, q). Product coordinate systems show that $\tilde{X}$ is smooth. Thus *the lift of* $X \in \mathfrak{X}(M)$ *to* $M \times N$ *is the unique element of* $\mathfrak{X}(M \times N)$ *that is* π*-related to* X *and* σ*-related to the zero vector field on* N.

The set of all such *horizontal lifts* $\tilde{X}$ is denoted by $\mathfrak{L}(M)$.

Functions, tangent vectors, and vector fields on N are lifted to $M \times N$ in the same way using the projection σ. Note that $\mathfrak{L}(M)$ and symmetrically the *vertical lifts* $\mathfrak{L}(N)$ are vector subspaces of $\mathfrak{X}(M \times N)$ but (except in trivial cases) neither is invariant under multiplication by arbitrary functions $f \in \mathfrak{F}(M \times N)$.

For example, on $\boldsymbol{R}^2$ the natural coordinate vector field $\partial_x = \partial/\partial x$ is the horizontal lift of the vector field d/dx on $\boldsymbol{R}^1$ (as x axis), but $y\,\partial_x$ is not a lift.

44. Corollary. (1) If $\tilde{X}, \tilde{Y} \in \mathfrak{L}(M)$ then $[\tilde{X}, \tilde{Y}] = [X, Y]^{\sim} \in \mathfrak{L}(M)$, and similarly for $\mathfrak{L}(N)$.

(2) If $\tilde{X} \in \mathfrak{L}(M)$ and $\tilde{V} \in \mathfrak{L}(N)$, then $[\tilde{X}, \tilde{V}] = 0$.

Proof. Both assertions follow from Lemma 22. In the case of (2) for example, $[\tilde{X}, \tilde{V}]$ is π-related to $[X, 0] = 0$ and σ-related to $[0, V] = 0$. Thus the result follows by Lemma 43. ∎

To establish these facts we have used a rather elaborate notation. In practice, Lemma 43 is usually rendered as $T_{(p,q)}(M \times N) = T_pM \times T_pN$, and the tilde ($\sim$) is omitted from lifts.

B. Vector Spaces as Manifolds

Let V be an n-dimensional vector space over $\boldsymbol{R}$. If ξ and η are linear isomorphisms from V to $\boldsymbol{R}^n$, then $\xi \circ \eta^{-1}: \boldsymbol{R}^n \to \boldsymbol{R}^n$ is a linear isomorphism—hence a diffeomorphism. Thus Proposition 42 is scarcely needed to show that there is a unique way to make V a manifold such that every linear isomorphism $\xi: V \to \boldsymbol{R}^n$ is a coordinate system.

The following convenient notation will be used frequently.

45. Definition. If $p, v \in V$, let $v_p \in T_p(V)$ be the initial velocity $\alpha'(0)$ of the curve $\alpha(t) = p + tv$.

We picture v_p as the arrow running from p to $p + v$.

46.   Lemma.   If $x^1, \ldots, x^n$ is a linear coordinate system on V, then

$$v_p = \sum x^i(v)\, \partial_i|_p.$$

Proof.   Since the coordinates x^i are linear,

$$x^i(\alpha(t)) = x^i(p) + tx^i(v).$$

Hence by the velocity formula,

$$v_p = \alpha'(0) = \sum \frac{d(x^i \circ \alpha)}{dt}(0)\, \partial_i|_p = \sum x^i(v)\, \partial_i|_p. \qquad \blacksquare$$

Thus v_p is the tangent vector at p with the same coordinates as $v \in V$. It follows immediately that

(1)   for fixed $p \in V$, the function $v \to v_p$ is a linear isomorphism $V \approx T_p(V)$;

(2)   for $p, q \in V$, the function $v_p \to v_q$ is a linear isomorphism $T_p(V) \approx T_q(V)$.

As in the familiar case $V = R^3$ these *canonical isomorphisms* allow free interchange between v (point of V), v_0 (arrow from 0 to v), and v_p (arrow from p to $p + v$).

The *position vector field* $P \in \mathfrak{X}(V)$ assigns to each $p \in V$ the tangent vector $p_p \in T_p(V)$ (intuitively a duplicate, starting at p, of the arrow from 0 to p). In terms of a linear coordinate system, $P = \sum x^i\, \partial_i$.

C.   The Tangent Bundle

For a manifold M, let TM be the set $\bigcup \{T_p(M) : p \in M\}$ of all tangent vectors to M. A technicality: For each $p \in M$ replace $0 \in T_p(M)$ by 0_p (otherwise the zero tangent vector is in every tangent space). Then each $v \in TM$ is in a unique $T_p(M)$, and the *projection* $\pi: TM \to M$ sends v to p. Thus $\pi^{-1}(p) = T_p(M)$,

There is a natural way to make TM a manifold, called the *tangent bundle* of M. Let ξ be a coordinate system on $\mathcal{U} \subset M$. If v is tangent to M at a point p of $\mathcal{U}$, then v is uniquely determined by the coordinates of p and the coordinates of v relative to $\partial_1, \ldots, \partial_n$ at p. To formalize this, let $\dot{x}^i$ be the real-valued function on $\pi^{-1}(\mathcal{U}) \subset M$ given by $\dot{x}^i(v) = v(x^i)$. Then define $\tilde{\xi}: \pi^{-1}(\mathcal{U}) \to R^{2n}$ by

$$\tilde{\xi} = (x^1 \circ \pi, \ldots, x^n \circ \pi, \dot{x}^1, \ldots, \dot{x}^n).$$

We shall now apply Proposition 42 to make TM a manifold with all such functions $\tilde{\xi}$ as coordinate systems.

By the basis theorem, if $v \in \pi^{-1}(\mathcal{U})$, then $v = \sum \dot{x}^i(v)\, \partial_i|_{\pi v}$. Thus $\tilde{\xi}$ is a one-to-one function from $\pi^{-1}(\mathcal{U})$ onto the open set $\xi(\mathcal{U}) \times R^n$ of R^{2n}.

Next we show that any two such functions $\tilde{\xi}$ and $\tilde{\eta}$ overlap smoothly. If $(a, b) \in \eta(\mathscr{U} \cap \mathscr{V}) \times \mathbf{R}^n$, then for $1 \leq i \leq n$

$$u^i \tilde{\xi}\tilde{\eta}^{-1}(a, b) = x^i \pi \tilde{\eta}^{-1}(a, b) = x^i \eta^{-1}(a).$$

Since $\partial/\partial y^i = \sum (\partial x^k/\partial y^i)\, \partial/\partial x^k$, we also have

$$u^{n+i} \tilde{\xi}\tilde{\eta}^{-1}(a, b) = \dot{x}^i \tilde{\eta}^{-1}(a, b) = \sum b^k \frac{\partial x^i}{\partial y^k}(\eta^{-1} a).$$

Thus $\tilde{\xi} \circ \tilde{\eta}^{-1}$ is Euclidean smooth.

It is easy to check the conditions in Proposition 42 that show TM is a Hausdorff and second countable.

A vector field $X \in \mathfrak{X}(M)$ is exactly a smooth section of TM, that is, a smooth function $X: M \to TM$ such that $\pi \circ X = \text{id}$. This suggests a useful generalization.

47. Definition. A *vector field Z on a smooth map $\phi: P \to M$* is mapping $Z: P \to TM$ such that $\pi \circ Z = \phi$, where π is the projection $TM \to M$.

Thus Z assigns to each point $p \in P$ a tangent vector to M at $\phi(p)$. (For example, its velocity α' is a vector field on a curve α in M.) Z is smooth as a mapping $P \to TM$ if and only if $f \in \mathfrak{F}(M)$ implies $Zf \in \mathfrak{F}(P)$, where $(Zf)(p) = Z(p)f$ for all $p \in P$. The set $\mathfrak{X}(\phi)$ of all smooth vector fields on $\phi: P \to M$ is, in a natural way, a module over $\mathfrak{F}(P)$.

INTEGRAL CURVES

A vector field on a manifold can be interpreted as a differential equation for which the appropriate notion of solution is as follows.

48. Definition. A curve $\alpha: I \to M$ is an *integral curve* of $V \in \mathfrak{X}(M)$ provided $\alpha' = V_\alpha$; that is, $\alpha'(t) = V_{\alpha(t)}$ for all $t \in I$.

Thus at each point the curve α has the velocity prescribed by V. If the equation above is expressed in terms of a coördinate system ξ, it yields a system of first-order ordinary differential equations

$$\frac{d(x^i \circ \alpha)}{dt} = F^i(x^1 \circ \alpha, \ldots, x^n \circ \alpha) \qquad (1 \leq i \leq n),$$

where F^i is the coordinate expression for Vx^i.

Since the parameter t does not appear on the right-hand side, we can think of V as giving the velocity of the steady state flow of a fluid through M—an idea pursued below. The fundamental existence and uniqueness theorem for the solutions of such systems has the following consequence in terms of manifold theory.

49. Proposition. If $V \in \mathfrak{X}(M)$ then for each $p \in M$ there is an interval I around 0 and a unique integral curve $\alpha: I \to M$ of V such that $\alpha(0) = p$.

Note that if α is an integral curve of V, then $t \to \alpha(t + c)$ is also.

50. Corollary. If $\alpha, \beta: I \to M$ are integral curves of V such that $\alpha(a) = \beta(a)$ for some $a \in I$, then $\alpha = \beta$.

Proof. By continuity the agreement set $A = \{t \in I : \alpha(t) = \beta(t)\}$ is closed. If A is also open, then since A is nonempty, $A = I$. Fix $t \in A$. Then $s \to \alpha(t + s)$ and $s \to \beta(t + s)$ are integral curves of V that agree at $s = 0$. Hence by Proposition 49 they agree for s sufficiently near 0. ∎

Consider the collection of all integral curves $\alpha: I_\alpha \to M$ of V that *start at* $p \in M$, that is, for which $\alpha(0) = p$. For any two such, the corollary shows that $\alpha = \beta$ on $I_\alpha \cap I_\beta$. Thus by a remark on page 5, all these curves define a single integral curve $\alpha_p: I_p \to M$ where $I_p = \bigcup I_\alpha$. We call α_p the *maximal* integral curve of V starting at p. Examples as in Exercise 12(c) show that this largest possible domain I_p need not be the entire real line.

51. Example. On the plane $\boldsymbol{R}^2$ let $V = x\,\partial_x - y\,\partial_y$. Then $\alpha(t) = (x(t), y(t))$ is an integral curve of V if and only if $dx/dt = x$ and $dy/dt = -y$. Hence $x(t) = Ae^t$ and $y(t) = Be^{-t}$. Thus the maximal integral curve starting at $p = (p_1, p_2)$ is

$$\alpha_p(t) = (p_1 e^t, p_2 e^{-t}) \qquad \text{for all} \quad t \in \boldsymbol{R}.$$

These curves parametrize the hyperbolas xy constant, except for the four coordinate semiaxes and a constant curve at the origin.

The following refinement of Corollary 50 shows that if maximal integral curves of V meet at one point then they differ only in parametrization.

52. Lemma. For $V \in \mathfrak{X}(M)$ let $q = \alpha_p(s)$. Then $s + I_q = I_p$, and $\alpha_p(s + t) = \alpha_q(t)$ for all $t \in I_q$.

Proof. Let $\beta(u) = \alpha_q(u - s)$; then β is an integral curve of V defined on $s + I_q$. Since $\beta(s) = q = \alpha_p(s)$, Corollary 50 implies that $\beta = \alpha_p$ on $(s + I_q) \cap I_p$. Thus β and α_p combine to give a single integral curve on $(s + I_q) \cup I_p$. Since I_p is maximal, $s + I_q \subset I_p$. Hence $\alpha_p(s + t) = \beta(s + t) = \alpha_q(t)$ for all $t \in I_q$. Then since I_q is maximal, $I_q \supset -s + I_p$. Thus $s + I_q = I_p$. ∎

A nonconstant curve $\gamma: \boldsymbol{R} \to M$ is *periodic* if there is a number $c > 0$ such that $\gamma(t + c) = \gamma(t)$ for all t. Since γ is nonconstant it follows that there is a smallest such $c > 0$, called the *period* of γ. If γ is one-to-one on some interval $[a, a + c)$, then γ is *simply periodic*. Thus a figure-8 curve is periodic but not simply periodic.

It follows from the preceding lemma that *every maximal integral curve is either one-to-one, simply periodic, or constant.*

A vector field is *complete* if each of its maximal integral curves is defined on the entire real line. We shall now assemble all the integral curves of a given vector field V into a single mapping, assuming at first that V is complete.

53. Definition. The *flow* of a complete vector field V on M is the mapping $\psi: M \times \boldsymbol{R}^1 \to M$ given by

$$\psi(p, t) = \alpha_p(t),$$

where α_p is the maximal integral curve starting at p.

If p is held constant, then $t \to \psi(p, t)$ is just the integral curve α_p. On the other hand, if t is held constant, then $p \to \psi(p, t)$ defines a function $\psi_t: M \to M$ that lets every point p of M flow for exactly time t. We call ψ_t the tth *stage* of the flow ψ, and sometime refer to $\{\psi_t : t \in \boldsymbol{R}\}$ as the flow of V.

54. Lemma. If ψ is the flow of a complete vector field, then:

(1) ψ_0 is the identity map id of M.
(2) $\psi_s \circ \psi_t = \psi_{s+t}$ for all $s, t \in \boldsymbol{R}$ (so the stages of ψ commute).
(3) Each stage is a diffeomorphism, with $\psi_t^{-1} = \psi_{-t}$.

Proof. (1) Obvious, since $\psi_0(p) = \alpha_p(0) = p$. (2) This follows immediately from Lemma 52. (3) By the preceding properties, $\psi_t \circ \psi_{-t} = \text{id} = \psi_{-t} \circ \psi_t$. ∎

If the vector field V is not complete, then for each point p of M we at least have a *local flow* $\psi: \mathscr{U} \times I \to M$ defined by the formula in Definition 53 but with $\mathscr{U}$ a neighborhood of p in M and I an interval around 0 in $\boldsymbol{R}^1$. By differential equations theory if $\mathscr{U}$ and I are sufficiently small then ψ is smooth. For such a local flow the analogue of Lemma 54 holds: (1) ψ_0 is the identity map of $\mathscr{U}$, (2) $\psi_{s+t} = \psi_s \circ \psi_t$ whenever s, t, and $s + t$ are in I, and (3) for $t \in I$, $\psi_t: \mathscr{U} \to \psi_t(\mathscr{U})$ is a diffeomorphism.

If V is not necessarily complete then its flow $\psi(p, t) = \alpha_p(t)$ has largest domain

$$\mathscr{D} = \{(p, t) \in M \times \boldsymbol{R}^1 : t \in I_p\}.$$

Patching together smooth local flows as above shows that $\mathscr{D}$ is an open set of $M \times \boldsymbol{R}^1$ and the flow $\psi\colon \mathscr{D} \to M$ is smooth. (See Theorem 5 in Chapter IV of [L].)

In considering the behavior of curves $\alpha\colon I \to M$ near the ends of the interval I it suffices to focus on the right end by taking $I = [0, B)$; the left end is dealt with analogously. (By convention $b < \infty$ but $B \leq \infty$.)

55. Definition. A piecewise smooth curve $\alpha\colon [0, B) \to M$ is *extendible* provided it has a continuous extension $\tilde{\alpha}\colon [0, B] \to M$. Then $q = \tilde{\alpha}(B)$ is called an *endpoint* of α.

Equivalently, there exists a point $q \in M$ such that for every sequence $\{s_i\}$ in $[0, B)$ approaching B, the sequence $\{\alpha(s_i)\}$ converges to q.

An extendible curve need not have a piecewise smooth extension, however this occurs in important special cases.

56. Lemma. Let $\alpha\colon [0, b) \to M$, $b < \infty$, be an integral curve of $V \in \mathfrak{X}(M)$. The following are equivalent:

(1) α is not maximal; that is, α is extendible as an integral curve of V to a larger interval $[0, b + \varepsilon)$.
(2) α is extendible.
(3) α lies in a compact set of M.
(4) There exists a sequence $\{s_i\} \to b$ such that $\{\alpha(s_i)\}$ converges.

Proof. Clearly (1) $\Rightarrow$ (2) $\Rightarrow$ (3) $\Rightarrow$ (4). To prove (4) $\Rightarrow$ (1), let $\mathscr{U}$ be a neighborhood of $\lim \alpha(s_i)$ such that a flow of V is defined on $\mathscr{U} \times (-\delta, \delta)$. There is an n such that $b - \delta < s_n$ and $\alpha(s_n) \in \mathscr{U}$. The integral curve of V starting at $\alpha(s_n)$ is defined on $[0, \delta)$. Hence by Lemma 52 it serves to extend α past b. ∎

Finally we give two applications of flows.

57. Lemma. If V is a vector field and p a point such that $V_p \neq 0$, then there is a coordinate system $x^1, \ldots, x^n$ at p such that $V = \partial/\partial x^1$ on the coordinate neighborhood.

Proof. Let $\psi\colon \mathscr{U} \times I \to M$ be a local flow for V, where $\mathscr{U}$ is a neighborhood of p on which V is never zero. Let S be a hypersurface through p in $\mathscr{U}$ such that $V_p \notin T_p(S)$, and consider the restriction $\psi\colon S \times I \to M$ of ψ.

The map $\psi | (S \times 0)$ is a trivial diffeomorphism to S, and $d\psi_{(p,0)}$ merely identifies $T_{(p,0)}(S)$ with $T_p(S)$. Since $d\psi(\partial/\partial t|_{(p,0)} = V_p \notin T_p(S)$, it follows that $d\psi_{(p,0)}$ is an isomorphism. Thus ψ is a diffeomorphism of some neighborhood $\mathscr{V} \times J$ of $(p, 0)$ onto a neighborhood $\mathscr{W}$ of p in M.

We can suppose that $y^2, \ldots, y^n$ is a coordinate system for S on $\mathscr{V}$. Let $y^1 = t$ be the natural coordinate on $J \subset \boldsymbol{R}^1$. Transferring the product coordinate system $y^1, \ldots, y^n$ to $\mathscr{W}$ by means of the diffeomorphism $\psi|(\mathscr{V} \times J)$ gives the required coordinate system $x^1, \ldots, x^n$. In fact, $\partial/\partial x^1 = d\psi(\partial/\partial t) = V$, since $\psi(p, t) = \alpha_p(t)$. ∎

The bracket of vector fields can be described in terms of flows. Recall first that a smooth function $F\colon I \to V$ into a finite-dimensional real vector space has derivative $F'\colon I \to V$ given by

$$F'(s) = \lim_{t \to 0} \frac{1}{t} [F(s+t) - F(s)] = \sum \frac{df^i}{dt}(s) e_i,$$

where $F = \sum f^i e_i$ relative to a basis $e_1, \ldots, e_n$ for V.

Now take the derivative of a vector field W with respect to a vector field V as follows. If $p \in M$ let α_p be the integral curve of V starting at p. Use the flow of V to move the value of W at $\alpha_p(s)$ backward along α_p into $T_p(M)$. Then take the vector derivative at $s = 0$. The result turns out to be $[V, W]_p$.

58. Proposition. If $V, W \in \mathfrak{X}(M)$, let ψ be a local flow of V near $p \in M$. Then

$$[V, W]_p = \lim_{t \to 0} \frac{1}{t} [d\psi_{-t}(W_{\psi_t p}) - W_p].$$

Proof. Write $F_p(t) = d\psi_{-t}(W_{\psi_t p})$. Then the right-hand side of the above equation is $F_p'(0)$.

Case 1. $V_p \neq 0$. Choose a coordinate system $x^1, \ldots, x^n$ as in Lemma 57 so that $V = \partial_1$. Thus the flow of V only changes the x^1 coordinate of points q near p:

$$x^1(\psi_t q) = x^1(q) + t, \qquad x^j(\psi_t q) = x^j(q) \qquad \text{for} \quad 2 \leq j \leq n.$$

It follows that $d\psi_t(\partial_i) = \partial_i$ for all i and t. Thus if $W = \sum W^i \partial_i$,

$$F_p(t) = \sum W^i(\psi_t p)\, \partial_i|_p.$$

Omitting the subscript p we compute

$$F_p'(0) = \sum \frac{d}{dt}(W^i \circ \alpha_p)(0)\, \partial_i = \sum V_p(W^i)\, \partial_i$$

$$= \sum \frac{\partial W^i}{\partial x^1}(0)\, \partial_i = [\partial_1, W]_p = [V, W]_p.$$

Case 2. $V = 0$ on a neighborhood of p. Then $[V, W]_p = 0$. Integral curves starting in the neighborhood are constant, hence $\psi_t = \text{id}$ for all t. Thus F_p is constant, hence $F_p'(0) = 0$.

Case 3. $V_p = 0$, but p is the limit of a sequence $\{p_i\}$ with $V_{p_i} \neq 0$ for all i. Their expressions in terms of coordinates show that both $F_p'(0)$ and $[V, W]_p$ depend continuously on p, hence the result follows by Case 1. ∎

Exercises

1. (a) A vector field V on M is smooth if for sufficiently many coordinate systems ξ to cover M the functions Vx^i are smooth. (b) A map $\phi\colon M \to N$ is smooth if for sufficiently many coordinate systems η to cover N the functions $y^j \circ \phi$ are smooth.

2. A linear transformation $\phi\colon V \to W$ is smooth, and $d\phi(v_p) = (\phi v)_{\phi_p}$.

3. (a) If id is the identity map of M, then $d(\text{id})_p$ is the identity map of $T_p(M)$. (b) If $\phi\colon M \to B$ is a diffeomorphism then for each $p \in M$, $d\phi_p$ is a linear isomorphism with inverse $d(\phi^{-1})_{\phi p}$.

4. On the domain of a coordinate system ξ, if $V = \sum V^i\, \partial_i$ and $W = \sum W^j\, \partial_j$, then

$$[V, W] = \sum_{ij} \left(V^i \frac{\partial W^j}{\partial x^i} - W^i \frac{\partial V^j}{\partial x^i} \right) \partial_j .$$

5. Given $v \in T_p(M)$, there is a vector field $V \in \mathfrak{X}(M)$ such that $V_p = v$ and a curve α such that $\alpha'(0) = v$.

6. (a) A map $\phi\colon M \to N$ is smooth if and only if $g \in \mathfrak{F}(N)$ implies $g \circ \phi \in \mathfrak{F}(M)$. (b) A smooth map $\phi\colon M \to N$, with M connected, is constant if and only if $d\phi_p = 0$ for all $p \in M$.

7. Smooth real-valued functions $y^1, \ldots, y^n$ on a neighborhood of p in M form a coordinate system on a (possibly smaller) neighborhood of p if and only if for one, hence every, coordinate system $x^1, \ldots, x^n$ at p,

$$\det\left(\frac{\partial y^i}{\partial x^j}(p) \right) \neq 0.$$

8. (a) If P and Q are submanifolds of M and $P \subset Q$ then P is a submanifold of Q. (b) If P is a submanifold of M such that $\dim P = \dim M$, then P is an open submanifold of M.

9. (a) A smooth map ψ of M onto B is a submersion if and only if ψ has *local sections*; that is, given $p \in M$ there is a neighborhood $\mathcal{U}$ of $\psi(p)$ in B and a smooth map $\lambda\colon \mathcal{U} \to M$ such that $\psi \circ \lambda = \text{id}$ and $\lambda(\psi p) = p$.

10. Let $\psi: M \to B$ be a submersion. (a) A vector $v \in T_p(M)$ is tangent to the fiber $\psi^{-1}(\psi p)$ if and only if $d\psi(v) = 0$. (b) A map $\phi: B \to N$ is smooth if and only if $\phi \circ \psi$ is smooth.

11. Let $\phi: M \to N$ be a smooth map and let $p \in M$ be a point at which $d\phi_p$ is a linear isomorphism. (a) Given a coordinate system ξ at p *or* η at $\phi(p)$ the other can be selected so that ϕ *preserves coordinates*: $y^i(\phi q) = x^i(q)$ for q near p. (b) Then $d\phi(\partial/\partial x^i) = \partial/\partial y^i$, and if $f \in \mathfrak{F}(N)$, $\partial(f \circ \phi)/\partial x^i = (\partial f/\partial y^i) \circ \phi$ near p.

12. In each case find a formula for the maximal integral curve α_p starting at an arbitrary point p. (a) $V = y\,\partial_x - x\,\partial_y$ on $\boldsymbol{R}^2$. (b) $V = x\,\partial_x + (x + y)\,\partial_y$ on $\boldsymbol{R}^2$. Sketch a few integral curves. (c) $V = u^2\, d/du$ on $\boldsymbol{R}^1$. Note that V is not complete; find the largest domain $\mathscr{D}$ of its flow.

13. For a function $f \in \mathfrak{F}(M)$, its differential $df: T(M) \to \boldsymbol{R}$ and its differential map $df: T(M) \to T(\boldsymbol{R}^1)$ differ only by a canonical isomorphism.

14. A *critical point* of $f \in \mathfrak{F}(M)$ is a point $p \in M$ at which $df_p = 0$. (a) At such a point there exists a Hessian function $H\colon T_pM \times T_pM \to \boldsymbol{R}$ such that $H(X_p, Y_p) = X_p(Yf) = Y_p(Xf)$ for all $X, Y \in \mathfrak{X}(M)$. (b) H is bilinear, symmetric, and satisfies $H(\partial_i|_p, \partial_j|_p) = (\partial^2 f/\partial x^i\,\partial x^j)(p)$ relative to a coordinate system. Also $H(v, v) = (d^2(f \circ \alpha)/ds^2)(0)$ if $\alpha'(0) = v$.

15. (a) Let $\phi: P \to M$ be a one-to-one immersion; make $\phi(P)$ a manifold so that $P \to \phi(P)$ is a diffeomorphism. Show that $\phi(P)$ is an immersed submanifold of M, and that if P is compact $\phi(P)$ is a submanifold. (b) By considering a subset of $\boldsymbol{R}^2$ shaped like a figure 8, show that an immersed submanifold need not have the induced topology and that Corollaries 29 and 30 both fail for immersed submanifolds.

16. Let ψ be the flow of $V \in \mathfrak{X}(M)$. (a) If $\psi_t(p) = p$ for a sequence of t-values approaching zero, then $V_p = 0$ (hence $\psi_t(p) = p$ for all t). (b) If an integral curve $\alpha\colon [0, \infty) \to M$ of V is extendible, with endpoint q, then $V_q = 0$.

17. (a) Let $a \in I \subset \boldsymbol{R}$. If $f \in \mathfrak{F}(I)$ and $f(a) = 0$ show that there exists a function $g \in \mathfrak{F}(I)$ such that $f(s) = (s - a)g(s)$ for all $s \in I$. (b) Prove the analogue with f replaced by a vector field on a curve.

18. Let $\phi: P \to M$ be an immersion and fix $p \in P$. (a) If $f \in \mathfrak{F}(P)$ there exists a function $\tilde{f} \in \mathfrak{F}(M)$ such that $f = \tilde{f} \circ \phi$ on some neighborhood of p in P. (Hint: See Proposition 33(3).) (b) If $Z \in \mathfrak{X}(\phi)$ there exists $X \in \mathfrak{X}(M)$ such that $Z = X_\phi$ on some neighborhood of p in P.

19. An atlas on a space S is *maximal* if it is not contained in any strictly larger atlas on S. Prove that an atlas is maximal if and only if it is complete.

2 TENSORS

The notion of tensor field on a manifold generalizes the notions of real-valued function, vector field, and one-form, and thus provides the mathematical means of describing more complicated objects on a manifold. Tensors occur in many different guises, but their characteristic property is always *multilinearity*. The definition we use stresses this and converts easily into the classical coordinate description of tensor.

The last part of the chapter deals with the generalization of *inner product* on which semi-Riemannian geometry is based.

BASIC ALGEBRA

The following definitions cover the two main cases we need: the module $\mathfrak{X}(M)$ over $\mathfrak{F}(M)$, and the vector space $T_p(M)$ over $\boldsymbol{R}$.

Let $V_1, \dots, V_s$ be modules over a ring K. Then $V_1 \times \cdots \times V_s$ is the set of all s-tuples $(v_1, \dots, v_s)$ with $v_i \in V_i$. The usual componentwise definitions of addition and of multiplication by an element of K make $V_1 \times \cdots \times V_s$ a module over K, called a *direct product* (or *direct sum* if the notation $\times$ is replaced by $\oplus$). If W is also a module over K, a function

$$A: V_1 \times \cdots \times V_s \to W$$

is *K-multilinear* provided A is K-linear *in each slot*, that is, for $1 \leq i \leq s$ and $v_j \in V_j$ $(j \neq i)$, the function

$$v \to A(v_1, \dots, v_{i-1}, v, v_{i+1}, \dots, v_s)$$

is K-linear.

If V is a module over K, let V^* be the set of all K-linear functions from V to K. The usual definition of addition of functions and multiplication by elements of K makes V^* a module over K, called the *dual module* of V.

If $V_i = V$ for $1 \leq i \leq s$, the notation $V_1 \times \cdots \times V_s$ is abbreviated to V^s.

1. Definition. For integers $r \geq 0$, $s \geq 0$ not both zero, a K-multilinear function $A: (V^*)^r \times V^s \to K$ is called a *tensor of type* (r, s) *over* V. (Here we understand $A: V^s \to K$ if $r = 0$, and $A: (V^*)^r \to K$ if $s = 0$.)

The set $\mathfrak{T}^r_s(V)$ of all tensors of type (r, s) over V is a module over K, again with the usual definitions of functional addition and multiplication by an element of K. A *tensor of type* (0, 0) *over* V is simply an element of K.

TENSOR FIELDS

A *tensor field* A on a manifold M is a tensor over the $\mathfrak{F}(M)$-module $\mathfrak{X}(M)$, as defined above. Thus if A has type (r, s) it is an $\mathfrak{F}(M)$-multilinear function

$$A: \mathfrak{X}^*(M)^r \times \mathfrak{X}(M)^s \to \mathfrak{F}(M).$$

So A is a multilinear machine which when fed r one-forms $\theta^1 \ldots, \theta^r$ and s vector fields $X_1, \ldots, X_s$ produces a real-valued function

$$f = A(\theta^1, \ldots, \theta^r, X_1, \ldots, X_s) \in \mathfrak{F}(M).$$

Here θ^i occupies the ith *contravariant slot*, X_j the jth *covariant slot* of A.

The set $\mathfrak{T}^r_s(M)$ of all tensor fields on M of type (r, s) is then a module over $\mathfrak{F}(M)$. In the exceptional case $r = s = 0$, a tensor field on M of type (0, 0) is just a function $f \in \mathfrak{F}(M)$; that is, $\mathfrak{T}^0_0(M) = \mathfrak{F}(M)$.

To show that a given function $A: \mathfrak{X}^*(M)^r \times \mathfrak{X}(M)^s \to \mathfrak{F}(M)$ is a tensor we must show that it is $\mathfrak{F}(M)$-linear in each slot (i.e., in each variable separately). Additivity in each slot is often obvious, so the crucial question is whether functions can be factored out of each slot:

$$A(\theta_1, \ldots, \theta^r, X_1, \ldots, fX_i, \ldots, X_s) = fA(\theta^1, \ldots, \theta^r, X_1, \ldots, X_i, \ldots, X_s).$$

Consider the following examples. (1) The evaluation function $E: \mathfrak{X}^*(M) \times \mathfrak{X}(M) \to \mathfrak{F}(M)$ is given by $E(\theta, X) = \theta X$. Clearly E is $\mathfrak{F}(M)$-linear in each slot, so it is a tensor field on M of type (1, 1).

(2) Fix a one-form $\omega \neq 0$ and define $F: \mathfrak{X}(M) \times \mathfrak{X}(M) \to \mathfrak{F}(M)$ by $F(X, Y) = X(\omega Y)$ for all X, Y. Then F is $\mathfrak{F}(M)$-linear in X but only additive in Y. In fact,

$$F(X, fY) = X\omega(fY) = X(f\omega Y) = (Xf)\omega Y + fF(X, Y).$$

Thus F is not a tensor field.

In these examples, note that the tensor (1) is algebraic, while the nontensor (2) involves differentiation.

While we only add tensors of the same type, any two tensors can be multiplied as follows: If $A \in \mathfrak{T}^r_s(M)$ and $B \in \mathfrak{T}^{r'}_{s'}(M)$, define

$$A \otimes B: \mathfrak{X}^*(M)^{r+r'} \times \mathfrak{X}(M)^{s+s'} \to \mathfrak{F}(M)$$

by

$$(A \otimes B)(\theta^1, \ldots, \theta^{r+r'}, X_1, \ldots, X_{s+s'})$$
$$= A(\theta^1, \ldots, \theta^r, X_1, \ldots, X_s)B(\theta^{r+1}, \ldots, \theta^{r+r'}, X_{s+1}, \ldots, X_{s+s'}).$$

Then $A \otimes B$ is a tensor of type $(r + r', s + s')$, called the *tensor product* of A and B. If $r' = s' = 0$, so B is a function $f \in \mathfrak{F}(M)$, define

$$A \otimes f = f \otimes A = fA.$$

Thus if A is also of type (0, 0), the tensor product reduces to ordinary multiplication in $\mathfrak{F}(M)$.

Evidently the tensor product is $\mathfrak{F}(M)$-bilinear, that is,

$$(fA + gA') \otimes B = fA \otimes B + gA' \otimes B,$$

with a similar identity for B. Furthermore it is immediate from the definition that the tensor product is associative; thus $A \otimes B \otimes C$ is well defined for tensors of any types. However the tensor product is generally not commutative. For example, on a coordinate neighborhood

$$(dx^1 \otimes dx^2)(\partial_1, \partial_2) = dx^1(\partial_1)\, dx^2(\partial_2) = 1,$$

$$(dx^2 \otimes dx^1)(\partial_1, \partial_2) = dx^2(\partial_1)\, dx^1(\partial_2) = 0,$$

so $dx^1 \otimes dx^2 \neq dx^2 \otimes dx^1$. On the other hand, functions commute with everything:

$$f(A \otimes B) = fA \otimes B = A \otimes fB.$$

INTERPRETATIONS

If ω is a smooth one-form on a manifold M, then the function $X \to \omega(X)$ is $\mathfrak{F}(M)$-linear from $\mathfrak{X}(M)$ to $\mathfrak{F}(M)$, hence is a (0, 1) tensor field. Every (0, 1) tensor field arises in this way from a unique one-form (Exercise 4), so we write simply

$$\mathfrak{T}^0_1(M) = \mathfrak{X}^*(M).$$

There are two less obvious interpretations that will be used frequently.

(1) If V is a (smooth) vector field on M, define

$$V(\theta) = \theta(V) \qquad \text{for all} \quad \theta \in \mathfrak{X}^*(M).$$

This function $V: \mathfrak{X}^*(M) \to \mathfrak{F}(M)$ is $\mathfrak{F}(M)$-linear, hence is a (1, 0) tensor field. Every (1, 0) tensor field on M arises in this way from a unique vector field (Exercise 5), so we write

$$\mathfrak{T}^1_0(M) = \mathfrak{X}(M).$$

(2) If $A: \mathfrak{X}(M)^s \to \mathfrak{X}(M)$ is $\mathfrak{F}(M)$-multilinear, define $\bar{A}: \mathfrak{X}^*(M) \times \mathfrak{X}(M)^s \to \mathfrak{F}(M)$ by

$$\bar{A}(\theta, X_1, \ldots, X_s) = \theta(A(X_1, \ldots, X_s)), \qquad \text{for all} \quad \theta \text{ and } X_i.$$

Evidently $\bar{A}$ is $\mathfrak{F}(M)$-multilinear, hence is a $(1, s)$ tensor field. We shall consider A itself to be a tensor field, using the formula above only when necessary.

Tensors of type $(0, s)$ are said to be *covariant*, while tensors of type $(r, 0)$ with $r \geq 1$ are *contravariant*. For example, real-valued functions and one-forms are covariant; vector fields are contravariant. An (r, s) tensor is *mixed* if neither r nor s is zero. Notice that the definition of tensor product shows that if A is covariant and B contravariant then $A \otimes B = B \otimes A$.

TENSORS AT A POINT

The goal of this section is to show that—just as for a vector field or one-form—any tensor field A on M can indeed be viewed as a *field* on M, assigning a value A_p at each point $p \in M$. The essential fact is that when A is evaluated on one-forms and vector fields to give a real-valued function

$$A(\theta^1, \ldots, \theta^r, X_1, \ldots, X_s),$$

the value of this function at a point $p \in M$ depends not on the entirety of each one-form and vector field—or even on their values on a neighborhood of p—but solely on their values at the point p itself. Formally:

2. Proposition. Let $p \in M$ and $A \in \mathfrak{T}^r_s(M)$. Let $\bar{\theta}^1, \ldots, \bar{\theta}^r$ and $\theta^1, \ldots, \theta^r$ be one-forms such that $\bar{\theta}^i|_p = \theta^i|_p$ $(1 \leq i \leq r)$; let $\bar{X}_1, \ldots, \bar{X}_s$ and $X_1, \ldots, X_s$ be vector fields such that $\bar{X}_j|_p = X_j|_p$ $(1 \leq j \leq s)$. Then

$$A(\bar{\theta}^1, \ldots, \bar{\theta}^r, \bar{X}_1, \ldots, \bar{X}_s)(p) = A(\theta^1, \ldots, \theta^r, X_1, \ldots, X_s)(p).$$

The proof will be easy once we establish the following fact.

3. Lemma. If any one of the one-forms $\theta^1, \ldots, \theta^r$ or vector fields $X_1, \ldots, X_s$ is zero at p, then $A(\theta^1, \ldots, \theta^r, X_1, \ldots, X_s)(p) = 0$.

Proof. Suppose that, say, $X_s|_p = 0$. Let $x^1, \ldots, x^n$ be a coordinate system on a neighborhood $\mathscr{U}$ of p. Then $X_s = \sum X^i \partial_i$ on $\mathscr{U}$, where $X^i = X_s x^i \in \mathfrak{F}(\mathscr{U})$. Let f be a bump function at p with support in $\mathscr{U}$ (Lemma 1.8). Then as usual fX^i is a smooth function on all of M, and similarly $f\partial_i \in \mathfrak{X}(M)$. Hence

$$f^2 A(\theta^1, \ldots, X_s) = A(\theta^1, \ldots, f^2 X_s) = A(\theta^1, \ldots, \sum fX^i f\,\partial_i)$$
$$= \sum_{i=1}^{n} fX^i A(\theta^1, \ldots, f\,\partial_i).$$

Since $X_s|_p = 0$, each $X^i(p) = 0$; also $f(p) = 1$. Hence evaluating the formula above at p yields $A(\theta^1, \ldots, X_s)(p) = 0$. ∎

Proof of Proposition 2. For clarity suppose $r = 1$ and $s = 2$. Consider the following telescoping identity (extendible in an obvious way to any r, s):

$$A(\bar\theta, \bar X, \bar Y) - A(\theta, X, Y) = A(\bar\theta - \theta, \bar X, \bar Y) + A(\theta, \bar X - X, \bar Y)$$
$$+ A(\theta, X, \bar Y - Y).$$

By hypothesis $\bar\theta - \theta$, $\bar X - X$, and $\bar Y - Y$ all vanish at the point p. Hence, by the lemma, $A(\bar\theta, \bar X, \bar Y)(p) = A(\theta, X, Y)(p)$. ∎

It follows immediately from Proposition 2 that a tensor field $A \in \mathfrak{T}^r_s(M)$ has a *value* A_p at each point p of M, namely, the function

$$A_p\colon (T_pM^*)^r \times (T_pM)^s \to \mathbf{R}$$

defined as follows. If $\alpha^1, \ldots, \alpha^r \in T_pM^*$ and $x_1, \ldots, x_s \in T_pM$, let

$$A_p(\alpha^1, \ldots, \alpha^r, x_1, \ldots, x_s) = A(\theta^1, \ldots, \theta^r, X_1, \ldots, X_s)(p),$$

where $\theta^1, \ldots, \theta^r$ are *any* one-forms on M such that $\theta^i|_p = \alpha^i$ $(1 \le i \le r)$ and $X_1, \ldots, X_s$ are *any* vector fields such that $X_j|_p = x_j$ $(1 \le j \le s)$.

It is easy to check that the function A_p is $\mathbf{R}$-multilinear; then by Definition 1 it is an (r, s) tensor over $T_p(M)$. *We can thus consider $A \in \mathfrak{T}^r_s(M)$ as a field smoothly assigning to each $p \in M$ the tensor A_p.* Just as a vector field is a smooth section of the tangent bundle TM, such a field $p \to A_p$ is a smooth section of the (r, s) tensor bundle—the latter obtained, roughly speaking, by replacing each $T_p(M)$ in TM by the space $T_p(M)^r_s$ of (r, s) tensors over $T_p(M)$.

Conversely, a smooth cross section, say $p \to B_p \in T_p(M)^1_2$ for simplicity, arises from the unique tensor $B \in \mathfrak{T}^1_2(M)$ given by

$$B(\theta, X, Y)(p) = B_p(\theta_p, X_p, Y_p)$$

for all points p, one-forms θ, and vector fields X, Y. (That the section is *smooth* means that the values of B are in $\mathfrak{F}(M)$.)

In particular the cross-sectional interpretation shows that if $A \in \mathfrak{T}^r_s(M)$ and $\mathscr{U}$ is an open set of M, then the *restriction* $A|\mathscr{U}$ of A to $\mathscr{U}$ is a well-defined tensor field on $\mathscr{U}$.

TENSOR COMPONENTS

The coordinate formulas $X = \sum X(x^i)\,\partial_i$ for a vector field and $\theta = \sum \theta(\partial_i)\,dx^i$ for a one-form extend readily to tensor fields of arbitrary type.

4. Definition. Let $\xi = (x^1, \ldots, x^n)$ be a coordinate system on $\mathscr{U} \subset M$. If $A \in \mathfrak{T}^r_s(M)$ the *components of A relative to* ξ are the real-valued functions

$$A^{i_1 \ldots i_r}_{j_1 \ldots j_s} = A(dx^{i_1}, \ldots, dx^{i_r}, \partial_{j_1}, \ldots, \partial_{j_s}) \quad \text{on} \quad \mathscr{U},$$

where all indices run from 1 to $n = \dim M$.

Evidently for a (0, 1) tensor, that is, a one-form, these components are exactly those of the formula $\theta = \sum \theta(\partial_i)\,dx^i$. To see that the corresponding agreement holds for a vector field X we must use its interpretation as a (1, 0) tensor field. By the definition above the ith component of X relative to ξ is $X(dx^i)$, which is interpreted as $dx^i(X) = Xx^i$.

Similarly, when a $(1, s)$ tensor field is given in the form $A\colon \mathfrak{X}(M)^s \to \mathfrak{X}(M)$ its components are determined directly by the equation

$$A(\partial_{i_1}, \ldots, \partial_{i_s}) = \sum_j A^j_{i_1 \ldots i_s}\,\partial_j,$$

since for its interpretation $\bar{A} \in \mathfrak{T}^1_s(M)$

$$\begin{aligned}\bar{A}(dx^j, \partial_{i_1}, \ldots, \partial_{i_s}) &= dx^j(A(\partial_{i_1}, \ldots, \partial_{i_s}) \\ &= \sum_k A^k_{i_1 \ldots i_s}\,dx^j(\partial_k) = A^j_{i_1 \ldots i_s}.\end{aligned}$$

The evaluation of a tensor field on one-forms and vector fields can be described in terms of coordinates by writing everything out in components. For example, suppose A is a (1, 2) tensor. Write $\theta = \sum \theta_k\,dx^k$ for an arbitrary one-form and $X = \sum X^i\,\partial_i$, $Y = \sum Y^j\,\partial_j$ for arbitrary vector fields. The $\mathfrak{F}(M)$-multilinearity of A then yields

$$A(\theta, X, Y) = \sum_{i,j,k} A(dx^k, \partial_i, \partial_j)\theta_k X^i Y^j = \sum_{i,j,k} A^k_{ij}\theta_k X^i Y^j.$$

For a fixed coordinate system, the components of a sum of tensors are just the sum of the components. (Recall that only tensors of the same type are added.) The components of a tensor product are given by

$$(A \otimes B)^{i_1 \ldots i_{r+r'}}_{j_1 \ldots j_{s+s'}} = A^{i_1 \ldots i_r}_{j_1 \ldots j_s} \cdot B^{i_{r+1} \ldots i_{r+r'}}_{j_{s+1} \ldots j_{s+s'}},$$

where, as usual, all indices run from 1 to $n = \dim M$. To check this formula, suppose that A has type (1, 2) and B has type (1, 1). Then $A \otimes B$ is a (2, 3) tensor with components:

$$\begin{aligned}(A \otimes B)^{kq}_{ijp} &= (A \otimes B)(dx^k, dx^q, \partial_i, \partial_j, \partial_p) \\ &= A(dx^k, \partial_i, \partial_j) \cdot B(dx^q, \partial_p) = A^k_{ij}B^q_p.\end{aligned}$$

Let ξ be a coordinate system on $\mathscr{U} \subset M$. Then just as for a vector field or one-form, any tensor has a unique expression on $\mathscr{U}$ in terms of its components relative to ξ. Suppose for example that $r = 1$ and $s = 2$. Then $\partial_k \otimes dx^i \otimes dx^j$ is a (1, 2) tensor on $\mathscr{U}$ for all $1 \leq i, j, k \leq n$. We assert that if A is any (1, 2) tensor, then

$$A = \sum A^k_{ij}\, \partial_k \otimes dx^i \otimes dx^j \quad \text{on} \quad \mathscr{U},$$

where each index is summed from 1 to n. Since both sides are $\mathfrak{F}(\mathscr{U})$-multilinear it suffices to check that they have the same value on $dx^m, \partial_p, \partial_q$ for all $1 \leq m, p, q \leq n$. This follows immediately from

$$(\partial_k \otimes dx^i \otimes dx^j)(dx^m, \partial_p, \partial_q) = dx^m(\partial_k)\, dx^i(\partial_p)\, dx^j(\partial_q) = \delta^m_k \delta^i_p \delta^j_q,$$

where for the sake of index balance we use the extended Kronecker delta

$$\delta_{ij} = \delta^j_i = \delta^{ij} = \begin{cases} 1 & \text{if} \quad i = j, \\ 0 & \text{if} \quad i \neq j. \end{cases}$$

In general:

5. Lemma. Let $x^1, \ldots, x^n$ be a coordinate system on $\mathscr{U} \subset M$. If A is an (r, s) tensor field, then on $\mathscr{U}$,

$$A = \sum A^{i_1 \cdots i_r}_{j_1 \cdots j_s}\, \partial_{i_1} \otimes \cdots \otimes \partial_{i_r} \otimes dx^{j_1} \otimes \cdots \otimes dx^{j_s},$$

where each index is summed from 1 to n.

CONTRACTION

There is a remarkable operation called *contraction* that shrinks (r, s) tensors to $(r - 1, s - 1)$ tensors. The general definition derives from the following special case.

6. Lemma. There is a unique $\mathfrak{F}(M)$-linear function $C: \mathfrak{T}^1_1(M) \to \mathfrak{F}(M)$, called (1, 1) *contraction*, such that $C(X \otimes \theta) = \theta X$ for all $X \in \mathfrak{X}(M)$ and $\theta \in \mathfrak{X}^*(M)$.

Proof. Since it is to be $\mathfrak{F}(M)$-linear, C will be a pointwise operation. On a coordinate neighborhood $\mathscr{U}$ a (1, 1) tensor field A can be written as $\sum A^i_j\, \partial_i \otimes dx^j$. Since $C(\partial_i \otimes dx^j)$ must be $dx^j(\partial_i) = \delta^j_i$, we have no choice but to define

$$C(A) = \sum A^i_i = \sum A(dx^i, \partial_i).$$

Then C has the required properties on $\mathscr{U}$. To obtain the required global function it suffices to show that this definition is independent of the choice of coordinate system. But

$$
\begin{aligned}
\sum_m A\left(dy^m, \frac{\partial}{\partial y^m}\right) &= \sum_m A\left(\sum \frac{\partial y^m}{\partial x^i}\, dx^i, \sum \frac{\partial x^j}{\partial y^m}\frac{\partial}{\partial x^j}\right) \\
&= \sum_{i,j,m} \frac{\partial y^m}{\partial x^i}\frac{\partial x^j}{\partial y^m} A\left(dx^i, \frac{\partial}{\partial x^j}\right) = \sum_{i,j} \delta^j_i A\left(dx^i, \frac{\partial}{\partial x^j}\right) \\
&= \sum_i A\left(dx^i, \frac{\partial}{\partial x^i}\right). \qquad \blacksquare
\end{aligned}
$$

Evidently (1, 1) contraction is closely related to the notion of *trace* (see Exercise 9).

To extend (1, 1) contraction C to tensors of higher type the scheme is to specify one covariant slot and one contravariant slot, and apply C to these.

Suppose $A \in \mathfrak{T}^r_s(M)$ and $1 \le i \le r$ and $1 \le j \le s$. Fix one-forms $\theta^1_1, \ldots, \theta^{r-1}$ and vector fields $X_1, \ldots, X_{s-1}$. Then the function

$$
(\theta, X) \to A(\theta^1, \ldots, \overset{\substack{i\text{th contravariant slot}\\ \downarrow}}{\theta}, \ldots, \theta^{r-1}, X_1, \ldots, \underset{\substack{\uparrow\\ j\text{th covariant slot}}}{X}, \ldots, X_{s-1})
$$

is a (1, 1) tensor that can be written as

$$A(\theta^1, \ldots, \cdot, \ldots, \theta^{r-1}, X_1, \ldots, \cdot, \ldots, X_{s-1}).$$

Applying the (1, 1) contraction to this tensor produces a real-valued function denoted by

$$(C^i_j A)(\theta^1, \ldots, \theta^{r-1}, X_1, \ldots, X_{s-1}).$$

Evidently $C^i_j A$ is $\mathfrak{F}(M)$-multilinear in its arguments. Hence it is a tensor of type $(r-1, s-1)$ called the *contraction* of A over i, j.

For example, if A is a (2, 3) tensor field then $C^1_3(A)$ is the (1, 2) tensor field given by

$$(C^1_3 A)(\theta, X, Y) = C\{A(\cdot, \theta, X, Y, \cdot)\}.$$

Relative to a coordinate system the components of $C^1_3 A$ are

$$
\begin{aligned}
(C^1_3 A)^k_{ij} &= (C^1_3 A)(dx^k, \partial_i, \partial_j) = C\{A(\cdot, dx^k, \partial_i, \partial_j, \cdot)\} \\
&= \sum_m A(dx^m, dx^k, \partial_i, \partial_j, \partial_m) = \sum_m A^{mk}_{ijm},
\end{aligned}
$$

where we use the coordinate formula for C from the preceding proof.

7. Corollary. Let $1 \le i \le r$ and $1 \le j \le s$. Relative to a coordinate system, if $A \in \mathfrak{T}_s^r(M)$ has components $A_{j_1 \dots j_s}^{i_1 \dots i_r}$, then $C_j^i A$ has components

$$\sum_m A_{j_1 \dots \underset{\underset{\text{—}j\text{th index}}{\uparrow}}{m} \dots j_s}^{i_1 \dots \overset{\overset{\text{—}i\text{th index}}{\downarrow}}{m} \dots i_r}.$$

COVARIANT TENSORS

By means of a mapping from M to N any covariant tensor on N can be pulled back to M. The following definition uses the cross-sectional view of tensors.

8. Definition. Let $\phi: M \to N$ be a smooth mapping. If $A \in \mathfrak{T}_s^0(N)$ with $s \ge 1$, let

$$(\phi^* A)(v_1, \dots, v_s) = A(d\phi v_1, \dots, d\phi v_s)$$

for all $v_i \in T_p(M)$, $p \in M$. Then $\phi^*(A)$ is called the *pullback* of A by ϕ.

At each point p on M, $\phi^*(A)$ gives an R-multilinear function from $T_p(M)^s$ to R, that is, an $(0, s)$ tensor over $T_p(M)$. Coordinate computations as in Exercise 8 show that $\phi^*(A)$ is a smooth covariant tensor field on M. In the special case of a $(0, 0)$ tensor $f \in \mathfrak{F}(N)$, the pullback to M is defined to be $\phi^*(f) = f \circ \phi \in \mathfrak{F}(M)$. Note that $\phi^*(df) = d(\phi^* f)$.

The following properties of the pullback operation are easily verified.

9. Lemma. (1) If $\phi: M \to N$ is a smooth mapping, then $\phi^*: \mathfrak{T}_s^0(N) \to \mathfrak{T}_s^0(M)$ is R-linear for each $s \ge 0$, and

$$\phi^*(A \otimes B) = \phi^*(A) \otimes \phi^*(B)$$

for covariant tensors of arbitrary types $(0, s)$ and $(0, t)$.

(2) If $\psi: N \to P$ is also a smooth map, then

$$(\psi \circ \phi)^* = \phi^* \circ \psi^*: \mathfrak{T}_s^0(P) \to \mathfrak{T}_s^0(M)$$

for all $s \ge 0$.

A tensor field of type (r, s) with $r \ge 1$ can generally be moved neither from M to N nor from N to M by an arbitrary mapping $M \to N$.

Let A be a covariant or contravariant tensor of type at least 2. A is *symmetric* if transposing any two of its argument leaves its value unchanged. A is *skew-symmetric* (or *alternate*) if each such reversal produces a sign change. Functions, one-forms, and vector fields are considered by convention to be both symmetric and skew symmetric. A *differential s-form* is a skew-symmetric covariant tensor field of type (0, s). For the calculus of differential forms, see, for example, [BG].

10. Lemma. Let μ be an n-form. If $V_i = \sum_{j=1}^{n} A_{ij} W_j$ for $1 \leq i \leq n$, then

$$\mu(V_1, \ldots, V_n) = (\det A)\mu(W_1, \ldots, W_n).$$

The proof is a standard combinatorial argument.

TENSOR DERIVATIONS

Previous sections have dealt with tensor algebra; we now consider some tensor calculus.

11. Definition. A *tensor derivation* $\mathscr{D}$ on a smooth manifold M is a set of $\boldsymbol{R}$-linear functions

$$\mathscr{D} = \mathscr{D}_s^r \colon \mathfrak{T}_s^r(M) \to \mathfrak{T}_s^r(M) \qquad (r \geq 0, s \geq 0)$$

such that for any tensors A and B:

(1) $\mathscr{D}(A \otimes B) = \mathscr{D}A \otimes B + A \otimes \mathscr{D}B$,
(2) $\mathscr{D}(CA) = C(\mathscr{D}A)$ for any contraction C.

Thus $\mathscr{D}$ is $\boldsymbol{R}$-linear, preserves tensor type, obeys the usual Leibnizian product rule, and commutes with all contractions. For a function $f \in \mathfrak{F}(M)$ recall that $fA = f \otimes A$; hence $\mathscr{D}(fA) = (\mathscr{D}f)A + f\mathscr{D}A$.

In the special case $t = s = 0$, $\mathscr{D}_0^0$ is a derivation on $\mathfrak{T}_0^0(M) = \mathfrak{F}(M)$ so, as discussed in Chapter 1, there is a unique vector field $V \in \mathfrak{X}(M)$ such that

$$\mathscr{D}f = Vf \qquad \text{for all} \quad f \in \mathfrak{F}(M).$$

Since tensor derivations are generally not $\mathfrak{F}(M)$-linear the value of $\mathscr{D}A$ at a point $p \in M$ cannot usually be found from A_p alone. However it can be found from the values of A on any arbitrarily small neighborhood of p. This local character of tensor derivations can be expressed as follows.

12. Proposition. If $\mathscr{D}$ is a tensor derivation on M and $\mathscr{U}$ is an open set of M, then there is a unique tensor derivation $\mathscr{D}_{\mathscr{U}}$ on $\mathscr{U}$ such that

$$\mathscr{D}_{\mathscr{U}}(A \mid \mathscr{U}) = (\mathscr{D}A) \mid \mathscr{U} \qquad \text{for all tensors} \quad A \quad \text{on} \quad M.$$

($\mathscr{D}_{\mathscr{U}}$ is called the *restriction* of $\mathscr{D}$ to $\mathscr{U}$, and henceforth we omit the subscript $\mathscr{U}$.)

Scheme of Proof. Let $B \in \mathfrak{T}^r_s(\mathscr{U})$. If $p \in \mathscr{U}$ let f be a bump function at p with support in $\mathscr{U}$. Thus $fB \in \mathfrak{T}^r_s(M)$. Define

$$(\mathscr{D}_{\mathscr{U}} B)_p = \mathscr{D}(fB)_p.$$

Then show: (1) This definition is independent of the choice of bump function. (2) $\mathscr{D}_{\mathscr{U}} B$ is a smooth tensor field on $\mathscr{U}$. (3) $\mathscr{D}_{\mathscr{U}}$ is a tensor derivation on $\mathscr{U}$. (4) $\mathscr{D}_{\mathscr{U}}$ has the stated restriction property. (5) $\mathscr{D}_{\mathscr{U}}$ is unique. ■

The Leibnizian formula $\mathscr{D}(A \otimes B) = \mathscr{D}A \otimes B + A \otimes \mathscr{D}B$ can be recast as follows.

13. Proposition (The Product Rule). Let $\mathscr{D}$ be a tensor derivation on M. If $A \in \mathfrak{T}^r_s(M)$ then

$$\begin{aligned}\mathscr{D}[A(\theta^1,\ldots,\theta^r, X_1,\ldots,X_s)] &= (\mathscr{D}A)(\theta^1,\ldots,\theta^r, X_1,\ldots,X_s)\\ &\quad + \sum_{i=1}^{r} A(\theta^1,\ldots,\mathscr{D}\theta^i,\ldots,\theta^r, X_1,\ldots,X_s)\\ &\quad + \sum_{j=1}^{s} A(\theta^1,\ldots,\theta^r, X_1,\ldots,\mathscr{D}X_j,\ldots,X_s).\end{aligned}$$

(The placement of parentheses is crucial here: on the left-hand side $\mathscr{D}$ is applied to a function, on the right-hand side to the tensor A, to one-forms, and to vector fields.)

Proof. For simplicity let $r = s = 1$. We assert that

$$A(\theta, X) = \bar{C}(A \otimes \theta \otimes X),$$

where $\bar{C}$ is a composition of two contractions. In fact, relative to a coordinate system $A \otimes \theta \otimes X$ has components $A^i_j \theta_k X^l$, while $A(\theta, X) = \sum A^i_j \theta_i X^j$. Thus

$$\begin{aligned}\mathscr{D}(A(\theta, X)) &= \mathscr{D}\,\bar{C}(A \otimes \theta \otimes X) = \bar{C}\,\mathscr{D}(A \otimes \theta \otimes X)\\ &= \bar{C}(\mathscr{D}A \otimes \theta \otimes X) + \bar{C}(A \otimes \mathscr{D}\theta \otimes X) + \bar{C}(A \otimes \theta \otimes \mathscr{D}X)\\ &= (\mathscr{D}A)(\theta, X) + A(\mathscr{D}\theta, X) + A(\theta, \mathscr{D}X).\end{aligned}$$ ■

For a $(1, s)$ tensor expressed as an $\mathfrak{F}(M)$-multilinear function $A: \mathfrak{X}(M)^s \to \mathfrak{X}(M)$ the tensor derivation obeys the same formal product rule, namely,

$$\mathscr{D}(A(X_1,\ldots,X_s)) = (\mathscr{D}A)(X_1,\ldots,X_s) + \sum_{i=1}^{s} A(X_1,\ldots,\mathscr{D}X_i,\ldots,X_s).$$

Both these versions of the product rule will frequently be solved for the term involving $\mathscr{D}A$. This gives a formula for $\mathscr{D}$ of an arbitrary tensor in terms of $\mathscr{D}$ applied solely to functions, vector fields, and one-forms. But for a one-form

$$(\mathscr{D}\theta)(X) = \mathscr{D}(\theta X) - \theta(\mathscr{D}X).$$

Thus functions and vector fields suffice.

14. Corollary. If tensor derivations $\mathscr{D}_1$ and $\mathscr{D}_2$ agree on functions $\mathfrak{F}(M)$ and vector fields $\mathfrak{X}(M)$, then $\mathscr{D}_1 = \mathscr{D}_2$.

Furthermore, from suitable data on $\mathfrak{F}(M)$ and $\mathfrak{X}(M)$ we can construct a tensor derivation.

15. Theorem. Given a vector field $V \in \mathfrak{X}(M)$ and an $\boldsymbol{R}$-linear function $\delta: \mathfrak{X}(M) \to \mathfrak{X}(M)$ such that

$$\delta(fX) = Vf\, X + f\, \delta(X) \qquad \text{for all} \quad f \in \mathfrak{F}(M), \quad X \in \mathfrak{X}(M),$$

there exists a unique tensor derivation $\mathscr{D}$ on M such that $\mathscr{D}_0^0 = V: \mathfrak{F}(M) \to \mathfrak{F}(M)$ and $\mathscr{D}_0^1 = \delta$.

Proof. $\mathscr{D}_0^0$ and $\mathscr{D}_0^1$ are given. The formula preceding Corollary 14 shows that $\mathscr{D}$ on a one-form θ must be defined by

$$(\mathscr{D}\theta)(X) = V(\theta X) - \theta(\delta X) \qquad \text{for all} \quad X \in \mathfrak{X}(M).$$

Using the formula given for δ it is easy to check that $\mathscr{D}\theta$ is $\mathfrak{F}(M)$-linear, hence is a one-form, and that $\mathscr{D} = \mathscr{D}_1^0: \mathfrak{X}^*(M) \to \mathfrak{X}^*(M)$ is $\boldsymbol{R}$-linear.

By the product rule (13), $\mathscr{D}$ on an (r, s) tensor A with $r + s \geq 2$ must be defined by

$$\begin{aligned}(\mathscr{D}A)(\theta^1, \ldots, \theta^r, X_1, \ldots, X_s) = {} & V(A(\theta^1, \ldots, \theta^r, X_1, \ldots, X_s)) \\ & - \sum_{i=1}^{r} A(\theta^1, \ldots, \mathscr{D}\theta^i, \ldots, \theta^r, X_1, \ldots, X_s) \\ & - \sum_{j=1}^{s} A(\theta^1, \ldots, \theta^r, X_1, \ldots, \delta X_j, \ldots, X_s).\end{aligned}$$

(On the right-hand side, $\mathscr{D}$ of a one-form is defined as above.)

Again it is easy to verify that $\mathscr{D}A$ is $\mathfrak{F}(M)$-multilinear hence is an (r, s) tensor, and that $\mathscr{D}: \mathfrak{T}_s^r(M) \to \mathfrak{T}_s^r(M)$ is $\boldsymbol{R}$-linear. Furthermore, a direct computation shows that $\mathscr{D}(A \otimes B) = \mathscr{D}A \otimes B + A \otimes \mathscr{D}B$. (Take A and B of type (1, 1) to see how this works.)

To prove that $\mathscr{D}$ commutes with contraction, consider first the case $C: \mathfrak{T}_1^1(M) \to \mathfrak{F}(M)$. That $\mathscr{D}C = C\mathscr{D}$ on tensor products $\theta \otimes X$ is immediate from the definition of $\mathscr{D}$ on one-forms. Hence $\mathscr{D}C = C\mathscr{D}$ on sums of terms of the form $\theta \otimes X$. Since $\mathscr{D}$ is local and C pointwise it suffices to prove $\mathscr{D}C = C\mathscr{D}$ on coordinate neighborhoods. But there Lemma 5 shows that every (1, 1) tensor can be written as such a sum.

The extension to arbitrary contractions is an exercise in parentheses. Taking $A \in \mathfrak{T}^1_2(M)$ for example,

$$\begin{aligned}(\mathscr{D}C^1_2 A)(X) &= \mathscr{D}((C^1_2 A)(X)) - (C^1_2 A)(\mathscr{D}X)\\ &= \mathscr{D}(C\{A(\cdot, X, \cdot)\}) - C\{A(\cdot, \mathscr{D}X, \cdot)\}\\ &= C\{\mathscr{D}(A(\cdot, X, \cdot)) - A(\cdot, \mathscr{D}X, \cdot)\}\\ &= C\{(\mathscr{D}A)(\cdot, X, \cdot)\} = (C^1_2 \mathscr{D}A)(X).\end{aligned}$$

Hence $\mathscr{D}C^1_2 A = C^1_2 \mathscr{D}A$. ∎

Here is an application of the theorem.

16. Definition. If $V \in \mathfrak{X}(M)$ the tensor derivation L_V such that

$$L_V(f) = Vf \qquad \text{for all} \quad f \in \mathfrak{F}(M),$$
$$L_V(X) = [V, X] \qquad \text{for all} \quad X \in \mathfrak{X}(M)$$

is called the *Lie derivative* relative to V.

The definition is valid since L_V on vector fields satisfies the hypothesis on δ in the theorem:

$$L_V(fX) = [V, fX] = Vf\,X + f[V, X] = Vf\,X + fL_V X.$$

SYMMETRIC BILINEAR FORMS

Semi-Riemannian geometry involves a particular kind of (0, 2) tensor on tangent spaces. To study these in general, let V be a real vector space (finite-dimensional where the context so indicates). A *bilinear form* on V is an $\boldsymbol{R}$-bilinear function $b\colon V \times V \to \boldsymbol{R}$, and we consider only the symmetric case: $b(v, w) = b(w, v)$ for all v, w.

17. Definition. A symmetric bilinear form b on V is

(1) *positive* [*negative*] *definite* provided $v \neq 0$ implies $b(v, v) > 0$ [<0],

(2) *positive* [*negative*] *semidefinite* provided $b(v, v) \geq 0$ [≤ 0] for all $v \in V$,

(3) *nondegenerate* provided $b(v, w) = 0$ for all $w \in V$ implies $v = 0$.

Also b is *definite* [*semidefinite*] provided either alternative in (1) [(2)] holds. If b is definite then it is obviously both semidefinite and nondegenerate; the converse follows from Exercise 12.

If b is a symmetric bilinear form on V then for any subspace W of V the restriction $b|(W \times W)$, denoted merely by $b|W$, is again symmetric and bilinear. If b is [semi-]definite, so is $b|W$.

18. Definition. The *index* ν of a symmetric bilinear form b on V is the largest integer that is the dimension of a subspace $W \subset V$ on which $b|W$ is negative definite.

Thus $0 \leq \nu \leq \dim V$, and $\nu = 0$ if and only if b is positive semidefinite.

The function $q: V \to \boldsymbol{R}$ given by $q(v) = b(v, v)$ is the *associated quadratic form* of b. It is often easier to deal with than b, and no information is lost since b can be reconstructed by the polarization identity

$$b(v, w) = \tfrac{1}{2}[q(v + w) - q(v) - q(w)].$$

If $e_1, \ldots, e_n$ is a basis for V, the $n \times n$ matrix $(b_{ij}) = b(e_i, e_j)$ is called the *matrix of* b *relative to* $e_1, \ldots, e_n$. Since b is symmetric, this matrix is symmetric. Clearly it determines b since

$$b(\textstyle\sum v_i e_i, \sum w_j e_j) = \sum b_{ij} v_i w_j.$$

19. Lemma. A symmetric bilinear form is nondegenerate if and only if its matrix relative to one (hence every) basis is invertible.

Proof. Let $e_1, \ldots, e_n$ be a basis for V. If $v \in V$, then $b(v, w) = 0$ for all $w \in V$ if and only if $b(v, e_i) = 0$ for $i = 1, \ldots, n$. Since (b_{ij}) is symmetric,

$$b(v, e_i) = b(\textstyle\sum v_j e_j, e_i) = \sum b_{ij} v_j.$$

Thus b is degenerate if and only if there exist numbers $v_1, \ldots, v_n$ not all zero such that $\sum b_{ij} v_j = 0$ for $i = 1, \ldots, n$. But this is equivalent to the linear dependence of the columns of (b_{ij}), that is, to (b_{ij}) being singular. ∎

SCALAR PRODUCTS

20. Definition. A *scalar product* g on a vector space V is a nondegenerate symmetric bilinear form on V.

An *inner product* is a positive definite scalar product, the canonical example being the *dot product* on $\boldsymbol{R}^n$, for which $v \cdot w = \sum v_i w_i$. Many properties of inner products carry over to scalar products, however some distinctive new phenomena arise when g is indefinite.

Changing one sign in the definition of the dot product on $\boldsymbol{R}^2$ gives the simplest example of an indefinite scalar product.

21. Example. Define $g: \boldsymbol{R}^2 \times \boldsymbol{R}^2 \to \boldsymbol{R}$ by

$$g(v, w) = v_1 w_1 - v_2 w_2 .$$

Obviously g is symmetric and bilinear. Taking w to be (1, 0) and then (0, 1) shows that g is nondegenerate. Thus g is a scalar product. The associated quadratic form has $q(v) = v_1^2 - v_2^2$, and g is indefinite.

Henceforth V will denote a *scalar product space*, that is, a (finite-dimensional, real) vector space furnished with a scalar product g. A vector $v \in V$ is *null* provided $q(v) = 0$ but $v \neq 0$. Evidently null vectors exist if and only if g is indefinite. In the example above, null vectors fill the two 45° lines with the origin omitted (0 is not null). For $c \neq 0$ the sets $q = c$ and $q = -c$ are hyperbolas asymptotic to the null lines (Figure 1).

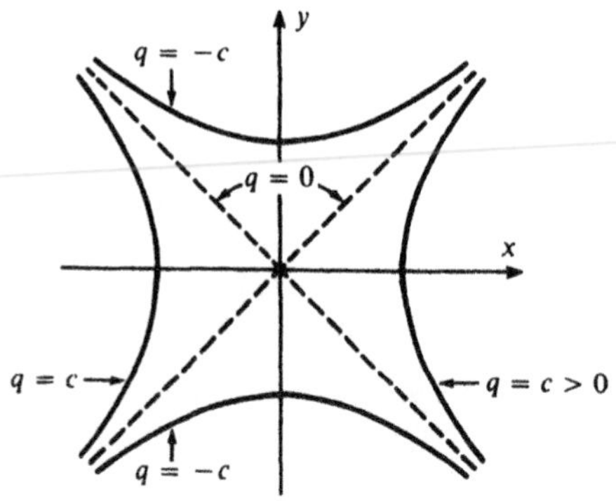

Figure 1

Vectors v, $w \in V$ are *orthogonal*, written $v \perp w$, provided $g(v, w) = 0$. Subsets A and B of V are *orthogonal*, written $A \perp B$, provided $v \perp w$ for all $v \in A$ and $w \in B$. When the scalar product g is indefinite we can no longer picture orthogonal vectors as being at right angles to each other. For Example 21, Figure 2 shows three pairs of orthogonal vectors $z \perp z'$, the last of which

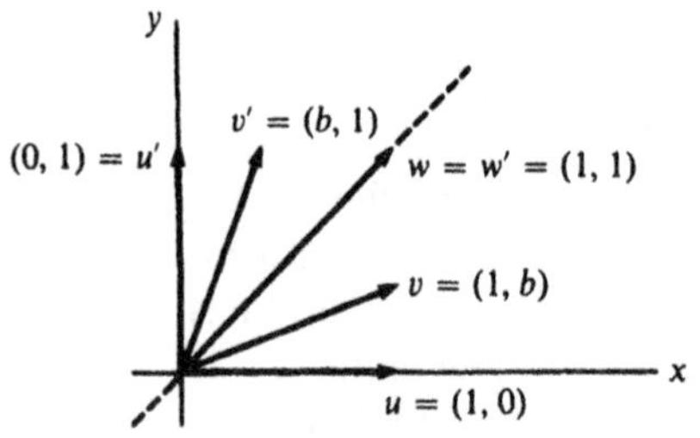

Figure 2

illustrates the fact that a null vector is a nonzero vector that is orthogonal to itself. If W is a subspace of V, let

$$W^{\perp} = \{v \in V : v \perp W\}.$$

$W^{\perp}$ is a subspace of V called W *perp*. We cannot call $W^{\perp}$ the *orthogonal complement* of W since $W + W^{\perp}$ is generally not all of V. (In Example 21 if W is the subspace spanned by (1, 1), then $W^{\perp} = W$.) However the perp operation does have two familiar properties:

22. Lemma. If W is a subspace of a scalar product space V, then

(1) $\dim W + \dim W^{\perp} = n = \dim V$,
(2) $(W^{\perp})^{\perp} = W$.

Proof. (1) Let $e_1, \ldots, e_n$ be a basis for V adapted to W, that is, for which $e_1, \ldots, e_k$ is a basis for W. Now $v \in W^{\perp}$ if and only if $g(v, e_i) = 0$ for $1 \leq i \leq k$, which in coordinate terms is

$$\sum_{j=1}^{n} g_{ij} v_j = 0 \qquad (1 \leq i \leq k).$$

This is k linear equations in n unknowns, and by Lemma 19 the rows of the coefficient matrix are linearly independent, so the matrix has rank k. Hence by linear algebra the space of solutions has dimension $n - k$. But by construction these n-tuple solutions $(v_1, \ldots, v_n)$ give exactly the vectors $v = \sum v_i e_i$ of $W^{\perp}$.

(2) Since $v \in (W^{\perp})^{\perp}$ means $v \perp W^{\perp}$, we have $W \subset (W^{\perp})^{\perp}$. By (1) these two subspaces have the same dimension, hence they are equal. ∎

Note that the nondegeneracy of g on the whole space V is equivalent to $V^{\perp} = 0$.

A subspace W of V is called *nondegenerate* if $g|W$ is nondegenerate. When V is an inner product space, every subspace W is again an inner product space (under $g|W$) hence is nondegenerate. However when g is indefinite there will always be degenerate subspaces; for example, any null vector spans one. Thus a subspace of a scalar product space need not be a scalar product space. This difficulty is linked to the earlier one involving the perp operation.

23. Lemma. A subspace W of V is nondegenerate if and only if V is the direct sum of W and $W^{\perp}$.

Proof. By a standard vector space identity

$$\dim(W + W^{\perp}) + \dim(W \cap W^{\perp}) = \dim W + \dim W^{\perp}.$$

According to Lemma 22, the right-hand side is $n = \dim V$. Hence $W + W^\perp = V$ if and only if $W \cap W^\perp = 0$. Thus either of these two conditions is equivalent to $V = W \oplus W^\perp$. But $W \cap W^\perp = \{w \in W : w \perp W\}$, so the vanishing of this subspace is equivalent to the nondegeneracy of W. ∎

Since $(W^\perp)^\perp = W$, it follows that W is nondegenerate if and only if $W^\perp$ is.

Because $q(v) = g(v, v)$ may be negative, the *norm* $|v|$ of a vector is defined to be $|g(v, v)|^{1/2}$. A *unit vector* u is a vector of norm 1, that is, $g(u, u) = \pm 1$. As usual a set of mutually orthogonal unit vectors is said to be *orthonormal*, and for $n = \dim V$, any set of n orthonormal vectors in V is necessarily a basis for V.

24. Lemma. A scalar product space $V \neq 0$ has an orthonormal basis.

Proof. Since g is nondegenerate there is a vector $v \in V$ such that $g(v, v) \neq 0$. Now $v/|v|$ is a unit vector. Thus it suffices by induction to show that *any orthonormal set* $e_1, \ldots, e_k$ *with* $k < n$ *can be enlarged by one*. By Lemma 19 these vectors span a (k-dimensional) nondegenerate subspace W. It remains only to find a unit vector in $W^\perp \neq 0$. But as noted above $W^\perp$ is also nondegenerate, so the preceding argument shows it contains a unit vector. ∎

The matrix of g relative to an orthonormal basis $e_1, \ldots, e_n$ for V is diagonal; in fact,

$$g(e_i, e_j) = \delta_{ij}\varepsilon_j, \qquad \text{where} \quad \varepsilon_j = g(e_j, e_j) = \pm 1.$$

Whenever convenient we shall order the vectors in an orthonormal basis so that the negative signs—if any—come first in the so-called *signature* $(\varepsilon_1, \ldots, \varepsilon_n)$.

Taking these signs into account *orthonormal expansion* is still available.

25. Lemma. Let $e_1, \ldots, e_n$ be an orthonormal basis for V, with $\varepsilon_i = g(e_i, e_i)$. Then each $v \in V$ has a unique expression

$$v = \sum \varepsilon_i g(v, e_i) e_i.$$

For the proof it suffices to check that v minus the sum is orthogonal to each e_i; thus by the nondegeneracy of g it is zero.

The *orthogonal projection* π of V onto a nondegenerate subspace W is the linear transformation that sends $W^\perp$ to 0 and leaves each vector of W fixed. An orthonormal basis $e_1, \ldots, e_k$ for W can always be enlarged to a basis for V; thus

$$\pi(v) = \sum_{j=1}^{k} \varepsilon_j g(v, e_j) e_j.$$

It is customary to refer to the index ν of the scalar product g of V as the *index* of V, writing $\nu = \operatorname{ind} V$.

26. Lemma. For any orthonormal basis $e_1, \ldots, e_n$ for V the number of negative signs in the signature $(\varepsilon_1, \ldots, \varepsilon_n)$ is the index ν of V.

Proof. Assume that the first m signs ε_i are the negative ones. The result is trivial if g is definite, so $0 < m < n$. Evidently g is negative definite on the subspace S spanned by $e_1, \ldots, e_m$; thus $\nu \geq m$.

To prove the reverse inequality, let W be an arbitrary subspace on which g is negative definite, and define $\pi\colon W \to S$ by

$$\pi(w) = -\sum_{i \leq m} g(w, e_i)e_i.$$

Evidently π is linear. Thus it suffices to show that π is one-to-one, for then $\dim W \leq \dim S = m$, hence $\nu \leq m$. If $\pi(w) = 0$, then by orthonormal expansion

$$w = \sum_{j>m} g(w, e_j)e_j.$$

But since $w \in W$,

$$0 \geq g(w, w) = \sum_{j>m} g(w, e_j)^2.$$

Hence $g(w, e_j) = 0$ for $j > m$, so $w = 0$. ∎

It follows that for a nondegenerate subspace W of V

$$\operatorname{ind} V = \operatorname{ind} W + \operatorname{ind} W^{\perp},$$

since the proof of Lemma 25 shows that there is an orthonormal basis for V adapted to the direct sum $V = W + W^{\perp}$.

Let V and $\bar{V}$ have scalar products g and $\bar{g}$. A linear transformation $T\colon V \to \bar{V}$ *preserves scalar products* provided

$$\bar{g}(Tv, Tw) = g(v, w) \qquad \text{for all} \quad v, w \in V.$$

In this case T is necessarily one-to-one, because if $Tv = 0$ then $g(v, w) = 0$ for all w, hence $v = 0$.

Note that T preserves scalar products if and only if it preserves their associated quadratic forms, that is,

$$\bar{q}(Tv) = q(v) \qquad \text{for all} \quad v \in V.$$

One implication is obvious; the other follows by polarization.

A linear isomorphism $T\colon V \to W$ that preserves scalar products is called a *linear isometry*. By the preceding remarks a linear transformation $T\colon V \to W$ is a linear isometry if and only if $\dim V = \dim W$ and T preserves scalar products (or equivalently their quadratic forms).

27. Lemma. Scalar product spaces V and W have the same dimension and index if and only if there exists a linear isometry from V to W.

Proof. Assuming the invariants are the same, pick orthonormal bases $e_1, \ldots, e_n$ for V and $e'_1, \ldots, e'_n$ for W. By Lemma 26 we can suppose that $\langle e_i, e_i \rangle = \langle e'_i, e'_i \rangle$ for all i. Let T be the linear transformation such that $Te_i = e'_i$ for all i. Then $\langle Te_i, Te_j \rangle = \langle e'_i, e'_j \rangle$ for all i, j. Hence by linearity T is a linear isometry.

Conversely, if $T: V \to W$ is a linear isometry, then T carries an orthonormal basis for V to an orthonormal basis for W. Hence dim V = dim W and, by Lemma 26, ind V = ind W. ∎

Exercises

1. A one-form θ is zero if and only if $\theta X = 0$ for all $X \in \mathfrak{X}(M)$. A vector field X is zero if and only if $\theta X = 0$ for all $\theta \in \mathfrak{X}^*(M)$.

2. (a) The bracket operation on vector fields is $\boldsymbol{R}$-linear $\mathfrak{X}(M) \times \mathfrak{X}(M) \to \mathfrak{X}(M)$, but cannot be interpreted as a (1, 2) tensor field. (b) If θ is a one-form its exterior derivative $d\theta$ is a tensor field, in fact a two-form. (By definition, $(d\theta)(X, Y) = X\theta Y - Y\theta X - \theta[X, Y]$.)

3. Tensor transformation rule. Let A be a tensor field, say of type (1, 2). Let ξ and η be coordinate systems on $\mathscr{U} \subset M$. Show that the components of A relative to η are determined as follows by the components of A relative to ξ:

$$^{\eta}A^c_{ab} = \sum_{i,j,k} \frac{\partial y^c}{\partial x^k} \frac{\partial x^i}{\partial y^a} \frac{\partial x^j}{\partial y^b} \, ^{\xi}A^k_{ij}.$$

4. Prove that the interpretation on page 36 gives an $\mathfrak{F}(M)$-linear isomorphism from $\mathfrak{X}^*(M)$ to $\mathfrak{T}^0_1(M)$. (Thus $\mathfrak{X}^*(M)$ is identified with the dual module of $\mathfrak{X}(M)$.)

5. (a) If V is a finite-dimensional vector space and $\phi \in (V^*)^*$ there is a unique $v \in V$ such that $\alpha(v) = \phi(\alpha)$ for all $\alpha \in V^*$. (b) The interpretation on page 37 gives an $\mathfrak{F}(M)$-linear isomorphism from $\mathfrak{X}(M)$ to $\mathfrak{T}^1_0(M)$. (Hint: If $Z \in \mathfrak{T}^1_0(M)$, then $Z_p \in T_p(M)^{**}$.) Thus, in view of the preceding exercise, $\mathfrak{X}(M)$ is identified with its double dual module $\mathfrak{T}^1_0(M) = (\mathfrak{X}^*(M))^* \approx \mathfrak{X}(M)^{**}$.

6. Let $\mathscr{D}$ be a tensor derivation on M. Relative to a coordinate system, (a) if $\mathscr{D}(\partial_i) = \sum F^j_i \partial_j$, show that $\mathscr{D}(dx^j) = -\sum F^j_i \, dx^i$. (b) If A is a (1, 2) tensor field, find a formula for the components of $\mathscr{D}A$ in terms of F^j_i and the components of A.

7. If $V \in \mathfrak{X}(M)$ and $A \in \mathfrak{T}_2^1(M)$, then relative to a coordinate system express the components of the Lie derivative $L_V A$ in terms of the components of A and V.

8. Let $\phi: M^m \to N^n$ be a smooth map that carries the coordinate neighborhood $\mathscr{U} \subset M$ of ξ into the coordinate neighborhood of η in N. If B is a covariant tensor field on N, with say $s = 2$, show that on $\mathscr{U}$,

$$(\phi^* B)\left(\frac{\partial}{\partial x^i}, \frac{\partial}{\partial x^j}\right) = \sum_{a,b=1}^{n} \frac{\partial(y^a \circ \phi)}{\partial x^i} \frac{\partial(y^b \circ \phi)}{\partial x^j} B\left(\frac{\partial}{\partial y^a}, \frac{\partial}{\partial y^b}\right) \circ \phi.$$

(Hint: Evaluate the left-hand side at $p \in \mathscr{U}$.)

9. (a) Interpret $A \in \mathfrak{T}_1^1(M)$ as a function smoothly assigning to each $p \in M$ a linear operator A_p on $T_p(M)$. (b) Show that $(CA)(p) = \text{trace } A_p$. (c) If $A, B \in \mathfrak{T}_1^1(M)$, express the function $p \to A_p \circ B_p$ as a element of $\mathfrak{T}_1^1(M)$.

10. (a) Prove that a tensor derivation has $\mathscr{D}_0^0 = 0$ if and only if $\mathscr{D}_0^1$ is $\mathfrak{F}(M)$-linear. Then by interpretation, $\mathscr{D}_0^1 = B \in \mathfrak{T}_1^1(M)$, and we write $\mathscr{D} = \mathscr{D}_B$. (b) If $\mathscr{D}$ is an arbitrary tensor derivation on M, show that there is a unique $V \in \mathfrak{X}(M)$ and a unique $B \in \mathfrak{T}_1^1(M)$ such that $\mathscr{D} = L_V + \mathscr{D}_B$.

11. Establish the following properties of Lie derivatives:

(a) $L_{aV+bW} = aL_V + bL_W$, $(a, b \in \boldsymbol{R})$

(b) $[L_V, L_W] = L_{[V,W]}$,

(c) $L_V(df) = d(Vf)$,

(d) $L_{fV} = fL_V - \mathscr{D}_{V \otimes df}$ (notation as in Exercise 10).

12. Let b be a symmetric bilinear form on V. The *nullspace* of b is $N = \{v: b(v, w) = 0 \text{ for all } w\}$. The *nullcone* of b is the set Λ of all null vectors in V. Let $A = \Lambda \cup 0$, so $A \supset N$. Prove : (a) N is a subspace, but A is not unless $A = 0$ or V. (b) b is nondegenerate $\Leftrightarrow N = 0$; b is definite $\Leftrightarrow A = 0$. (c) b is semidefinite $\Leftrightarrow N = A$.

13. Let g be a scalar product of index ν on an n-dimensional vector space V. Prove that there exists a subspace W of dimension $\min(\nu, n - \nu)$, and no larger, on which $g = 0$.

14. (Dajczer, Nomizu, and others.) Let V have indefinite scalar product g, and let b be a symmetric bilinear form on V with corresponding quadratic form q. Show that the following conditions are equivalent: (i) $b = cg$ for some $c \in \boldsymbol{R}$, (ii) $q = 0$ on null vectors, (iii) $|q|$ is bounded on timelike unit vectors, and (iv) $|q|$ is bounded on spacelike unit vectors. (Hint: Polarize.)

3 SEMI-RIEMANNIAN MANIFOLDS

The familiar geometry of the Euclidean space $\boldsymbol{R}^3$ can be traced back to its natural inner product, the dot product. By means of the natural isomorphism $T_p(\boldsymbol{R}^3) \approx \boldsymbol{R}^3$ the dot product can be deployed on each tangent space. Then one can perform such basic geometric operations as measuring the length of a tangent vector or the angle between two tangent vectors.

The theory of surfaces in $\boldsymbol{R}^3$ attained its classical form in the work of Gauss, who showed in 1827 that the *intrinsic geometry* of a surface S in $\boldsymbol{R}^3$ (roughly, the geometry perceived by the inhabitants of S) derives solely from the dot product as applied to tangent vectors to S.

As long ago as 1854 Riemann saw what was needed to generalize these two special cases and introduce geometry on an arbitrary n-dimensional manifold: an inner product must be given on each tangent space. This is thought of as providing, in particular, an infinitesimal measurement of distance. Crudely, if p and $p + dp$ are nearby points, the distance between them is the norm of the "tangent vector" dp.

Under the impetus of Einstein's general theory of relativity (1915) a further generalization, technical but far-reaching, appeared: the positive definiteness of the inner product was weakened to nondegeneracy.

1. Definition. A *metric tensor* $\boldsymbol{g}$ on a smooth manifold M is a symmetric nondegenerate (0, 2) tensor field on M of constant index.

In other words $\boldsymbol{g} \in \mathfrak{T}^0_2(M)$ smoothly assigns to each point p of M a scalar product $\boldsymbol{g}_p$ on the tangent space $T_p(M)$, and the index of $\boldsymbol{g}_p$ is the same for all p.

2. Definition. A *semi-Riemannian manifold* is a smooth manifold M furnished with a metric tensor $\boldsymbol{g}$.

Thus strictly speaking a semi-Riemannian manifold is an ordered pair (M, g): two different metric tensors on the same manifold constitute different semi-Riemannian manifolds. Nevertheless we usually denote a semi-Riemannian manifold by the name of its smooth manifold $M, N, \ldots$.

The common value ν of index g_p on a semi-Riemannian manifold M is called the *index* of M: $0 \leq \nu \leq n = \dim M$. If $\nu = 0$, M is a *Riemannian manifold*; each g_p is then a (positive definite) inner product on $T_p(M)$. If $\nu = 1$ and $n \geq 2$, M is a *Lorentz manifold.*

Semi-Riemannian manifolds are often called *pseudo-Riemannian manifolds*, or even—in older terminology—Riemannian manifolds, but we reserve the latter term for the distinctive positive definite case.

We use $\langle\ ,\ \rangle$ as an alternative notation for g, writing $g(v, w) = \langle v, w\rangle \in \mathbf{R}$ for tangent vectors, and $g(V, W) = \langle V, W\rangle \in \mathfrak{F}(M)$ for vector fields.

If $x^1, \ldots, x^n$ is a coordinate system on $\mathscr{U} \subset M$ the components of the metric tensor g on $\mathscr{U}$ are

$$g_{ij} = \langle \partial_i, \partial_j \rangle \qquad (1 \leq i, j \leq n).$$

Thus for vector fields $V = \sum V^i\, \partial_i$ and $W = \sum W^j\, \partial_j$,

$$g(V, W) = \langle V, W\rangle = \sum g_{ij} V^i W^j.$$

Since g is nondegenerate, at each point p of $\mathscr{U}$ the matrix $(g_{ij}(p))$ is invertible, and its inverse matrix is denoted by $(g^{ij}(p))$. The usual formula for the inverse of a matrix shows that the functions g^{ij} are smooth on $\mathscr{U}$.

Since g is symmetric, $g_{ij} = g_{ji}$ and hence $g^{ij} = g^{ji}$ for $1 \leq i, j \leq n$. Finally on $\mathscr{U}$ the metric tensor can be written as

$$g = \sum g_{ij}\, dx^i \otimes dx^j.$$

Recall from Chapter 1 that for each $p \in \mathbf{R}^n$ there is a canonical linear isomorphism from $\mathbf{R}^n$ to $T_p(\mathbf{R}^n)$ that, in terms of natural coordinates, sends v to $v_p = \sum v^i\, \partial_i$. Thus the dot product on $\mathbf{R}^n$ gives rise to a metric tensor on $\mathbf{R}^n$ with

$$\langle v_p, w_p \rangle = v \cdot w = \sum v^i w^i.$$

Henceforth in any geometric context $\mathbf{R}^n$ will denote the resulting Riemannian manifold, called *Euclidean n-space.*

For an integer ν with $0 \leq \nu \leq n$, changing the first ν plus signs above to minus gives a metric tensor

$$\langle v_p, w_p \rangle = -\sum_{i=1}^{\nu} v^i w^i + \sum_{j=\nu+1}^{n} v^j w^j$$

of index ν. The resulting *semi-Euclidean space* $\mathbf{R}^n_\nu$ reduces to $\mathbf{R}^n$ if $\nu = 0$. For $n \geq 2$, $\mathbf{R}^n_1$ is called *Minkowski n-space*; if $n = 4$ it is the simplest example of a relativistic spacetime.

Fix the notation

$$\varepsilon_i = \begin{cases} -1 & \text{for} \quad 1 \leq i \leq \nu, \\ +1 & \text{for} \quad \nu + 1 \leq i \leq n. \end{cases}$$

Then the metric tensor of $\boldsymbol{R}^n_\nu$ can be written

$$\boldsymbol{g} = \sum \varepsilon_i \, du^i \otimes du^i.$$

The geometric significance of the index of a semi-Riemannian manifold derives from the following trichotomy.

3. **Definition.** A tangent vector v to M is

$$\begin{array}{lllll} \textit{spacelike} & \text{if} & \langle v, v \rangle > 0 & \text{or} & v = 0, \\ \textit{null} & \text{if} & \langle v, v \rangle = 0 & \text{and} & v \neq 0, \\ \textit{timelike} & \text{if} & \langle v, v \rangle < 0. & & \end{array}$$

The set of all null vectors in $T_p(M)$ is called the *nullcone* at $p \in M$. The category into which a given tangent vector falls is called its *causal character*. This terminology derives from relativity theory, and particularly in the Lorentz case, null vectors are also said to be *lightlike*.

Let $\boldsymbol{q}(v) = \langle v, v \rangle$ for each tangent vector v to M. At each point p of M, $\boldsymbol{q}$ gives the associated quadratic form of the scalar product at p; thus $\boldsymbol{q}$ determines the metric tensor.

If $V \in \mathfrak{X}(M)$ and $f \in \mathfrak{F}(M)$, then $\boldsymbol{q}(fV) = f^2\boldsymbol{q}(V) \in \mathfrak{F}(M)$, so $\boldsymbol{q}$ is not a tensor field. Classically $\boldsymbol{q}$ is called the *line element* of M, and denoted by ds^2. In terms of a coordinate system,

$$\boldsymbol{q} = ds^2 = \sum g_{ij} \, dx^i \, dx^j.$$

Here the juxtaposition of differentials denotes ordinary multiplication of functions (on each tangent space), so

$$\boldsymbol{q}(V) = \sum g_{ij} \, dx^i(V) \, dx^j(V) = \sum g_{ij} V^i V^j.$$

As in Chapter 2, the *norm* $|v|$ of a tangent vector is $|\boldsymbol{q}(v)|^{1/2} = |\langle v, v \rangle|^{1/2}$, and unit vectors, orthogonality, and orthonormality are as before.

The origin of the unusual notation ds^2 can be seen intuitively as follows. Assume for simplicity that M is Riemannian. If p and p' are nearby points with coordinates $(x^1, \ldots, x^n)$ and $(x^1 + \Delta x^1, \ldots, x^n + \Delta x^n)$ relative to some coordinate system, then the tangent vector $\Delta p = \sum \Delta x^i \, \partial_i$ at p points approximately to p'. Thus we expect the square of the distance Δs from p to p' to be approximately

$$|\Delta p|^2 = \langle \Delta p, \Delta p \rangle = \sum g_{ij}(p) \, \Delta x^i \, \Delta x^j,$$

as in the formula $ds^2 = \sum g_{ij} \, dx^i \, dx^j$.

Given a way to get new smooth manifolds from old, there is often a corresponding way to derive a metric tensor on the new manifold from metric tensors on the old.

For example, suppose first that P is a submanifold of a Riemannian manifold M. Since each tangent space $T_p(P)$ of P is regarded as a subspace of $T_p(M)$, we obtain a Riemannian metric tensor g_P on P merely by applying the metric tensor g of M to each pair of tangent vectors to P. Formally, g_P is the pullback $j^*(g)$, where $j: P \subset M$ is the inclusion map. For example, the *standard n-sphere of radius* $r > 0$ is the Riemannian submanifold

$$S^n(r) = \{p : |p| = r\} \quad \text{of} \quad \boldsymbol{R}^{n+1}.$$

However when the metric tensor g of M is indefinite, then $j^*(g)$ need not be a metric on P. It is a smooth symmetric (0, 2) tensor field, hence it is a metric if and only if each $T_p(P)$ is nondegenerate in $T_p(M)$ relative to g—and the index of $T_p(P)$ is the same for all p (see Exercise 10(b)).

4. Definition. Let P be a submanifold of a semi-Riemannian manifold M. If the pullback $j^*(g)$ (as above) is a metric tensor on P it makes P a *semi-Riemannian submanifold* of M.

(If P is known to be Riemannian or Lorentz, these terms replace semi-Riemannian.)

Now we consider product manifolds.

5. Lemma. Let M and N be semi-Riemannian manifolds with metric tensors g_M and g_N. If π and σ are the projections of $M \times N$ onto M and N, respectively, let

$$g = \pi^*(g_M) + \sigma^*(g_N).$$

Then g is a metric tensor on $M \times N$ making it a *semi-Riemannian product manifold*.

Proof. Translating from the pullback notation: if $v, w \in T_{(p,q)}(M \times N)$, then

$$g(v, w) = g_M(d\pi(v), d\pi(w)) + g_N(d\sigma(v), d\sigma(w)).$$

Thus g is symmetric. To show nondegeneracy, suppose $g(v, w) = 0$ for all $w \in T_{(p,q)}(M \times N)$. Then, in particular, for all $w \in T_{(p,q)}M$ we have $g_M(d\pi(v), d\pi(w)) = 0$, since $d\sigma(w) = 0$. But such $d\pi(w)$ fill $T_p(M)$, hence $d\pi(v) = 0$. Similarly $d\sigma(v) = 0$; hence $v = 0$.

Orthonormal bases for $T_p(M)$ and $T_q(N)$ combine to give an orthonormal basis for $T_{(p,q)}(M \times N)$. Hence the index of g has constant value ind M + ind N. ∎

The same scheme extends in an obvious way to any finite product of semi-Riemannian manifolds. For example, the semi-Euclidean space $\mathbf{R}^n_\nu$ is

$$\underbrace{\mathbf{R}^1_1 \times \cdots \times \mathbf{R}^1_1}_{\nu \text{ factors}} \times \underbrace{\mathbf{R}^1 \times \cdots \times \mathbf{R}^1}_{n-\nu \text{ factors}} = \mathbf{R}^\nu_\nu \times \mathbf{R}^{n-\nu},$$

where by definition $\mathbf{R}^1_1$ is the real line with metric tensor the negative of the usual dot product on $\mathbf{R}^1$.

ISOMETRIES

An isometry is the special type of mapping that expresses the notion of isomorphism for semi-Riemannian manifolds.

6. Definition. Let M and N be semi-Riemannian manifolds with metric tensors g_M and g_N. An *isometry* from M to N is a diffeomorphism $\phi: M \to N$ that *preserves metric tensors*: $\phi^*(g_N) = g_M$.

Explicitly, $\langle d\phi(v), d\phi(w)\rangle = \langle v, w\rangle$ for all $v, w \in T_p(M)$, $p \in M$. Since ϕ is a diffeomorphism, each differential map $d\phi_p$ is a linear isomorphism; thus the metric condition means that each $d\phi_p$ is a linear isometry. The pullback operates in the usual way on line elements, and since these determine their metric tensors, preservation of metrics is equivalent to $\phi^*(q_N) = q_M$.

It is easy to see that

(1) The identity map of a semi-Riemannian manifold is an isometry.
(2) A composition of isometries is an isometry.
(3) The inverse map of an isometry is an isometry.

The interpretation of $q = ds^2$ as the square of infinitesimal distance suggests thinking of an isometry as a rigid motion, by contrast with an arbitrary diffeomorphism which can deform M in applying it to N.

An object preserved in an appropriate sense by all isometries is called an *isometric invariant*; and *semi-Riemannian geometry* is traditionally described as the study of such invariants. If there exists an isometry between M and N, they are said to be *isometric*; roughly speaking, isometric manifolds are geometrically the same.

Let V be a scalar product space, that is, a real vector space furnished with a scalar product. Then V is a manifold, and just as in the case $V = \mathbf{R}^n$ the formula $\langle v_p, w_p\rangle = \langle v, w\rangle$ defines a metric tensor on V, making it a semi-Riemannian manifold.

7. Lemma. If $\psi\colon V \to W$ is a linear isometry of scalar product spaces, then (for V and W semi-Riemannian as above) $\psi\colon V \to W$ is an isometry.

Proof. Since linear maps are smooth, the linear isomorphism ψ is a diffeomorphism. If $v_p \in T_p(V)$, then Exercise 1.2 gives $d\psi(v_p) = (\psi(v))_{\psi(p)}$. Thus ψ preserves metric tensors, since

$$\begin{aligned}\langle d\psi(v_p), d\psi(w_p)\rangle &= \langle(\psi(v))_{\psi(p)}, (\psi(w))_{\psi(p)}\rangle \\ &= \langle\psi(v), \psi(w)\rangle = \langle v, w\rangle = \langle v_p, w_p\rangle. \qquad \blacksquare\end{aligned}$$

It follows that if V is a scalar product space of dimension n and index ν, then as a semi-Riemannian manifold, V is isometric to $\mathbf{R}^n_\nu$. In fact, the coordinate isomorphism of any orthonormal basis for V is a (linear) isometry.

If M is an arbitrary semi-Riemannian manifold, its metric tensor makes each of its tangent spaces a semi-Euclidean space of the same dimension and index as M itself. This is one view of how semi-Riemannian geometry generalizes semi-Euclidean geometry.

THE LEVI-CIVITA CONNECTION

Let V and W be vector fields on a semi-Riemannian manifold M. The goal of this section is to show how to define a new vector field $D_V W$ on M whose value at each point p is the vector rate of change of W in the V_p direction. There is a natural way to do this on $\mathbf{R}^n_\nu$.

8. Definition. Let $u^1, \ldots, u^n$ be the natural coordinates on $\mathbf{R}^n_\nu$. If V and $W = \sum W^i\,\partial_i$ are vector fields on $\mathbf{R}^n_\nu$, the vector field

$$D_V W = \sum V(W^i)\,\partial_i$$

is called the *natural covariant derivative* of W with respect to V.

Since this definition uses the distinctive coordinates of $\mathbf{R}^n_\nu$ it is not obvious how to extend it to an arbitrary semi-Riemannian manifold. We begin, therefore, by axiomatizing its key properties.

9. Definition. A *connection* D on a smooth manifold M is a function $D\colon \mathfrak{X}(M) \times \mathfrak{X}(M) \to \mathfrak{X}(M)$ such that

(D1) $D_V W$ is $\mathfrak{F}(M)$-linear in V,
(D2) $D_V W$ is $\mathbf{R}$-linear in W,
(D3) $D_V(fW) = (Vf)W + f\,D_V W$ for $f \in \mathfrak{F}(M)$.

$D_V W$ is called the *covariant derivative* of W with respect to V for the connection D.

Axiom (D1) asserts that $D_V W$ is *tensor in* V; hence by Proposition 2.2, for an individual tangent vector $v \in T_p(M)$ we have a well-defined tangent vector $D_v W \in T_p(M)$, namely, $(D_V W)_p$ where V is any vector field such that $V_p = v$. On the other hand, (D3) shows that $D_V W$ is not tensor in W.

We can now state our goal more precisely: it is to show that on every semi-Riemannian manifold there is a unique connection sharing two further properties ((D4) and (D5) below) of the natural connection on $\boldsymbol{R}^n_\nu$.

The next step is algebraic.

10. Proposition. Let M be a semi-Riemannian manifold. If $V \in \mathfrak{X}(M)$ let V^* be the one-form on M such that

$$V^*(X) = \langle V, X \rangle \qquad \text{for all} \quad X \in \mathfrak{X}(M).$$

Then the function $V \to V^*$ is an $\mathfrak{F}(M)$-linear isomorphism from $\mathfrak{X}(M)$ to $\mathfrak{X}^*(M)$.

Proof. Since V^* is $\mathfrak{F}(M)$-linear it is indeed a one-form, and the function $V \to V^*$ is also $\mathfrak{F}(M)$-linear. That it is an isomorphism follows from two facts:

(a) If $\langle V, X \rangle = \langle W, X \rangle$ for all $X \in \mathfrak{X}(M)$, then $V = W$.

(b) Given any one-form $\theta \in \mathfrak{X}^*(M)$ there is a unique vector field $V \in \mathfrak{X}(M)$ such that $\theta(X) = \langle V, X \rangle$ for all X.

Let $U = V - W$. Then assertion (a) amounts to showing that if $\langle U_p, X_p \rangle = 0$ for all $X \in \mathfrak{X}(M)$ and all $p \in M$, then $U = 0$. Since every element of $T_p(M)$ has the form X_p, the result follows by the nondegeneracy of the metric tensor.

Now (a) is exactly the uniqueness assertion in (b), hence to prove (b) it suffices to find V on an arbitrary coordinate neighborhood $\mathscr{U}$. (All these local Vs will be consistent on overlaps.) If $\theta = \sum \theta_i\, dx^i$ on $\mathscr{U}$, let $V = \sum_{i,j} g^{ij}\theta_i\, \partial_j$. Then since (g_{ij}) and (g^{ij}) are inverse matrices,

$$\begin{aligned}\langle V, \partial_k \rangle &= \sum_{i,j} g^{ij}\theta_i \langle \partial_j, \partial_k \rangle = \sum_{i,j} \theta_i g^{ij} g_{jk} \\ &= \sum_i \theta_i \delta_{ik} = \theta_k = \theta(\partial_k).\end{aligned}$$

It follows by $\mathfrak{F}(M)$-linearity that $\langle V, X \rangle = \theta(X)$ for all X on $\mathscr{U}$. ∎

Thus in semi-Riemannian geometry we can freely transform a vector field into a one-form, and vice versa. Corresponding pairs $V \leftrightarrow \theta$ contain exactly the same information, and are said to be *metrically equivalent*.

The following fundamental result has been called the miracle of semi-Riemannian geometry:

11. Theorem. On a semi-Riemannian manifold M there is a unique connection D such that

(D4) $[V, W] = D_V W - D_W V$, and
(D5) $X\langle V, W\rangle = \langle D_X V, W\rangle + \langle V, D_X W\rangle$,

for all $X, V, W \in \mathfrak{X}(M)$. D is called the *Levi-Civita connection* of M, and is characterized by the *Koszul formula*

$$\begin{aligned}2\langle D_V W, X\rangle = {} & V\langle W, X\rangle + W\langle X, V\rangle - X\langle V, W\rangle \\ & - \langle V, [W, X]\rangle + \langle W, [X, V]\rangle + \langle X, [V, W]\rangle.\end{aligned}$$

Proof. Suppose that D is a connection on M satisfying axioms (D4) and (D5). On the right-hand side of the Koszul formula use (D5) on the first three terms and (D4) on the last three. Most terms cancel in pairs leaving $2\langle D_V W, X\rangle$. Thus D satisfies the Koszul formula, hence by assertion (a) in the preceding proof it is unique.

For the existence define $F(V, W, X)$ to be the right-hand side of the Koszul formula. For fixed $V, W \in \mathfrak{X}(M)$ a straightforward computation shows that the function $X \to F(V, W, X)$ is $\mathfrak{F}(M)$-linear, hence is a one-form. By Proposition 10, there is a unique vector field, which we denote by $D_V W$, such that $2\langle D_V W, X\rangle = F(V, W, X)$ for all X. Thus the Koszul formula holds and from it we can deduce (D1)–(D5).

For example, let us prove (D3). For an arbitrary X,

$$\begin{aligned}2\langle D_V(fW), X\rangle = {} & V\langle fW, X\rangle + fW\langle X, V\rangle - X\langle V, fW\rangle \\ & - \langle V, [fW, X]\rangle + \langle fW, [X, V]\rangle + \langle X, [V, fW]\rangle.\end{aligned}$$

Functions can be factored out of the tensor $\langle\, ,\rangle$, and for the bracket operation we have, for example, $[fW, X] = -XfW + f[W, X]$. Thus the expression on the right-hand side above becomes

$$\begin{aligned}& Vf\langle W, X\rangle + Vf\langle X, W\rangle + Xf\langle V, W\rangle - Xf\langle V, W\rangle + fF(V, W, X) \\ & \quad = 2\langle VfW + fD_V W, X\rangle.\end{aligned}$$

Then by the preceding proof, $D_V(fW) = (Vf)W + f D_V W$.

To prove (D4) start from

$$2\langle D_V W - D_W V, X\rangle = F(V, W, X) - F(W, V, X).$$

The right-hand side reduces to

$$\langle X, [V, W]\rangle - \langle X, [W, V]\rangle = 2\langle [V, W], X\rangle.$$

Hence the result follows. The other verifications are similar. ∎

12. Definition. Let $x^1, \ldots, x^n$ be a coordinate system on a neighborhood $\mathscr{U}$ in a semi-Riemannian manifold M. The *Christoffel symbols* for this coordinate system are the real-valued functions Γ_{ij}^k on $\mathscr{U}$ such that

$$D_{\partial_i}(\partial_j) = \sum_k \Gamma_{ij}^k \, \partial_k \qquad (1 \leq i, j \leq n).$$

Since $[\partial_i, \partial_j] = 0$, it follows from (D4) that $D_{\partial_i}(\partial_j) = D_{\partial_j}(\partial_i)$, hence $\Gamma_{ij}^k = \Gamma_{ji}^k$.

The connection D is not a tensor, so the Christoffel symbols do not obey the usual tensor transformation rule under change of coordinates.

13. Proposition. For a coordinate system $x^1, \ldots, x^n$ on $\mathscr{U}$,

(1) $$D_{\partial_i}\left(\sum W^j \partial_j\right) = \sum_k \left\{ \frac{\partial W^k}{\partial x^i} + \sum_j \Gamma_{ij}^k W^j \right\} \partial_k,$$

where the Christoffel symbols are given by

(2) $$\Gamma_{ij}^k = \frac{1}{2} \sum_m g^{km} \left\{ \frac{\partial g_{jm}}{\partial x^i} + \frac{\partial g_{im}}{\partial x^j} - \frac{\partial g_{ij}}{\partial x^m} \right\}.$$

Proof. (1) is an immediate consequence of (D3). To derive (2) set $V = \partial_i$, $W = \partial_j$, $X = \partial_m$ in the Koszul formula. The brackets are zero, leaving

$$2\langle D_{\partial_i}(\partial_j), \partial_m \rangle = \frac{\partial}{\partial x^i}(g_{jm}) + \frac{\partial}{\partial x^j}(g_{im}) - \frac{\partial}{\partial x^m}(g_{ij}).$$

But by the definition of Christoffel symbols,

$$2\langle D_{\partial_i}(\partial_j), \partial_m \rangle = 2 \sum_a \Gamma_{ij}^a g_{am}.$$

Attacking both equations above with $\sum_m g^{mk}$ gives the required result. ∎

Using (D1) we can compute any $D_V W$ on coordinate neighborhoods by the first formula above, while the second formula is the coordinate description of how the metric tensor determines the Levi-Civita connection.

14. Lemma. The natural connection D of Definition 8 is the Levi-Civita connection of the semi-Euclidean space $\boldsymbol{R}_\nu^n$ for every $\nu = 0, 1, \ldots, n$. Relative to natural coordinates on $\boldsymbol{R}_\nu^n$

(1) $$g_{ij} = \delta_{ij}\varepsilon_j, \quad \text{where} \quad \varepsilon_j = \begin{cases} -1 & \text{for} \quad 1 \leq j \leq \nu, \\ +1 & \text{for} \quad \nu + 1 \leq j \leq n, \end{cases}$$

(2) $\Gamma_{ij}^k = 0$,

for all $1 \leq i, j, k \leq n$.

Proof. (1) is essentially the definition of the metric tensor of $\boldsymbol{R}^n_\nu$. To prove that D is the Levi-Civita connection of $\boldsymbol{R}^n_\nu$ one must check that it satisfies (D1)–(D5). Take (D5), for example. Since $\langle V, W\rangle = \sum \varepsilon_i V^i W^i$,

$$\begin{aligned} X\langle V, W\rangle &= \sum \varepsilon_i X(V^i)W^i + \sum \varepsilon_i V^i X(W^i) \\ &= \langle D_X V, W\rangle + \langle V, D_X W\rangle. \end{aligned}$$

Then (2) follows from Proposition 13(2), since the g_{ij}s are constant. ∎

A vector field V is *parallel* provided its covariant derivatives $D_X V$ are zero for all $X \in \mathfrak{X}(M)$. Thus the vanishing of Christoffel symbols in the lemma means that the natural coordinate vector fields on $\boldsymbol{R}^n_\nu$ are parallel. In general the Christoffel symbols of a coordinate system measure the failure of its coordinate vector fields to be parallel.

15. Example. Cylindrical Coordinates in $\boldsymbol{R}^3$. Let r, φ, z be the usual cylindrical coordinates in $\boldsymbol{R}^3$ as indicated in Figure 1. Actually (r, φ, z) is a coordinate system only on $\boldsymbol{R}^3 - H$, where H is, for example, the half-plane $x \geq 0$, $y = 0$. There the coordinate functions are well defined and an inverse mapping exists given by $x = r\cos\varphi$, $y = r\sin\varphi$, $z = z$.

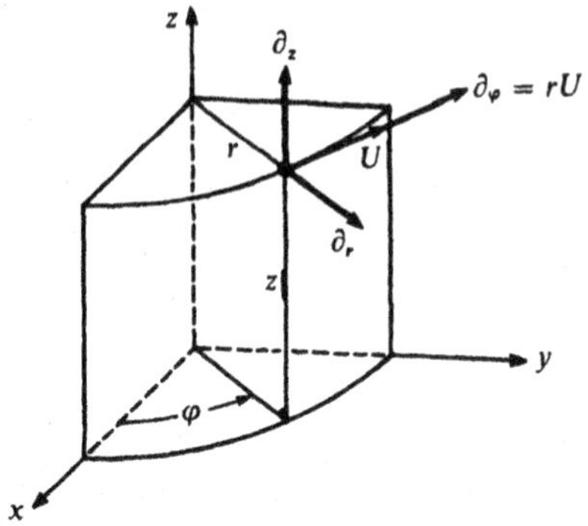

Figure 1

Hence by the basis theorem

$$\partial_r = \cos\varphi\,\partial_x + \sin\varphi\,\partial_y;$$

$$\partial_\varphi = rU, \qquad \text{where} \quad U = -\sin\varphi\,\partial_x + \cos\varphi\,\partial_y;$$

$$\partial_z = \partial_z.$$

For the sake of indexing, let $y^1 = r$, $y^2 = \varphi$, $y^3 = z$. Then $g_{11} = g_{33} = 1$, $g_{22} = r^2$, and $g_{ij} = 0$ for $i \neq j$, hence

$$ds^2 = dr^2 + r^2\, d\varphi^2 + dz^2.$$

In particular this is an *orthogonal* coordinate system; that is, the coordinate vector fields are mutually orthogonal.

A direct computation shows that the Christoffel symbols of cylindrical coordinates are all zero except $\Gamma^1_{22} = -r$ and $\Gamma^2_{21} = \Gamma^2_{12} = 1/r$. Hence all coordinate covariant derivatives are zero, except

$$D_{\partial_\varphi}(\partial_\varphi) = -r\,\partial_r, \qquad D_{\partial_\varphi}(\partial_r) = D_{\partial_r}(\partial_\varphi) = U.$$

These formulas are consistent with what we can visualize from Figure 1. Since ∂_z is also a natural coordinate vector field it must be parallel. Similarly we expect $D_{\partial_z}(\partial_r) = D_{\partial_z}(\partial_\varphi) = 0$, since ∂_r and ∂_φ remain parallel as the point p moves in the z direction.

The covariant derivative D_V can be extended to operate on arbitrary tensor fields. In fact, axioms (D2) and (D3) are exactly what is needed to apply Theorem 2.15.

16. Definition. Let V be a vector field on a semi-Riemannian manifold M. The (Levi-Civita) *covariant derivative* D_V is the unique tensor derivation on M such that

$$D_V f = Vf \qquad \text{for} \quad f \in \mathfrak{F}(M),$$

and $D_V W$ is the Levi-Civita covariant derivative for all $W \in \mathfrak{X}(M)$.

If $A \in \mathfrak{T}^r_s(M)$ then the (r, s) tensor field $D_V A$ is $\mathfrak{F}(M)$-linear in $V \in \mathfrak{X}(M)$. In fact, by Corollary 2.14 it suffices to check that the tensor derivations D_{fV+gW} and $f\,D_V + g\,D_W$ agree on $\mathfrak{F}(M)$ and $\mathfrak{X}(M)$. But the former is definitional and the latter is (D1). This remark justifies the following definition.

17. Definition. The *covariant differential* of an (r, s) tensor A on M is the $(r, s+1)$ tensor DA such that

$$(DA)(\theta^1, \ldots, \theta^r, X_1, \ldots, X_s, V) = (D_V A)(\theta^1, \ldots, \theta^r, X_1, \ldots, X_s)$$

for all $V, X_i \in \mathfrak{X}(M)$ and $\theta^j \in \mathfrak{X}^*(M)$.

In the exceptional case $r = s = 0$ the covariant differential of a function f is its usual differential $df \in \mathfrak{X}^*(M)$, since

$$(Df)(V) = D_V f = Vf = df(V) \qquad \text{for all} \quad V \in \mathfrak{X}(M).$$

DA is simply a convenient way to collect all the covariant derivatives of A. The fact that the covariant type of DA is one larger than that of A accounts for the term *covariant* as applied to both derivatives and differentials.

Just as for a vector field, a tensor field A is *parallel* provided its covariant differential is zero, that is, $D_V A = 0$ for all $V \in \mathfrak{X}(M)$. For example, using the product rule (2.13) it follows that (D5) is equivalent to the parallelism of the metric tensor g.

If $A \in \mathfrak{T}^r_s(M)$ the components of DA relative to a coordinate system are denoted by $A^{i_1 \cdots i_r}_{j_1 \cdots j_s;k}$. Exercise 2 shows how to express these components in terms of the components of A and the Christoffel symbols of the coordinate system. The general formula is somewhat complicated, but as we shall see its use can be avoided. In the special case of natural coordinates on R^n_ν, since the coordinate vector fields and hence the differentials $du^1, \ldots, du^n$ are parallel it follows that $A^{i_1 \cdots i_r}_{j_1 \cdots j_s;k} = (\partial/\partial u^k) A^{i_1 \cdots i_r}_{j_1 \cdots j_s}$.

PARALLEL TRANSLATION

The simplest case of a vector field on a mapping (Definition 1.47) is a vector field Z on a curve $\alpha: I \to M$. Z smoothly assigns to each $t \in I$ a tangent vector to M at $\alpha(t)$. For example, the velocity α' is a vector field on α, as is the restriction V_α of any $V \in \mathfrak{X}(M)$. The set $\mathfrak{X}(\alpha)$ of all (smooth) vector fields on α is a module over $\mathfrak{F}(I)$.

When M is a semi-Riemannian manifold there is a natural way to define the vector rate of change Z' of a vector field $Z \in \mathfrak{X}(\alpha)$.

18. Proposition. Let $\alpha: I \to M$ be a curve in a semi-Riemannian manifold M. Then there is a unique function $Z \to Z' = DZ/dt$ from $\mathfrak{X}(\alpha)$ to $\mathfrak{X}(\alpha)$, called the *induced covariant derivative*, such that

(1) $(aZ_1 + bZ_2)' = aZ_1' + bZ_2'$ $\quad (a, b \in R)$,
(2) $(hZ)' = (dh/dt)Z + hZ'$ $\quad (h \in \mathfrak{F}(I))$,
(3) $(V_\alpha)'(t) = D_{\alpha'(t)}(V)$ $\quad (t \in I, \quad V \in \mathfrak{X}(M))$.

Furthermore,

(4) $(d/dt)\langle Z_1, Z_2\rangle = \langle Z_1', Z_2\rangle + \langle Z_1, Z_2'\rangle$.

Proof. Uniqueness. Suppose an induced connection exists satisfying only the first three properties. We can assume that α lies in the domain of a single coordinate system $x^1, \ldots, x^n$. By the basis theorem, if $Z \in \mathfrak{X}(\alpha)$, then at $\alpha(t)$,

$$Z(t) = \sum Z(t)x^i\, \partial_i = \sum (Zx^i)(t)\, \partial_i.$$

Denote the component function $Zx^i: I \to R$ by Z^i. By properties (1) and (2)

$$Z' = \sum \frac{dZ^i}{dt} \partial_i|_\alpha + \sum Z^i(\partial_i|_\alpha)'.$$

But by (3), $(\partial_i|_\alpha)' = D_{\alpha'}(\partial_i)$; thus

$$Z' = \sum \frac{dZ^i}{dt} \partial_i + \sum Z^i D_{\alpha'}(\partial_i).$$

Thus Z' is completely determined by the Levi-Civita connection D.

Existence. On any subinterval J of I such that $\alpha(J)$ lies in a coordinate neighborhood, define Z' by the formula above. Then straightforward computations show that all four properties hold. By the uniqueness these local definitions of Z' constitute a single vector field in $\mathfrak{X}(\alpha)$. ∎

In the special case $Z = \alpha'$ the derivative $Z' = \alpha''$ is called the *acceleration* of the curve α. More elaborate notations are sometimes used to emphasize that α'' involves geometry while α' does not.

For a vector field Z on α it is tempting to write $Z' = D_{\alpha'} Z$ and hence also $\alpha'' = D_{\alpha'}(\alpha')$. Though Z and α' are not vector fields on M, these formulas can be justified—but only at points $\alpha(t)$ where $\alpha'(t) \neq 0$ (see Exercise 12).

Introducing Christoffel symbols into the coordinate formula above yields

$$Z' = \sum_k \left\{ \frac{dZ^k}{dt} + \sum_{i,j} \Gamma^k_{ij} \frac{d(x^i \circ \alpha)}{dt} Z^j \right\} \partial_k.$$

If $Z' = 0$, then Z is said to be *parallel*. This formula shows that the equation $Z' = 0$ is equivalent to a system of linear ordinary differential equations. Thus the fundamental existence and uniqueness theorem for such systems gives:

19. Proposition. For a curve $\alpha: I \to M$, let $a \in I$ and $z \in T_{\alpha(a)}(M)$. Then there is a unique parallel vector field Z on α such that $Z(a) = z$.

Here we take advantage of the fact that solutions of a linear system are defined on the entire interval for which its coefficient functions are given.

In the notation of the proposition, if $b \in I$ then the function

$$P = P_a^b(\alpha): T_p(M) \to T_q(M)$$

sending each z to $Z(b)$ is called *parallel translation along* α from $p = \alpha(a)$ to $q = \alpha(b)$.

20. Lemma. Parallel translation is a linear isometry.

Proof. With notation as above, let $v, w \in T_p(M)$ correspond as in the proposition to parallel vector fields V, W. Since $V + W$ is also parallel,

$P(v + w) = (V + W)(b) = V(b) + W(b) = P(v) + P(w)$. Similarly, $P(cv) = cP(v)$. Thus P is linear.

If $P(v) = 0$ then by the uniqueness in the proposition, V can only be the identically zero vector field on α. Hence $v = V(a) = 0$. Thus P is one-to-one, and since tangent spaces to M have the same dimension, P is a linear isomorphism.

Finally, for V, W as above,

$$\frac{d}{dt}\langle V, W\rangle = \langle V', W\rangle + \langle V, W'\rangle = 0.$$

Hence $\langle V, W\rangle$ is constant, so

$$\langle P(v), P(w)\rangle = \langle V(b), W(b)\rangle = \langle V(a), W(a)\rangle = \langle v, w\rangle. \quad \blacksquare$$

In general, parallel translation from p to q depends on the particular curve joining p to q. On $\mathbf{R}^n_\nu$ the natural coordinate vector fields are parallel and hence so are their restrictions to any curve. Hence parallel translation from p to q along any curve is just the canonical isomorphism $v_p \to v_q$. This phenomenon is called *distant parallelism.*

GEODESICS

We now generalize the Euclidean notion of straight line. A *geodesic* in a semi-Riemannian manifold M is a curve $\gamma: I \to M$ whose vector field γ' is parallel. Equivalently, geodesics are the curves of acceleration zero: $\gamma'' = 0$.

21. Corollary. Let $x^1, \ldots, x^n$ be a coordinate system on $\mathscr{U} \subset M$. A curve γ in $\mathscr{U}$ is a geodesic of M if and only if its coordinate functions $x^k \circ \gamma$ satisfy

$$\frac{d^2(x^k \circ \gamma)}{dt^2} + \sum_{i,j} \Gamma^k_{ij}(\gamma)\frac{d(x^i \circ \gamma)}{dt}\frac{d(x^j \circ \gamma)}{dt} = 0$$

for $1 \leq k \leq n$.

In fact, these expressions are the components of γ'' relative to the coordinate vector fields $\partial_1, \ldots, \partial_n$.

In dealing with curves it is often convenient to use a common abbreviation, writing the coordinate functions of γ as x^i rather than $x^i \circ \gamma$. In any reasonable context there should be no confusion between these functions on the domain I of γ and the coordinate functions on $\mathscr{U} \subset M$. The geodesic equations then become

$$\frac{d^2(x^k)}{dt^2} + \sum_{i,j} \Gamma^k_{ij}\frac{dx^i}{dt}\frac{dx^j}{dt} = 0 \qquad (1 \leq k \leq n).$$

The existence and uniqueness theorem for ordinary differential equations gives the following local result.

22. Lemma. If $v \in T_p(M)$ there exists an interval I about 0 and a unique geodesic $\gamma: I \to M$ such that $\gamma'(0) = v$.

The last equation implies, of course, that $\gamma(0) = p$; we say that γ is a geodesic *starting at p* with *initial velocity v*.

23. Lemma. Let $\alpha, \beta: I \to M$ be geodesics. If there is a number $a \in I$ such that $\alpha'(a) = \beta'(a)$, then $\alpha = \beta$.

Proof. Suppose the conclusion is false; then there is a $t_0 \in I$ such that $\alpha(t_0) \neq \beta(t_0)$, with say $t_0 > a$. Thus the set $\{t \in I : t > a \text{ and } \alpha(t) \neq \beta(t)\}$ has a greatest lower bound b, for which $b \geq a$. We assert that $\alpha'(b) = \beta'(b)$. This is given if $b = a$. If $b > a$, then α and β agree on the interval (a, b). Coordinate expressions show that the functions $t \to \alpha'(t)$ and $t \to \alpha'(t)$ from (a, b) into the tangent manifold TM are continuous (in fact, smooth). Thus as t approaches b from below

$$\alpha'(b) = \lim \alpha'(t) = \lim \beta'(t) = \beta'(b).$$

Since $t \to \alpha(t + b)$ and $t \to \beta(t + b)$ are also geodesics, Lemma 22 shows that $\alpha = \beta$ on some interval around b. But this contradicts the definition of b. ∎

24. Proposition. Given any tangent vector $v \in T_p(M)$ there is a unique geodesic γ_v in M such that

(1) The initial velocity of γ_v is v; that is, $\gamma_v'(0) = v$.

(2) The domain I_v of γ_v is the largest possible. Hence, if $\alpha: J \to M$ is a geodesic with initial velocity v, then $J \subset I$ and $\alpha = \gamma_v | J$.

Proof. Let $\mathscr{G}$ be the collection of all geodesics $\gamma: I_\gamma \to M$ with initial velocity v. (By Lemma 22 there are some.) Lemma 23 shows that α and β in $\mathscr{G}$ agree on $I_\alpha \cap I_\beta$. Hence the collection $\mathscr{G}$ consistently defines a single curve γ_v on the interval $I = \bigcup I_\gamma$. Evidently γ_v has the required properties. ∎

Because of (2) the geodesic γ_v is said to be *maximal* or *geodesically inextendible*. The notation γ_v will be used frequently.

Picture M as a surface in $\boldsymbol{R}^3$ and p as a penny constrained to remain on M. Once p is given an initial velocity its motion is completely determined and it traces out a geodesic of M.

A semi-Riemannian manifold M for which every maximal geodesic is defined on the entire real line is said to be *geodesically complete*—or merely *complete*. (See Exercise 7). Note that if even a single point p is removed from a complete manifold M then $M - p$ is no longer complete, since geodesics that formerly went through p are now obliged to stop.

25. Example. Geodesics of Semi-Euclidean Space. For natural coordinates the Christoffel symbols vanish, so the geodesic equations become

$$\frac{d^2(u^i \circ \gamma)}{dt^2} = 0 \qquad (1 \leq i \leq n).$$

Thus $u^i(\gamma(t)) = p^i + tv^i$ for all t, where p^i and v^i are arbitrary constants. In vector notation, $\gamma(t) = p + tv$. Hence the geodesics of $\boldsymbol{R}^n_\nu$ are straight lines. In particular, $\boldsymbol{R}^n_\nu$ is geodesically complete.

Since its velocity vector field is parallel, a geodesic γ has quite uniform behavior. Every constant curve in M is trivially geodesic, but if $\gamma'(t) \neq 0$ for one single t then γ' never vanishes. Thus a geodesic cannot slow down and stop.

A curve α in M is *spacelike* if all of its velocity vectors $\alpha'(s)$ are spacelike; similarly for *timelike* and *null*. An arbitrary curve need not have one of these *causal characters*, but a geodesic γ always does since γ' is parallel, and parallel translation preserves causal character of vectors.

26. Lemma. Let $\gamma: I \to M$ be a nonconstant geodesic. A reparametrization $\gamma \circ h: J \to M$ is a geodesic if and only if h has the form $h(t) = at + b$.

Proof. For any curve γ, $(\gamma \circ h)'(t) = (dh/dt)(t)\gamma'(h(t))$. Hence by Exercise 3,

$$(\gamma \circ h)''(t) = \frac{d^2h}{dt^2}\gamma'(h(t)) + \left(\frac{dh}{dt}\right)^2 \gamma''(h(t)).$$

Since γ is a geodesic, $\gamma'' = 0$, and γ nonconstant implies γ' never zero. Thus $\gamma \circ h$ geodesic $\Leftrightarrow (\gamma \circ h)'' = 0 \Leftrightarrow d^2h/dt^2 = 0 \Leftrightarrow h(t) = at + b$. ∎

This result shows that that geodesic parametrizations have geometric significance. If a curve has a reparametrization as a geodesic we call it a *pregeodesic*.

If a system of second-order ordinary differential equations is given by smooth functions, then its solutions are smooth not just in the parameter but simultaneously in the parameter, initial values, and initial first derivatives. Applying this fact to the geodesic differential equations gives

27. Lemma. Let v be a tangent vector to M, that is, an element of the tangent bundle TM. Then there exists a neighborhood $\mathscr{N}$ of v in TM and an interval I around 0 such that $(w, s) \to \gamma_w(s)$ is a well-defined smooth function from $\mathscr{N} \times I$ into M.

A second-order differential equation for y can be converted into a pair of first-order equations by taking y' as a new variable. By essentially the same device, geodesics in M can be represented by integral curves in the tangent bundle TM.

28. Proposition. There is a vector field G on TM such that the projection $\pi: TM \to M$ establishes a one-to-one correspondence between [maximal] integral curves of G and [maximal] geodesics of M.

Proof. If $v \in TM$ let G_v be the initial velocity of the curve $s \to \gamma_v'(s)$ in TM. It follows using the preceding lemma that G is a smooth vector field on TM.

(a) *If γ is a geodesic in M, then γ' is an integral curve of G.*

For all s, let $\alpha(s) = \gamma'(s)$. For arbitrary fixed t, let $w = \gamma'(t)$ and $\beta(s) = \gamma_w'(s)$. By Lemma 23, $\gamma(t + s) = \gamma_w(s)$. Taking velocities in M gives $\alpha(t + s) = \gamma_w'(s) = \beta(s)$. Then taking velocities in TM gives $\alpha'(t + s) = \beta'(s)$. In particular,

$$\alpha'(t) = \beta'(0) = G_w = G_{\alpha(t)}.$$

(b) *If α is an integral curve of G, then $\pi \circ \alpha$ is a geodesic in M.*

If $v = \alpha(0)$ then by (a), $s \to \gamma_v'(s)$ is also an integral curve of G. Like α it starts at v, hence the uniqueness of integral curves implies that, initially at least, $\pi \circ \alpha = \pi \circ \gamma_v' = \gamma_v$. For arbitrary t let δ be the integral curve of G starting at $\alpha(t)$. By Lemma 1.50, $\alpha(t + s) = \delta(s)$, hence $\pi\alpha(t + s) = \pi\delta(s) = \gamma_{\delta(0)}(s)$.

The identities $\pi \circ \gamma' = \gamma$ and $(\pi \circ \alpha)' = \alpha$ show that the maps $\alpha \to \pi \circ \alpha$ and $\gamma \to \gamma'$ are inverses, and the result follows. ∎

THE EXPONENTIAL MAP

At each point o of a semi-Riemannian manifold M we collect the geodesics starting at o into a single mapping.

29. Definition. If $o \in M$, let $\mathscr{D}_o$ be the set of vectors v in $T_o(M)$ such that the inextendible geodesic γ_v is defined at least on [0, 1]. The *exponential map* of M at o is the function

$$\exp_o: \mathscr{D}_o \to M$$

such that $\exp_o(v) = \gamma_v(1)$ for all $v \in \mathscr{D}_o$.

Obviously $\mathscr{D}_o$ is the largest subset of $T_o(M)$ on which $\exp_o$ can be defined. If M is complete, then $\mathscr{D}_o = T_o(M)$ for every point o of M.

Fix $v \in T_o(M)$ and $t \in \boldsymbol{R}$; then the geodesic $s \to \gamma_v(ts)$ has initial velocity $t\gamma_v'(0) = tv$. Hence $\gamma_{tv}(s) = \gamma_v(ts)$ for all s and t such that either side (hence both) is well defined. In particular, if $v \in \mathscr{D}_o$ then

$$\exp_o(tv) = \gamma_{tv}(1) = \gamma_v(t).$$

Thus *the exponential map* $\exp_o$ *carries lines through the origin of* $T_o(M)$ *to geodesics of* M *through* o.

30. Proposition. For each point $o \in M$ there exists a neighborhood $\tilde{\mathscr{U}}$ of 0 in $T_o(M)$ on which the exponential map $\exp_o$ is a diffeomorphism onto a neighborhood $\mathscr{U}$ of o in M.

Proof. It follows from Lemma 27 that $\exp_o$ is a well-defined smooth mapping on some neighborhood of 0 in $T_o(M)$. We assert that *the differential map*

$$d\exp_o\colon T_0(T_oM) \to T_o(M)$$

is the canonical isomorphism $v_0 \to v$ (page 26). By definition $v_0 = \rho'(0)$, where $\rho(t) = tv$; as noted above, $\exp_o(tv) = \gamma_v(t)$. Thus

$$d\exp_o(v_0) = d\exp_o(\rho'(0)) = (\exp_o \circ \rho)'(0) = \gamma_v'(0) = v.$$

The result then follows by the inverse function theorem (1.16). ∎

A subset S of a vector space is *starshaped* about 0 if $v \in S$ implies $tv \in S$ for all $0 \le t \le 1$. Then S is a union of radial line segments. If $\mathscr{U}$ and $\tilde{\mathscr{U}}$ are as in the preceding proposition and $\tilde{\mathscr{U}}$ is starshaped about 0, then $\mathscr{U}$ is called a *normal neighborhood* of o (see Figure 2). Now we show that $\mathscr{U}$ deserves to be called starshaped about o.

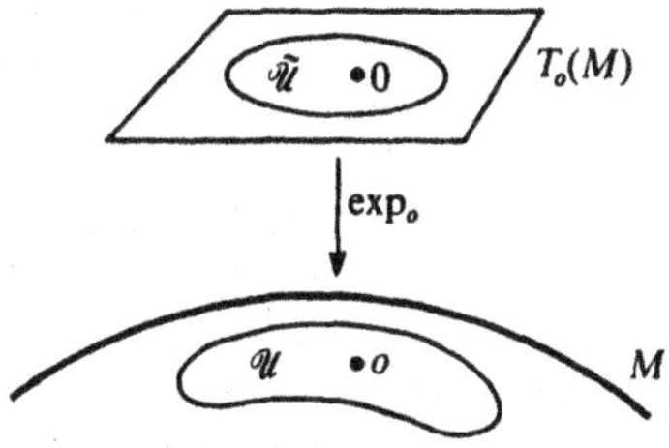

Figure 2

31. Proposition. If $\mathscr{U}$ is a normal neighborhood of $o \in M$, then for each point $p \in \mathscr{U}$ there is a unique geodesic $\sigma\colon [0, 1] \to \mathscr{U}$ from o to p in $\mathscr{U}$. Furthermore $\sigma'(0) = \exp_o^{-1}(p) \in \tilde{\mathscr{U}}$.

Proof. By definition $\tilde{\mathscr{U}}$ is a starshaped neighborhood of 0 in $T_o(M)$ such that $\exp_o|\tilde{\mathscr{U}}$ is a diffeomorphism onto $\mathscr{U}$. For $p \in \mathscr{U}$ let $v = \exp_o^{-1}(p) \in \tilde{\mathscr{U}}$. Since $\tilde{\mathscr{U}}$ is starshaped the ray $\rho(t) = tv\,(0 \leq t \leq 1)$ lies in $\tilde{\mathscr{U}}$. Thus the geodesic segment $\sigma = \exp_o \circ \rho$ lies in $\mathscr{U}$ and runs from o to p. (σ is said to be *radial.*) At the origin of $T_o(M)$, $d\exp_o$ is the canonical isomorphism $T_0(T_o M) \approx T_o(M)$. But $\rho'(0) = v_0$, hence

$$\sigma'(0) = d\exp_o(\rho'(0)) = d\exp_o(v_0) = v.$$

Suppose $\tau\colon [0, 1] \to \mathscr{U}$ is an arbitrary geodesic in $\mathscr{U}$ from o to p. If $w = \tau'(0)$, then the geodesics $t \to \exp_o(tw)$ and τ have the same initial velocities, hence are equal.

The radial segment $t \to tw\,(0 \leq t \leq 1)$ does not leave $\tilde{\mathscr{U}}$, for if it does there is a $0 < t_0 < 1$ such that $t_0 w \in \tilde{\mathscr{U}}$ but $\exp_o(t_0 w) \in \mathscr{U} - \tau([0, 1])$. Thus $w \in \tilde{\mathscr{U}}$. But $\exp_o(w) = \tau(1) = p = \exp_o(v)$ and $\exp_o$ is one-to-one on $\tilde{\mathscr{U}}$, so $w = v$. Hence by the uniqueness of geodesics $\tau = \sigma$. ∎

This proof shows that a normal neighborhood $\mathscr{U}$ of o uniquely determines the neighborhood $\tilde{\mathscr{U}}$ in $T_o(M)$.

A *broken geodesic* is a piecewise smooth curve segment whose smooth subsegments are geodesics. For example, a broken geodesic in $\boldsymbol{R}^2$ is just a polygonal curve.

32. Lemma. A semi-Riemannian manifold M is connected if and only if any two points of M can be joined by a broken geodesic.

Proof. Assume M is connected and fix $p \in M$. Let $\mathscr{C}$ be the set of points that can be connected to p by a broken geodesic. For $q \in M$ let $\mathscr{U}$ be a normal neighborhood. If $q \in \mathscr{C}$ then clearly $\mathscr{U} \subset \mathscr{C}$. But also if $q \in M - \mathscr{C}$ then $\mathscr{U} \subset M - \mathscr{C}$. Thus by connectedness, $M = \mathscr{C}$. The converse is obvious. ∎

On any normal neighborhood $\mathscr{U}$ of $o \in M$ there is a special type of coordinate system that is particularly simple. Let $e_1, \ldots, e_n$ be an orthonormal basis for $T_o(M)$, so $\langle e_i, e_j\rangle = \delta_{ij}\varepsilon_j$. The *normal coordinate system* $\xi = (x^1, \ldots, x^n)$ determined by $e_1, \ldots, e_n$ assigns to each point $p \in \mathscr{U}$ the vector coordinates relative to $e_1, \ldots, e_n$ of the corresponding point

$$\exp_o^{-1}(p) \in \tilde{\mathscr{U}} \subset T_o(M).$$

In short,

$$\exp_o^{-1}(p) = \sum x^i(p) e_i \qquad (p \in \mathscr{U}).$$

Hence if $f^1, \ldots, f^n$ is the dual basis to $e_1, \ldots, e_n$, then $x^i \circ \exp_o = f^i$ on $\tilde{\mathscr{U}}$.

33. Proposition. If $x^1, \ldots, x^n$ is a normal coordinate system at $o \in M$, then for all i, j, k

(1) $g_{ij}(o) = \delta_{ij}\varepsilon_j$; (2) $\Gamma_{ij}^k(o) = 0$.

Proof. With notation as above, if $v \in T_o(M)$ write $v = \sum a^i e_i$. Since $\exp_o(tv) = \gamma_v(t)$,

$$x^i(\gamma_v(t)) = f^i(tv) = tf^i(v) = ta^i.$$

Hence $v = \gamma_v'(0) = \sum a^i \partial_i|_o$. Taking $a^i = \delta_{ij}$ shows that $e_j = \partial_j|_o$, and (1) follows.

The expression for $x^i \circ \gamma_v$ shows that the geodesic differential equations for γ_v reduce to

$$\sum_{ij} \Gamma_{ij}^k(\gamma_v(t))a^i a^j = 0 \qquad \text{for all} \quad k.$$

In particular, $\sum \Gamma_{ij}^k(o)a^i a^j = 0$ holds for all $a = (a^1, \ldots, a^n) \in \boldsymbol{R}^n$. For fixed k, this expresses the fact that a certain quadratic form on $\boldsymbol{R}^n$ is identically zero. Hence by polarization the corresponding symmetric bilinear form is identically zero, that is, $\Gamma_{ij}^k(o) = 0$. ∎

Comparison with Lemma 14 shows that at the point o—though in general not elsewhere—the metric tensor and Christoffel symbols of a normal coordinate system are semi-Euclidean. Since tensors are pointwise creatures this fact is surprisingly powerful in computations. For example, suppose a problem involves the covariant differential DA of $A \in \mathfrak{T}_2^1(M)$. For arbitrary coordinates the components of DA (Exercise 2) are rather clumsy, but if for each point $o \in M$ we use normal coordinates at o, then

$$A_{jk;m}^i(o) = ((\partial/\partial x^m)A_{jk}^i)(o)$$

just as for natural coordinates in semi-Euclidean space.

Near o the formulas in the proposition are at least approximately true: the closer we get to o the more nearly M resembles $T_o(M) \approx \boldsymbol{R}_\nu^n$. But this approximation cannot be pushed too far. For example, at o the first derivatives of Γ_{ij}^k generally do not vanish (although those of g_{ij} do).

34. Example. Exponential Maps for $\boldsymbol{R}_\nu^n$. According to Example 25 the geodesic with initial velocity $v_p \in T_p(\boldsymbol{R}_\nu^n)$ is the straight line $t \to p + tv$. Thus the exponential map at p sends v_p to $p + v$ (the tip of the arrow v_p). It follows that $\exp_p: T_p(\boldsymbol{R}_\nu^n) \to \boldsymbol{R}_\nu^n$ is a diffeomorphism, since it is the composition of the canonical isomorphism $T_p(\boldsymbol{R}_\nu^n) \approx \boldsymbol{R}_\nu^n$ and the translation $x \to p + x$. In fact, when the scalar product space $T_p(\boldsymbol{R}_\nu^n)$ is given its usual metric tensor, both these maps are isometries, so $\exp_p$ is an isometry.

CURVATURE

In the theory of surfaces in $\mathbf{R}^3$ evolving during the late 1700s, a notion of curvature was defined that gives a very reasonable description of the way the surface is shaped in $\mathbf{R}^3$. It was Gauss who showed ("theorema egregium") that this *Gaussian curvature* is an isometric invariant of the surface itself, independent of the fact that the surface happens to be in 3-space. This theorem led Riemann to his invention of Riemannian geometry, whose dominant feature is the generalization of Gaussian curvature to arbitrary Riemannian manifolds. No significant changes are required in extending to semi-Riemannian manifolds.

Lie derivatives satisfy the identity $L_{[X,Y]} = [L_X, L_Y]$, where as usual the right-hand side means $L_X L_Y - L_Y L_X$. Hence if $[X, Y] = 0$ (as, for example, for coordinate vector fields) then L_X and L_Y commute. By contrast, these results fail in general for the covariant derivative D_X. This failure is measured by a tensor field that plays a central role in all differential geometry.

35. Lemma. Let M be a semi-Riemannian manifold with Levi-Civita connection D. The function $R: \mathfrak{X}(M)^3 \to \mathfrak{X}(M)$ given by

$$R_{XY}Z = D_{[X,Y]}Z - [D_X, D_Y]Z$$

is a (1, 3) tensor field on M called the *Riemannian curvature tensor* of M.

Proof. As on page 37, R can be interpreted as an element of $\mathfrak{T}^1_3(M)$ provided it is $\mathfrak{F}(M)$-multilinear. Since $\mathbf{R}$-linearity is obvious, this amounts to showing we can "factor out functions." For example, since $[X, fY] = Xf \cdot Y + f[X, Y]$,

$$\begin{aligned} R_{X,fY}Z &= D_{[X,fY]}Z - D_X D_{fY}Z + D_{fY}D_X Z \\ &= Xf \cdot D_Y Z + fD_{[X,Y]}Z - D_X(fD_Y Z) + fD_Y D_X Z \\ &= Xf \cdot D_Y Z - Xf \cdot D_Y Z + fR_{XY}Z = fR_{XY}Z. \quad \blacksquare \end{aligned}$$

The bracket operation on vector fields is not a tensor and the covariant derivative is not a tensor, but in the combination above they produce the tensor R. The alternative notation $R(X, Y)Z$ for $R_{XY}Z$ is convenient when X and Y are replaced by more complicated expressions.

As shown in Chapter 2 the tensor R can be considered as an $\mathbf{R}$-multilinear function on individual tangent vectors. If $x, y \in T_p(M)$ the linear operator

$$R_{xy}: T_p(M) \to T_p(M)$$

sending each z to $R_{xy}z$ is called a *curvature operator*. The following identities are the *symmetries* of curvature.

36. Proposition. If $x, y, z, v, w \in T_p(M)$, then

(1) $R_{xy} = -R_{yx}$,
(2) $\langle R_{xy}v, w\rangle = -\langle R_{xy}w, v\rangle$,
(3) $R_{xy}z + R_{yz}x + R_{zx}y = 0$,
(4) $\langle R_{xy}v, w\rangle = \langle R_{vw}x, y\rangle$.

The first two identities show that the curvature tensor contains considerable skew-symmetry. In particular (2) says that curvature operators are skew-adjoint. Equation (3) is called the *first Bianchi identity*—note that its vectors are cyclically permuted. *Symmetry by pairs*, (4), will follow from the earlier identities.

Proof. Since both the covariant derivative D_X and the bracket operation on vector fields are local operations, it suffices to work on any neighborhood of the point p. Because the identities to be proved are tensor equations, the tangent vectors $x, y, \ldots$ can be extended to vector fields $X, Y, \ldots$ on some neighborhood in any convenient way. In the case at hand, we choose the extensions so that *all their brackets are zero.* (This is accomplished by taking them to have constant components relative to a coordinate system.) In particular, $R_{XY}Z$ then reduces to $D_Y(D_XZ) - D_X(D_YZ)$.

(1) Whenever the bracket $[A, B] = AB - BA$ makes sense, it is skew-symmetric in A and B. Thus (1) is immediate from the definition of curvature.

(2) By polarization we need only show that $\langle R_{xy}v, v\rangle = 0$. But using (D5) from Theorem 11,

$$\begin{aligned}\langle R_{XY}V, V\rangle &= \langle D_YD_XV, V\rangle - \langle D_XD_YV, V\rangle \\ &= Y\langle D_XV, V\rangle - \langle D_XV, D_YV\rangle - X\langle D_YV, V\rangle + \langle D_YV, D_XV\rangle \\ &= \tfrac{1}{2}YX\langle V, V\rangle - \tfrac{1}{2}XY\langle V, V\rangle = 0,\end{aligned}$$

since $[X, Y] = 0$.

(3) Suppose $F: \mathfrak{X}(M)^3 \to \mathfrak{X}(M)$ is a function that is merely **R**-linear, and let $\mathfrak{S}F(X, Y, Z)$ be the sum over the cyclic permutations of X, Y, Z:

$$F(X, Y, Z) + F(Y, Z, X) + F(Z, X, Y).$$

A cyclic permutation of X, Y, Z leaves $\mathfrak{S}F(X, Y, Z)$ unchanged. Consequently,

$$\begin{aligned}\mathfrak{S}R_{XY}Z &= \mathfrak{S}D_YD_XZ - \mathfrak{S}D_XD_YZ \\ &= \mathfrak{S}D_XD_ZY - \mathfrak{S}D_XD_YZ = \mathfrak{S}D_X[Z, Y] = 0.\end{aligned}$$

(4) The proof is a combinatorial exercise. By (3), $\langle \mathfrak{S}R_{YV}X, W\rangle = 0$, where now $\mathfrak{S}$ acts on whichever three vectors fields are attached to R. Sum over the four cyclic permutations of Y, V, X, W, and then expand each $\mathfrak{S}$ to

obtain twelve terms. Using (1) and (2), eight of these will cancel in pairs, leaving

$$2\langle R_{XY}V, W\rangle + 2\langle R_{WV}X, Y\rangle = 0.$$

Hence $\langle R_{XY}V, W\rangle = \langle R_{VW}X, Y\rangle$. ∎

The symmetries of the curvature tensor R lead to a less obvious symmetry of its covariant differential DR, called the *second Bianchi identity*. By definition DR is a (1, 4) tensor that we interpret as assigning to four vector fields the (vector field) value $(D_Z R)_{XY}V = (D_Z R)(X, Y)V$. As always, this makes sense for individual tangent vectors, so the summands below are linear operators on $T_p(M)$.

37. Proposition (Second Bianchi Identity). If $x, y, z \in T_p(M)$, then

$$(D_z R)(x, y) + (D_x R)(y, z) + (D_y R)(z, x) = 0.$$

Proof. As in the previous proof extend the tangent vectors x, y, z to vector fields X, Y, Z on a neighborhood of p. This time we choose more carefully: for a *normal* coordinate system at p, let the extensions have constant components. Thus not only do all brackets vanish identically, but at the point p (where the Christoffel symbols are zero) it follows from the formula in Proposition 13(1) that all nine covariant derivatives involving only X, Y, Z are zero. By the product rule, applying $(D_Z R)(X, Y)$ to an arbitrary vector field V yields

$$D_Z(R(X, Y)V) - R(D_Z X, Y)V - R(X, D_Z Y)V - R(X, Y)(D_Z V).$$

At the point p the two middle terms are zero, hence dropping the now superfluous vector field V we have

$$(D_Z R)(X, Y) = [D_Z, R(X, Y)] = [D_Z, [D_Y, D_X]] \quad \text{at} \quad p.$$

But the *Jacobi identity* (as one can see by writing it out) is valid here just as for brackets of vector fields. Thus summing the above formula over the cyclic permutations of X, Y, Z gives the required result $\mathfrak{S}(D_Z R)(X, Y) = 0$ at p. ∎

38. Lemma. On the coordinate neighborhood of a coordinate system $x^1, \ldots, x^n$,

$$R_{\partial_k \partial_l}(\partial_j) = \sum_i R^i_{jkl}\, \partial_i,$$

where the components of R are given by

$$R^i_{jkl} = \frac{\partial}{\partial x^l}\Gamma^i_{kj} - \frac{\partial}{\partial x^k}\Gamma^i_{lj} + \sum_m \Gamma^i_{lm}\Gamma^m_{kj} - \sum_m \Gamma^i_{km}\Gamma^m_{lj}.$$

Proof. For coordinate vector fields

$$R_{\partial_k \partial_l}(\partial_j) = D_{\partial_l}(D_{\partial_k} \partial_j) - D_{\partial_k}(D_{\partial_l} \partial_j).$$

The first term on the right-hand side is

$$D_{\partial_l}\left(\sum \Gamma_{kj}^{m} \partial_m\right) = \sum_m \frac{\partial}{\partial x^l} \Gamma_{kj}^{m} \partial_m + \sum \Gamma_{kj}^{m} \Gamma_{lm}^{r} \partial_r.$$

Relabeling two pairs of indices gives

$$\left\{\sum_i \frac{\partial}{\partial x^l} \Gamma_{kj}^{i} + \sum_m \Gamma_{lm}^{i} \Gamma_{kj}^{m}\right\} \partial_i.$$

Then subtracting the corresponding expression with k and l reversed gives the result. ∎

Substituting from Proposition 13(2) into the formula above gives an explicit formula for curvature in terms of the metric tensor. Even in simple cases such computations are tedious and give minimum information. To compute the curvature of a given manifold M the practical way is to use theoretical results to exploit the distinctive features of M. Many examples of this general approach will appear later on.

SECTIONAL CURVATURE

The Riemannian curvature tensor R is fairly complicated; we now consider a simpler real-valued function which completely determines R.

A two-dimensional subspace Π of the tangent space $T_p(M)$ is called a *tangent plane* to M at p. For tangent vectors v, w define

$$Q(v, w) = \langle v, v\rangle\langle w, w\rangle - \langle v, w\rangle^2.$$

By Lemma 2.19 a tangent plane Π is nondegenerate if and only if $Q(v, w) \neq 0$ for one—hence every—basis v, w for Π. The absolute value $|Q(v, w)|$ is the square of the area of the parallelogram with sides v and w. $Q(v, w)$ is positive if $g|\Pi$ is definite, negative if it is indefinite (use an orthonormal basis).

39. Lemma. Let Π be a nondegenerate tangent plane to M at p. The number

$$K(v, w) = \langle R_{vw}v, w\rangle / Q(v, w)$$

is independent of the choice of basis v, w for Π, and is called the *sectional curvature* $K(\Pi)$ of Π.

Proof. Any two bases for Π are related by equations

$$v = ax + by,$$
$$w = cx + dy,$$

where the determinant of coefficients $ad - bc$ is not zero. A direct computation shows that

$$\langle R_{vw}v, w\rangle = (ad - bc)^2\langle R_{xy}x, y\rangle,$$

and

$$Q(v, w) = (ad - bc)^2 Q(x, y). \quad \blacksquare$$

Thus the sectional curvature K of M is a real-valued function on the set of all nondegenerate tangent planes to M.

By definition, R determines K; to show that K determines R, a technicality about indefinite scalar products is needed.

40. Lemma. Given vectors v, w in a scalar product space, there exist vectors $\bar{v}$ and $\bar{w}$, arbitrarily close to v and w, respectively, that span a nondegenerate plane.

Proof. We can assume that v and w are linearly independent since any pair of vectors can be approximated by independent vectors. Obviously the plane spanned by v and w can be assumed to be degenerate—hence the scalar product is indefinite. If v is null, let x be a vector such that $\langle v, x\rangle \neq 0$; if v is nonnull, pick $x \neq 0$ of opposite causal character. In both cases $Q(v, x) < 0$.

It suffices now to show that for all sufficiently small $\delta \neq 0$ the vectors v and $w + \delta x$ span a nondegenerate plane. Expansion of $Q(v, w + \delta x)$ gives an expression of the form

$$2\delta b + \delta^2 Q(v, x).$$

If $b \neq 0$, this will be nonzero since δ small dominates δ^2. If $b = 0$, it is nonzero since $Q(v, x) < 0$. $\blacksquare$

41. Proposition. If $K = 0$ at $p \in M$, then $R = 0$ at p.

Explicitly, if $K(\Pi) = 0$ for every nondegenerate plane in $T_p(M)$, then $R_{xy}z = 0$ for all x, y, z in $T_p(M)$.

Proof. (1) $\langle R_{vw}v, w\rangle = 0$ *for all* $v, w \in T_p(M)$.

If v and w span a nondegenerate plane, then $\langle R_{vw}v, w\rangle = 0$. But by the lemma any pair of vectors is a limit of such vectors. Since $\langle R_{vw}x, y\rangle$ is multilinear it is continuous on $T_p(M)^4$, so (1) is true.

(2) $R_{vw}v = 0$ *for all* $v, w \in T_p(M)$.

For arbitrary x, polarize thus:

$$\langle R_{v,w+x}v, w + x\rangle = \langle R_{vw}v, w\rangle + \langle R_{vx}v, w\rangle + \langle R_{vw}v, x\rangle + \langle R_{vx}v, x\rangle.$$

Three of these terms vanish by (1). Symmetry by pairs asserts that the remaining two are equal, so $\langle R_{vw}v, x\rangle = 0$ for all x.

(3) $R_{vw}x = R_{wx}v$ *for all* $v, w, x \in T_p(M)$.

Polarize a second time:

$$R_{v+x,w}(v + x) = R_{vw}v + R_{xw}v + R_{vw}x + R_{xw}x.$$

Again three terms vanish, by (2); hence (3) follows by the skew-symmetry of R in its subscripts.

According to (3), $R_{vw}x$ is unchanged by a cyclic permutation of the vectors v, w, x. Thus the first Bianchi identity implies $R_{vw}x = 0$ for all v, w, x; so $R = 0$ at p. ■

A semi-Riemannian manifold M for which the curvature tensor R is zero at every point is said to be *flat*. By the proposition, M is flat if and only if the sectional curvature function K is identically zero. For example, every semi-Euclidean space $\mathbf{R}^n_\nu$ is flat: for natural coordinates the Christoffel symbols all vanish, hence $R = 0$ by Lemma 38.

The preceding proof actually shows somewhat more. Let us say that a multilinear function $F: T_p(M)^4 \to \mathbf{R}$ is *curvaturelike* provided F has the symmetries stated in Proposition 36 for the function $(v, w, x, y) \to \langle R_{vw}x, y\rangle$. The preceding proof used only these abstract properties; thus $F(v, w, v, w) = 0$ for all $v, w \in T_p(M)$ spanning a nondegenerate plane implies $F = 0$. It follows that K determines R in this sense:

42. Corollary. Let F be a curvaturelike function on $T_p(M)$ such that

$$K(v, w) = \frac{F(v, w, v, w)}{\langle v, v\rangle\langle w, w\rangle - \langle v, w\rangle^2}$$

whenever v and w span a nondegenerate plane. Then

$$\langle R_{vw}x, y\rangle = F(v, w, x, y)$$

for all v, w, x, y in $T_p(M)$.

Proof. The difference function $\Delta(v, w, x, y) = F(v, w, x, y) - \langle R_{vw}x, y\rangle$ is also curvaturelike. By hypothesis, $\Delta(v, w, v, w) = 0$ if v and w span a nondegenerate plane. Thus by the remark preceding this corollary, $\Delta = 0$. ■

A semi-Riemannian manifold M has *constant curvature* if its sectional curvature function is constant. In Chapter 4 we shall find many such manifolds, for example, the sphere $S^n(r)$.

The preceding corollary leads to a simple formula for R when K is constant.

43. Corollary. If M has constant curvature C, then

$$R_{xy}z = C\{\langle z, x\rangle y - \langle z, y\rangle x\}.$$

Proof. A routine computation shows that the formula

$$F(x, y, v, w) = C\{\langle v, x\rangle\langle y, w\rangle - \langle v, y\rangle\langle x, w\rangle\}$$

defines a curvaturelike function at each point, and $F(x, y, x, y) = CQ(x, y)$. So if x and y span a nondegenerate plane,

$$K(x, y) = C = \frac{F(x, y, x, y)}{Q(x, y)},$$

and the result follows by Corollary 42. ∎

In the Riemannian case, this curvature formula has a simple geometric meaning: if x, y is an orthonormal basis for a plane Π, then R_{xy} is zero on $\Pi^\perp$, and on Π is the rotation sending x to y and y to $-x$, followed by scalar multiplication by C.

SEMI-RIEMANNIAN SURFACES

Let M be a semi-Riemannian surface, that is, a semi-Riemannian manifold of dimension 2. For a coordinate system u, v in M the components of the metric tensor are traditionally denoted by

$$E = g_{11} = \langle\partial_u, \partial_u\rangle, \qquad F = g_{12} = g_{21} = \langle\partial_u, \partial_v\rangle, \qquad G = g_{22} = \langle\partial_v, \partial_v\rangle,$$

where for indexing purposes, $u = u^1$, and $v = u^2$. The line element is thus

$$ds^2 = E\,du^2 + 2F\,du\,dv + G\,dv^2,$$

and $Q = Q(\partial_u, \partial_v) = EG - F^2$.

Then by Proposition 13 or by differentiation of E, F, G the Christoffel symbols are as follows:

$$Q\Gamma^1_{11} = \begin{vmatrix} E_u/2 & F \\ F_u - (E_v/2) & G \end{vmatrix}, \qquad Q\Gamma^2_{11} = \begin{vmatrix} E & E_u/2 \\ F & F_u - (E_v/2) \end{vmatrix},$$

$$Q\Gamma^1_{12} = \begin{vmatrix} E_v/2 & F \\ G_u/2 & G \end{vmatrix}, \qquad Q\Gamma^2_{12} = \begin{vmatrix} E & E_v/2 \\ F & G_u/2 \end{vmatrix},$$

$$Q\Gamma^1_{22} = \begin{vmatrix} F_v - (G_u/2) & F \\ G_v/2 & G \end{vmatrix}, \qquad Q\Gamma^2_{22} = \begin{vmatrix} E & F_v - (G_u/2) \\ F & G_v/2 \end{vmatrix}.$$

Coordinate geodesic equations then follow by substitution in

$$u'' + \Gamma^1_{11}u'^2 + 2\Gamma^1_{12}u'v' + \Gamma^1_{22}v'^2 = 0,$$
$$v'' + \Gamma^2_{11}u'^2 + 2\Gamma^2_{12}u'v' + \Gamma^2_{22}v'^2 = 0.$$

Since M is two-dimensional, $T_p(M)$ is the only tangent plane at p. Thus the sectional curvature K becomes a real-valued function on M, called the *Gaussian curvature* of M. General formulas for K are complicated so we consider a useful special case.

44. Proposition. Let u, v be an orthogonal coordinate system in a semi-Riemannian surface, so $F = \langle \partial_u, \partial_v \rangle = 0$.

(1) $$D_{\partial_u}\partial_u = \frac{E_u}{2E}\partial_u - \frac{E_v}{2G}\partial_v, \qquad D_{\partial_v}\partial_v = -\frac{G_u}{2E}\partial_u + \frac{G_v}{2G}\partial_v,$$

$$D_{\partial_u}\partial_v = D_{\partial_v}\partial_u = \frac{E_v}{2E}\partial_u + \frac{G_u}{2G}\partial_v.$$

(2) Let $e = |E|^{1/2}$ and $g = |G|^{1/2}$, and let $\varepsilon_1 = \pm 1$ be the sign of E and ε_2 the sign of G. Then

$$K = \frac{-1}{eg}\left[\varepsilon_1\left(\frac{g_u}{e}\right)_u + \varepsilon_2\left(\frac{e_v}{g}\right)_v\right].$$

Proof. (1) Set $F = 0$ in the formulas above for the Christoffel symbols. (2) By definition, $K = \langle R_{\partial_u\partial_v}(\partial_u), \partial_v\rangle/EG$. In computing the numerator, use

$$\langle D_{\partial_u}D_{\partial_v}\partial_u, \partial_v\rangle = \frac{\partial}{\partial u}\langle D_{\partial_v}\partial_u, \partial_v\rangle - \langle D_{\partial_v}\partial_u, D_{\partial_u}\partial_v\rangle,$$

which by (1) becomes $G_{uu}/2 - (E_v)^2/4E - (G_u)^2/4G$. There is one other analogous term. Then the curvature formula can be verified by calculating its right-hand side. ∎

For some examples, see Exercise 8.

TYPE-CHANGING AND METRIC CONTRACTION

In tensor language, Proposition 10 asserts that for a semi-Riemannian manifold M there is a natural $\mathfrak{F}(M)$-linear isomorphism $\mathfrak{T}^1_0(M) \approx \mathfrak{T}^0_1(M)$. This isomorphism is readily extended to higher types as follows. Fix integers

$1 \le a \le r$ and $1 \le b \le s$. If $A \in \mathfrak{T}^r_s(M)$ then the value of $\downarrow^a_b A \in \mathfrak{T}^{r-1}_{s+1}(M)$ on arbitrary one-forms and vector fields is defined by

$$(\downarrow^a_b A)(\theta^1, \ldots, \theta^{r-1}, X_1, \ldots, X_{s+1})$$

$$= A(\theta^1, \ldots, \underset{a\text{th slot}}{X_b^*}, \ldots, \theta^{r-1}, X_1, \ldots, X_{b-1}, X_{b+1}, \ldots, X_{s+1}),$$

where X_b^* is the one-form metrically equivalent to X_b. Thus on the right-hand side we extract the bth vector field and insert its metrically equivalent one-form in the ath slot among the one-forms.

For example, let A be a (2, 2) tensor field. Then $B = \downarrow^1_2 A$ is the (1, 3) tensor field such that $B(\theta, X, Y, Z) = A(Y^*, \theta, X, Z)$ for all one-forms θ and vector fields X, Y, Z. In coordinate terms, the one-form dual to ∂_i is $\sum g_{ij}\, dx^j$. Hence

$$B^i_{jkl} = B(dx^i, \partial_j, \partial_k, \partial_l) = A\Big(\sum_m g_{km}\, dx^m, dx^i, \partial_j, \partial_l\Big)$$

$$= \sum_m g_{km} A^{mi}_{jl}$$

Thus $\downarrow^1_2$ uses the metric tensor to turn first superscripts into second subscripts.

The operation $\downarrow^a_b : \mathfrak{T}^r_s(M) \to \mathfrak{T}^{r-1}_{s+1}(M)$ is known classically as *lowering an index*. It is clearly $\mathfrak{F}(M)$-linear, and is in fact an isomorphism, since there is a inverse operation $\uparrow^a_b$ which, with notation essentially as above, extracts the ath one-form and inserts its metrically equivalent vector field in the bth slot among the vector fields. In coordinates, the vector field metrically equivalent to dx^i is $\sum g^{ij}\,\partial_j$. If B is a (1, 3) tensor, then

$$(\uparrow^1_2 B)^{ij}_{kl} = \sum_q g^{iq} B^j_{kql},$$

with the metric tensor turning the second subscript into the first superscript. Thus the operation $\uparrow^a_b$ is classically called *raising an index*. As an example, in coordinate terms, of the inverse nature of the two operations,

$$(\uparrow^1_2 \downarrow^1_2 A)^{ij}_{kl} = \sum_p g^{ip}(\downarrow^1_2 A)^j_{kpl} = \sum_{pm} g^{ip} g_{pm} A^{mj}_{kl} = \sum_m \delta^i_m A^{mj}_{kl} = A^{ij}_{kl}.$$

Type-changing of tensors is so natural that it is apt to occur in practice without even being noticed. An important case is that of a $(1, s)$ tensor A given as an $\mathfrak{F}(M)$-multilinear function $A: \mathfrak{X}(M)^s \to \mathfrak{X}(M)$. It is then particularly simple to lower the unique contravariant slot to, say, the first covariant position:

$$(\downarrow^1_1 A)(V, X_1, \ldots, X_s) = \langle V, A(X_1, \ldots, X_s)\rangle.$$

In fact, the left-hand side is by definition $A(V^*, X_1, \ldots, X_s)$, which by the interpretation on page 37 is $V^*(A(X_1, \ldots, X_s) = \langle V, A(X_1, \ldots, X_s)\rangle$.

All tensors obtained from a given tensor by the raising and lowering operations are said to be *metrically equivalent.* They all contain the same information, and hence can be viewed as different manifestations of a single object.

The classical coordinatized version of multidimensional differential geometry was developed long before the invariant version, and in harmonizing the two approaches one point requires care. When the curvature tensor $R\colon \mathfrak{X}(M)^3 \to \mathfrak{X}(M)$ is written in the ordinary way as a function of three vector fields, the classical index pattern in Lemma 38 demands $R(Z, X, Y) = R_{XY}Z$. The components of the (0, 4) tensor $\downarrow_1^1 R$ are then given by

$$R_{ijkl} = (\downarrow_1^1 R)(\partial_i, \partial_j, \partial_k, \partial_l) = \langle \partial_i, R_{\partial_k \partial_l}(\partial_j)\rangle = \sum g_{im} R^m_{jkl}.$$

This is the usual way to lower the contravariant index of R—an agreement expressed classically by writing $R^i_{\cdot jkl}$ for R^i_{jkl}.

On a smooth manifold, contraction operates on one contravariant and one covariant slot to reduce an (r, s) tensor to an $(r-1, s-1)$ tensor. But on a semi-Riemannian manifold we can *metrically contract* two covariant indices by first raising either one of them and then contracting in the usual way. Thus for $1 \le a < b \le s$ and arbitrary r, the *metric contraction* $C_{ab}\colon \mathfrak{T}^r_s(M) \to \mathfrak{T}^r_{s-2}(M)$ is given in coordinates by

$$(C_{ab}A)^{i_1 \dots i_r}_{j_1 \dots j_{s-2}} = \sum_{p,q} g^{pq} A^{i_1 \dots i_r}_{j_1 \dots \underset{a\text{th position}}{p} \dots \underset{b\text{th position}}{q} \dots j_{s-2}}.$$

For example if A is a (1, 3) tensor, then

$$(C_{12}A)^i_j = \sum_{p,q} g^{pq} A^i_{pqj}.$$

Similarly, in the contravariant case, for $1 \le a < b \le r$ and arbitrary s, we get

$$C^{ab}\colon \mathfrak{T}^r_s(M) \to \mathfrak{T}^{r-2}_s(M)$$

with a coordinate formula reversing covariant and contravariant indices above (so g_{ij} replaces g^{ij}). When specific indices are not important, all contractions will be denoted by C.

45. Lemma. Covariant derivatives D_V and the covariant differential D commute with both type-changing and contraction.

Proof. Since a raising operation is the inverse of a lowering, it suffices to consider the latter. By a simple permutation argument we need only consider $\downarrow_1^a$. The coordinate expression for $\downarrow_1^a A$ shows that this tensor is the ordinary contraction C_1^a applied to $g \otimes A$. As a tensor derivation, D_V commutes with ordinary contraction. Since the metric tensor is parallel,

$$D_V(\downarrow_1^a A) = D_V(C_1^a(g \otimes A)) = C_1^a(g \otimes D_V A) = \downarrow_1^a (D_V A).$$

Hence D_V also commutes with metric contraction.

Formal computations then give the corresponding results for D. ∎

FRAME FIELDS

An orthonormal basis for a tangent space $T_p(M)$ is called a *frame on M at p*. If $n = \dim M$ then a set $E_1, \ldots, E_n$ of n mutually orthogonal unit vector fields is called a *frame field*, since it assigns a frame at each point. For example, on $\mathbf{R}^n$ the natural coordinate vector fields form a frame field.

In general there may not be a frame field on all of M, but we shall see in a moment that they always exist locally. By orthonormal expansion (2.25) any vector field V can be expressed in terms of a frame field as

$$V = \sum \varepsilon_i \langle V, E_i \rangle E_i, \qquad \text{where} \quad \varepsilon_i = \langle E_i, E_i \rangle.$$

Thus

$$\langle V, W \rangle = \sum \varepsilon_i \langle V, E_i \rangle \langle W, E_i \rangle.$$

At the origin o of a normal coordinate system the coordinate vectors are orthonormal. It follows that as long as only *pointwise* operations are involved, frame field formulas are (simpler) consequences of corresponding coordinate formulas. For example, consider the metric contraction C_{ab} of $A \in \mathfrak{T}_s^0(M)$. Relative to a frame field,

$$(C_{ab}A)(X_1, \ldots, X_{s-2}) = \sum \varepsilon_m A(X_1, \ldots, \underset{a\text{th slot}}{E_m}, \ldots, \underset{b\text{th slot}}{E_m}, \ldots, X_{s-2}).$$

To prove this tensor equation it suffices to work at a single point o, origin of normal coordinates such that $\partial_i|_o = E_i|_o$. By multilinearity it suffices to let the X_is be the coordinate vector fields ∂_i. But then the formula follows from the coordinate formula for C_{ab}, since at the point o both g_{ij} and g^{ij} become $\delta_{ij}\varepsilon_j$.

Similarly, for a $(1, s)$ tensor field $A: \mathfrak{X}(M)^s \to \mathfrak{X}(M)$,

$$(C_b^1 A)(X_1, \ldots, X_{s-1}) = \sum_m \varepsilon_m \langle E_m, A(X_1, \ldots, \underset{b\text{th slot}}{E_m}, \ldots, X_{s-1}) \rangle.$$

These remarks apply only to tensor algebra: for tensor calculus the advantage of $\langle E_i, E_j \rangle = \delta_{ij}\varepsilon_j$ over $\langle \partial_i, \partial_j \rangle = g_{ij}$ must be balanced against the disadvantage that, unlike $[\partial_i, \partial_j]$, the brackets $[E_i, E_j]$ are generally not zero.

A *frame field on a curve* $\alpha: I \to M$ is a set of mutually orthogonal unit vector fields $E_1, \ldots, E_n$ on α. Not only can such a frame field be defined on the entire curve, but we can choose the vector fields $E_i \in \mathfrak{X}(\alpha)$ to be parallel.

46. Corollary. If $\alpha: I \to M$ is a curve and $e_1, \ldots, e_n$ is a frame at $\alpha(0)$, then there is a unique parallel frame field $E_1, \ldots, E_n$ on α such that $E_i(0) = e_i$ for $1 \leq i \leq n$.

Proof. By Proposition 19 there is a unique parallel vector field E_i on α such that $E_i(0) = e_i$. But since parallel translation to any $t \in I$ is a linear isometry, $E_1, \ldots, E_n$ is in fact a (parallel) frame field. ∎

The triple advantages of orthonormality, parallelism, and global definition give the use of parallel frame fields on a curve a decisive superiority over coordinate methods.

It follows that on M frame fields exist locally: given any frame $e_1, \ldots, e_n$ in a tangent space $T_o(M)$, choose a normal neighborhood $\mathscr{U}$ of o and extend the frame to a frame field $E_1, \ldots, E_n$ on $\mathscr{U}$ by parallel translation along radial geodesics. Differential equations theory guarantees that the vector fields E_i are smooth.

SOME DIFFERENTIAL OPERATORS

On a semi-Riemannian manifold M there are natural generalizations of the well-known differential operators of vector calculus on $\boldsymbol{R}^3$: *gradient*, *divergence*, and *Laplacian*.

47. Definition. The *gradient* grad f of a function $f \in \mathfrak{F}(M)$ is the vector field metrically equivalent to the differential $df \in \mathfrak{X}^*(M)$. Thus

$$\langle \operatorname{grad} f, X \rangle = df(X) = Xf \qquad \text{for all} \quad X \in \mathfrak{X}(M).$$

In terms of a coordinate system $df = \sum (\partial f/\partial x^i)\, dx^i$, hence

$$\operatorname{grad} f = \sum_{i,j} g^{ij} \frac{\partial f}{\partial x^i} \partial_j.$$

In particular, for natural coordinates on semi-Euclidean space we have $\operatorname{grad} f = \sum \varepsilon_i (\partial f/\partial u^i)\, \partial_i$, which reduces to the usual formula on $\boldsymbol{R}^3$.

For a tensor A the contraction of the new covariant slot in its covariant differential DA with one of its original slots is called a *divergence* div A of A. We mostly use two special cases where there is a unique divergence:

(1) If V is a vector field, then div $V = C(DV) \in \mathfrak{F}(M)$. Thus for a frame field

$$\operatorname{div} V = \sum \varepsilon_i \langle D_{E_i} V, E_i \rangle,$$

and for a coordinate system

$$\operatorname{div} V = \sum V^i_{;i} = \sum_i \left\{ \frac{\partial V^i}{\partial x^i} + \sum_j \Gamma^i_{ij} V^j \right\}.$$

Hence for natural coordinates on $\boldsymbol{R}^n_\nu$, div $V = \sum \partial V^i / \partial u^i$, which on $\boldsymbol{R}^3$ is the usual formula.

(2) If A is a symmetric (0, 2) tensor, then div $A = C_{13}(DA) = C_{23}(DA) \in \mathfrak{X}^*(M)$.

For a frame field, $(\operatorname{div} A)(X) = \sum \varepsilon_i (D_{E_i} A)(E_i, X)$, while for coordinates

$$(\operatorname{div} A)_i = \sum_{r,s} g^{rs} A_{ri;s} = \sum_s A^s_{i;s}.$$

48. Definition. The *Hessian* of a function $f \in \mathfrak{F}(M)$ is its second covariant differential $H^f = D(Df)$.

49. Lemma. The Hessian H^f of f is the symmetric (0, 2) tensor field such that

$$H^f(X, Y) = XYf - (D_X Y)f = \langle D_X(\operatorname{grad} f), Y \rangle.$$

Proof. Since $Df = df$,

$$\begin{aligned} H^f(X, Y) = D(df)(X, Y) = D_Y(df)(X) &= Y(df(X)) - df(D_Y X) \\ &= YXf - (D_Y X)f. \end{aligned}$$

Because $XY - YX = [X, Y] = D_X Y - D_Y X$, we can reverse X and Y in the preceding formula—showing also that H^f is symmetric. Finally,

$$\langle D_X(\operatorname{grad} f), Y \rangle = X\langle \operatorname{grad} f, Y \rangle - \langle \operatorname{grad} f, D_X Y \rangle = H^f(X, Y). \quad \blacksquare$$

50. Definition. The *Laplacian* Δf of a function $f \in \mathfrak{F}(M)$ is the divergence of its gradient: $\Delta f = \operatorname{div}(\operatorname{grad} f) \in \mathfrak{F}(M)$.

Since the covariant differential commutes with type-changing, it follows that *the Laplacian of f is the contraction of its Hessian.* In fact,

$$\begin{aligned} \Delta f &= \operatorname{div}(\operatorname{grad} f) = \mathbf{C}D(\operatorname{grad} f) = \mathbf{C}D(\uparrow^1_1 df) \\ &= \mathbf{C} \uparrow^1_1 Ddf = (\mathbf{C} \uparrow^1_1) H^f = C_{12}(H^f). \end{aligned}$$

For a coordinate system the components of the Hessian can be read from the lemma above. Thus the Laplacian of f has coordinate expression

$$\Delta f = \sum g^{ij} H_{ij} = \sum_{ij} g^{ij} \left\{ \frac{\partial^2 f}{\partial x^i \, \partial x^j} - \sum_k \Gamma^k_{ij} \frac{\partial f}{\partial x^k} \right\}.$$

For natural coordinates in $\boldsymbol{R}^n_\nu$ the components of H^f are just the second partials $\partial^2 f/\partial u^i \, \partial u^j$, and $\Delta f = \sum \varepsilon_i \, \partial^2 f/\partial (u^i)^2$, which reduces to usual formula on $\boldsymbol{R}^3$. See Exercise 7.5 for another formula for Δf. (In special contexts, Δ acquires other names, and it is sometimes defined with opposite sign.)

RICCI AND SCALAR CURVATURE

Contraction of Riemannian curvature yields simpler invariants.

51. Definition. Let R be the Riemannian curvature tensor of M. The *Ricci curvature tensor* Ric of M is the contraction $C^1_3(R) \in \mathfrak{T}^0_2(M)$, whose components relative to a coordinate system are $R_{ij} = \sum R^m_{ijm}$.

Because of the symmetries of R the only nonzero contractions of R are $\pm$Ric.

52. Lemma. The Ricci curvature tensor Ric is symmetric, and is given relative to a frame field by

$$\mathrm{Ric}(X, Y) = \sum_m \varepsilon_m \langle R_{XE_m} Y, E_m \rangle,$$

where as usual $\varepsilon_m = \langle E_m, E_m \rangle$.

Proof. As pointed out earlier, classical indexing demands the notation $R(X, Y, E_m) = R_{YE_m} X$. Thus

$$\mathrm{Ric}(X, Y) = (C^1_3 R)(X, Y) = \sum \varepsilon_m \langle E_m, R(X, Y, E_m) \rangle = \sum \varepsilon_m \langle R_{YE_m} X, E_m \rangle.$$

Symmetry by pairs then gives the required formula and shows that Ric is symmetric. ∎

If its Ricci tensor is identically zero, M is said to be *Ricci flat*. A flat manifold is certainly Ricci flat, but we shall see later that the converse does not hold. Note the trace formula $\mathrm{Ric}(X, Y) = \mathrm{trace}\{V \to R_{XV} Y\}$.

Since sectional curvature determines the curvature tensor R it also determines Ric—and in a rather simple way. By polarization and scalar multiplication, Ric can be reconstructed at each point p from its values $\mathrm{Ric}(u, u)$ on the unit vectors at p. But if $e_1, \ldots, e_n$ is a frame at p such that $u = e_1$, then by the preceding lemma,

$$\mathrm{Ric}(u, u) = \sum \varepsilon_m \langle R_{ue_m}(u), e_m \rangle = \langle u, u \rangle \sum K(u, e_m).$$

Thus $\mathrm{Ric}(u, u)$ is, but for the sign $\langle u, u \rangle = \pm 1$, the sum of the sectional curvatures of any $n - 1$ orthogonal nondegenerate planes through u.

53. Definition. The *scalar curvature* S of M is the contraction $C(\mathrm{Ric}) \in \mathfrak{F}(M)$ of its Ricci tensor.

In coordinates,

$$S = \sum g^{ij} R_{ij} = \sum g^{ij} R^k_{ijk}.$$

Contracting relative to a frame field yields

$$S = \sum_{i \neq j} K(E_i, E_j) = 2 \sum_{i < j} K(E_i, E_j).$$

The following consequence of the second Bianchi identity (37) is crucial in the foundations of general relativity.

54. Corollary. $dS = 2 \operatorname{div} \mathrm{Ric}$.

Proof. To express the second Bianchi identity $\mathfrak{S}(D_Z R)_{XY} = 0$ in terms of coordinates apply it to

$$(D_{\partial_r} R)_{\partial_k \partial_l}(\partial_j) = \sum R^i_{jkl;r}\, \partial_i$$

to obtain

$$R^i_{jkl;r} + R^i_{jlr;k} + R^i_{jrk;l} = 0.$$

Reverse r and k in the third term, with change of sign, and contract on i and r. Thus

$$\sum_r R^r_{jkl;r} + \sum_r R^r_{jlr;k} - \sum R^r_{jkr;l} = 0,$$

which is just

$$\sum_r R^r_{jkl;r} + R_{jl;k} - R_{jk;l} = 0.$$

Now metrically contract on j and k:

$$\sum_{r,j,k} g^{jk} R^r_{jkl;r} + \sum g^{jk} R_{jl;k} - S_{;l} = 0.$$

In the first term write R^r_{jkl} as $\sum g^{rm} R_{mjkl}$ and use the coordinate symmetries of R (Exercise 4) to express this term as $\sum R^m_{l;m}$. Thus

$$2 \sum R^m_{l;m} = S_{;l}.$$

The left-hand side is the coordinate expression for 2 div(Ric); the right-hand side is that of $DS = dS$. ∎

55. Remark. Curvature Sign. Our definition of the Riemannian curvature tensor is not the only reasonable possibility; one can also change its sign, defining curvature to be $\mathcal{R} = -R$. By contrast the notions of sectional, Ricci, and scalar curvature are inviolate. To convert the formulas of this book from R to $\mathcal{R}$, simply change the sign of Riemannian curvature and its *four index* components. Do not change the signs of K, Ric, R_{ij}, or S. Replacing R by $-\mathcal{R}$ will adjust definitions as required. For example,

$$Q(X, Y)K(X, Y) = \langle R_{XY}X, Y\rangle = -\langle \mathcal{R}_{XY}X, Y\rangle = \langle \mathcal{R}_{XY}Y, X\rangle,$$

$$R_{ij} = \sum R^m_{ijm} = -\sum \mathcal{R}^m_{ijm} = \sum \mathcal{R}^m_{imj}.$$

SEMI-RIEMANNIAN PRODUCT MANIFOLDS

To see how the geometry of a semi-Riemannian product $M \times N$ depends on that of M and N, an essential tool is the notion of *lift* as discussed in Chapter 1. If $X \in \mathfrak{X}(M)$ we will use the same notation for its horizontal lift $X \in \mathfrak{L}(M) \subset \mathfrak{X}(M \times N)$; similarly for the vertical lift $V \in \mathfrak{L}(N)$ of $V \in \mathfrak{X}(N)$.

56. Proposition. If $X, Y \in \mathfrak{L}(M)$ and $V, W \in \mathfrak{L}(N)$, then

(1) $D_X Y$ is the lift of ${}^M D_X Y \in \mathfrak{X}(M)$.
(2) $D_V W$ is the lift of ${}^N D_V W \in \mathfrak{X}(N)$.
(3) $D_V X = 0 = D_X V$.

These assertions can be easily verified using the Koszul formula. (Compare the generalization in Proposition 7.35.)

57. Corollary. (1) A curve $\gamma(s) = (\alpha(s), \beta(s))$ in $M \times N$ is a geodesic if and only if its projections α in M and β in N are both geodesics. (2) $M \times N$ is complete if and only if both M and N are complete.

The first assertion follows from a more general result in Chapter 7; evidently (1) implies (2).

Applying Proposition 56 to the definition of curvature gives:

58. Corollary. On $M \times N$ if $X, Y, Z \in \mathfrak{L}(M)$ and $U, V, W \in \mathfrak{L}(N)$, then: (1) $R_{XY}Z$ is the lift of ${}^M R_{XY}Z$ on M. (2) $R_{VW}U$ is the lift on ${}^N R_{VW}U$ on N. (3) R is zero on any other choices from $X, \ldots, W$.

These tensor results are valid for individual tangent vectors. It follows that the sectional curvature of a nondegenerate horizontal plane is the same as that of its projection into M, and analogously for a vertical plane. A nondegenerate plane spanned by a vertical and a horizontal vector has $K = 0$ since $R_{xv} = 0$. Thus there is always some flatness in a semi-Riemannian product manifold.

LOCAL ISOMETRIES

The various features of semi-Riemannian geometry defined so far are isometric invariants: each is, in an appropriate sense, preserved by isometries. This can hardly be doubted since each is constructed, using the tools of manifold theory, from the metric tensor, and an isometry preserves both the tools and the tensor (Definition 6). For example, Levi-Civita connections are preserved in the following sense.

59. Proposition. If $\phi: M \to N$ is an isometry, then $d\phi(D_X Y) = D_{d\phi X}(d\phi Y)$ for all $X, Y \in \mathfrak{X}(M)$.

Proof. Recall that the value of the transferred vector field $d\phi(X)$ at a point $\phi(p)$ is $d\phi(X_p)$.

Because ϕ is a diffeomorphism, for any $p \in M$ there are coordinate systems ξ at p and η at $\phi(p)$ that are preserved by ϕ, that is, $y^i(\phi q) = x^i(q)$ for q near p. It follows at once that ϕ preserves coordinate vector fields and partial derivatives (as in Exercise 1.11). Then the components of X and $Y = d\phi(X)$ are preserved since

$$Y^i(\phi q) = Y_{\phi q}(y^i) = (d\phi X_q)y^i = X_q(y^i \circ \phi) = X_q(x^i) = X^i(q).$$

Because ϕ is also an isometry, metric tensor components are preserved: $g^N_{ij}(\phi q) = g^M_{ij}(q)$. Hence the formula in Proposition 13(2) shows that Christoffel symbols are similarly preserved. The result then follows from formula (1) of that proposition. ∎

The local character of this proof suggests that such invariance results can be extended to a broader class of mappings.

60. Definition. A smooth map $\phi: M \to N$ of semi-Riemannian manifolds is a *local isometry* provided each differential map $d\phi: T_p(M) \to T_{\phi p}(N)$ is a linear isometry.

In view of the inverse function theorem an equivalent formulation, justifying the term *local isometry*, is this: Each point p of M has a neighborhood $\mathscr{U}$ such that $\phi | \mathscr{U}$ is an isometry of $\mathscr{U}$ onto a neighborhood of $\phi(p)$ in N.

61. Example. Let S^1 be the unit circle in $\boldsymbol{R}^2$. The *exponential map* exp: $\boldsymbol{R}^1 \to S^1$ wraps the line evenly around the circle via $t \to (\cos t, \sin t)$. Consider S^1 as a Riemanian submanifold of $\boldsymbol{R}^2$; then exp is a local isometry. (It suffices to check that, as a curve in $\boldsymbol{R}^2$, exp has unit speed.)

By the above criterion for local isometries, every object of local (or pointwise) character that is preserved by isometries will automatically be preserved by local isometries $\phi: M \to N$. Here are some examples:

(1) *The induced covariant derivative on a curve.* If Y is a vector field on a curve α in M, then $(d\phi Y)' = d\phi(Y')$, where $(d\phi Y)(s) = d\phi(Y(s))$ for all s. (The proof is a mild variant of that of Proposition 59.) Hence:

(2) *Parallel translation.* Let P be parallel translation from $\alpha(a)$ to $\alpha(b)$ along a curve α in M. Let $\bar{P}$ be parallel translation from $\phi\alpha(a)$ to $\phi\alpha(b)$ along $\phi \circ \alpha$ in N. Then $d\phi_{\alpha b} \circ P = \bar{P} \circ d\phi_{\alpha a}$. Hence:

(3) *Geodesics.* If γ is a geodesic in M, then $\phi \circ \gamma$ is a geodesic in N. Thus $\phi \circ \gamma_v = \gamma_{d\phi v} | I_v$, since both sides are geodesics with the same initial velocity. (The domain of $\gamma_{d\phi v}$ may be larger than I_v; for example, consider the inclusion map ϕ of an open disk into $\boldsymbol{R}^2$.) Hence:

(4) *Exponential maps.* $\phi \circ \exp_p = \exp_{\phi p} \circ d\phi_p$, wherever the left-hand—hence right-hand—side is defined.

(5) *Riemannian curvature tensors.* $d\phi(R(x, y)z) = R(d\phi x, d\phi y)(d\phi z)$. (This is immediate from the definition of R, since a local isometry locally preserves both brackets and covariant derivatives.) Hence:

(6) *Sectional curvature.* $K_N(d\phi\Pi) = K_M(\Pi)$ for all nondegenerate tangent planes on M. *Ricci curvature*: $\phi^*(\mathrm{Ric}_N) = \mathrm{Ric}_M$. *Scalar curvature*: $S_N \circ \phi = S_M$. (The latter two use the fact that the pointwise operation of contraction is also preserved.)

A local isometry is uniquely determined by its differential map at a single point.

62. Proposition. Let $\phi, \psi: M \to N$ be local isometries of a connected semi-Riemannian manifold M. If there is a point $p \in M$ such that $d\phi_p = d\psi_p$ (hence $\phi(p) = \psi(p)$), then $\phi = \psi$.

Proof. Let $A = \{q \in M : d\phi_q = d\psi_q\}$. By continuity, A is closed in M. Since A is nonempty it suffices to show that A is open. We assert that if $q \in A$ then any normal neighborhood $\mathscr{U}$ of q is contained in A. If $r \in \mathscr{U}$ there is a vector $v \in T_q(M)$ such that $\gamma_v(1) = \exp_q(v) = r$. Hence

$$\phi(r) = \phi(\gamma_v(1)) = \gamma_{d\phi v}(1) = \gamma_{d\psi v}(1) = \psi(\gamma_v(1)) = \psi(r).$$

Thus $\phi = \psi$ on $\mathscr{U}$; hence $d\phi_r = d\psi_r$ for all $r \in \mathscr{U}$. ∎

A smooth mapping $\psi: M \to N$ of semi-Riemannian mappings is *conformal* provided $\psi^*(g_N) = hg_M$ for some function $h \in \mathfrak{F}(M)$ such that $h > 0$ or $h < 0$. The following special case will prove quite useful.

63. Definition. A diffeomorphism $\psi: M \to N$ of semi-Riemannian manifolds such that $\psi^*(g_N) = cg_M$ for some constant $c \neq 0$ is called a *homothety* of *coefficient* c.

Thus $\langle d\psi(v), d\psi(w)\rangle = c\langle v, w\rangle$ for all $v, w \in T_p(M)$, $p \in M$: all scalar products are stretched by the same constant c. If $c > 0$ [$c < 0$] then ψ is called a positive [negative] homothety of *scale factor* $|c|^{1/2}$. For example, in $\boldsymbol{R}^{n+1}$ scalar multiplication by b/a is a positive homothety from $S^n(a)$ to $S^n(b)$ of scale factor b/a.

An isometry is just a homothety with $c = 1$. If $c = -1$ we call ψ an *anti-isometry*.

64. Lemma. Homotheties preserve Levi-Civita connections.

Proof. If $\psi: M \to N$ is a homothety with coefficient c, let N' be the smooth manifold of N furnished with the new metric tensor cg_N. Then $\psi: M \to N'$ is an isometry, hence preserves connections. Thus it remains to show that cg_N and g_N determine the same Levi-Civita connection. This is clear from the Koszul formula, since the metric tensor appears exactly once in each of its terms, hence the coefficient c cancels. ∎

65. Remark. Effect of a Homothety. (1) Since a homothety preserves Levi-Civita connections it also preserves all geometric notions that derive solely from D, notably the induced covariant derivative on a curve, parallel translation, geodesics, Riemannian curvature R, and Ricci curvature—the latter since Ric is a nonmetric contraction of R.

(2) By contrast, sectional and scalar curvature are not invariant under homotheties. In fact, if $\psi: M \to N$ has coefficient c, it is easy to check that

$$K_N(d\psi\Pi) = \frac{1}{c} K_M(\Pi) \qquad \text{and} \qquad S_N \circ \psi = \frac{1}{c} S_M.$$

(3) Clearly a homothety of coefficient $c > 0$ preserves causal character of tangent vectors (hence of curves). However if $c < 0$ then causal character is *reversed*: v timelike $\Rightarrow d\phi v$ spacelike; v spacelike $\Rightarrow d\phi v$ timelike, and v null $\Rightarrow d\phi v$ null.

The operation of changing the semi-Riemannian manifold M with metric tensor $\boldsymbol{g}$ to the same smooth manifold with metric tensor $-\boldsymbol{g}$ is called *reversing the metric* of M. The effect on geometry is clear from the remarks above. In particular we will largely neglect manifolds with negative-definite metrics in favor of Riemannian manifolds. Analogously, the case of index $\nu = n - 1 \geq 1$ could also be considered as Lorentz geometry.

LEVELS OF STRUCTURE

Basic to all mathematics is the notion—here used quite informally—of a *set with structure*. For every type of structure there is a notion of *equivalence* (or *isomorphism*)—a one-to-one onto function that, in an appropriate sense, *preserves* the structure. A particular type of structure defines a branch of mathematics: the study of those concepts preserved by equivalence. For example, a *group* is a set furnished with the structure *group operation*. The notion of equivalence is the usual notion of isomorphism of groups.

In the case of semi-Riemannian geometry there is a hierarchy of structures (see Table 1).

TABLE 1

Branch of mathematics	Set with structure	Structure	Equivalence
Semi-Riemannian geometry	Semi-Riemannian manifold	Metric tensor, atlas, topology	Isometry
Manifold theory	Manifold	Atlas, topology	Diffeomorphism
Topology	Topological space	Topology	Homeomorphism
Set theory	Set	(None)	One-to-one onto function

Exercises

1. For an arbitrary connection D on a manifold M show that $T(X, Y) = [X, Y] - D_X Y + D_Y X$ defines a (1, 2) tensor field on M. T is called the *torsion tensor* of D. (Thus for a semi-Riemannian manifold, property (D4) asserts that the Levi-Civita connection has torsion zero.)

2. For a (1, 2) tensor field A, derive the formula

$$A^i_{jk;l} = \frac{\partial}{\partial x^l} A^i_{jk} + \sum_m \{A^m_{jk}\Gamma^i_{ml} - A^i_{mk}\Gamma^m_{jl} - A^i_{jm}\Gamma^m_{kl}\}.$$

(Hint: See Exercise 2.6.)

3. Let $h: J \to I$ reparametrize a curve $\alpha: I \to M$. If $Z \in \mathfrak{X}(\alpha)$ then $Z \circ h \in \mathfrak{X}(\alpha \circ h)$ and (a) $(Z \circ h)' = (dh/dt)Z' \circ h$; (b) $(Z \circ h)'' = (d^2h/dt^2)Z' \circ h + (dh/dt)^2 Z'' \circ h$.

4. (a) Symmetries are preserved under covariant differentiation; in particular $D_X R$ has the same symmetries as R, and $D_X \text{Ric}$ is symmetric. (b) With indices as on page 83, $R_{ijkl} = -R_{jikl} = -R_{ijlk} = R_{klij}$ and $R_{ijkl} + R_{iklj} + R_{iljk} = 0$.

5. If M^n has constant sectional curvature C, then $\text{Ric} = (n-1)Cg$ and $S = n(n-1)C$.

6. For a semi-Riemannian surface, (a) $R_{XY}Z = K[\langle Z, X\rangle Y - \langle Z, Y\rangle X]$. (b) $\text{Ric} = Kg$ and $S = 2K$. (c) $K = -R_{1212}$ (see Remark 55).

7. Let P be an open submanifold of a semi-Riemannian manifold M. If P is complete and M is connected, then $P = M$. (Hint: Assume not, and consider a boundary point of P.)

8. (a) The *Poincaré half-plane* is the region $v > 0$ in $\mathbf{R}^2$, but with line element $ds^2 = (du^2 + dv^2)/v^2$. Show that P has constant curvature $K = -1$. (b) Exercise 13 will imply that the unit sphere S^2 in $\mathbf{R}^3$ has $ds^2 = d\vartheta^2 + \sin^2 \vartheta \, d\varphi^2$. Show that S^2 has constant curvature $K = +1$. (These examples will be extensively generalized in the next chapter.)

9. If $f, g \in \mathfrak{F}(M)$ and $V \in \mathfrak{X}(M)$, then

(a) $D(fA) = A \otimes df + f\,DA \quad (A \in \mathfrak{T}^r_s(M))$.

(b) $\text{grad}(fg) = f \text{ grad } g + g \text{ grad } f$.

(c) $\text{div}(fV) = Vf + f \text{ div } V$.

(d) $H^{fg} = fH^g + gH^f + df \otimes dg + dg \otimes df$.

(e) $\Delta(fg) = f\Delta g + g\Delta f + 2\langle \text{grad } f, \text{grad } g\rangle$.

10. For the metric tensor g of M: (a) The contraction of g is dim M. (b) If M is connected the hypothesis that g has constant index is superfluous. (c) $\text{div}(fg) = df$. Prove using coordinates; using frame fields.

11. Let $\phi: M \to N$ be a smooth map of semi-Riemannian manifolds. For coordinate systems ξ at p in M and η at ϕp in N prove that $d\phi_p$ preserves scalar products if and only if for all i, j

$$g_{ij}(p) = \sum_{k,l} \bar{g}_{kl}(\phi p) \frac{\partial(y^k \circ \phi)}{\partial x^i}(p) \frac{\partial(y^l \circ \phi)}{\partial x^j}(p).$$

12. Let Z be a vector field on a curve α in M. (a) If $\alpha'(s_0) \neq 0$ there exists a smooth vector field $\tilde{Z}$ on a neighborhood of $\alpha(s_0)$ in M such that $\tilde{Z}_{\alpha(s)} = Z(s)$ for s near s_0. In particular, α *is locally an integral curve.* (Hint: For some interval J around s_0, $\alpha(J)$ is a submanifold of M.) (b) $Z'(s) = D_{\alpha'(s)}\tilde{Z}$ for s near s_0. (c) If α is an integral curve of X then $Xf \circ \alpha = d(f \circ \alpha)/ds$ for $f \in \mathfrak{F}(M)$.

13. We use *spherical coordinates* r, ϑ, φ in $\mathbf{R}^3$ as suggested by Figure 3 so that r, φ give polar coordinates in the plane $\vartheta = \pi/2$. (Some mathematicians prefer to reverse the notations ϑ and φ.) (a) Prove that (r, ϑ, φ) is a coordinate system on $\mathbf{R}^3 - H$, where H is a suitable half-plane. (b) Express $\partial_r, \partial_\vartheta, \partial_\varphi$ in

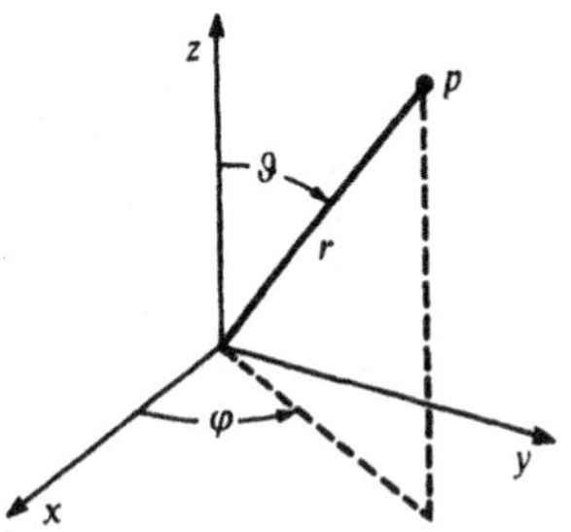

Figure 3

terms of $\partial_x, \partial_y, \partial_z$. (c) Deduce $ds^2 = dr^2 + r^2(d\vartheta^2 + \sin^2 \vartheta\, d\varphi^2)$. (d) Compute

$$D_{\partial_r} \partial_r = 0, \qquad D_{\partial_r} \partial_\varphi = (1/r)\, \partial_\varphi, \qquad D_{\partial_r} \partial_\vartheta = (1/r)\, \partial_\vartheta,$$

$$D_{\partial_\vartheta} \partial_\vartheta = -r\, \partial_r, \qquad D_{\partial_\vartheta} \partial_\varphi = \cot \vartheta\, \partial_\varphi, \qquad D_{\partial_\varphi} \partial_\varphi = -r \sin \vartheta\, U,$$

where $U = \cos \varphi\, \partial_x + \sin \varphi\, \partial_y = \sin \vartheta\, \partial_r + (\cos \vartheta)/r\, \partial_\vartheta$.

14. (a) Let Π be a tangent plane. If $A\colon \Pi \to \Pi$ is a linear operator of determinant 1, then $R(Av, Aw) = R(v, w)$. (b) For a semi-Riemannian product $M \times N$,

$$\mathrm{Ric}(x + v, y + w) = \mathrm{Ric}_M(x, y) + \mathrm{Ric}_N(v, w),$$

where $x, y \in T_p(M)$ and $v, w \in T_q(N)$. Hence $S = S_M + S_N$.

15. For an orthogonal coordinate system the geodesic differential equations become

$$\frac{d}{ds}\left(g_{kk} \frac{dx^k}{ds}\right) = \frac{1}{2} \sum_i \frac{\partial g_{ii}}{\partial x^k} \left(\frac{dx^i}{ds}\right)^2 \qquad (1 \le k \le n).$$

16. (a) If γ is a geodesic and $f \in \mathfrak{F}(M)$, then $H^f(\gamma', \gamma') = d^2(f \circ \gamma)/ds^2$. (b) At a critical point of f, H^f agrees with the Hessian in Exercise 1.14.

17. (a) If Y is a vector field on a curve α, then $Y'(s) = \lim_{t \to 0} (1/t)[P_t Y(s + t) - Y(s)]$, where P_t is parallel translation along α from $\alpha(s + t)$ to $\alpha(s)$. (b) Parallel translation along geodesics uniquely determines the Levi-Civita connection.

18. The *curl* of $V \in \mathfrak{X}(M)$ is defined by $(\mathrm{curl}\ V)(X, Y) = \langle D_X V, Y\rangle - \langle D_Y V, X\rangle$. (a) curl V is a skew-symmetric (0, 2) tensor field (that is, two-form) with coordinate components $\partial V_j/\partial x^i - \partial V_i/\partial x^j$. (b) $\mathrm{curl}(\mathrm{grad}\ f) = 0$. (c) $\mathrm{curl}\ V = d\theta$, where θ is the one-form metrically equivalent to V. (d) On $\mathbf{R}^3$, $(\mathrm{curl}\ V)(X, Y) = (X \times Y)\cdot(\nabla \times V)$.

19. To show that a regular curve α with α'' and α' collinear is a pregeodesic, write $\alpha''(s) = f(s)\alpha'(s)$ and prove: (a) $\beta = \alpha \circ h$ is a geodesic if and only if

$h'' + f(h)h'^2 = 0$. (b) If $\langle \alpha', \alpha' \rangle$ is never zero, then any constant speed reparametrization of α is a geodesic. (c) $\langle \alpha', \alpha' \rangle$ is always zero or never zero. (d) α is pregeodesic in the case $\langle \alpha', \alpha' \rangle = 0$.

20. The Riemannian curvature tensor on M^n has $n^2(n^2 - 1)/12$ independent components. Formally, if V is an n-dimensional scalar product space, the vector space of all curvaturelike (page 79) functions $V^4 \to \boldsymbol{R}$ has dimension $n^2(n^2 - 1)/12$.

21. A semi-Riemannian manifold M is an *Einstein manifold* provided $\text{Ric} = cg$ for some constant c. (a) If M is connected, $n = \dim M \geq 3$, and $\text{Ric} = fg$, then M is Einstein. (b) (Schur.) If M is connected, $n \geq 3$, and for each $p \in M$, K is constant on the nondegenerate planes in $T_p(M)$, then K is constant. (Hint: For (a), contract and use div.)

22. If $\mathscr{U}$ is an open set in $T_p(M)$ then the curvature tensor R at p is completely determined by $K(\Pi)$ for all nondegenerate planes Π that meet $\mathscr{U}$. (Hint: If $u \in \mathscr{U}$ and $x \in T_p(M)$ then for δ small, $u + \delta x \in \mathscr{U}$.)

4 SEMI-RIEMANNIAN SUBMANIFOLDS

If M is a semi-Riemannian submanifold of $\bar{M}$ (Definition 3.4) then for vectors $v, w \in T_p(M) \subset T_p(\bar{M})$ the notation $\langle v, w \rangle$ is unambiguous. Though they agree on this measurement it will soon be clear that inhabitants of the submanifold see their world M differently than do the inhabitants of the outside world $\bar{M}$. Comparing the Levi-Civita connections of M and $\bar{M}$ gives a tensor II that provides an infinitesimal description of the shape of M in $\bar{M}$. Then II is used systematically to compare the geometries of M and $\bar{M}$. In particular, the geometry of M can be derived from II and the geometry of $\bar{M}$.

Overbars are used to distinguish corresponding geometrical objects on M and $\bar{M}$. Thus for connection and curvature we have D, R on M and $\bar{D}, \bar{R}$ on $\bar{M}$. In the case of a vector field Y on a curve, we use the notations $\dot{Y} = \bar{D}Y/ds$ and $Y' = DY/ds$. (But for a curve α in M, $\dot{\alpha} = \alpha' = d\alpha/ds$.) The abbreviation $M \subset \bar{M}$ means that M is a semi-Riemannian submanifold of $\bar{M}$.

TANGENTS AND NORMALS

If M is merely a smooth submanifold of $\bar{M}$, a vector field X on the inclusion map $j: M \subset \bar{M}$ (Definition 1.47) is called an $\bar{M}$ *vector field on* M. Thus X assigns to each point p of M a tangent vector X_p to $\bar{M}$ at p, and X is smooth if $f \in \mathfrak{F}(\bar{M})$ implies $Xf \in \mathfrak{F}(M)$. The set $\bar{\mathfrak{X}}(M)$ of all such (smooth) vector fields is a module over $\mathfrak{F}(M)$. For any $Y \in \mathfrak{X}(\bar{M})$ the restriction $Y|M$ is in $\bar{\mathfrak{X}}(M)$. Since we have agreed to ignore differential maps of j, $\mathfrak{X}(M)$ is a submodule of $\bar{\mathfrak{X}}(M)$.

Now let M be a semi-Riemannian submanifold of $\bar{M}$. Each tangent space $T_p(M)$ is, by definition, a nondegenerate subspace of $T_p(\bar{M})$. Hence Lemma 2.23 gives the direct sum decomposition

$$T_p(\bar{M}) = T_p(M) + T_p(M)^\perp,$$

and $T_p(M)^\perp$ is also nondegenerate. Its dimension is k, the *codimension* of M^n in $\bar{M}^{n+k}$. Similarly the index of ($\bar{g}$ restricted to) $T_p(M)^\perp$ is called the *co-index* of M in $\bar{M}$, and by Lemma 2.26, ind $\bar{M}$ = ind M + coind M.

Vectors in $T_p(M)^\perp$ are said to be *normal* to M, while those in $T_p(M)$ are of course *tangent* to M. That the sum above is direct means that every vector $x \in T_p(\bar{M})$, $p \in M$, has a unique expression

$$x = \tan x + \operatorname{nor} x,$$

where $\tan x \in T_p(M)$ and $\operatorname{nor} x \in T_p(M)^\perp$. The resulting orthogonal projections

$$\tan\colon T_p(\bar{M}) \to T_p(M) \qquad \text{and} \qquad \operatorname{nor}\colon T_p(\bar{M}) \to T_p(M)^\perp$$

are obviously $\boldsymbol{R}$-linear.

A vector field $Z \in \bar{\mathfrak{X}}(M)$ is *normal* to M provided each value Z_p is normal to M. The set $\mathfrak{X}(M)^\perp$ of all such is a submodule of $\bar{\mathfrak{X}}(M)$.

For $X \in \bar{\mathfrak{X}}(M)$, applying *tan* and *nor* at each point of M gives vector fields $\tan X \in \mathfrak{X}(M)$ and $\operatorname{nor} X \in \mathfrak{X}(M)^\perp$. (A computation with adapted coordinate systems shows that these vector fields are smooth.) The resulting *orthogonal projections*

$$\tan\colon \bar{\mathfrak{X}}(M) \to \mathfrak{X}(M) \qquad \text{and} \qquad \operatorname{nor}\colon \bar{\mathfrak{X}}(M) \to \mathfrak{X}(M)^\perp$$

are $\mathfrak{F}(M)$-linear, and the identity $X = \tan X + \operatorname{nor} X$ is that of the direct sum $\bar{\mathfrak{X}}(M) = \mathfrak{X}(M) + \mathfrak{X}(M)^\perp$.

In this chapter, V and W will always denote vector fields tangent to M, and Z will be normal to M.

THE INDUCED CONNECTION

If M is a semi-Riemannian submanifold of $\bar{M}$ the Levi-Civita connection $\bar{D}$ of $\bar{M}$ gives rise in a natural way to a function $\mathfrak{X}(M) \times \bar{\mathfrak{X}}(M) \to \bar{\mathfrak{X}}(M)$ called the *induced connection* on $M \subset \bar{M}$. If $V \in \mathfrak{X}(M)$ and $X \in \bar{\mathfrak{X}}(M)$ then $\bar{D}_V X$ is not yet meaningful, since V and X are not in $\mathfrak{X}(\bar{M})$. However for each $p \in M$ let $\bar{V}$ and $\bar{X}$ be smooth local extensions of V and X over a coordinate neighborhood $\mathscr{U}$ of p in $\bar{M}$ (Exercise 1.18). Then define $\bar{D}_V X$ on each $\mathscr{U} \cap M$ to be the restriction of $\bar{D}_{\bar{V}} \bar{X}$ to $\mathscr{U} \cap M$.

1. Lemma. $\bar{D}_V X$ is a well-defined smooth $\bar{M}$ vector field on M.

Proof. As the restriction of a smooth vector field, $\bar{D}_{\bar{V}} \bar{X} | \mathscr{U} \cap M$ is smooth. Thus it suffices to show that it is independent of the choice of extensions. In terms of a coordinate system on $\mathscr{U}$, write $\bar{X} = \sum f^i \partial_i$. Then $\bar{D}_{\bar{V}} \bar{X} = \sum \bar{V}(f^i) \partial_i + \sum f^i \bar{D}_{\bar{V}}(\partial_i)$. But at $q \in \mathscr{U} \cap M$, $(\bar{V} f^i)(q) = V_q(f^i) = V_q(f^i | \mathscr{U} \cap M)$ and $\bar{D}_{\bar{V}}(\partial_i)|_q = \bar{D}_{V_q}(\partial_i)$. Thus the restriction of $\bar{D}_{\bar{V}} \bar{X}$ depends solely on V and X. ∎

Since the induced connection

$$\bar{D}: \mathfrak{X}(M) \times \bar{\mathfrak{X}}(M) \to \bar{\mathfrak{X}}(M)$$

is so closely related to the Levi-Civita connection of $\bar{M}$ we have used the same notation for both. In particular the induced connection has the five Levi-Civita properties, as follows.

2. Corollary. Let $\bar{D}$ be the induced connection of $M \subset \bar{M}$. If V, $W \in \mathfrak{X}(M)$ and $X, Y \in \bar{\mathfrak{X}}(M)$, then

(1) $\bar{D}_V X$ is $\mathfrak{F}(M)$-linear in V.
(2) $\bar{D}_V X$ is $\mathbf{R}$-linear in X.
(3) $\bar{D}_V(fX) = VfX + f \bar{D}_V X$ for $f \in \mathfrak{F}(M)$.
(4) $[V, W] = \bar{D}_V W - \bar{D}_W V$.
(5) $V\langle X, Y \rangle = \langle \bar{D}_V X, Y \rangle + \langle X, \bar{D}_V Y \rangle$.

Proof. For each point $p \in M$ extend all the vector fields and functions over a neighborhood of p in $\bar{M}$. The corresponding five properties hold for the Levi-Civita connection of $\bar{M}$; then restriction to M gives the results above, since (a) $(\bar{D}_{\bar{V}} \bar{X})|M = \bar{D}_V X$; (b) $\bar{V} \bar{f} | M = Vf$; (c) $\langle \bar{X}, \bar{Y} \rangle | M = \langle X, Y \rangle$; and by Proposition 1.32, (d) $[\bar{V}, \bar{W}] | M = [V, W]$. ∎

A basic fact here is that for V, W both tangent to M, the covariant derivative $\bar{D}_V W$ need not be tangent to M. So it is natural to ask what tan $\bar{D}_V W$ and nor $\bar{D}_V W$ are.

3. Lemma. For $M \subset \bar{M}$, if $V, W \in \mathfrak{X}(M)$, then

$$D_V W = \tan \bar{D}_V W,$$

where D is the Levi-Civita connection of M.

Proof. For an arbitrary vector field $X \in \mathfrak{X}(M)$, locally extend the vector fields X, V, W and write out the Koszul equation $2\langle \bar{D}_{\bar{V}} \bar{W}, \bar{X} \rangle = F(\bar{V}, \bar{W}, \bar{X})$. Upon restriction to M, $\langle \bar{D}_{\bar{V}} \bar{W}, \bar{X} \rangle$ becomes $\langle \bar{D}_V W, X \rangle$, and properties (a)–(d) in the previous proof show that $F(\bar{V}, \bar{W}, \bar{U})|M = F(V, W, U)$. Thus $\langle \bar{D}_V W, X \rangle = \langle D_V W, X \rangle$. Since X is tangent to M, we can replace $\bar{D}_V W$ by tan $\bar{D}_V W$, and the result follows. ∎

But nor $\bar{D}_V W$ is something new.

4. Lemma. The function $II\colon \mathfrak{X}(M) \times \mathfrak{X}(M) \to \mathfrak{X}(M)^\perp$ such that

$$II(V, W) = \text{nor } \bar{D}_V W$$

is $\mathfrak{F}(M)$-bilinear and symmetric. II is called the *shape tensor* (or *second fundamental form tensor*) of $M \subset \bar{M}$.

Proof. Since $\bar{D}_V W$ is $\mathfrak{F}(M)$-linear in V and $\boldsymbol{R}$-linear in W, so is II. For $f \in \mathfrak{F}(M)$,

$$\bar{D}_V(f W) = Vf\, W + f\, \bar{D}_V W.$$

But W is tangent to M, and the projection *nor* is $\mathfrak{F}(M)$-linear, hence

$$II(V, f W) = \text{nor } \bar{D}_V(f W) = f \text{ nor } \bar{D}_V W = f\, II(V, W).$$

Finally,

$$II(V, W) - II(W, V) = \text{nor}(\bar{D}_V W - \bar{D}_W V) = \text{nor}[V, W] = 0. \qquad \blacksquare$$

II is a more general tensor field than those defined in Chapter 2, since its values lie in $\mathfrak{X}(M)^\perp$ rather than $\mathfrak{X}(M)$. But evidently its $\mathfrak{F}(M)$-bilinearity means that it has the pointwise character expressed by Proposition 2.2 (see page 199). Thus at each point $p \in M$, II determines an $\boldsymbol{R}$-bilinear function

$$T_p(M) \times T_p(M) \to T_p(M)^\perp$$

sending (v, w) to $II(v, w)$.

The two preceding lemmas are summarized by

$$\bar{D}_V W = \underset{\text{tangent to } M}{D_V W} + \underset{\text{normal to } M}{II(V, W)} \qquad \text{for} \quad V, W \in \mathfrak{X}(M).$$

This decomposition leads to a fundamental curvature result called the *Gauss equation.*

5. Theorem. Let M be a semi-Riemannian submanifold of $\bar{M}$, with R and $\bar{R}$ their Riemannian curvature tensors and II the shape tensor. Then for vector fields V, W, X, Y all tangent to M,

$$\begin{aligned}\langle R_{VW} X, Y\rangle = \langle \bar{R}_{VW} X, Y\rangle &+ \langle II(V, X), II(W, Y)\rangle \\ &- \langle II(V, Y), II(W, X)\rangle.\end{aligned}$$

Proof. As usual we can suppose $[V, W] = 0$. Thus $\langle \bar{R}_{VW} X, Y\rangle = -(VW) + (WV)$, where

$$(VW) = \langle \bar{D}_V \bar{D}_W X, Y\rangle = \langle \bar{D}_V D_W X, Y\rangle + \langle \bar{D}_V(II(W, X)), Y\rangle.$$

Since Y is tangent to M the projection *tan* can be introduced in the left side of the first summand, giving $\langle D_V D_W X, Y \rangle$. The second summand can be written as

$$V\langle II(W, X), Y\rangle - \langle II(W, X), \bar{D}_V Y\rangle.$$

Since $II(W, X)$ is normal to M this expression reduces to

$$-\langle II(W, X), \text{nor}\, \bar{D}_V Y\rangle = -\langle II(W, X), II(V, Y)\rangle.$$

We conclude that

$$(VW) = \langle D_V D_W X, Y\rangle - \langle II(W, X), II(V, Y)\rangle.$$

The result then follows by evaluating $-(VW) + (WV)$. ∎

Because it is a tensor equation the Gauss equation remains valid if its vector fields are replaced by individual tangent vectors. Thus substitution in the formula of Lemma 3.39 gives the following relation between the sectional curvatures K of M and $\bar{K}$ of $\bar{M}$ (also called the *Gauss equation*).

6. Corollary. If the vectors v and w form a basis for a nondegenerate tangent plane to M, then

$$K(v, w) = \bar{K}(v, w) + \frac{\langle II(v, v), II(w, w)\rangle - \langle II(v, w), II(v, w)\rangle}{\langle v, v\rangle\langle w, w\rangle - \langle v, w\rangle^2}.$$

As an application let us show that the sphere $S^n(r)$ has constant sectional curvature $K = 1/r^2$ if $n \geq 2$. (All one-dimensional manifolds are trivially flat.) The position vector field $P = \sum u^i\, \partial_i$ of $\mathbf{R}^{n+1}$ is normal to the standard sphere $S^n(r) \subset \mathbf{R}^{n+1}$ at each point. If $\bar{D}$ is the connection of $\mathbf{R}^{n+1}$, note that $\bar{D}_X P = \sum X u^i\, \partial_i = X$ for every vector field X. We assert that the shape tensor of the sphere is given by

$$II(V, W) = (-1/r)\langle V, W\rangle U,$$

where $U = P/r$ is the outward unit normal on $S^n(r)$. In fact,

$$\begin{aligned}\langle II(V, W), U\rangle &= \langle \text{nor}\, \bar{D}_V W, U\rangle = \langle \bar{D}_V W, P\rangle/r \\ &= -\langle W, \bar{D}_V P\rangle/r = -\langle V, W\rangle/r.\end{aligned}$$

Since $\mathbf{R}^{n+1}$ is flat, substitution in Corollary 6 gives the result $K = 1/r^2$.

Since it is a $(0, 2)$ tensor field with values in $\mathfrak{X}(M)^\perp$, II can be metrically contracted to give a normal vector field on M. Dividing by $n = \dim M$ gives the *mean curvature vector field* H of $M \subset \bar{M}$. Explicitly, at $p \in M$,

$$H_p = \frac{1}{n}\sum_{i=1}^{n} \varepsilon_i II(e_i, e_i),$$

where $e_1, \ldots, e_n$ is any frame on M at p.

If M is a semi-Riemannian submanifold of $\bar{M}$, the usual geometry of M is sometimes called its *intrinsic geometry* to emphasize that it is independent of the fact that M happens to be in $\bar{M}$. Roughly speaking, the *extrinsic geometry* of M is that seen by observers in $\bar{M}$. Formally, a *pair isometry* from $M \subset \bar{M}$ to $N \subset \bar{N}$ is an isometry $\phi\colon \bar{M} \to \bar{N}$ such that $\phi | M$ is an isometry from M to N. (When $\bar{M} = \bar{N}$, ϕ is also called a *congruence* from M to N.) Features of M that are preserved by all such pair isometries—and do not belong to its intrinsic geometry—constitute the *extrinsic geometry* of M. For example the shape tensor of $M \subset \bar{M}$ is an extrinsic invariant.

7. Lemma. A pair isometry ϕ from $M \subset \bar{M}$ to $N \subset \bar{N}$ preserves shape tensors, that is,

$$d\phi(II(v, w)) = II(d\phi v, d\phi w)$$

for all $v, w \in T_p(M)$, $p \in M$.

Proof. Let $V, W \in \mathfrak{X}(M)$. Since $\phi | M$ is in particular a diffeomorphism $M \to N$, $d\phi(V)$ and $d\phi(W)$ are in $\mathfrak{X}(N)$. Because $\phi\colon \bar{M} \to \bar{N}$ preserves connections it follows that $d\phi(\bar{D}_V W) = \bar{D}_{d\phi V}(d\phi W)$. For each $p \in M$, the linear isometry $d\phi\colon T_p(\bar{M}) \to T_p(\bar{N})$ carries $T_p(M)$ to $T_p(N)$, hence $T_p(M)^\perp$ to $T_p(N)^\perp$. Thus $d\phi$ preserves tangential and normal components; hence

$$\begin{aligned} d\phi(II(V, W)) &= d\phi(\operatorname{nor} \bar{D}_V W) = \operatorname{nor}(d\phi(\bar{D}_V W)) \\ &= \operatorname{nor}(\bar{D}_{d\phi V}(d\phi W)) = II(d\phi V, d\phi W). \qquad \blacksquare \end{aligned}$$

The shape tensor is not intrinsic to M. For example, if different shapes of a piece of paper in space represent isometric submanifolds of $\boldsymbol{R}^3$ then $II = 0$ when the paper is flat but not when it is curved.

GEODESICS IN SUBMANIFOLDS

The decomposition $\bar{D}_V W = D_V W + II(V, W)$ is adapted to vector fields on a curve as follows.

8. Proposition. Let Y be a vector field—always tangent to M—on a curve α in $M \subset \bar{M}$. Then

$$\dot{Y} = \underset{\substack{\text{tangent}\\ \text{to } M}}{Y'} + \underset{\text{normal to } M}{II(\alpha', Y)},$$

where $\dot{Y} = \bar{D}Y/ds$ and $Y' = DY/ds$.

Proof. As usual we can assume α lies in a single coordinate neighborhood of M and write $Y = \sum Y^i \partial_i$. Relative to the geometry of $\bar{M}$,

$$\dot{Y} = \sum \frac{dY^i}{ds} \partial_i + \sum Y^i (\partial_i|_\alpha)^{\cdot}.$$

But

$$(\partial_i|_\alpha)^{\cdot} = \bar{D}_{\alpha'}(\partial_i) = D_{\alpha'}(\partial_i) + II(\alpha', \partial_i).$$

Substituting in the previous equation then gives the required result. ∎

9. Corollary. If α is a curve in $M \subset \bar{M}$, then

$$\ddot{\alpha} = \alpha'' + II(\alpha', \alpha'),$$

where $\ddot{\alpha}$ is the acceleration of α in $\bar{M}$, and α'' its acceleration in M.

To see how the tensor II describes shape, fix a point $p \in M$ and for $v \in T_p(M)$ let γ be the M geodesic with initial velocity v. In M, γ is "straight," thus *its curving in $\bar{M}$ is that forced by the curving of M itself in $\bar{M}$*. By the corollary, $\ddot{\gamma}(0) = II(v, v)$. Thus for all v, II describes the shape at p of M in $\bar{M}$ (see Figure 1). If $\bar{M}$ is a Euclidean space, then γ has the quadratic approximation $s\dot{\gamma}(s) + \frac{1}{2}s^2\ddot{\gamma}(s)$, and in fact it is true in general that the hypersurface $\{v + \frac{1}{2}II(v, v): v \in T_p(M)\}$ in $T_p(\bar{M})$ is the best quadratic approximation of M in $\bar{M}$ near p.

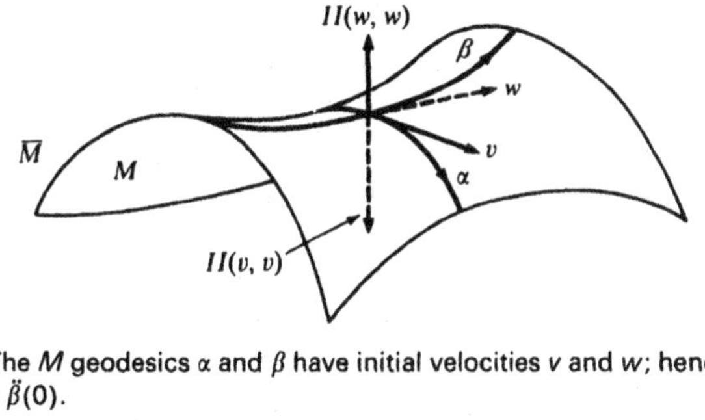

Figure 1. The M geodesics α and β have initial velocities v and w; hence $II(v, v) = \ddot{\alpha}(0)$ and $II(w, w) = \ddot{\beta}(0)$.

10. Corollary. A curve α in $M \subset \bar{M}$ is a geodesic of M if and only if its $\bar{M}$ acceleration is everywhere normal to M.

This is obvious since, by the preceding corollary, $\ddot{\alpha}$ normal to M implies that the M acceleration of α is zero.

11. Corollary. The nonconstant geodesics of the sphere $S^n(r)$ are precisely all constant speed parametrizations of the great circles of $S^n(r) \subset \boldsymbol{R}^{n+1}$.

Proof. A *great circle* of the sphere $S = S^n(r)$ is a circle $\Pi \cap S$ cut from S by a plane Π through the origin of $\mathbf{R}^{n+1}$. If α is a constant speed parameterization of $\Pi \cap S$ then $\dot{\alpha} = d\alpha/dt$ and $\ddot{\alpha} = d^2\alpha/dt^2$ are orthogonal to each other and tangent to the plane Π. But on α the position vector field P_α is also tangent to Π and orthogonal to $\dot{\alpha} \neq 0$. Thus $\ddot{\alpha}$ and P_α are collinear at each point, so that $\ddot{\alpha}$ is normal to S and hence α is a geodesic of the sphere.

To show that every nonconstant geodesic γ can be so obtained, let Π be the plane through the origin and $\gamma(0)$ to which $\gamma'(0)$ is tangent. Then a suitable constant speed parametrization of $\Pi \cap S$ has $\alpha'(0) = \gamma'(0)$. Thus $\gamma = \alpha$ by the uniqueness of geodesics. ∎

In particular, every geodesic of $S^n(r)$ is periodic, with period $2\pi r$. It is now easy to see what the exponential maps of the sphere are like. Fix a point p of $S = S^n(r)$. For each unit vector $v \in T_p(S)$, $\exp_p$ wraps the radial line through v around the great circle to which v is tangent. Explicitly,

$$\exp_p(tv) = r[\cos(t/r)\, p + \sin(t/r)\, v].$$

Thus the $(n-1)$-spheres t constant in $T_p(S)$ are typically carried to "latitudinal" $(n-1)$-spheres in S cut by hyperplanes orthogonal to the diameter $p, -p$. Exceptionally, the spheres $t = k\pi$ collapse to the north pole p when k is an even integer and to the south pole $-p$ when k is odd. Note that $\exp_p$ is a diffeomorphism from the ball $0 \leq t < \pi r$ to $S - \{-p\}$, hence the latter is a normal neighborhood of p.

TOTALLY GEODESIC SUBMANIFOLDS

These are the submanifolds with the simplest possible shape.

12. Definition. A semi-Riemannian submanifold M of $\bar{M}$ is *totally geodesic* provided its shape tensor vanishes: $II = 0$.

Thus a totally geodesic submanifold M is extrinsically flat: observers in $\bar{M}$ see no curving. That is not to say that M is intrinsically flat; indeed by the Gauss equation (6) it has the same intrinsic curvature as $\bar{M}$.

13. Proposition. For $M \subset \bar{M}$ the following are equivalent.

(1) M is totally geodesic in $\bar{M}$.

(2) Every geodesic of M is also a geodesic of $\bar{M}$.

(3) If $v \in T_p(\bar{M})$ is tangent to M, then the $\bar{M}$ geodesic γ_v lies initially in M.

(4) If α is a curve in M and $v \in T_{\alpha(0)}(M)$, then parallel translation of v along α is the same for M and for $\bar{M}$.

Proof. (2) ⇒ (3). If $\alpha\colon I \to M$ is the geodesic of M with initial velocity v, then since α is also a geodesic of $\bar{M}$ it follows by the uniqueness of geodesics that $\alpha = \gamma_v | I$. (However, γ_v may later on leave M; for example, let M be an open disk in the xy-plane of $\bar{M} = \boldsymbol{R}^3$.)

(3) ⇒ (1). For every v tangent to M, applying Corollary 9 to γ_v shows that $II(v, v) = 0$. Then the polarization

$$II(v + w, v + w) = II(v, v) + 2II(v, w) + II(w, w)$$

shows that $II = 0$.

(1) ⇒ (4). Let V be the M parallel vector field on α such that $V(0) = v$. By Proposition 8, V is $\bar{M}$ parallel. Thus the two parallel translations agree.

(4) ⇒ (2). If γ is a geodesic of M, then γ' is M parallel hence $\bar{M}$ parallel, so γ is a geodesic of $\bar{M}$. ∎

If W is a [nondegenerate] k-dimensional subspace of $\boldsymbol{R}^n_\nu$ then any translate $x + W$ is called a [nondegenerate] *k-plane* in $\boldsymbol{R}^n_\nu$ ($0 \le k \le n$). It is easy to see that nondegenerate k-planes are totally geodesic semi-Riemannian submanifolds of $\boldsymbol{R}^n_\nu$. The following lemma will imply that they are the only complete connected ones.

14. Lemma. Let M and N be complete, connected, totally geodesic semi-Riemannian submanifolds of $\bar{M}$. If there is a point $p \in M \cap N$ at which $T_p(M) = T_p(N)$, then $M = N$.

Proof. It suffices to show that if M is connected and N is complete, then $M \subset N$. Let σ be a geodesic segment in M running from p to q. Then σ is a geodesic of $\bar{M}$, and by the hypothesis on tangent spaces, $\sigma'(0)$ is tangent to N. Thus σ is a geodesic of N as long as it remains in N. But since N is complete, σ is entirely contained in N. By the preceding proposition, M parallel translation of $T_p(M) = T_p(N)$ along σ will give $T_q(M) = T_q(N)$. Thus the argument can be repeated to show that every broken geodesic of M starting at p lies also in N. Since M is connected it follows that $M \subset N$. (In fact M is an open submanifold of N, by Exercise 1.8.) ∎

It follows that the complete, connected, k-dimensional totally geodesic Riemannian submanifolds of the sphere $S^n(r)$ are exactly its *great k-spheres*: the submanifolds $W \cap S^n(r)$ where W is a $(k + 1)$-plane through the origin.

These results about totally geodesic submanifolds M can be extended to the case where M is not semi-Riemannian (Exercise 9).

15. Definition. A point p of $M \subset \bar{M}$ is *umbilic* provided there is a normal vector $z \in T_p(M)^\perp$ such that

$$II(v, w) = \langle v, w \rangle z \qquad \text{for all} \quad v, w \in T_p(M).$$

Then z is called the *normal curvature vector* of M at p.

In the Riemannian case, $II(u, u) = z$ for all unit vectors: M bends the same way in all directions at an umbilic point. However for indefinite metrics, the formula $II(u, u) = \langle u, u \rangle z$ shows that *M bends toward z in spacelike directions, away from z in timelike directions.*

A semi-Riemannian submanifold M of $\bar{M}$ is *totally umbilic* provided every point of M is umbilic. Then there is a smooth normal vector field $\mathbf{z}$ on M, called the *normal curvature vector field* of M, such that $II(V, W) = \langle V, W \rangle \mathbf{z}$ for all $V, W \in \mathfrak{X}(M)$. Thus a totally geodesic submanifold is a totally umbilic submanifold for which $\mathbf{z} = 0$. Our earlier computation of the shape tensor of the sphere $S^n(r) \subset \mathbf{R}^{n+1}$ shows that it is totally umbilic with $\mathbf{z} = -U/r$, where U is the outward unit normal.

SEMI-RIEMANNIAN HYPERSURFACES

A *semi-Riemannian hypersurface M of $\bar{M}$* is just a semi-Riemannian submanifold of codimension 1. Thus the co-index of M—the common index of all (one-dimensional) normal spaces $T_p(M)$—must be 0 or 1. It is more efficient to separate these two cases by a sign as follows.

16. Definition. The *sign* ε of a semi-Riemannian hypersurface M of $\bar{M}$ is

$+1$ if the co-index of M is 0, that is, $\langle z, z \rangle > 0$ for every normal vector $z \neq 0$;

-1 if the co-index of M is 1, that is, $\langle z, z \rangle < 0$ for every normal vector $z \neq 0$.

Note that ind M = ind $\bar{M}$ if $\varepsilon = 1$, but ind M = ind $\bar{M} - 1$ if $\varepsilon = -1$. For a Riemannian manifold every hypersurface is Riemannian with sign $+1$, but in the indefinite case, sign -1 is as natural as $+1$.

17. Proposition. Let c be a value of $f \in \mathfrak{F}(\bar{M})$. Then $M = f^{-1}(c)$ is a semi-Riemannian hypersurface of $\bar{M}$ if and only if $\langle \text{grad } f, \text{grad } f \rangle$ is >0 or <0 on M. In this case the sign of M is the (constant) sign of $\langle \text{grad } f, \text{grad } f \rangle$, and $U = \text{grad } f / |\text{grad } f|$ is a unit normal vector field on M.

Proof. Since grad f is metrically equivalent to the differential df, it follows that M is a hypersurface, and the condition on $\langle \text{grad } f, \text{grad } f \rangle$ ensures that M is semi-Riemannian. (If M is connected, nonvanishing of the scalar product will suffice.) Finally grad f is normal to M, since for any $v \in T_p(M)$,

$$\langle \text{grad } f, v \rangle = v(f) = v(f \mid M) = 0,$$

because f is constant on M. ∎

In R^{n+1}, for example, if $f = \sum (u^i)^2$ then $f^{-1}(r^2)$ is the standard sphere $S^n(r)$.

Not every semi-Riemannian hypersurface $M \subset \bar{M}$ can be obtained from this proposition, since in general there does not exist a smooth unit normal on all of M: the Möbius band in R^3 is one example. However it is easy to show that there is always a unit normal on some neighborhood of any point of M.

For hypersurfaces the shape tensor can be reduced to simpler tensors as follows.

18. Definition. Let U be a unit normal vector field on a semi-Riemannian hypersurface $M \subset \bar{M}$. The (1, 1) tensor field S on M such that

$$\langle S(V), W\rangle = \langle II(V, W), U\rangle \qquad \text{for all} \quad V, W \in \mathfrak{X}(M)$$

is called the *shape operator of* $M \subset \bar{M}$ *derived from* U.

As usual, S determines a linear operator $S: T_p(M) \to T_p(M)$ at each point $p \in M$.

19. Lemma. If S is the shape operator derived from U, then $S(v) = -\bar{D}_v U$, and at each point the linear operator S on $T_p(M)$ is self-adjoint.

Proof. Since $\langle U, U\rangle$ is constant, $\langle \bar{D}_V U, U\rangle = 0$. Hence $\bar{D}_V U$ is tangent to M for all $V \in \mathfrak{X}(M)$. But if $W \in \mathfrak{X}(M)$ then

$$\langle S(V), W\rangle = \langle II(V, W), U\rangle = \langle \bar{D}_V W, U\rangle = -\langle \bar{D}_V U, W\rangle.$$

Hence $S(V) = -\bar{D}_V U$. The symmetry of II implies that S is self-adjoint. ∎

This characterization of S shows how it describes the shape of M in $\bar{M}$: S measures the $\bar{M}$ rate of change of U in all tangent directions, and since $U_p^{\perp} = T_p(M)$ it thereby records the turning of $T_p(M)$ in $\bar{M}$ as p traverses M.

The symmetric (0, 2) tensor B metrically equivalent to S is traditionally called the *second fundamental form* of $M \subset \bar{M}$. (The *first fundamental form* is just the metric tensor of M.)

If a unit normal U, perhaps defined only locally, is replaced by $-U$ then S changes sign. Thus even if M does not admit a global unit normal, S is globally defined *up to sign*. This sign ambiguity is something of a nuisance, but in intrinsic formulas it must cancel out.

For hypersurfaces the Gauss equation takes the following form.

20. Corollary. Let S be the shape operator of a semi-Riemannian hypersurface $M \subset \bar{M}$. If v, w span a nondegenerate tangent plane on M, then

$$K(v, w) = \bar{K}(v, w) + \varepsilon \, \frac{\langle Sv, v\rangle\langle Sw, w\rangle - \langle Sv, w\rangle^2}{\langle v, v\rangle\langle w, w\rangle - \langle v, w\rangle^2},$$

where ε is the sign of $M \subset \bar{M}$.

This follows immediately from Corollary 6 since $II(v, w) = \varepsilon\langle Sv, w\rangle U$ and $\langle U, U\rangle = \varepsilon$.

21. Lemma. A semi-Riemannian hypersurface $M \subset \bar{M}$ is totally umbilic if and only if its shape operator is scalar.

Proof. Suppose first that M is totally umbilic, with normal curvature vector field z. Let S be the shape operator derived from a unit normal U, perhaps only locally defined. Then

$$\langle SV, W\rangle = \langle II(V, W), U\rangle = \langle V, W\rangle\langle \mathrm{z}, U\rangle$$

for all tangent vector fields V, W. Thus $SV = \langle U, \mathrm{z}\rangle V$ for all V, so S is scalar.

Conversely, suppose that for every choice of U the derived S is scalar, that is, there is a function k_U on the domain of U such that $SV = k_U V$ for all V. Then

$$II(V, W) = \varepsilon\langle SV, W\rangle U = \varepsilon k_U\langle V, W\rangle U.$$

Since $k_{-U} = -k_U$ the vector field $\mathrm{z} = \varepsilon k_U U$ is globally well defined and the above equation becomes $II(V, W) = \langle V, W\rangle \mathrm{z}$. ∎

The (sign ambiguous) function k in this proof is called the *normal curvature function* of $M \subset \bar{M}$.

HYPERQUADRICS

The same methods used above to deal with spheres in $\mathbf{R}^{n+1}$ can be applied in $\mathbf{R}^{n+1}_\nu$ to an important family of semi-Riemannian hypersurfaces. Let $q \in \mathfrak{F}(R^{n+1}_\nu)$ as usual be the function $q(v) = \langle v, v\rangle$. Relative to natural coordinates,

$$q = \sum \varepsilon_i (u^i)^2 = -\sum_{i=1}^{\nu} (u^i)^2 + \sum_{j=\nu+1}^{n+1} (u^j)^2.$$

If P is the position vector field of $\mathbf{R}^n_\nu$, then $q = \langle P, P\rangle$. Consequently grad $q = 2P$, since for all V,

$$\langle \operatorname{grad} q, V\rangle = Vq = V\langle P, P\rangle = 2\langle D_V P, P\rangle = 2\langle V, P\rangle.$$

Thus $\langle \operatorname{grad} q, \operatorname{grad} q\rangle = 4q$. By Proposition 17 we conclude that *for* $r > 0$ *and* $\varepsilon = \pm 1$, $Q = q^{-1}(\varepsilon r^2)$ *is a semi-Riemannian hypersurface of* $\mathbf{R}^{n+1}_\nu$ *with unit normal* $U = P/r$ *and* sign ε. These hypersurfaces are called the (*central*) *hyperquadrics* of $\mathbf{R}^{n+1}_\nu$. The two families $\varepsilon = 1$ and $\varepsilon = -1$ fill all of $\mathbf{R}^{n+1}_\nu$ except the set $q^{-1}(0)$, which consists of the nullcone $\Lambda = q^{-1}(0) - \{0\}$ and the origin 0. (See Figure 2.)

22. Proposition. The nullcone Λ of $\boldsymbol{R}_\nu^{n+1}$ is a hypersurface invariant under scalar multiplication and diffeomorphic to $(\boldsymbol{R}^\nu - 0) \times S^{n-\nu}$. The position vector field P of $\boldsymbol{R}_\nu^{n+1}$ is both tangent to Λ and normal to Λ, hence Λ is not semi-Riemannian.

Proof. Since P is zero only at the origin 0 it follows as in the proof of Proposition 17 that Λ is a hypersurface and P is normal to Λ. P is also tangent to Λ since at $v \in \Lambda$ the vector $P_v = v_v$ is tangent to the radial null geodesic $t \to tv$ in Λ. The scalar multiplication assertion is obvious, so it remains to determine the structure of Λ as a smooth manifold. Let $S^{n-\nu}$ be the unit sphere in $\boldsymbol{R}^{n-\nu+1}$, and define a mapping ϕ of $(\boldsymbol{R}^\nu - 0) \times S^{n-\nu}$ into $\boldsymbol{R}_\nu^{n+1}$ by

$$\phi(x, p) = (x_1, \ldots, x_\nu, |x|p_1, \ldots, |x|p_{n-\nu+1}).$$

Since

$$\langle \phi(x, p), \phi(x, p) \rangle = -\sum_1^\nu (x_i)^2 + |x|^2 = 0,$$

the image of ϕ lies in Λ.

If $v \in \Lambda$, let $\psi(v) = (x, p) \in (\boldsymbol{R}^\nu - 0) \times S^{n-\nu}$, where $x = (v_1, \ldots, v_\nu)$, $p = (v_{\nu+1}, \ldots, v_{n+1})/\kappa(v)$, and $\kappa(v) = [\sum_{\nu+1}^{n+1} (v_j)^2]^{1/2}$. Then ϕ and ψ are inverse mappings, hence diffeomorphisms. ∎

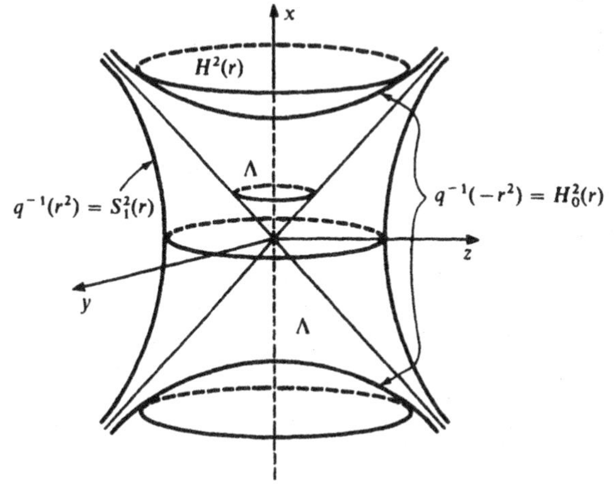

Figure 2. Hyperquadrics in $\boldsymbol{R}_1^3$.

23. Definition. Let $n \geq 2$ and $0 \leq \nu \leq n$. Then

(1) The *pseudosphere* of radius $r > 0$ in $\mathbf{R}^{n+1}_\nu$ is the hyperquadric

$$S^n_\nu(r) = q^{-1}(r^2) = \{p \in \mathbf{R}^{n+1}_\nu : \langle p, p\rangle = r^2\},$$

with dimension n and index ν.

(2) The *pseudohyperbolic space* of radius $r > 0$ in $\mathbf{R}^{n+1}_{\nu+1}$ is the hyperquadric

$$H^n_\nu(r) = q^{-1}(-r^2) = \{p \in \mathbf{R}^{n+1}_{\nu+1} : \langle p, p\rangle = -r^2\},$$

with dimension n and index ν.

Since pseudospheres have sign $+1$, by a previous remark their index in $\mathbf{R}^n_\nu$ is ν. Pseudohyperbolic spaces have sign -1, thus in $\mathbf{R}^n_\nu$ have index $\nu - 1$ (hence the shift to $\mathbf{R}^n_{\nu+1}$ in the definition above).

For $\nu = 0$, $S^n_0(r)$ is just the standard sphere $S^n(r)$ in Euclidean space $\mathbf{R}^{n+1}_0 = \mathbf{R}^{n+1}$.

The study of hyperquadrics is simplified by the fact that any hyperquadric is homothetic to a suitable *unit pseudosphere* $S^n_\nu = S^n_\nu(1)$. First, for any $r > 0$, scalar multiplication by r is a (positive) homothety $\mathbf{R}^{n+1}_\nu \to \mathbf{R}^{n+1}_\nu$ of scale factor r; hence so is its restriction $S^n_\nu \to S^n_\nu(r)$. Then $S^n_\nu(r)$ is homothetic to a pseudohyperbolic space as follows.

24. Lemma. The mapping $\sigma: \mathbf{R}^{n+1}_\nu \to \mathbf{R}^{n+1}_{n-\nu+1}$ given by

$$\sigma(p_1, \ldots, p_{n+1}) = (p_{\nu+1}, \ldots, p_{n+1}, p_1, \ldots, p_\nu)$$

is an anti-isometry that carries each $S^n_\nu(r)$ anti-isometrically onto $H^n_{n-\nu}(r)$, and vice versa.

Proof. Since σ is a linear isomorphism and

$$\langle \sigma(p), \sigma(p)\rangle = -\sum_{\nu+1}^{n+1} (p_j)^2 + \sum_{1}^{\nu} (p_i)^2 = -\langle p, p\rangle,$$

it follows as in the proof of Lemma 3.7 that σ is an anti-isometry. The formula also shows that σ carries $S^n_\nu(r)$ into $H^n_{n-\nu}(r)$, and vice versa. Thus $\sigma | S^n(r)$ is a diffeomorphism, hence an anti-isometry. ∎

In view of Remark 3.65 these homotheties reduce the intrinsic geometry of hyperquadrics to the case of the unit pseudosphere.

25. Lemma. The pseudosphere $S^n_\nu(r)$ is diffeomorphic to $\mathbf{R}^\nu \times S^{n-\nu}$; the pseudohyperbolic space $H^n_\nu(r)$ is diffeomorphic to $S^\nu \times \mathbf{R}^{n-\nu}$.

Proof. By the preceding results it suffices to deal with the unit pseudosphere S^n_ν. If $x \in \mathbf{R}^\nu$ and $p \in S^{n-\nu}$, let

$$\phi(x, p) = (x, (1 + |x|^2)^{1/2}p) \in \mathbf{R}^\nu_\nu \times \mathbf{R}^{n+1-\nu} \approx \mathbf{R}^{n+1}_\nu.$$

Because

$$\langle \phi(x, p), \phi(x, p) \rangle = -|x|^2 + (1 + |x|^2) = 1,$$

the mapping ϕ carries $\mathbf{R}^\nu \times S^{n-\nu}$ into S^n_ν. Since ϕ has inverse mapping $(x, q) \to (x, (1 + |x|^2)^{-1/2}q)$, it is a diffeomorphism. ∎

By convention $\mathbf{R}^0$ is a single point; by definition S^0 consists of two points.

Aside from spheres the only Riemannian manifolds among the hyperquadrics are the hypersurfaces $H^n_0(r)$ in Minkowski space $\mathbf{R}^{n+1}_1$. By the lemma, $H^n_0(r)$ consists of two connected components, each diffeomorphic to $\mathbf{R}^n$. In fact, the components are congruent under the isometry

$$(p_1, \ldots, p_{n+1}) \to (-p_1, p_2, \ldots, p_{n+1})$$

of $\mathbf{R}^{n+1}_1$ (see Figure 2).

26. Definition. The component of $H^n_0(r)$ through $(r, 0, \ldots, 0)$ is called the *upper imbedding*—the component through $(-r, 0, \ldots, 0)$ the *lower imbedding*—of *hyperbolic n-space* $H^n(r)$ in $\mathbf{R}^{n+1}_1$.

Like the sphere, every hyperquadric is totally umbilic.

27. Lemma. The hyperquadric $Q = q^{-1}(\varepsilon r^2) \subset \mathbf{R}^{n+1}_\nu$ of sign ε is totally umbilic, with shape operator $S = -I/r$ derived from the outward unit normal P/r.

Proof. If $V \in \mathfrak{X}(Q)$, then $S(V) = -\bar{D}_V(P/r) = -V/r$. ∎

It follows that the normal curvature vector field of Q is $z = -(\varepsilon/r)U$. Since U is outward (away from the origin), z is inward on pseudospheres and outward on pseudohyperbolic spaces. This is reasonable enough on the Riemannian hyperquadrics $S^n(r)$ and $H^n(r)$ since the sphere bends inward at all points, while hyperbolic space bends outward (Figure 2). But in the indefinite case, where $II(u, u) = \langle u, u \rangle z$, pseudospheres bend inward in spacelike directions but outward in timelike directions (Figure 2) while on pseudohyperbolic spaces this bending pattern is reversed.

The following proof will show that, just as for the sphere, the geodesics of any hyperquadric $Q \subset \mathbf{R}^{n+1}_\nu$ are the curves sliced from Q by planes Π through the origin of $\mathbf{R}^{n+1}_\nu$. ($\Pi \cap Q$ may have two components instead of one as for the sphere.) First consider pseudospheres.

28. Proposition. Let γ be a nonconstant geodesic of $S^n_\nu(r) \subset \mathbf{R}^{n+1}_\nu$.

(1) If γ is timelike it is a parametrization of one branch of a hyperbola in $\mathbf{R}^{n+1}_\nu$.

(2) If γ is null, it is a straight line, that is, a geodesic of $\mathbf{R}^{n+1}_\nu$.

(3) If γ is spacelike it is a periodic parametrization of an ellipse in $\mathbf{R}^{n+1}_\nu$ (see Figure 3).

In particular $S^n_\nu(r)$ is complete.

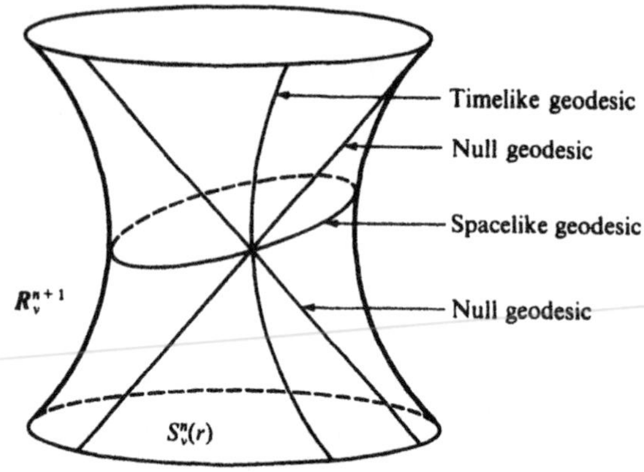

Figure 3

Proof. Let $p \in \mathbf{S} = S^n_\nu$ and let Π be a plane in $\mathbf{R}^{n+1}_\nu$ through the origin 0 and p. If g is the scalar product of $\mathbf{R}^{n+1}_\nu$, then since p is spacelike, there are three possibilities for the restriction of g to Π.

Case 1. $g|\Pi$ is positive definite. Then $\mathbf{S} \cap \Pi$ is a circle in $\Pi \approx \mathbf{R}^2$. In fact, if e_1, e_2 is an orthonormal basis for Π then a point $ae_1 + be_2$ of Π is in $\mathbf{S} = \{v \in \mathbf{R}^{n+1}_\nu : \langle v, v \rangle = r^2\}$ if and only if $a^2 + b^2 = r^2$. Thus $\alpha(t) = r \cos t\, e_1 + r \sin t\, e_2$ is a constant speed parametrization of $\Pi \cap \mathbf{S}$. In fact, $\langle \dot\alpha, \dot\alpha \rangle = r^2$, so α is spacelike. Furthermore $\ddot\alpha = -P_\alpha$, so $\ddot\alpha$ is normal to $\mathbf{S}$, hence α is a geodesic of $\mathbf{S}$.

Case 2. $g|\Pi$ is nondegenerate, with index 1. Let e_0, e_1 be an orthonormal basis for Π such that $p = re_1$, so e_0 is timelike. A point $ae_0 + be_1$ of Π is in $\mathbf{S}$ if and only if $-a^2 + b^2 = r^2$. Thus $\Pi \cap \mathbf{S}$ consists of the two branches of a hyperbola in $\Pi \approx \mathbf{R}^2_1$. The branch through p has constant speed parametrization

$$\alpha(t) = r \sinh t\, e_0 + r \cosh t\, e_1 \qquad (t \in \mathbf{R}).$$

Then $\langle \dot{\alpha}, \dot{\alpha} \rangle = -r^2 \cosh^2 t + r^2 \sinh t = -r^2$, hence α is timelike. Since $\ddot{\alpha} = P_\alpha$, $\ddot{\alpha}$ is normal to S; hence α is a geodesic of S.

Case 3. $g|\Pi$ is degenerate with nullspace of dimension 1. Thus if $v \neq 0$ is in the nullspace, v is in particular a null vector and p, v is a basis for Π. A point $ap + bv$ of Π is in S if and only if $a^2 = 1$, that is, $a = \pm 1$. Thus $\Pi \cap S$ consists of two parallel straight lines. The one through p is parametrized by $\alpha(t) = p + tv$. Since α is a geodesic of $\boldsymbol{R}^{n+1}_\nu$ lying in S it is a geodesic of S. Furthermore $\dot{\alpha}(0) = v_p$, so α is a null geodesic.

A proof such as the one for the sphere shows that every geodesic of S is a reparametrization of one of the three types above. ■

The geodesics of pseudohyperbolic spaces can now easily be derived. Since the mapping σ in Lemma 24 is an anti-isometry it follows that *the preceding proposition holds also for $H^n_\nu(r)$ provided the words* spacelike *and* timelike *are reversed.* Thus in particular, every geodesic γ of hyperbolic space $H^n(r)$ is one-to-one, since γ is spacelike in $\boldsymbol{R}^{n+1}_1$.

Substituting from Lemma 27 into Corollary 20 shows that hyperquadrics have constant curvature.

29. Proposition. Let $n \geq 2$ and $0 \leq \nu \leq n$.

(1) The pseudosphere $S^n_\nu(r)$ is a complete semi-Riemannian manifold of constant positive curvature $K = 1/r^2$.

(2) The pseudohyperbolic space $H^n_\nu(r)$ is a complete semi-Riemannian manifold of constant negative curvature $K = -1/r^2$.

In the Riemannian case the hyperbolic space $H^n(r)$ contrasts sharply with the sphere $S^n(r)$. Where the sphere is compact, with periodic geodesics and positive curvature, $H^n(r)$ is noncompact (indeed diffeomorphic to $\boldsymbol{R}^n$), with one-to-one geodesics and negative curvature.

Intuitively speaking all points on the unit sphere S^2 in $\boldsymbol{R}^3$, and all directions, are geometrically the same. Figure 2 may seem to suggest that this uniformity fails for pseudospheres, but in fact it holds in the following strong form.

30. Proposition. Let $e_1, \ldots, e_n$ and $f_1, \ldots, f_n$ be (tangent) frames on $S^n_\nu(r)$ at points p and q, respectively. Then there is a unique isometry $\phi\colon \boldsymbol{R}^{n+1}_\nu \to \boldsymbol{R}^{n+1}_\nu$ carrying $S^n_\nu(r)$ isometrically to itself, with $\phi(p) = q$ and $d\phi(e_i) = f_i$ for $1 \leq i \leq n$.

Proof. The position vector p_p at $p \in S = S^n(r)$ is normal to S, hence orthogonal to each e_i. Thus if $\tilde{e}_i$ denotes the element of $\boldsymbol{R}^{n+1}_\nu$ canonically corresponding to e_i, then $\tilde{e}_1, \ldots, \tilde{e}_n$, p/r is an orthonormal basis for $\boldsymbol{R}^{n+1}_\nu$. The same holds for the other frame.

The proof of Lemma 2.27 shows that there is a unique linear isometry $\phi: \boldsymbol{R}_\nu^{n+1} \to \boldsymbol{R}_\nu^{n+1}$ such that $\phi(e_i) = f_i$ for all i, and $\phi(p/r) = q/r$ hence $\phi(p) = q$. By Lemma 3.7, ϕ is an isometry of $\boldsymbol{R}_\nu^{n+1}$.

It is clear from the definition of $S = S_\nu^n(r)$ that $\phi(S) = S$. Since S is a semi-Riemannian submanifold, ϕ restricts to an isometry $S \to S$. Since ϕ is linear on $\boldsymbol{R}_\nu^{n+1}$ it differs only by canonical isomorphism from its differential map. Thus $d\phi(e_i) = f_i$ for all i.

To show that ϕ is unique, let ψ be another isometry of $\boldsymbol{R}_\nu^{n+1}$ with the required properties. Clearly $d\phi$ and $d\psi$ agree on $T_p(S)$. Since ϕ and ψ are pair isometries, it follows from Lemma 7 that $d\phi$ and $d\psi$ preserve normal curvature vectors. Thus they agree on $T_p(\boldsymbol{R}_\nu^{n+1})$. Proposition 3.62 then gives $\phi = \psi$. ∎

Evidently the same result holds for pseudohyperbolic spaces.

THE CODAZZI EQUATION

For $M \subset \bar{M}$, if the geometry of M is considered to be that of vectors tangent to M then there is an analogous geometry of vectors normal to M.

31. Definition. The *normal connection* of $M \subset \bar{M}$ is the function $D^\perp: \mathfrak{X}(M) \times \mathfrak{X}(M)^\perp \to \mathfrak{X}(M)^\perp$ given by

$$D_V^\perp Z = \text{nor } \bar{D}_V Z \qquad \text{for} \quad V \in \mathfrak{X}(M), \quad Z \in \mathfrak{X}(M)^\perp.$$

$D_V^\perp Z$ is called the *normal covariant derivative* of Z with respect to V. Evidently it registers the strictly normal rate of change of Z_p when p moves as prescribed by V. The following properties are immediate:

(1) $D_V^\perp Z$ is $\mathfrak{F}(M)$-linear in V and $\boldsymbol{R}$-linear in Z.
(2) $D_V^\perp(fZ) = f\, D_V^\perp Z + Vf\, Z$, where $f \in \mathfrak{F}(M)$.
(3) $V\langle Y, Z\rangle = \langle D_V^\perp Y, Z\rangle + \langle Y, D_V^\perp Z\rangle$, where $Y, Z \in \mathfrak{X}(M)^\perp$.

Deeper results about semi-Riemannian submanifolds often require information about the rate of change of the shape tensor. To avoid tensor complications we use a standard definition modeled on the product rules in Chapter 2.

32. Definition. Let II be the shape tensor of $M \subset \bar{M}$. If $V, X, Y \in \mathfrak{X}(M)$, let

$$(\nabla_V II)(X, Y) = D_V^\perp(II(X, Y)) - II(D_V X, Y) - II(X, D_V Y).$$

It is easily verified that, like II itself, $\nabla_V II$ is a symmetric $\mathfrak{F}(M)$-bilinear function $\mathfrak{X}(M) \times \mathfrak{X}(M) \to \mathfrak{X}(M)^\perp$.

The Gauss equation describes tan $\bar{R}_{VW}X$, where V, W, X are all tangent to M, in terms of the shape tensor. The following analogue for nor $\bar{R}_{VW}X$ is called the *Codazzi equation.*

33. Proposition. Let M be a semi-Riemannian submanifold of $\bar{M}$. If $V, W, X \in \mathfrak{X}(M)$ then

$$\text{nor } \bar{R}_{VW}X = -(\nabla_V II)(W, X) + (\nabla_W II)(V, X).$$

Proof. As usual we can assume $[V, W] = 0$. Then nor $\bar{R}_{VW}X = -(VW) + (WV)$, where $(VW) = \text{nor } \bar{D}_V \bar{D}_W X$. Now

$$(VW) = \text{nor } \bar{D}_V(D_W X) + \text{nor } \bar{D}_V(II(W, X)).$$

The first term here is just $II(V, D_W X)$. The second is, by definition, $D_V^\perp(II(W, X))$; hence reading from the definition of $\nabla_V II$ gives

$$(VW) = II(V, D_W X) + (\nabla_V II)(W, X) + II(D_V W, X) + II(W, D_V X).$$

Since $D_V W - D_W V = [V, W] = 0$, cancellations in $-(VW) + (WV)$ give the required result. ∎

A vector field $Z \in \mathfrak{X}(M)^\perp$ is *normal parallel* provided $D_V^\perp Z = 0$ for all $V \in \mathfrak{X}(M)$.

34. Corollary. Let $\bar{M}$ have constant curvature.

(1) The Codazzi equation for $M \subset \bar{M}$ becomes

$$(\nabla_V II)(W, X) = (\nabla_W II)(V, X) \qquad \text{for all} \quad V, W, X \in \mathfrak{X}(M).$$

(2) If M is a hypersurface in $\bar{M}$ with shape operator S, then

$$(D_V S)(W) = (D_W S)(V) \qquad \text{for all} \quad V, W \in \mathfrak{X}(M).$$

Proof. (1) By the formula in Corollary 3.43, $\bar{R}_{VW}X$ is tangent to M, hence nor $\bar{R}_{VW}X = 0$.

(2) Suppose S derives from the unit normal U. Note that U is normal parallel, that is, $D_X^\perp U = 0$ for all $X \in \mathfrak{X}(M)$. In fact, $\langle U, U\rangle = \text{const} \Rightarrow \bar{D}_X U \perp U \Rightarrow \bar{D}_X U$ tangent to $M \Rightarrow D_X^\perp U = 0$. As usual in a tensor proof we can suppose that the vector fields $V, W, X \in \mathfrak{X}(M)$ have all M covariant derivatives zero at a point p. There

$$\begin{aligned}\langle (D_V S)(W), X\rangle &= \langle D_V(SW), X\rangle = V\langle SW, X\rangle = V\langle II(W, X), U\rangle \\ &= \langle D_V(II(W, X)), U\rangle = \langle (\nabla_V II)(W, X), U\rangle.\end{aligned}$$

The result then follows from (1). ∎

TOTALLY UMBILIC HYPERSURFACES

As an application of the methods of this chapter we will find all connected totally umbilic hypersurfaces of semi-Euclidean space $\mathbf{R}^n_\nu$. We have already seen that every connected totally geodesic hypersurface of $\mathbf{R}^n_\nu$ is an open set in a nondegenerate hyperplane. Thus it remains to determine the totally umbilic hypersurfaces that are not totally geodesic. This class includes, of course, all hyperquadrics. The essential step is the following simple application of the Codazzi equation.

35. Lemma. Let $M \subset \bar{M}$ be a connected semi-Riemannian hypersurface of sign ε and dimension ≥ 2. If M is totally umbilic and $\bar{M}$ has constant curvature $\bar{C}$, then the normal curvature k is constant (mod sign) and M has constant curvature $\bar{C} + \varepsilon k^2$.

Proof. The curvature assertion is clear from the Gauss equation once k is known to be constant. Since M is connected it suffices to show that $v(k) = 0$ for every tangent vector to M. If $v \in T_p(M)$ choose $V, W \in \mathfrak{X}(M)$ so that $V_p = v$ and W_p is independent of v_p. Since $S = kI$,

$$(D_V S)(W) = D_V(SW) - S(D_V W) = D_V(kW) - kD_V W = Vk\, W.$$

By the choice of V and W it follows from Corollary 34 that $v(k) = (Vk)(p) = 0$. ∎

For example, a flat totally umbilic hypersurface in $\mathbf{R}^n_\nu$ is totally geodesic.

A central hyperquadric in $\mathbf{R}^n_\nu$, given by $\langle p, p \rangle = c \neq 0$, is carried by the translation $p \to p - x_0$ to a *hyperquadric centered at* x_0, given by $\langle p - x_0, p - x_0 \rangle = c$. Evidently the translation is a pair isometry; hence, in particular, all hyperquadrics are totally umbilic.

A semi-Riemannian hypersurface of a two-dimensional manifold is trivially totally umbilic, so we deal with $\mathbf{R}^n_\nu$ only for $n \geq 3$.

36. Proposition. If M is a connected semi-Riemannian hypersurface of $\mathbf{R}^n_\nu$, $n \geq 3$, that is totally umbilic but not totally geodesic, then M is an open set in a hyperquadric. Hence if M is also complete, M is a component of a hyperquadric.

(This reference to components is operative only in the two nonconnected cases H^n_0 and S^n_n.)

Proof. If U is a locally defined unit normal on M then $S = kI$, where by the preceding lemma, k is constant. The normal vector field $U/k = (-U)/-k$ is well defined on all of M.

Identify the tangent spaces of $\boldsymbol{R}^n_\nu$ with $\boldsymbol{R}^n_\nu$ itself by canonical isomorphism, and define a mapping $\phi: M \to \boldsymbol{R}^n$ by

$$\phi(p) = p + U_p/k.$$

The goal is to show that ϕ is a constant map. Since M is connected it suffices to show that each differential map of ϕ is zero.

With the identifications above we assert that for any tangent vector v to M,

$$d\phi(v) = v + (\bar{D}_v U)/k.$$

It suffices to show that both sides have the same components relative to the natural coordinates $u^1, \ldots, u^n$ of $\boldsymbol{R}^n_\nu$. If $U = \sum f^i \partial_i$, then

$$u^i(\phi(p)) = u^i(p) + f^i(p)/k \qquad \text{for all} \quad p,$$

hence

$$(d\phi(v))u^i = v(u^i \circ \phi) = vu^i + vf^i/k.$$

Since $\bar{D}_v U = \sum vf^i \partial_i$, the assertion is true.

Now $\bar{D}_v U = -S(v) = -kv$, hence $d\phi(v) = v + (-kv)/k = 0$. Thus ϕ is a constant map, say to the point x_0 of $\boldsymbol{R}^n$. But then for every point p of M, $p - x_0 = -U_p/k$, hence $\langle p - x_0, p - x_0 \rangle = \varepsilon/k^2$. Thus M is contained in this hyperquadric Q. Since M is connected it is contained in a single component C of Q. By Exercise 1.8, M is an open submanifold of C. If M is complete, then $M = C$, by Exercise 3.7. ∎

Thus *the complete, connected, totally umbilic hypersurfaces of* $\boldsymbol{R}^n_\nu$ $(n \geq 3)$ *exactly the nondegenerate hyperplanes of* $\boldsymbol{R}^n_\nu$ *and the components of hyperquadrics.* On the basis of this extrinsic information it turns out to be easy to characterize intrinsically the semi-Riemannian hypersurfaces of $\boldsymbol{R}^n_\nu$ that have constant nonzero curvature.

37. Proposition. If M is a connected semi-Riemannian hypersurface of $\boldsymbol{R}^n_\nu$ $(n \geq 4)$ with constant curvature $C \neq 0$, then M is an open set in a hyperquadric. Hence if M is complete, M is a component of a hyperquadric.

This follows immediately from Proposition 36 and the following algebraic result.

38. Lemma. Let M be a semi-Riemannian hypersurface of $\bar{M}$. If M and $\bar{M}$ have constant curvatures $C \neq \bar{C}$ and dim $M \geq 3$, then M is totally umbilic.

Proof. Let $\Delta = \varepsilon(C - \bar{C}) \neq 0$, where ε is the sign of M. By the Gauss equation

(a) $\langle Sv, v\rangle\langle Sw, w\rangle - \langle Sv, w\rangle^2 = \Delta[\langle v, v\rangle\langle w, w\rangle - \langle v, w\rangle^2]$,

whenever v and w span a nondegenerate tangent plane to M. By Lemma 3.40 the relation is valid for arbitrary v, w.

At each point $p \in M$ the shape operator S on $T_p(M)$ is invertible. In fact, if $v \neq 0$ then by an assertion in the proof of Lemma 3.40, there is a vector $w \in T_p(M)$ such that $\langle v, v\rangle\langle w, w\rangle - \langle v, w\rangle^2 \neq 0$. Hence by (a), $Sv \neq 0$.

We assert that for all v, w, x, y in $T_p(M)$,

(b) $\langle Sv, x\rangle\langle Sw, y\rangle - \langle Sv, y\rangle\langle Sw, x\rangle = \Delta[\langle v, x\rangle\langle w, y\rangle - \langle v, y\rangle\langle w, x\rangle]$.

To prove this, note that each side defines a curvaturelike function. Thus Corollary 3.42 shows that (a) implies (b). Since $\dim T_p(M) \geq 3$ there is a vector $y \neq 0$ orthogonal to both v and Sv. Thus (b) becomes

$$\langle Sv, x\rangle\langle Sw, y\rangle = \Delta\langle v, x\rangle\langle w, y\rangle \qquad \text{for all} \quad v, w, x.$$

Because S is invertible, the image of S is not contained in $y^\perp$. Hence there is a vector w such that $\langle Sw, y\rangle \neq 0$. Thus for some constant k, we have $\langle Sv, x\rangle = k\langle v, x\rangle$ for all v, x. Hence S is scalar. ∎

In Proposition 37, if $C = \bar{C}$ the shape of the submanifold M can not always be specified.

THE NORMAL CONNECTION

For semi-Riemannian submanifolds of codimension greater than one the normal connection $D^\perp$ becomes more important. Just as the shape tensor of $M \subset \bar{M}$ measures the difference between D and $\bar{D}$ it also measures the difference between $D^\perp$ and $\bar{D}$—but in a different formulation, as follows.

39. Remarks. The Tensor $\widetilde{II}$ of $M \subset \bar{M}$. (1) If $M \subset \bar{M}$ then for $V \in \mathfrak{X}(M)$ and $Z \in \mathfrak{X}(M)^\perp$ define $\widetilde{II}(V, Z) = \tan \bar{D}_V Z$. Thus

$$\bar{D}_V Z = \underbrace{\widetilde{II}(V, Z)}_{\text{tangent to } M} + \underbrace{D_V^\perp W}_{\text{normal to } M}.$$

It is easy to check that the function $\widetilde{II}\colon \mathfrak{X}(M) \times \mathfrak{X}(M)^\perp \to \mathfrak{X}(M)$ is $\mathfrak{F}(M)$-bilinear. Hence at each point $p \in M$, $\widetilde{II}$ gives an $\mathbf{R}$-bilinear map $T_p(M) \times T_p(M)^\perp \to T_p(M)$.

(2) The tensor $\widetilde{II}$ contains no new information, since

$$\langle \widetilde{II}(V, Z), W\rangle = -\langle II(V, W), Z\rangle)$$

for all $V, W \in \mathfrak{X}(M)$ and $Z \in \mathfrak{X}(M)^\perp$. This identity is derived by differentiating $\langle Z, W \rangle = 0$ to get $\langle \bar{D}_V Z, W \rangle = -\langle Z, \bar{D}_V W \rangle$.

(3) When a particular $Z \in \mathfrak{X}(M)^\perp$ is important, the notation $S_Z V = -\tilde{II}(V, Z)$ is sometimes convenient. Since

$$\langle S_Z(V), W \rangle = \langle II(V, W), Z \rangle,$$

S_Z gives at each point $p \in M$ a self-adjoint linear operator on $T_p(M)$. (Compare [KN] where II is α, and S_Z is A_Z.) In the case of a unit normal to a hypersurface, S_Z is consistent with Definition 18 so we call it a *shape operator*.

The normal connection $D^\perp$ is adapted as follows to normal vector fields on curves in $M \subset \bar{M}$. If Y is a vector field on α always normal to M, then its *normal covariant derivative* $Y' = D^\perp Y/ds$ is defined to be the normal component of its $\bar{M}$ covariant derivative $\dot{Y} = \bar{D}Y/ds$. The properties in Proposition 3.18 imply at once the analogous properties for Y'. Corresponding to Proposition 8 is

$$\dot{Y} = \underset{\substack{\text{tangent}\\ \text{to } M}}{\tilde{II}(\alpha', Y)} + \underset{\substack{\text{normal}\\ \text{to } M}}{Y'}.$$

If $Y' = 0$, Y is said to be *normal parallel.*

40. Lemma. Let α be a curve in $M \subset \bar{M}$. If y is a vector normal to M at $\alpha(a)$ there is a unique normal parallel vector field Y on α such that $Y(\alpha) = y$.

Proof. Using an adapted coordinate systems and the projection *nor* it is easy to find, on subintervals of α, normal vector fields $E_1, \ldots, E_k$ that at each point give a basis for the normal space. With somewhat more work these can be patched together smoothly to give such fields defined on the entire curve α.

Write $Y = \sum f_i E_i$, $y = \sum c_i E_i|_{\alpha(a)}$, and $E_i' = \sum h_{ji} E_j$ $(1 \le i \le k)$. Then $Y' = \sum (f_i' + \sum h_{ij} f_j) E_i$. Let $f_1, \ldots, f_k$ be the unique solutions of the linear differential equations $f_i' + \sum h_{ij} f_j = 0$ such that $f_i(a) = c_i$. The solutions are defined on the entire interval I, hence $Y = \sum f_i E_i$ is the required vector field. ∎

In the notation of the lemma, $Y(b)$ is the *normal parallel translate* of y, and a proof like that for Lemma 3.20 shows that *normal parallel translation* $y \to Y(b)$ is a linear isometry $P\colon T_{\alpha(a)}(M)^\perp \to T_{\alpha(b)}(M)^\perp$. Hence normal parallel translation of a normal frame gives a normal parallel frame field on α.

A proof like that of Lemma 7 shows that the normal connection is preserved by pair isometries, hence $D^\perp$ and its subsidiary notions above belong to the extrinsic geometry of $M \subset \bar{M}$.

A CONGRUENCE THEOREM

As in elementary Euclidean geometry two objects in a semi-Riemannian manifold $\overline{M}$ are *congruent* if there is an isometry of $\overline{M}$ carrying one to the other. In particular, submanifolds M_1 and M_2 of $\overline{M}$ are *congruent* provided there is an isometry ψ of $\overline{M}$ such that $\psi | M_1$ is an isometry from M_1 to M_2. Then by definition M_1 and M_2 have the same intrinsic and extrinsic geometry. Evidently congruence is a useful notion only when $\overline{M}$ has many isometries. Taking $\overline{M} = \boldsymbol{R}^n_\nu$ we prove that *isometric submanifolds are congruent if and only if they have "the same shape tensor."*

41. Theorem. Let $\phi \colon M_1 \to M_2$ be an isometry of connected semi-Riemannian submanifolds of $\boldsymbol{R}^n_\nu$. There is an isometry $\tilde{\phi}$ of $\boldsymbol{R}^n_\nu$ such that $\tilde{\phi}|M_1 = \phi$ if and only if at a point $o \in M_1$ there is a linear isometry $F_o \colon T_o(M_1)^\perp \to T_{\phi o}(M_2)^\perp$ with this property: If α is any curve in M_1 starting at o, then for each s the linear isometry

$$F_{\alpha(s)} = P_{\phi\alpha} \circ F_o \circ P_\alpha^{-1} \colon T_{\alpha(s)}(M_1)^\perp \to T_{\phi\alpha(s)}(M_2)^\perp$$

(P normal parallel translation) preserves shape tensors, that is,

$$F_{\alpha(s)}(II_1(v, w)) = II_2(d\phi v, d\phi w) \qquad \text{for all} \quad v, w \in T_{\alpha(s)}(M_1).$$

Proof. The condition on shape tensors is necessary since we have seen that normal parallel translation and shape tensors belong to extrinsic geometry. To prove sufficiency, let α be a curve as above.

(a) If Z is a vector field on α normal to M_1, then $(F_\alpha Z)(s) = F_{\alpha(s)}Z(s)$ defines a vector field on $\phi \circ \alpha$ normal to M_2. The definition of $F_{\alpha(s)}$ shows that if Z is normal parallel then so is $F_\alpha Z$. By expressing an arbitrary Z in terms of a normal parallel frame field on α it follows that $(F_\alpha Z)' = F_\alpha(Z')$.

(b) Since F_α preserves II, it also preserves $\widetilde{II}$, that is,

$$d\phi(\widetilde{II}_1(V, Z)) = \widetilde{II}_2(d\phi V, F_\alpha Z),$$

where V and Z are vector fields on α tangent and normal, respectively, to M. To check this take the scalar product with $d\phi(W)$ of both sides of the II equation in the theorem and use Remark 39(2).

(c) For each s, $d\phi + F_{\alpha(s)}$ is a well-defined linear isometry from $T_{\alpha(s)}(\boldsymbol{R}^n_\nu)$ to $T_{\phi\alpha(s)}(\boldsymbol{R}^n_\nu)$. If X is an $\boldsymbol{R}^n_\nu$ vector field on α, $(d\phi + F_\alpha)X$ is an $\boldsymbol{R}^n_\nu$ vector field on $\phi \circ \alpha$. We assert that *if X is $\boldsymbol{R}^n_\nu$ parallel, so is $(d\phi + F_\alpha)X$*. Let $\dot{X}$ denote the $\boldsymbol{R}^n_\nu$ covariant derivative along α; then $0 = \dot{X} = (\tan X)^{\cdot} + (\text{nor}\, X)^{\cdot}$ gives

$$\begin{cases} (\tan X)' + \widetilde{II}_1(\alpha', \text{nor}\, X) = 0, \\ II_1(\alpha', \tan X) + (\text{nor}\, X)' = 0. \end{cases}$$

Now

$$((d\phi + F_\alpha)X)^{\cdot} = (d\phi \tan X)^{\cdot} + (F_\alpha \operatorname{nor} X)^{\cdot}$$
$$= (d\phi \tan X)' + II_2(d\phi\alpha', d\phi \tan X)$$
$$+ \tilde{II}_2(d\phi\alpha', F_\alpha \operatorname{nor} X) + (F_\alpha \operatorname{nor} X)'.$$

By (a), (b), and properties of isometries this becomes

$$d\phi[(\tan X)' + \tilde{II}_1(\alpha', \operatorname{nor} X)] + F_\alpha[II_1(\alpha', \tan X) + (\operatorname{nor} X)'].$$

The equations for $\dot{X} = 0$ show that this expression is zero.

(d) *If curves* $\alpha, \beta: I \to \boldsymbol{R}^n_\nu$ *have* $\alpha(0) = \beta(0)$, $\alpha'(0) = \beta'(0)$, *and for all* s, $\alpha''(s) \| \beta''(s)$ (distant parallelism), *then* $\alpha = \beta$. In fact, the parallelism means that the natural coordinates of α and β satisfy $d^2(u^i \circ \alpha)/ds^2 = d^2(u^i \circ \beta)/ds^2$. Hence $u^i \circ \alpha - u^i \circ \beta = a_i s + b_i$ for all i. But the initial conditions imply $a_i = b_i = 0$.

(e) Since a pair map preserves the intrinsic and extrinsic geometry of submanifolds we can assume without loss of generality that the point $o \in M_1$ in the theorem and its image $\phi(o) \in M_2$ are both at the origin of $\boldsymbol{R}^n_\nu$. In fact it suffices to translate each manifold, with corresponding obvious changes in ϕ and F_o.

(f) Let $\tilde{\phi}$ be the linear isometry of $\boldsymbol{R}^n_\nu$ canonically corresponding to $d\phi_o + F_o: T_o(\boldsymbol{R}^n_\nu) \to T_o(\boldsymbol{R}^n_\nu)$; thus $d\tilde{\phi} = d\phi_o + F_o$. We will show that $\tilde{\phi} | M_1 = \phi$. If $p \in M_1$ let α be a curve in M_1 from o to p. It suffices to prove $\phi \circ \alpha = \tilde{\phi} \circ \alpha$. By (e), $\phi(o) = o = \tilde{\phi}(o)$, and since $d\tilde{\phi} | T_o(M) = d\phi$ it follows that the two curves have the same initial position and velocity. As usual $\ddot{\alpha}$ denotes $\boldsymbol{R}^n_\nu$ acceleration. Then

$$(\phi \circ \alpha)^{\cdot\cdot} = (\phi \circ \alpha)'' + II_2(d\phi\alpha', d\phi\alpha')$$
$$= d\phi(\alpha'') + F_\alpha II_1(\alpha', \alpha') = (d\phi + F_\alpha)(\ddot{\alpha}).$$

Let A_s be the vector in $T_o(\boldsymbol{R}^n)$ that is distantly parallel to $\ddot{\alpha}(s)$. By (c),

$$(d\phi + F_{\alpha(s)})(\ddot{\alpha}(s)) \| (d\phi_o + F_o)(A_s).$$

The latter vector is just $d\tilde{\phi}(A_s)$. Since $\tilde{\phi}$ is a linear isometry, $d\tilde{\phi}(A_s) \| d\tilde{\phi}(\ddot{\alpha}(s))$ where the second vector is $(\tilde{\phi} \circ \alpha)^{\cdot\cdot}(s)$. Consequently $(\phi \circ \alpha)^{\cdot\cdot}(s) \| (\tilde{\phi} \circ \alpha)^{\cdot\cdot}(s)$ for all s. Hence (d) shows that $\phi \circ \alpha = \tilde{\phi} \circ \alpha$. ∎

ISOMETRIC IMMERSIONS

The range of applicability of the results of this chapter can be enlarged in the following way.

42. Definition. Let M and $\overline{M}$ be semi-Riemannian manifolds with metric tensors g and $\bar{g}$. An *isometric immersion* of M into $\overline{M}$ is a smooth immersion such that $\phi^*(\bar{g}) = g$. An *isometric imbedding* is a one-to-one isometric immersion.

The latter notion includes that of semi-Riemannian submanifold $M \subset \bar{M}$ in the sense that the inclusion map j of M into $\bar{M}$ is an isometric imbedding. The machinery for the study of semi-Riemannian submanifolds is readily adapted to the more general case, mostly by everywhere replacing j by ϕ. In terms of the notion of vector field on a mapping, $\bar{\mathfrak{X}}(M)$ is just $\mathfrak{X}(j)$. Thus $\bar{\mathfrak{X}}(M)$ is replaced by $\mathfrak{X}(\phi)$, and similarly $\mathfrak{X}^{\perp}(M)$ is replaced by

$$\mathfrak{X}^{\perp}(\phi) = \{Z \in \mathfrak{X}(\phi) : Z_p \perp d\phi(T_pM) \quad \text{for all} \quad p \in M\}.$$

We could continue as before to get the induced connection and shape tensor, and thereby generalize the basic results of this chapter. While this continuation has technical advantages, all local results and thereby many global results can be obtained automatically from the submanifold case by means of the observation that *locally an isometric immersion is essentially a semi-Riemannian submanifold*:

43. Lemma. Let $\phi: M \to \bar{M}$ be an isometric imbedding. Then

(1) Each point $p \in M$ has a neighborhood $\mathcal{U}$ such that $\phi|\mathcal{U}$ is an imbedding, and

(2) If $\phi(\mathcal{U})$ is assigned the metric tensor such that the induced map $\mathcal{U} \to \phi(\mathcal{U})$ is an isometry, then $\phi(\mathcal{U})$ is a semi-Riemannian manifold of $\bar{M}$.

Proof. (1) As noted in Chapter 1 this is an immediate consequence of Lemma 1.33. (2) If $v, w \in T_p(M), p \in \mathcal{U}$, then $\langle d\phi(v), d\phi(w)\rangle$ has the same value, $\langle v, w\rangle$, whether computed relative to $\phi(\mathcal{U})$ or $\bar{M}$. ∎

TWO-PARAMETER MAPS

Let $\mathcal{D}$ be an open subset of the plane $\mathbf{R}^2$ satisfying this interval condition: horizontal or vertical lines intersect $\mathcal{D}$ in intervals (if at all). A *two-parameter map* is a smooth map $\mathbf{x}: \mathcal{D} \to M$. Thus $\mathbf{x}$ is composed of two interwoven families of *parameter curves*:

The *u-parameter curve* $v = v_0$ of $\mathbf{x}$ is $u \to \mathbf{x}(u, v_0)$.
The *v-parameter curve* $u = u_0$ of $\mathbf{x}$ is $v \to \mathbf{x}(u_0, v)$. The *partial velocities*

$$\mathbf{x}_u = d\mathbf{x}(\partial_u), \qquad \mathbf{x}_v = d\mathbf{x}(\partial_v)$$

are vector fields on $\mathbf{x}$ (Definition 1.47). Evidently $\mathbf{x}_u(u_0, v_0)$ is the velocity vector at u_0 of the u-parameter curve $v = v_0$, and symmetrically for $\mathbf{x}_v(u_0, v_0)$.

If $\mathbf{x}$ lies in the domain of a coordinate system $x^1, \ldots, x^n$, then its *coordinate functions* $\mathbf{x}^i = x^i \circ \mathbf{x}$ $(1 \le i \le n)$ are real-valued functions on $\mathscr{D}$, and

$$\mathbf{x}_u = \sum \frac{\partial \mathbf{x}^i}{\partial u} \partial_i, \qquad \mathbf{x}_v = \sum \frac{\partial \mathbf{x}^i}{\partial v} \partial_i.$$

So far M could be a smooth manifold; now suppose it is semi-Riemannian. If Z is a smooth vector field on $\mathbf{x}$, its *partial covariant derivatives* are

$Z_u = DZ/\partial u$, the covariant derivative of Z along u-parameter curves, and
$Z_v = DZ/\partial v$, the covariant derivative of Z along v-parameter curves.

Explicitly, $Z_u(u_0, v_0)$ is the covariant derivative at u_0 of the vector field $u \to Z(u, v_0)$ on the curve $u \to \mathbf{x}(u, v_0)$.

In terms of coordinates, $Z = \sum Z^i \partial_i$, where each $Z^i = Z(\mathbf{x}^i)$ is a real-valued function on $\mathscr{D}$. Then by the formula following Proposition 3.18,

$$Z_u = \sum_k \left\{ \frac{\partial Z^k}{\partial u} + \sum_{i,j} \Gamma^k_{ij} Z^i \frac{\partial \mathbf{x}^j}{\partial u} \right\} \partial_k.$$

In the special case $Z = \mathbf{x}_u$ the derivative $Z_u = \mathbf{x}_{uu}$ gives the accelerations of u-parameter curves, while $\mathbf{x}_{vv}$ gives v-parameter accelerations.

44. Proposition. (1) If $\mathbf{x}$ is a two-parameter map into a semi-Riemannian manifold M, then $\mathbf{x}_{uv} = \mathbf{x}_{vu}$. (2) If Z is a vector field on $\mathbf{x}$, then

$$Z_{uv} - Z_{vu} = R(\mathbf{x}_u, \mathbf{x}_v)Z,$$

where R is the Riemannian curvature tensor of M.

Proof. (1) With coordinate notation as above,

$$\mathbf{x}_{uv} = \sum_k \left\{ \frac{\partial^2 \mathbf{x}^k}{\partial v\, \partial u} + \sum_{i,j} \Gamma^k_{ij} \frac{\partial \mathbf{x}^i}{\partial u} \frac{\partial \mathbf{x}^j}{\partial v} \right\} \partial_k.$$

This formula is symmetric in u and v, since Γ_{ij} is symmetric in i and j.

(2) A coordinate computation of $Z_{uv} - Z_{vu}$ produces curvature as in Lemma 3.38. ∎

Here (1) expresses for a two-parameter map the axiom (D4) (see Theorem 3.11) on the Levi-Civita connection, while in (2) curvature arises as usual from the failure of commutativity of covariant differentiation.

Exercises

1. Let $x^1, \ldots, x^{n+k}$ be an adapted coordinate system for $M \subset \bar{M}$. Show (a) the mean curvature vector field H is $(1/n)\sum g^{ij} II(\partial_i, \partial_j)$ for $1 \le i, j \le n$. (b) If $\partial_{n+1}, \ldots, \partial_{n+k}$ are normal to M, then

$$II(\partial_i, \partial_j) = \sum_{r>n} \bar{\Gamma}^r_{ij} \partial_r.$$

2. Let Π be a nondegenerate tangent plane to M at p. If P is a small enough neighborhood of 0 in Π, prove that $\exp_p(P)$ is a semi-Riemannian submanifold of M whose Gaussian curvature at p is $K(\Pi)$, where K is the sectional curvature of M.

3. Let S derive from the unit normal $U = \text{grad } f/|\text{grad } f|$ on a semi-Riemannian hypersurface $M = f^{-1}(c)$ of $\bar{M}$. (a) Show that

$$\langle Sv, w\rangle = -H^f(v, w)/|\text{grad } f|$$

for v, w tangent to M, where H^f is the Hessian of f. (b) Find S at $p = (1, 0, 0)$ for the hyperboloid $x^2 + y^2 = 1 + z^2$ in $\boldsymbol{R}^3$ first by using (a) then by using $S(v) = -\bar{D}_v U$.

4. Let M^n be a spacelike hypersurface of sign ε in $\bar{M}$ (Riemannian or Lorentz). Let S derive from unit normal U, and let $k_1, \ldots, k_n$ be the eigenvalues of S. Prove: (a) $\langle H, U\rangle = (1/n)\sum k_i$. (b) $K(\Pi) = \varepsilon k_i k_j + \bar{K}(\Pi)$, where Π is spanned by the eigenvectors of k_i and k_j. (c) $p \in M$ is an umbilic $\Leftrightarrow k_1 = \cdots = k_n$ at p.

5. (a) Show that the Ricci curvature of a hypersurface $M \subset \boldsymbol{R}^{n+1}_v$ is given by $\text{Ric}(V, W) = \varepsilon[\langle SV, W\rangle \text{ trace } S - \langle SV, SW\rangle]$.

6. Generalize Lemma 35 as follows: For arbitrary codimension, if M is connected, has dim $M \geq 2$, and is totally umbilic in a manifold of constant curvature $\bar{C}$, then (a) the normal curvature vector z is normal parallel, and (b) M has constant curvature $\bar{C} + \langle z, z\rangle$.

7. Classical surface theory [O1]. Let the two-parameter map $\mathbf{x}\colon \mathcal{D} \to \boldsymbol{R}^3$ be an immersion. Use the dot product on $\boldsymbol{R}^3$; $\mathbf{x}$ is an isometric immersion relative to the pulled-back metric on $\mathcal{D}$. (a) Check that

$$U = \mathbf{x}_u \times \mathbf{x}_v/|\mathbf{x}_u \times \mathbf{x}_v|$$

is a unit normal vector field on $\mathbf{x}$. Define

$$E = \mathbf{x}_u \cdot \mathbf{x}_u, \qquad F = \mathbf{x}_u \cdot \mathbf{x}_v, \qquad G = \mathbf{x}_v \cdot \mathbf{x}_v,$$

$$L = \mathbf{x}_{uu} \cdot U, \qquad M = \mathbf{x}_{uv} \cdot U, \qquad N = \mathbf{x}_{vv} \cdot U.$$

Prove: (b) $L = S\mathbf{x}_u \cdot \mathbf{x}_u$, $M = S\mathbf{x}_u \cdot \mathbf{x}_v$, $N = S\mathbf{x}_v \cdot \mathbf{x}_v$. (c) The Gaussian curvature of $\mathcal{D}$ is $K = (LN - M^2)/(EG - F^2) = \det S$. (d) If H is the mean curvature vector field of $\mathbf{x}$, then $H = hU$, where $h = \frac{1}{2}$ trace $S = (GL + EN - 2FM)/2(EG - F^2)$. *Note*: if $\mathbf{x}$ parametrizes a surface $\Sigma \subset \boldsymbol{R}^3$, these results transfer to Σ since $\mathbf{x}$ is then a local isometry $\mathcal{D} \to M$. (See also Exercise 9.22.)

8. Total shape tensor. For $M \subset \bar{M}$ define $T\colon \mathfrak{X}(M) \times \bar{\mathfrak{X}}(M) \to \bar{\mathfrak{X}}(M)$ by $T_V X = \tan \bar{D}_V(\text{nor } X) + \text{nor } \bar{D}_V(\tan X)$. Prove: (a) T is $\mathfrak{F}(M)$-bilinear;

(b) T_V is skew-adjoint; (c) T_V reverses $\mathfrak{X}(M)$ and $\mathfrak{X}(M)^\perp$; (d) $T_V W = II(V, W)$ and $T_V Z = \tilde{II}(V, Z)$ where $V, W \in \mathfrak{X}(M)$ and $Z \in \mathfrak{X}(M)^\perp$; (e) T is completely determined by II.

9. Let M be merely a smooth submanifold of a semi-Riemannian manifold $\bar{M}$. If $v \in T_p(\bar{M})$ let $\{v\}$ be the element of the quotient vector space $T_p(\bar{M})/T_p(M)$ containing v. If $V, W \in \mathfrak{X}(M)$ define $II(V, W) = \{\bar{D}_V W\}$. Prove: (a) II is symmetric and $\mathfrak{F}(M)$-bilinear. (b) The following are equivalent: (i) M is totally geodesic (that is, $II = 0$). (ii) if γ is an $\bar{M}$ geodesic such that $\gamma'(0)$ is tangent to M then γ remains initially in M. (iii) If a tangent vector to M is $\bar{M}$ parallel translated along a curve in M it remains tangent to M. (Hint: To prove (i) $\Rightarrow$ (iii) write the coordinate expression for $\bar{M}$ parallelism in terms of an adapted coordinate system.) (c) This notion of totally geodesic agrees with the previous notion if M is semi-Riemannian.

10. (a) If $\phi\colon M \to M$ is a homothety then each component of its fixed point set $\{p \in M : \phi(p) = p\}$ is a totally geodesic submanifold of M in the sense of the preceding exercise. (b) If $\psi\colon M \to N$ is a homothety, then its graph

$$\{(p, \psi p) : p \in M\}$$

is a totally geodesic submanifold of $M \times N$, and is semi-Riemannian if ψ is not an anti-isometry.

11. If $M \subset \bar{M}$ the function $R^\perp\colon \mathfrak{X}(M) \times \mathfrak{X}(M) \times \mathfrak{X}(M)^\perp \to \mathfrak{X}(M)^\perp$ given by

$$R^\perp_{VW} X = D^\perp_{[V,W]} X - [D^\perp_V, D^\perp_W] X$$

is called the *normal curvature tensor* of $M \subset \bar{M}$. Check that $R^\perp$ is $\mathfrak{F}(M)$-multilinear, and prove the *Ricci equation*:

$$\begin{aligned}\langle R^\perp_{VW} X, Y\rangle = \langle \bar{R}_{VW} X, Y\rangle &+ \langle \tilde{II}(V, X), \tilde{II}(W, Y)\rangle \\ &- \langle \tilde{II}(V, Y), \tilde{II}(W, X)\rangle\end{aligned}$$

where $X, Y \in \mathfrak{X}(M)^\perp$. (Since $\bar{R}_{VW}$ is skew-adjoint, the Gauss, Codazzi, and Ricci equations cover all four tan/nor choices in $\langle \bar{R}_{VW}\cdot, \cdot\rangle$.)

12. The sectional curvature K of M is constant at p if and only if every unit vector in $T_p(M)$ is normal to a hypersurface totally geodesic at p. (Hint: Use Codazzi to get $R_{xy}x = \langle x, x\rangle K(x, y)y$ for nonnull $x \perp y$.)

13. Show that for $0 < \nu < n$ there are no compact semi-Riemannian hypersurfaces in $\boldsymbol{R}^n_\nu$. (Hint: Deduce a contradiction to constancy of index.)

5 RIEMANNIAN AND LORENTZ GEOMETRY

After some general preliminaries we consider special features of the two most important geometries determined by the *index*: Riemannian geometry, $\nu = 0$, and Lorentz geometry, $\nu = 1$.

The metric tensor of a Riemannian manifold M makes each tangent space an inner product space, linearly isometric to Euclidean space $\boldsymbol{R}^n$. Then the notion of arc length leads to a notion of distance between points of M that generalizes the usual Euclidean distance in $\boldsymbol{R}^n$. Riemannian distance makes every Riemannian manifold a metric space and simplifies the study of its geometry.

Each tangent space to a Lorentz manifold is linearly isometric to Minkowski space $\boldsymbol{R}^n_1$, and Lorentz geometry begins with the study of the causal character of vectors in such a space.

The emphasis in this chapter is on local geometry and on geodesics. In particular, a useful analogy appears between timelike geodesics in a Lorentz manifold and arbitrary geodesics in a Riemannian manifold.

THE GAUSS LEMMA

The key to the local geometry of a semi-Riemannian manifold near a point $o \in M$ is the comparison with the semi-Euclidean space $T_o(M) \approx \boldsymbol{R}^n_\nu$ afforded by the exponential map. We have seen that $\exp_o$ carries rays $t \to tx$ in $T_o(M)$ to radial geodesics γ_x. The following result, called the *Gauss lemma*, implies in particular that orthogonality to radial directions is also preserved.

1. Lemma. Let $o \in M$ and $0 \neq x \in T_o(M)$. If $v_x, w_x \in T_x(T_o M)$ with v_x radial, then

$$\langle d\exp_o(v_x), d\exp_o(w_x)\rangle = \langle v_x, w_x\rangle.$$

Proof. That v_x is radial means that v is a scalar multiple of x. Hence we can suppose that $v = x$ (and we elect to replace x by v).

Consider the two-parameter map $\tilde{\mathbf{x}}(t, s) = t(v + sw)$ in $T_o(M)$, and its exponential image in M:

$$\mathbf{x}(t, s) = \exp_o(t(v + sw)).$$

Now $\tilde{\mathbf{x}}_t(1, 0) = v_v$ and $\tilde{\mathbf{x}}_s(1, 0) = w_v$, hence

$$\mathbf{x}_t(1, 0) = d\exp_o(v_v), \qquad \mathbf{x}_s(1, 0) = d\exp_o(w_v).$$

So we must show that $\langle \mathbf{x}_t(1, 0), \mathbf{x}_s(1, 0)\rangle = \langle v, w\rangle$.

The longitudinal curve $t \to \mathbf{x}(t, s)$ is a geodesic with initial velocity $v + sw$. Hence $\mathbf{x}_{tt} = 0$ and $\langle \mathbf{x}_t, \mathbf{x}_t\rangle = \langle v + sw, v + sw\rangle$. Proposition 4.44(1) says that $\mathbf{x}_{st} = \mathbf{x}_{ts}$. Thus

$$\frac{\partial}{\partial t}\langle \mathbf{x}_t, \mathbf{x}_s\rangle = \langle \mathbf{x}_t, \mathbf{x}_{st}\rangle = \langle \mathbf{x}_t, \mathbf{x}_{ts}\rangle = \frac{1}{2}\frac{\partial}{\partial s}\langle \mathbf{x}_t, \mathbf{x}_t\rangle.$$

The formula above for $\langle \mathbf{x}_t, \mathbf{x}_t\rangle$ then shows that

$$\left(\frac{\partial}{\partial t}\langle \mathbf{x}_t, \mathbf{x}_s\rangle\right)(t, 0) = \langle v, w\rangle \qquad \text{for all} \quad t.$$

Since $\mathbf{x}(0, s) = \exp_o(0) = o$ for all s, $\langle \mathbf{x}_t, \mathbf{x}_s\rangle(0, 0) = 0$. Thus by elementary calculus, $\langle \mathbf{x}_t, \mathbf{x}_s\rangle(t, 0) = t\langle v, w\rangle$. Setting $t = 1$ gives the required result. ∎

In particular the length of radial vectors is preserved. Thus the Gauss lemma describes the exponential map $\exp_o$ as a kind of partial isometry whose principal distortions are in directions orthogonal to radial directions in $T_o(M)$.

Using the Gauss lemma we now set up a detailed comparison between a normal neighborhood of a point o in M and the corresponding neighborhood of 0 in $T_p(M)$. A tilde ($\sim$) is used systematically to distinguish objects in $T_o(M)$ from the corresponding objects in $\mathscr{U} \subset M$. In this context $\exp_o^{-1}$ will always mean the diffeomorphism $(\exp_o|\tilde{\mathscr{U}})^{-1}: \mathscr{U} \to \tilde{\mathscr{U}}$.

(1) If $\tilde{q}$ is the function $v \to \langle v, v\rangle$ on $T_o(M)$ the corresponding function $\tilde{q} \circ \exp_o^{-1}$ is denoted by q.

(2) Hyperquadrics appear in $T_o(M) \approx \boldsymbol{R}^n_\nu$, just as in $\boldsymbol{R}^n_\nu$ itself, as level hypersurfaces $\tilde{Q} = \tilde{q}^{-1}(c)$, $c \neq 0$. The diffeomorphic image of $\tilde{Q} \cap \tilde{\mathscr{U}}$ under $\exp_o$ is the hypersurface $Q = q^{-1}(c)$ in $\mathscr{U} \subset M$, called a *local hyperquadric* at o. (See Figure 1.)

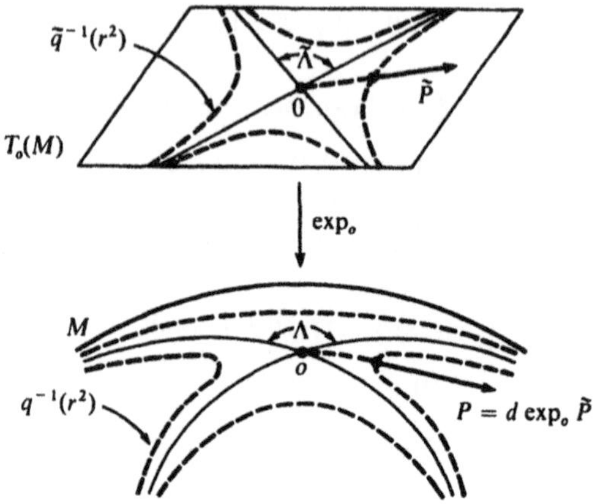

Figure 1

(3) For indefinite metrics, if $\tilde{\Lambda}$ is the nullcone $\tilde{q}^{-1}(0) - 0$ in $T_o(M)$, then the diffeomorphic image of $\tilde{\Lambda} \cap \tilde{\mathscr{U}}$ under $\exp_o$ is the *local nullcone* $\Lambda(o) = q^{-1}(0) - o$. Thus $\Lambda(o)$ consists of initial segments of all null geodesics starting at o, and in $\mathscr{U}$ as in $T_o(M)$ the two families of hyperquadrics ($c < 0$, $c > 0$) are separated by the nullcone. The simpler Riemannian situation is spelled out later.

(4) If $\tilde{P}$ is the position vector field $v \to v_v$ on $\tilde{\mathscr{U}} \subset T_o(M)$ then the transferred vector field $P = d\exp_o(\tilde{P})$ is called the *local position vector field* at o. Like $\tilde{P}$ it is radial, that is, tangent to all radial geodesics emanating from o. Applying the Gauss lemma to known properties of $\tilde{P}$ gives

2. Corollary. The local position vector field P at o is orthogonal to every local hyperquadric of M at o. Furthermore P is both orthogonal and tangent to the local nullcone $\Lambda(o)$.

Again by the Gauss lemma, $\langle P, P\rangle \circ \exp_o = \langle \tilde{P}, \tilde{P}\rangle$. Thus by the corollary each local hyperquadric Q is semi-Riemannian with the same sign (Definition 4.16) as the corresponding $\tilde{Q}$. (Though Q and $\tilde{Q}$ are diffeomorphic they are generally not isometric, since tangent directions to $\tilde{Q}$ are the distortion directions of $\exp_o$.)

3. Corollary. grad $q = 2P$.

Proof. Recall from Chapter 4 that grad $\tilde{q} = 2\tilde{P}$. If v is tangent to $\mathscr{U}$, let $\tilde{v}$ be the tangent vector to $\tilde{\mathscr{U}}$ such that $d\exp_o(\tilde{v}) = v$. Then

$$\begin{aligned}\langle \text{grad } q, v\rangle &= v(q) = (d\exp_o \tilde{v})(q) = \tilde{v}(q \circ \exp_o)\\ &= \tilde{v}(\tilde{q}) = \langle \text{grad } \tilde{q}, \tilde{v}\rangle = 2\langle \tilde{P}, \tilde{v}\rangle = 2\langle P, v\rangle,\end{aligned}$$

where the last step is by the Gauss lemma. ∎

CONVEX OPEN SETS

Let M be a semi-Riemannian manifold, assumed for the moment to be complete. The exponential maps $\exp_p\colon T_p(M) \to M$ for all $p \in M$ constitute a single mapping $\exp\colon TM \to M$ of the tangent bundle TM. If π is the natural projection of TM onto M, define $E\colon TM \to M \times M$ by $E(v) = (\pi(v), \exp(v))$. Explicitly,

$$E(v) = (p, \exp_p(v)) \qquad \text{if} \quad v \in T_p(M) \subset TM.$$

When M is not complete, exp and E have the same largest domain, namely the set $\mathscr{D}$ of all vectors $v \in TM$ such that the geodesic γ_v is defined at least on the interval $[0, 1]$. Thus $\mathscr{D}_p = \mathscr{D} \cap T_p(M)$ is the largest domain of the exponential map $\exp_p$.

4. Corollary. The domain $\mathscr{D}$ of exp is open in TM. The domain $\mathscr{D}_p$ of $\exp_p$ is an open set of $T_p(M)$ starshaped about 0.

Proof. Let $\tilde{\mathscr{D}}$ be the domain of the mapping $\psi(v, s) = \gamma_v'(s)$. Proposition 3.28 shows that ψ is the flow of a vector field on TM, hence $\tilde{\mathscr{D}}$ is an open set of $TM \times \boldsymbol{R}$. $\tilde{\mathscr{D}}$ is also the domain of the map $\pi \circ \psi$ sending (v, s) to $\gamma_v(s)$. Thus $\mathscr{D}$ is open in TM, since it corresponds to $\tilde{\mathscr{D}} \cap (TM \times 1)$ under the diffeomorphism $v \leftrightarrow (v, 1)$. Then $\mathscr{D}_p$ is open in $T_p(M)$.

If $v \in \mathscr{D}_p$ then γ_v is defined on $[0, 1]$. But $\gamma_{tv}(1) = \gamma_v(t)$, so $tv \in \mathscr{D}_p$ for $t \in [0, 1]$. Hence $\mathscr{D}_p$ is starshaped. ∎

5. Definition. An open set $\mathscr{C}$ in a semi-Riemannian manifold is *convex* provided $\mathscr{C}$ is a normal neighborhood of each of its points.

In particular for any two points p, q of $\mathscr{C}$ there is a unique geodesic segment $\sigma_{pq}\colon [0, 1] \to M$ from p to q that lies entirely in $\mathscr{C}$. Unlike the situation in $\boldsymbol{R}^n$ there may well be other geodesics from p to q that do not remain in $\mathscr{C}$.

6. Lemma. If $\exp_p\colon \mathscr{D}_p \to M$ is nonsingular at $x \in \mathscr{D}_p$, then $E\colon \mathscr{D} \to M \times M$ is nonsingular at x.

Proof. Suppose $dE(v) = 0$ for $v \in T_x(TM)$. We must show that $v = 0$. Let $\pi\colon T(M) \to M$ be the usual projection and let π_1 be the projection of $M \times M$ on its first factor. Then $\pi_1 \circ E = \pi$, so that $d\pi(v) = d\pi_1(dE(v)) = 0$. It follows that v is vertical, that is, tangent to $T_p(M)$, where $p = \pi(x)$. But $E \mid T_p(M)$ differs from $\exp_p$ only by the obvious diffeomorphism from $p \times M$ to M. Thus $d\exp_p(v) = 0$, which by hypothesis implies $v = 0$. ∎

Since $d\exp_p$ is always nonsingular at $0 \in T_p(M) \subset TM$ it follows by the inverse function theorem that *E maps some neighborhood of 0 in TM diffeomorphically onto a neighborhood of (p, p) in $M \times M$.*

7. Proposition. Each point o of M has a convex neighborhood.

Proof. Let $\xi = (x^1, \ldots, x^n)$ be a normal coordinate system on a neighborhood $\mathscr{V}$ of $o \in M$. If $N = \sum (x^i)^2$, then for $\delta > 0$ sufficiently small, $\mathscr{V}(\delta) = \{p \in \mathscr{V} : N(p) < \delta\}$ is a neighborhood of o diffeomorphic under ξ to an open ball in $\boldsymbol{R}^n$.

By a remark above, if δ is sufficiently small then E is a diffeomorphism of a neighborhood $\mathscr{W}$ of $0 \in T_0(M)$ in TM onto $\mathscr{V}(\delta) \times \mathscr{V}(\delta)$.

Consider the symmetric (0, 2) tensor B whose components are $\delta_{ij} - \sum_k \Gamma_{ij}^k x^k$. Obviously B is positive definite at o. Thus a further reduction of δ, if necessary, ensures that B is positive definite on $\mathscr{V}(\delta)$. We assert that *this neighborhood $\mathscr{U} = \mathscr{V}(\delta)$ is a normal neighborhood of each point $p \in \mathscr{U}$.*

Let $\mathscr{W}_p = \mathscr{W} \cap T_p(M)$. By construction $E|\mathscr{W}_p$ is a diffeomorphism onto $p \times \mathscr{U}$, hence $\exp_p|\mathscr{W}_p$ is a diffeomorphism onto $\mathscr{U}$. Also we must show that $\mathscr{W}_p$ is starshaped about 0. If $q \in \mathscr{U}, q \neq p$, let $v = E^{-1}(p, q)$. Then v is in $\mathscr{W}_p$, and $\sigma = \gamma_v|[0, 1]$ is a geodesic from p to q. If σ lies in $\mathscr{U}$ the proof of Proposition 3.30 shows that $tv \in \mathscr{W}_p$ for $0 \leq t \leq 1$; hence, $\mathscr{W}_p$ is starshaped.

Assume that σ leaves $\mathscr{U} = \mathscr{V}(\delta)$. We derive a contradiction as follows. Since $N(p)$, $N(q) < \delta$, the function $N \circ \sigma$ has a maximum at some t_0 with $0 < t_0 < 1$. Abbreviating $x^i \circ \sigma$ to x^i, we compute

$$\frac{d^2(N \circ \sigma)}{dt^2} = 2\sum_i \left\{\left(\frac{dx^i}{dt}\right)^2 + x^i \frac{d^2(x^i)}{dt^2}\right\}.$$

By the geodesic equations (3.21) this becomes

$$2\sum_{i,j} \left\{\delta_{ij} - \sum_k \Gamma_{ij}^k x^k\right\} \frac{dx^i}{dt}\frac{dx^j}{dt}.$$

Thus

$$\frac{d^2(N \circ \sigma)}{dt^2}(t_0) = 2B(\sigma'(t_0), \sigma'(t_0)) > 0.$$

This contradicts the fact that t_0 is a maximum point. ■

A first consequence is the following partial analogue of Lemma 1.56.

8. Lemma. A geodesic $\gamma: [0, b) \to M$ is extendible as a geodesic if and only if it is (continuously) extendible.

Proof. The "only if" assertion is obvious, so suppose that γ has a continuous extension $\tilde{\gamma}: [0, b] \to M$. Let $\mathscr{C}$ be a convex neighborhood of $\tilde{\gamma}(b)$. There is a number a with $0 \leq a < b$ such that $\tilde{\gamma}[a, b] \subset \mathscr{C}$. Then $\mathscr{C}$ is a normal neighborhood of $o = \gamma(a)$ and $\gamma|[a, b)$ is a radial geodesic. Like any radial geodesic, $\gamma|[a, b)$ can be geodesically extended until it approaches bd $\mathscr{C}$—or until its domain is $[a, \infty)$. But $\tilde{\gamma}(b)$ is not in bd $\mathscr{C}$, hence γ can be extended past b. ∎

If p and q are points of a convex open set $\mathscr{C}$ and σ_{pq} is the geodesic in $\mathscr{C}$ from $p = \sigma_{pq}(0)$ to $q = \sigma_{pq}(1)$, the *displacement vector* $\overrightarrow{pq}$ is $\sigma'_{pq}(0) \in T_p(M)$.

9. Lemma. If $\mathscr{C}$ is a convex open set then the map $\Delta: \mathscr{C} \times \mathscr{C} \to TM$ sending (p, q) to $\overrightarrow{pq}$ is smooth.

In fact, arguments as above show that $\Delta(\mathscr{C} \times \mathscr{C})$ is open and that Δ is the inverse map of the diffeomorphism $E: \Delta(\mathscr{C} \times \mathscr{C}) \to \mathscr{C} \times \mathscr{C}$.

In $\boldsymbol{R}^n$ an intersection of convex sets is convex, but in the circle S^1 such an intersection need not even be connected. In general, if convex open sets $\mathscr{C}$ and $\mathscr{C}'$ are contained in a convex open set $\mathscr{E}$, then $\mathscr{C} \cap \mathscr{C}'$ is convex (if nonempty). In fact, if $p, q \in \mathscr{C} \cap \mathscr{C}'$ then both $\mathscr{C}$ and $\mathscr{C}'$ provide the same geodesic σ_{pq} as $\mathscr{E}$; hence $\overrightarrow{pq}$ is unambiguously defined. Thus the diffeomorphism $\exp_p: \tilde{\mathscr{C}} \to \mathscr{C}$ restricts to a diffeomorphism of $\{\overrightarrow{pq}: q \in \mathscr{C} \cap \mathscr{C}'\}$ onto $\mathscr{C} \cap \mathscr{C}'$.

A *convex covering* $\mathfrak{K}$ of a semi-Riemannian manifold M is a covering of M by convex open sets such that if elements $\mathscr{U}$, $\mathscr{V}$ of $\mathfrak{K}$ meet then $\mathscr{U} \cap \mathscr{V}$ is convex.

10. Lemma. Given any open covering $\mathfrak{C}$ of M there is a convex covering $\mathfrak{K}$ such that each element of $\mathfrak{K}$ is contained in some element of $\mathfrak{C}$.

Proof. Let $\mathfrak{C}^*$ be the open covering of M consist of all convex open sets contained in any element of $\mathfrak{C}$. Since M is second countable, hence paracompact, there is an open covering $\mathfrak{B}$ such that if two of its elements meet then their union is contained in some element of $\mathfrak{C}^*$. Let $\mathfrak{K}$ consist of all convex open sets contained in elements of $\mathfrak{B}$. By a remark above, the covering $\mathfrak{K}$ has the required intersection property. ∎

ARC LENGTH

The familiar notion of arc length of a curve segment in Euclidean space generalizes in a natural way.

11. Definition. Let $\alpha: [a, b] \to M$ be a piecewise smooth curve segment in a semi-Riemannian manifold M. The *arc length* of α is

$$L(\alpha) = \int_a^b |\alpha'(s)|\,ds.$$

By definition $|\alpha'| = |\langle\alpha', \alpha'\rangle|^{1/2}$, hence in coordinates

$$|\alpha'| = \left|\sum_{i,j} g_{ij} \frac{d(x^i \circ \alpha)}{ds}\frac{d(x^j \circ \alpha)}{ds}\right|^{1/2}.$$

On a Riemannian manifold arc length behaves much as in Euclidean space, but for indefinite metrics the term *length* can be misleading since, for example, a null curve has length zero.

We consider now the effect of a change of parametrization. A *reparametrization function* h: $[c, d] \to [a, b]$ is a piecewise smooth function such that either $h(c) = a$, $h(d) = b$ (h *orientation-preserving*), or $h(c) = b$, $h(d) = a$ (h *orientation-reversing*). If its derivative does not change sign, h is *monotone*.

The same proofs as in elementary calculus then show the following.

12. Lemma. (1) The length of a piecewise smooth curve segment is unchanged by monotone reparametrization. (2) If α is a curve segment with $|\alpha'| > 0$, there is a strictly increasing reparameterization function h such that $\beta = \alpha(h)$ has $|\beta'| = 1$.

In the latter case β is said to have *unit speed* or *arc length parametrization*.

Let $\mathscr{U}$ be a normal neighborhood of a point o of M. The function r on $\mathscr{U}$ given by $r(p) = |\exp_o^{-1}(p)|$ is the *radius function* of M at o. In terms of normal coordinates,

$$r = \left| -\sum_{1}^{\nu} (x^i)^2 + \sum_{\nu+1}^{n} (x^j)^2 \right|^{1/2}.$$

Thus r is smooth except where it is zero, namely at the point o and on the local nullcone at o.

13. Lemma. Let r be the radius function on a normal neighborhood $\mathscr{U}$ of o. If σ is the radial geodesic from o to $p \in \mathscr{U}$, then $L(\sigma) = r(p)$.

Proof. If v is the initial velocity $\sigma'(0)$ of σ, we know from Proposition 3.30 that $v = \exp_o^{-1}(p)$. Since $|\sigma'|$ is constant,

$$L(\sigma) = \int_0^1 |\sigma'|\,ds = \int_0^1 |v|\,ds = |v| = r(p). \qquad \blacksquare$$

RIEMANNIAN DISTANCE

In this section M will be a Riemannian manifold; thus its metric tensor makes each tangent space an inner product space. The local geometry of M near a point o is relatively simple. If $\tilde{r}$ is the norm function $\tilde{r}(v) = |v| = \langle v, v\rangle^{1/2}$ on $T_o(M)$, then on any normal neighborhood $\mathscr{U}$ of o the radius function $r = \tilde{r} \circ \exp_o^{-1}$ is smooth except at o. There is only a single family of hyperquadrics at o, namely, the spheres r constant, which are the images under $\exp_o$ of the standard spheres $\tilde{r}$ constant in $T_o(M) \approx \boldsymbol{R}^n$. If as usual $q = \tilde{q} \circ \exp_o^{-1}$ on $\mathscr{U}$, and P is the local position vector field near o, then

$$|P| = \langle P, P\rangle^{1/2} = \sqrt{q} = r.$$

Hence $U = P/r$ is the outward radial unit vector field on $\mathscr{U} - o$, and U is normal to all the hyperspheres at o. Since $r = \sqrt{q}$, it follows from Corollary 3 that grad $r = P/r = U$ on $\mathscr{U} - \{o\}$.

The local geometry of Riemannian geodesics is dominated by the following result.

14. Lemma. Let $\mathscr{U}$ be a normal neighborhood of a point o in a Riemannian manifold M. If $p \in \mathscr{U}$ then the radial geodesic segment

$$\sigma\colon [0, 1] \to \mathscr{U}$$

from o to p is the unique shortest curve in $\mathscr{U}$ from o to p.

Proof. In view of Lemma 12 the uniqueness of σ must be interpreted as uniqueness up to monotone reparametrization. Thus if $\alpha\colon [0, b] \to \mathscr{U}$ is a curve from o to p we must show that

(a) $L(\alpha) \geq L(\sigma)$.
(b) If $L(\alpha) = L(\sigma)$, then α is a monotone reparametrization of σ.

For (a), restricting the radial unit vector field U to α we can write

$$\alpha' = \langle \alpha', U\rangle U + N,$$

where N is a vector field on α orthogonal to U. (At $t = 0$, let U be $\alpha'(0)$ and N zero.) Then

$$|\alpha'| = \langle \alpha', \alpha'\rangle^{1/2} = [\langle \alpha', U\rangle^2 + \langle N, N\rangle]^{1/2} \geq |\langle \alpha', U\rangle| \geq \langle \alpha', U\rangle.$$

But $U = \operatorname{grad} r$ implies $\langle \alpha', U\rangle = d(r \circ \alpha)/dt$. Hence by the fundamental theorem of calculus

$$L(\alpha) = \int_0^b |\alpha'|\, dt \geq r(\alpha(b)) - r(\alpha(0)) = r(p).$$

But by Lemma 13, $r(p) = L(\sigma)$.

For (b), if $L(\alpha) = L(\sigma)$, then the inequalities above all become equalities; hence, writing r for $r \circ \alpha$,

$$N = 0, \qquad \text{and} \qquad dr/dt = \langle \alpha', U\rangle = |\langle \alpha', U\rangle| \geq 0.$$

Thus $\alpha' = (dr/dt)U$, showing that α travels monotonically along the radial geodesic from o to p, namely, σ. In fact, since r measures radial distance, α is the monotone reparametrization $\alpha(t) = \sigma(r(t)/r(b))$ of σ. ∎

Refinements: (1) The proof remains valid if α is only required to be piecewise smooth. Then dr/dt has only a finite number of double values and the reparametrization function $t \to r(t)/b$ is piecewise smooth.

(2) Dropping the normal neighborhood $\mathscr{U}$ from the hypotheses, assertions (a) and (b) still hold provided $\alpha = \exp_o \circ \tau$, where τ is a curve in $T_o(M)$ from 0 to $\sigma'(0)$ such that $d \exp_o$ is nonsingular at each point of τ.

In Euclidean space the distance $d(p, q) = |p - q|$ between two points p and q could also be defined to be the length of the shortest curve segment joining them, namely the straight line segment from p to q. But in $\mathbf{R}^2 - (0, 0)$, for example, there is no shortest curve from $p = (-1, 0)$ to $q = (1, 0)$. However the following modification works in general.

15. Definition. For any points p and q of a connected Riemannian manifold M, the *Riemannian distance* $d(p, q)$ from p to q is the greatest lower bound of $\{L(\alpha) : \alpha \in \Omega(p, q)\}$,where $\Omega(p, q)$ is the set of all piecewise smooth curve segments in M from p to q.

If $o \in M$ and $\varepsilon > 0$ the set $\mathscr{N}_\varepsilon(o)$ of points $p \in M$ such that $d(o, p) < \varepsilon$ is called the *ε-neighborhood* of o in M. Using this notion the preceding lemma can be strengthened as follows.

16. Proposition. Let o be a point of a Riemannian manifold M. (1) For $\varepsilon > 0$ sufficiently small, the ε-neighborhood $\mathscr{N}_\varepsilon(o)$ is normal; (2) For a normal ε-neighborhood $\mathscr{N}_\varepsilon(o)$ the radial geodesic σ from o to $p \in \mathscr{N}_\varepsilon(0)$ is the unique shortest curve in M from o to p. In particular,

$$L(\sigma) = r(p) = d(o, p).$$

Proof. (1) Let $\mathscr{U}$ be a normal neighborhood of o in M, with $\tilde{\mathscr{U}}$ the corresponding neighborhood of 0 in $T_o(M)$. For $\varepsilon > 0$ sufficiently small, $\tilde{\mathscr{U}}$ contains the starshaped open set

$$\tilde{\mathscr{N}} = \tilde{\mathscr{N}}_\varepsilon(0) = \{v \in T_0(M) : |v| < \varepsilon\}.$$

Thus $\mathscr{N} = \exp_o(\tilde{\mathscr{N}})$ is also a normal neighborhood of o.

If $p \in \mathscr{N}$ then by Lemma 14 the radial geodesic σ from o to p is the unique shortest curve in $\mathscr{N}$ from o to p, and $L(\sigma) = r(p)$. Since $\exp_0^{-1}(p) = v$ is in $\tilde{\mathscr{N}}$, $r(p) = |v| < \varepsilon$. It now suffices to prove:

If α is a curve in M starting at o and leaving $\mathscr{N}$, then $L(\alpha) \geq \varepsilon$.

In fact, this will show that σ is the unique shortest curve in all M from o to p, hence $L(\sigma) = r(p)$ is $d(o, p)$. Thus $d(o, p) < \varepsilon$. But if $q \notin \mathscr{N}$ then $d(o, q) \geq \varepsilon$, showing that $\mathscr{N}$ is the ε-neighborhood of o.

To prove the assertion above, note first that since α leaves $\mathscr{N}$ it meets every sphere $S(a)$, $r = a$, with $a < \varepsilon$. If α_1 is the shortest initial segment of α from o to $S(a)$ then α lies in $\mathscr{N}$; so Lemma 14 gives $L(\alpha) \geq L(\alpha_1) \geq a$ for all $a < \varepsilon$. Thus $L(\alpha) \geq \varepsilon$. ∎

The following simple example illustrates many properties involving geodesics, in particular, the distinction between arbitrary normal neighborhoods and normal ε-neighborhoods.

17. Example. Geodesics on a Cylinder. The Riemannian product manifold $S^1 \times \boldsymbol{R}^1$ can be viewed as the cylinder $M\colon x^2 + y^2 = 1$ in $\boldsymbol{R}^3$. The mapping $\phi(u, v) = (\cos u, \sin u, v)$, which wraps the plane around the cylinder, is a local isometry. Thus the geodesics of M are all curves of the form

$$\gamma(t) = (\cos(at + b), \sin(at + b), ct + d).$$

These are typically helixes, reducing to cross-sectional circles for $c = 0$ and to vertical lines L for $a = 0$.

(a) For any such line L, the open set $M - L$ is convex; hence, if L does not pass through the point $o \in M$, then $M - L$ is a normal neighborhood of o. Thus by Lemma 14, the radial geodesic σ from o to $p \in M - L$ is the unique shortest curve in $M - L$ from o to p. But evidently the geodesic τ in Figure 2 is a shorter curve from o to p that does not remain in $M - L$.

(b) For any point o, its largest normal ε-neighborhood is $\mathscr{N}_\pi(o)$. Thus by Proposition 16 a radial geodesic segment from o to a point q of this neighborhood is the unique shortest curve in all of M from o to q. For a point r outside $\mathscr{N}_\pi(o)$, there is always a shortest curve from o to r, but uniqueness is lost if r lies on the vertical line opposite o, that is, through $-o$.

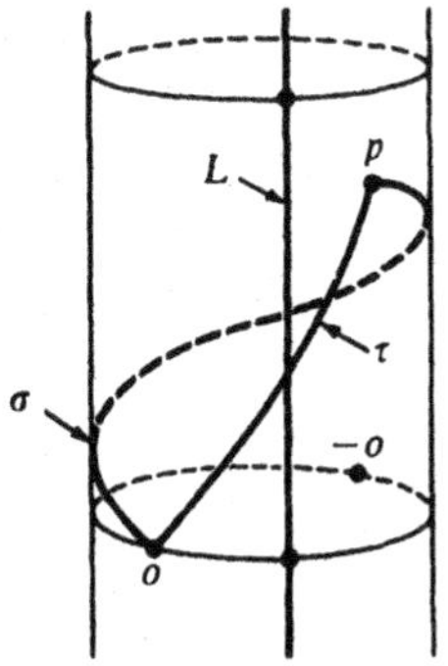

Figure 2

It is now easy to show that d has the properties expected of a distance function.

18. Proposition. For a connected Riemannian manifold M the Riemannian distance function $d: M \times M \to \boldsymbol{R}$ is a *metric* on M, that is, for all $p, q, r \in M$:

(1) $d(p, q) \geq 0$, and $d(p, q) = 0$ if and only if $p = q$ (positive definiteness)

(2) $d(p, q) = d(p, q)$ (symmetry).

(3) $d(p, q) + d(p, r) \geq d(p, r)$ (triangle inequality).

Furthermore d is compatible with the topology of M.

Proof. To prove the triangle inequality, for example, given any $\varepsilon > 0$ choose $\alpha \in \Omega(p, q)$ and $\beta \in \Omega(q, r)$ such that

$$L(\alpha) < d(p, q) + \varepsilon \quad \text{and} \quad L(\beta) < d(q, r) + \varepsilon.$$

Joining α and β produces a curve γ in $\Omega(p, r)$ for which

$$d(p, r) \leq L(\gamma) = L(\alpha) + L(\beta) < d(p, q) + d(q, r) + 2\varepsilon.$$

Since ε is arbitrary the result follows. The other metric properties are trivial except for $d(p, q) = 0 \Rightarrow p = q$. But if $p \neq q$ then since M is Hausdorff there is a normal neighborhood $\mathscr{U}$ of p that does not contain q. The proof of the preceding proposition shows that $\mathscr{U}$ contains an ε-neighborhood of p, so $d(p, q) \geq \varepsilon > 0$.

Since every neighborhood of a point of M contains an ε-neighborhood, and (also by the preceding proof) ε-neighborhoods are open sets of M, the metric d is compatible with the topology of M, that is, a subset $\mathscr{V}$ of M is open if and only if for each $p \in \mathscr{V}$ there is an $\varepsilon > 0$ such that $\mathscr{N}_\varepsilon(p) \subset \mathscr{V}$. ∎

Compatibility means that ε-neighborhoods can be used in the same familiar way as in Euclidean space. Hence, for example, a sequence of points $\{p_i\}$ in M converges to $p \in M$ if and only if the numerical sequence $\{d(p, p_i)\}$ converges to 0 in $\boldsymbol{R}$.

By definition of Riemannian distance, a curve segment σ from p to q in a Riemannian manifold M is a shortest curve segment from p to q if and only if $L(\sigma) = d(p, q)$. (There may be many or none such segments.) In this case we also say that σ *minimizes arc length from p to q*, or merely that σ is *minimizing*. Note that any subsegment of a minimizing segment is also minimizing.

Picturing a minimizing segment as a tightly stretched string correctly suggests that it is geodesic:

19. Corollary. In a Riemannian manifold a minimizing curve segment α from p to q is a monotone reparametrization of an (unbroken) geodesic segment from p to q.

Proof. The domain I of α can be decomposed into subintervals I_i such that each $\alpha_i = \alpha | I_i$ lies in a convex open set. (Each α_i can be assumed to be nonconstant, for otherwise I_i could be adjoined to an adjacent subinterval.) Since α_i is minimizing it follows from the uniqueness feature of Lemma 14 that it is a monotone reparametrization of a unit speed geodesic σ_i. Joining these gives a possibly broken geodesic σ from p to q. Similarly, the reparametrization functions can be patched together to exhibit α as a monotone reparametrization of σ.

Since $L(\sigma) = L(\alpha) = d(p, q)$, the following useful fact will imply that σ is unbroken: *If a geodesic segment γ_1 ending at p and a geodesic segment γ_2 starting at p combine to give a minimizing curve segment γ, then γ is an (unbroken) geodesic.* (Intuitively, if there were a corner at p we could round it off to get a shorter curve.) Assume γ has constant speed and let $\mathscr{C}$ be a convex neighborhood of p. Then a final segment of γ_1 and an initial segment in γ_2 combine to give a minimizing curve segment $\bar{\gamma}$ in $\mathscr{C}$. Since $\mathscr{C}$ is a normal neighborhood of the initial point of $\bar{\gamma}$ it follows from Lemma 14 that $\bar{\gamma}$ is a constant speed reparametrization of a radial geodesic. Consequently, $\bar{\gamma}$ and hence γ are unbroken geodesics. ∎

20. Example. The Sphere $S^n(r)$. If p and q are distinct nonantipodal points ($q \neq \pm p$) there is a unique great circle through p and q. Its shorter arc thus provides the unique minimizing geodesic σ from p to q. If ϑ is the angle, $0 < \vartheta < \pi$, between p and q (as vectors in $\mathbf{R}^{n+1}$), it follows that

$$d(p, q) = L(\sigma) = r\vartheta.$$

By continuity, $d(p, -p) = r\pi$. Thus the normal neighborhood $S^n(r) - \{-p\}$ is exactly $\mathscr{N}_{r\pi}(p)$.

Each semicircle from p to $-p$ gives a minimizing geodesic from p to $-p$, but a slightly longer geodesic is not minimizing (the other arc of its great circle being shorter). Thus a geodesic of $S^n(r)$ is minimizing if and only if its length is at most $r\pi$.

The hemisphere $\mathscr{N}_{r\pi/2}(p)$ is the largest ε-neighborhood of p that is convex.

RIEMANNIAN COMPLETENESS

The fundamental theorem of complete Riemannian manifolds is this:

21. Theorem (Hopf–Rinow). For a connected Riemannian manifold M the following conditions are equivalent:

(MC) As a metric space under Riemannian distance d, M is complete; that is, every Cauchy sequence converges.
(C_1) There exists a point $p \in M$ from which M is geodesically complete, that is, $\exp_p$ is defined on the entire tangent space $T_p(M)$.
(C) M is geodesically complete.
(HB) Every closed bounded subset of M is compact.

The following companion result will be proved simultaneously.

22. Proposition. If a connected Riemannian manifold is complete then any two of its points be joined by a minimizing geodesic segment.

The open disk in $\mathbf{R}^2$ shows that the converse of the proposition is false.

From the viewpoint of semi-Riemannian geometry the most striking feature of the proposition is that arbitrary points can be joined by any geodesic at all, much less a minimizing one. This property, called *geodesic connectedness*, is equivalent to all exponential maps of M being onto. We shall soon see (after Proposition 38) that for indefinite metrics, connectedness and geodesic completeness do not imply geodesic connectedness.

In general the proposition permits the free use of geodesic constructions throughout M, while the Hopf–Rinow theorem itself links the geodesics of a complete Riemannian manifold firmly to its structure as a metric space; for example:

23. Corollary. A compact Riemannian manifold is complete.

Proof. The Heine–Borel condition (HB) holds trivially. ∎

This result too fails for indefinite metrics; see Example 7.16. (Even if an indefinite M is compact, the set of unit vectors in TM is not.)

The following is the essential step in the proof of (21) and (22).

24. Lemma. If $\exp_p$ is defined on all of $T_p(M)$ (condition (C_1)), then for any $q \in M$, there is a minimizing geodesic segment from p to q.

Proof. Let $\mathscr{U}$ be a normal ε-neighborhood of p. We suppose $q \notin \mathscr{U}$, for otherwise the result is trivial. If r is the radius function at p, then for $\delta > 0$ sufficiently small, $r = \delta$ defines an entire $(n - 1)$-sphere S in $\mathscr{U}$. The function

$s \to d(s, q)$ is continuous on S (compact), hence has a minimum point $m \in S$. We assert that

$$(*) \qquad d(p, m) + d(m, q) = d(p, q).$$

Let $\alpha: [0, b] \to M$ be any curve from p to q. Since M is Hausdorff, α meets S at some parameter value a, with $0 < a < b$. Let α_1 and α_2 be the restrictions of α to $[0, a]$ and $[a, b]$, respectively. Then by Proposition 16,

$$L(\alpha) = L(\alpha_1) + L(\alpha_2) \geq \delta + d(m, q).$$

Hence $d(p, q) \geq \delta + d(m, q) = d(p, m) + d(m, q)$. The reverse inequality is the triangle inequality, so (*) is proved.

Now let $\gamma: [0, \infty) \to M$ be the unit speed geodesic whose initial segment runs radially from p through m (thus γ is aimed at q). Let $d = d(p, q)$ and let T be the set of all $t \in [0, d]$ such that

$$t + d(\gamma(t), q) = d.$$

It suffices to show that $d \in T$, for then $d(\gamma(d), q) = 0$ hence $\gamma(d) = q$, and since $\gamma|[0, d]$ has length $d = d(p, q)$ it is minimizing. In fact $\gamma|[0, t]$ is minimizing for any $t \in T$ since its length is t, and

$$d \leq d(p, \gamma(t)) + d(\gamma(t), q) = d(p, \gamma(t)) + d - t,$$

hence $d(p, \gamma(t)) = t$.

The set T is closed by continuity and nonempty by (*); thus it contains a largest number $t_0 \leq d$. Assuming $t_0 < d$, we deduce a contradiction as follows. In a normal neighborhood $\mathscr{U}'$ of $\gamma(t_0)$ the same procedure as above produces a unit speed radial geodesic segment $\sigma: [0, \delta'] \to \mathscr{U}'$ from $\gamma(t_0)$ to a point $m' \in \mathscr{U}'$ such that

$$\delta' + d(m', q) = d(\gamma(t_0), q).$$

(See Figure 3.) Because $t_0 \in T$,

$$t_0 + \delta' + d(m', q) = d.$$

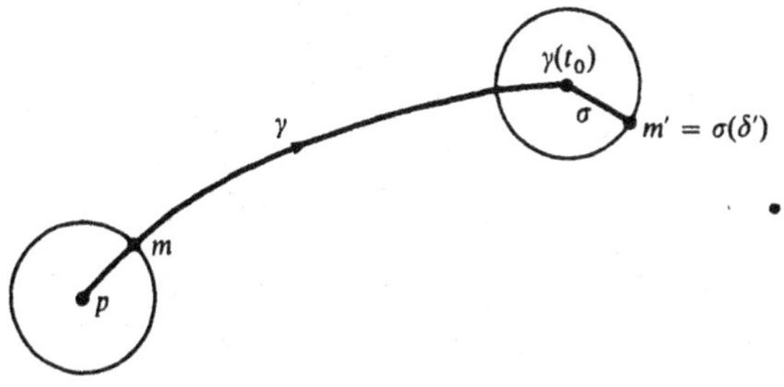

Figure 3

Since $d \le d(p, m') + d(m', q)$, another application of the triangle inequality yields $t_0 + \delta' = d(p, m')$. Thus the sum of $\gamma|[0, t_0]$ and σ minimizes, so by the proof of Corollary 19 there is no corner at their meeting point $\gamma(t_0) = \sigma(0)$. This means that $m' = \sigma(\delta')$ is $\gamma(t_0 + \delta')$. But then

$$t_0 + \delta' + d(\gamma(t_0 + \delta'), q) = d(p, m') + d(m', q) = d,$$

giving the contradiction $t_0 < t_0 + \delta' \in T$. ∎

Proof of (21) *and* (22). It suffices to prove the Hopf–Rinow theorem, for Proposition 22 then follows from the lemma.

(MC) ⇒ (C). Let $\gamma: [0, b) \to M$ be a unit speed geodesic. If $\{t_i\} \to b$ in $[0, b)$ then $\{\gamma(t_i)\}$ is Cauchy, since $d(\gamma t_i, \gamma t_j) \le |t_i - t_j|$. Hence $\{\gamma(t_i)\}$ converges to some point $q \in M$. For another such sequence $\{s_i\}$ the sequence $\{\gamma(s_i)\}$ converges to the same point q, since $d(\gamma t_i, \gamma s_i) \le |t_i - s_i|$. Thus γ is continuously extendible; hence by Lemma 8 it is geodesically extendible.

(C) ⇒ (C_1). Trivial.

(C_1) ⇒ (HB). Let A be a closed bounded subset of M. By the preceding lemma, if $q \in A$ there is a minimizing geodesic $\sigma_q: [0, 1] \to M$ from p to q. But $|\sigma_q'(0)| = L(\sigma_q) = d(p, q)$, and by the triangle inequality the set of these numbers for all $q \in A$ is bounded, say by r. Thus each $\sigma_q'(0)$ is in the compact ball $B_r = \{v \in T_p(p): |v| \le b\}$. Since $\exp_p(B_r)$ is compact and contains the closed set A, the latter is compact. ∎

(HB) ⇒ (MC). The point set of a Cauchy sequence is bounded, hence its closure is compact. Thus the sequence contains a convergent subsequence and, being Cauchy, must itself converge.

We conclude with another distinctively Riemannian property.

25. Lemma. Every (second countable) smooth manifold admits a Riemannian metric tensor.

Proof. Let $\{f_\alpha\}$ be a partition of unity subordinate to the covering of M by coordinate neighborhoods. For each α, pick a coordinate system $x^1, \ldots, x^n$ whose domain $\mathscr{U}$ contains supp f_α, and let g_α be the metric tensor $\sum dx^i \otimes dx^i$ on $\mathscr{U}$. A linear combination of (positive definite) inner products, with positive coefficients, is again an inner product. Thus $\sum f_\alpha g_\alpha$ is a Riemannian metric on M. ∎

LORENTZ CAUSAL CHARACTER

To study the tangent spaces of a Lorentz manifold in abstract terms, define a *Lorentz vector space* to be a scalar product space of index 1 and dimension ≥2. The notion of causal character of vectors has in this context a natural generalization to vector subspaces.

Let W be a subspace of a Lorentz vector space V, and let g be the scalar product of V. There are three mutually exclusive possibilities for W:

(1) $g|W$ is positive definite; that is, W is an inner product space. Then W is said to be *spacelike*.
(2) $g|W$ is nondegenerate of index 1. Then W is *timelike*.
(3) $g|W$ is degenerate. Then W is *lightlike*.

The type into which W falls is called its *causal character*. This definition is consistent with Definition 3.3 in the sense that the causal character of an individual vector v is the same as the causal character of the subspace $\mathbf{R}v$ it generates. (The zero subspace, like the zero vector, is spacelike.)

The following simple result is widely useful.

26. Lemma. If z is a timelike vector in a Lorentz vector space, then the subspace $z^\perp$ is spacelike and V is the direct sum $\mathbf{R}z + z^\perp$.

Proof. The subspace $\mathbf{R}z$ is nondegenerate with index 1. Hence by Lemma 2.23, $z^\perp$ is nondegenerate and $V = \mathbf{R}z + z^\perp$ is a direct sum. Thus ind $V =$ ind $\mathbf{R}z +$ ind $z^\perp$, which implies ind $z^\perp = 0$. Hence $z^\perp$ is spacelike. ∎

This argument shows, more generally, that a *subspace W is timelike if and only if $W^\perp$ is spacelike*. Since $(W^\perp)^\perp = W$ the words *timelike* and *spacelike* can be reversed in this assertion. It follows then that W is lightlike if and only if $W^\perp$ is lightlike.

Spacelike subspaces W are the easiest to deal with since, for example, every subspace of W is also spacelike and the *Schwarz inequality* is available:

$|\langle v, w\rangle| \leq |v||w|$, with equality if and only if v and w are dependent (collinear).

Now we consider some criteria for a subspace to be timelike, omitting the trivial case dim $W = 1$.

27. Lemma. Let W be a subspace of dimension ≥ 2 in a Lorentz vector space. Then the following are equivalent:

(1) W is timelike, hence is itself a Lorentz vector space.
(2) W contains two linearly independent null vectors.
(3) W contains a timelike vector.

Proof. (1) ⇒ (2). Let $e_1, \ldots, e_m$ be an orthonormal basis for W with e_1 the timelike vector. Then $e_1 \pm e_2$ are independent null vectors.

(2) ⇒ (3). By Exercise 2, if u, v are independent null vectors, then $g(u, v) \neq 0$. Hence one of the vectors $u \pm v$ is timelike.

(3) ⇒ (1). If z is a timelike vector in W, then $W^\perp \subset z^\perp$ and the latter is spacelike. Hence $W^\perp$ is spacelike. But then $W = (W^\perp)^\perp$ is timelike. ∎

28. Lemma. For a subspace W of a Lorentz vector space the following are equivalent

(1) W is lightlike, that is, degenerate.
(2) W contains a null vector but not a timelike vector.
(3) $W \cap \Lambda = L - 0$, where L is a one-dimensional subspace and Λ is the nullcone of V.

Proof. (1) ⇒ (2). Since W is degenerate it contains a null vector. By the previous lemma it cannot contain a timelike vector.

(2) ⇒ (3). Since W contains a null vector, $W \cap \Lambda$ is nonempty. By the previous lemma, two independent null vectors would imply that W contains timelike vector.

(3) ⇒ (1). W cannot be spacelike, and again by the preceding lemma cannot be timelike. Hence W is lightlike. (L is in fact the nullspace $W \cap W^{\perp}$ of W.) ■

Causal characters for a subspace W are illustrated in Figure 4.

Let P be a submanifold of a Lorentz manifold. If for every $p \in P$ the subspace $T_p(P)$ has the same causal character in $T_p(M)$, then that causal character is attributed to P itself. Thus semi-Riemannian submanifolds of M are either spacelike or timelike. The nullcone in $\boldsymbol{R}_1^n$ is an example of a lightlike submanifold. Of course an arbitrary submanifold need not have a causal character.

The causal character of tangent vectors, curves, and submanifolds is preserved not just by local isometries but by conformal maps with $h > 0$

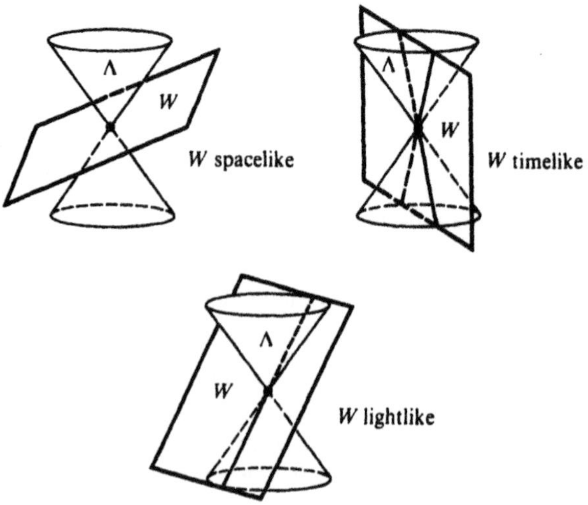

Figure 4. Causal character of subspace W.

(see the description preceding Definition 3.63). If $h < 0$ then *spacelike* and *timelike* are reversed.

Relativity theory merges three space dimensions and one time dimension into a four-dimensional Lorentz manifold. Instead of making time "the fourth dimension" it is more consistent in dealing with Lorentz manifolds of various different dimensions to make it the zero-th dimension [MTW]. Thus for the natural coordinates of $\boldsymbol{R}^n_1$ we frequently use the *relativistic indexing* $t = u^0$, $u^1, \ldots, u^{n-1}$. Similarly an orthonormal basis for a Lorentz vector space will be denoted by $e_0, e_1, \ldots, e_{n-1}$ with e_0 the timelike vector. (However in the diagrams of relativity theory, time axes are customarily drawn to be vertical.)

TIMECONES

Let $\mathscr{T}$ be the set of all timelike vectors in a Lorentz vector space V. For $u \in \mathscr{T}$

$$C(u) = \{v \in \mathscr{T} : \langle u, v \rangle < 0\}$$

is the *timecone* of V containing u. The *opposite* timecone is

$$C(-u) = -C(u) = \{v \in \mathscr{T} : \langle u, v \rangle > 0\}.$$

Since $u^\perp$ is spacelike, $\mathscr{T}$ is the disjoint union of these two timecones.

29. Lemma. Timelike vectors v and w in a Lorentz vector space are in the same timecone if and only if $\langle v, w \rangle < 0$.

Proof. We show that if $v \in C(u)$ and w is timelike, then $w \in C(u)$ if and only if $\langle v, w \rangle < 0$. Since $C(u/|u|) = C(u)$, we can assume u is a (timelike) unit vector.

Write $v = au + \vec{v}$, $w = bu + \vec{w}$, where $\vec{v}, \vec{w} \in u^\perp$. Since these are timelike vectors, $|a| > |\vec{v}|$ and $|b| > |\vec{w}|$. Now $\langle v, w \rangle = -ab + \langle \vec{v}, \vec{w} \rangle$, where by the Schwarz inequality $|\langle \vec{v}, \vec{w} \rangle| \leq |\vec{v}|\,|\vec{w}| < |ab|$.

Since $v \in C(u)$, $a > 0$. Hence $\operatorname{sgn}\langle v, w \rangle = \operatorname{sgn}(-ab) = -\operatorname{sgn}(b)$, which gives the result. ∎

It follows that for timelike vectors

$$u \in C(v) \Leftrightarrow v \in C(u) \Leftrightarrow C(v) = C(u).$$

Furthermore, *timecones are convex*, for if $v, w \in C(u)$ and $a \geq 0$, $b \geq 0$ (not both zero), then it is easy to check that $av + bw$ is in $C(u)$.

Many features of inner product spaces have novel analogues in the Lorentz case. For example, in an inner product space the Schwarz inequality permits the definition of the angle ϑ between v and w as the unique number $0 \leq \vartheta \leq \pi$ such that $\cos \vartheta = \langle v, w \rangle / |v|\,|w|$. An analogous Lorentz result is as follows.

30. Proposition. Let v and w be timelike vectors in a Lorentz vector space. Then

(1) $|\langle v, w\rangle| \geq |v|\,|w|$, with equality if and only if v and w are collinear.

(2) If v and w are in the same timecone of V, there is a unique number $\varphi \geq 0$, called the *hyperbolic angle* between v and w, such that

$$\langle v, w\rangle = -|v|\,|w| \cosh \varphi.$$

Proof. (1) Write $w = av + \vec{w}$, with $\vec{w} \in v^{\perp}$. Since w is timelike,

$$\langle w, w\rangle = a^2\langle v, v\rangle + \langle \vec{w}, \vec{w}\rangle < 0.$$

Then

$$\begin{aligned}\langle v, w\rangle^2 &= a^2\langle v, v\rangle^2 = (\langle w, w\rangle - \langle \vec{w}, \vec{w}\rangle)\langle v, v\rangle \\ &\geq \langle w, w\rangle\langle v, v\rangle = |v|^2|w|^2,\end{aligned}$$

since $\langle \vec{w}, \vec{w}\rangle \geq 0$ and $\langle v, v\rangle < 0$. Evidently equality holds if and only if $\langle \vec{w}, \vec{w}\rangle = 0$, which is equivalent to $\vec{w} = 0$, that is, to $w = av$.

(2) If v and w are in the same timecone, then $\langle v, w\rangle < 0$, hence

$$-\langle v, w\rangle/|v|\,|w| \geq 1,$$

and the result follows from properties of the hyperbolic cosine. ∎

Since the Schwarz inequality runs backwards in this context, so does the triangle inequality.

31. Corollary. If v and w are (timelike) vectors in the same timecone, then $|v| + |w| \leq |v + w|$, with equality if and only if v and w are collinear.

Proof. Since $\langle v, w\rangle < 0$, the backwards Schwarz inequality gives $|v|\,|w| \leq -\langle v, w\rangle$. Hence

$$(|v| + |w|)^2 = |v|^2 + 2|v|\,|w| + |w|^2 \leq -\langle v + w, v + w\rangle = |v + w|^2.$$

This becomes an equality if and only if $|v|\,|w| = -\langle v, w\rangle$. But the latter term is $|\langle v, w\rangle|$, so the previous proposition gives the collinearity criterion. ∎

To our Euclidean intuitions it can only seem distressing at first that a straight line segment is no longer the shortest route between two points, that cutting across a corner makes a trip *longer* rather than shorter. But the preceding result is fundamental in Lorentz geometry and its applications to relativity theory.

The existence of timecones raises a fundamental global question about an arbitrary Lorentz manifold M. In each (Lorentz) tangent space $T_p(M)$ there are two timecones, and there is no intrinsic way to distinguish one from the other. To choose one of them is to *time-orient* $T_p(M)$. The question is: Can every tangent space of M be time-oriented in a suitably continuous way?

Let τ be a function on M that assigns to each point p a timecone τ_p in $T_p(M)$. τ is *smooth* if for each $p \in M$ there is a (smooth) vector field V on some neighborhood $\mathscr{U}$ of p such that $V_q \in \tau_q$ for each $q \in \mathscr{U}$. Such a smooth function is called a *time-orientation* of M. If M admits a time-orientation, then M is said to be *time-orientable.* Then to choose a specific time-orientation on M is to *time-orient M.*

For example, Minkowski space $\boldsymbol{R}^n_1$ is time-orientable; its *usual time-orientation* is the one containing the coordinate vector field ∂_0 of natural coordinates $u^0, \ldots, u^{n-1}$.

32. Lemma. A Lorentz manifold M is time-orientable if and only if there exists a timelike vector field $X \in \mathfrak{X}(M)$.

Proof. If such an X exists then (as above for $\boldsymbol{R}^n_1$) assigning to each $p \in M$ the timecone containing X_p gives a time-orientation.

Conversely, let τ be a time-orientation of M. Since τ is smooth, each point of M has a neighborhood $\mathscr{U}$ on which is defined a timelike vector field $X_{\mathscr{U}}$ whose value at each $p \in \mathscr{U}$ is in τ_p. Now let $\{f_\alpha | \alpha \in A\}$ be a smooth partition of unity subordinate to the covering of M by all such neighborhoods. Thus each supp f_α is contained in some member $\mathscr{U}(\alpha)$ of the covering. The functions f_α are nonnegative and timecones are convex. Thus the vector field $X = \sum f_\alpha X_{\mathscr{U}(\alpha)}$ is timelike. ∎

For example, *all Lorentz spheres S^n_1 are time-orientable,* because if ∂_0 is the (timelike) natural coordinate vector field on $\boldsymbol{R}^{n+1}_1$, then $X = \tan \partial_0$ is a timelike vector field on S^n_1. Also, *all Lorentz hyperbolic spaces $H^n_1 \subset \boldsymbol{R}^{n+1}_2$ are time-orientable.* In fact, since $P = \sum u^i \partial_i$ is normal to H^n_1, it is easy to check that $u^2 \partial_1 - u^1 \partial_2$ is tangent to H^n_1 and timelike.

For a Lorentz manifold there is no relation between orientability (1.41) and time-orientability. For example, it is easy to assign a time-orientable Lorentz metric to the orientable band $S^1 \times I$ and a not time-orientable metric to the (nonorientable) Möbius band. The reverse situation is suggested in Figure 5.

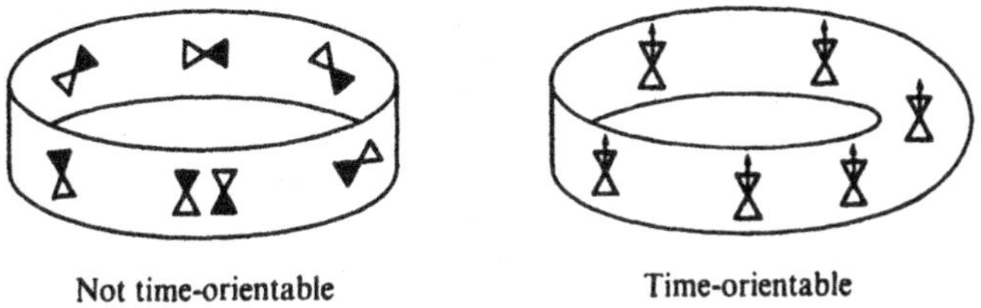

Figure 5

In a Lorentz vector space a vector that is nonspacelike (hence either null or timelike) is also said to be *causal*. For a timelike vector v the set $\overline{C}(v)$ of all causal vectors w such that $\langle v, w\rangle < 0$ is the *causal cone* containing v. Causal cones have properties quite similar to timecones (see Exercise 3).

In a Lorentz manifold a *causal curve* is one whose velocity vectors are all nonspacelike.

LOCAL LORENTZ GEOMETRY

Since Lorentz manifolds have index 1, it is natural to focus attention on their timelike curves. In the case of a piecewise smooth curve, α *timelike* means not only that every $\alpha'(t)$ is timelike, but that at each break t_i of α

$$\langle \alpha'(t_i^-), \alpha'(t_i^+)\rangle < 0.$$

Here the first vector derives from $\alpha|[t_{i-1}, t_i]$ and the second from $\alpha|[t_i, t_{i+1}]$. Thus α' does not switch timecones at a break. Similarly, we require that a piecewise smooth causal curve does not switch causal cones at a break.

33. Lemma. Let o be a point of a Lorentz manifold M. Suppose that $\beta\colon [0, b] \to T_o(M)$ is a piecewise smooth curve starting at 0 such that $\alpha = \exp_o \circ \beta$ is timelike. Then β remains in a single timecone of $T_o(M)$.

Proof. Suppose first that β (hence α) is smooth. In the following argument, *initially* will mean: *for all* $0 < t < \varepsilon$ *with* $\varepsilon > 0$ *sufficiently small.* Since $\beta'(0)$ is timelike and timecones are convex open sets, initially β is in a single timecone C. Since the position vector field $\tilde{P}$ is outward radial and timelike on C, initially $\langle \beta', \tilde{P}\rangle$ is negative.

Since grad $\tilde{q} = 2\tilde{P}$,

$$\frac{d(\tilde{q} \circ \beta)}{dt} = 2\langle \beta', \tilde{P}\rangle,$$

and by the Gauss lemma

$$\langle \alpha', P\rangle = \langle \beta', \tilde{P}\rangle.$$

It follows that initially $\langle \alpha', P\rangle$ and $d(\tilde{q} \circ \beta)/dt$ are negative.

So long as β remains in C, $\tilde{P}_\beta$ and (by the Gauss lemma) P_α remain timelike. Thus $\langle \alpha', P\rangle$ hence $\langle \beta', \tilde{P}\rangle$ hence $d(\tilde{q} \circ \beta)/dt$ remain negative. But β can leave C only by reaching either 0 or the nullcone, on both of which $\tilde{q}$ is 0. Thus β must remain in C.

Now suppose that β (hence α) is merely piecewise smooth. We know from above that on its first smooth segment β stays in C; thus at the first break, $\langle \beta'(t_1^-), \tilde{P}\rangle < 0$. Hence by the Gauss lemma $\langle \alpha'(t_1^-), P_1\rangle < 0$, where

$P_1 = d\,\exp_0(P_{\beta(t_1)})$. The additional condition on α at breaks keeps $\alpha'(t_1^+)$ in the same timecone, namely that of P_1. So, again by the Gauss lemma, $\langle \beta'(t_i^+), \tilde{P} \rangle < 0$. Thus it follows as above that $d(\tilde{q} \circ \beta)/dt$ cannot change signs at breaks, hence the argument for the smooth case remains valid. ∎

Minor changes in this proof show that the lemma remains true if the words *timelike* and *timecone* are replaced by *causal* and *causal cone.*

We can now prove a Lorentz analogue of Lemma 14 for timelike curves.

34. Proposition. Let $\mathscr{U}$ be a normal neighborhood of o in a Lorentz manifold M. If there exists a timelike curve in $\mathscr{U}$ from o to p, then the radial geodesic segment σ from o to p is the unique longest timelike curve in $\mathscr{U}$ from o to p.

Proof. As before, uniqueness is up to monotone reparametrization. If α is any timelike curve in $\mathscr{U}$ from o to p, then by the lemma, $\beta = \exp_o^{-1} \circ \alpha$ lies in a single timecone C. Hence (but for $\alpha(0) = o$), α lies in a region on which $U = P/r$ is a unit timelike vector field.

In particular σ is timelike. The argument is now a straightforward variant of that in the Riemannian case. Write

$$\alpha' = -\langle \alpha', U \rangle U + N,$$

where N is a (spacelike) vector field on α orthogonal to U. Then

$$|\alpha'| = (-\langle \alpha', \alpha' \rangle)^{1/2} = [\langle \alpha', U \rangle^2 - \langle N, N \rangle]^{1/2} \le |\langle \alpha', U \rangle|.$$

Since $r = \sqrt{-q}$ along α we have $\operatorname{grad} r = -P/r = -U$. By the lemma, $\langle \beta, \tilde{P} \rangle$, hence $\langle \alpha', U \rangle$, is negative, so

$$|\langle \alpha', U \rangle| = -\langle \alpha', U \rangle = d(r \circ \alpha)/dt.$$

Consequently,

$$L(\alpha) = \int_0^b |\alpha'|\, dt \le r(p) = L(\sigma).$$

If $L(\alpha) = L(\sigma)$, then as in the Riemannian case, $N = 0$ and α is a monotone reparametrization of σ; in fact, $\alpha(t) = \sigma(r(\alpha(t))/r(p))$. ∎

This proof works because if the timelike curve α strays from the radial geodesic σ its velocity acquires a spacelike component N that serves to reduce $|\alpha'|$ relative to $|\sigma'|$, hence to reduce $L(\alpha)$ relative to $L(\sigma)$.

As before the result can be refined by omitting the normal neighborhood and comparing σ with timelike curves of the form $\alpha = \exp_o \circ \beta$, where β is a curve in $T_o(M)$ from 0 to v such that $\exp_o(v) = p$ and $d\,\exp_o$ is nonsingular at each point of β.

In Minkowski space R^n_1 the entire manifold is a normal neighborhood of each point. Thus the proposition implies that any timelike geodesic segment in R^n_1 is the unique longest timelike curve joining its endpoints.

35. Example. Lorentz Cylinders. (1) The Lorentz surface $M = S^1_1 \times R^1$ can be viewed as a cylinder in R^3_2 (see Figure 6). It has essentially the same connection, hence same geodesics, as its Riemannian analogue in Example 17. Nullcones on M are marked out by intersecting null geodesics, for example $(\pm\cos s, \sin s, s + c)$.

Assertion (a) in Example 17 has the following Lorentz analogue: By Proposition 34 the timelike geodesic σ shown in Figure 6 is the unique longest timelike curve in $M - L$ from o to p. There are longer curves in $M - L$ that are not timelike, and there are longer timelike curves that do not remain in $M - L$.

Assertion (b) has no Lorentz analogue. Explicitly, there exists no normal neighborhood of o whose timelike radial geodesic segments are the longest in M joining their endpoints. In fact, any two points of M can be joined by arbitrarily long timelike geodesics (spiraling repeatedly around the cylinder).

(2) The Lorentz cylinder $R^1_1 \times S^1$ can be gotten by reversing the metric tensor of $S^1_1 \times R^1$, hence reversing the causal character of geodesics. It is not hard to see that the analogue of (b) mentioned above is valid now.

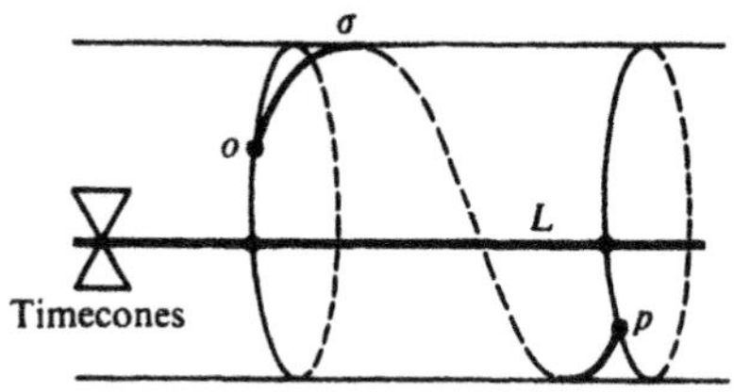

Figure 6. $S^1_1 \times R^1$.

In favorable cases a Riemannian manifold can be turned into a Lorentz manifold as follows.

36. Lemma. Suppose U is a unit vector field on a Riemannian manifold M with metric tensor g. Then $\tilde{g} = g - 2U^* \otimes U^*$ is a Lorentz metric on M. Furthermore, U becomes timelike so the resulting Lorentz manifold is time-orientable.

Proof. Recall that $U^*(X) = g(U, X)$ for vector fields X. Locally there exist vector fields E_j such that $U, E_2, \ldots, E_n$ is a frame field relative to g.

Then $\tilde{g}(E_i, E_j) = g(E_i, E_j) = \delta_{ij}$, and $\tilde{g}(U, E_j) = g(U, E_j) = 0$, but

$$\tilde{g}(U, U) = -1.$$

The final assertion follows from Lemma 32. ∎

For example, both Lorentz cylinders in Example 35 can be gotten in this way from the Riemannian cylinder $S^1 \times \boldsymbol{R}^1$.

By contrast with the Riemannian case, not every smooth manifold can be made a Lorentz manifold.

37. Proposition. For a smooth manifold M the following are equivalent:

(1) There exists a Lorentz metric on M.
(2) There exists a time-orientable Lorentz metric on M.
(3) There is a nonvanishing vector field on M.
(4) Either M is noncompact, or M is compact and has Euler number $\chi(M) = 0$.

Proof. (3) $\leftrightarrow$ (4). This is a standard topological result [V]. (2) $\rightarrow$ (1) is trivial. (3) $\rightarrow$ (2). By Lemma 25 there is a Riemannian metric tensor on M. For $X \in \mathfrak{X}(M)$ nonvanishing, apply the preceding lemma to $X/|X|$. (2) $\rightarrow$ (3) is immediate from Lemma 32. Thus it suffices to prove (1) $\rightarrow$ (4).

If M is time-orientable the preceding results establish (4). If M is not time-orientable we will see in Chapter 7 that it has a double-covering Lorentz manifold $\tilde{M}$ that is time-orientable. Thus $\tilde{M}$ is either noncompact or has $\chi(\tilde{M}) = 0$. Because the covering map $\tilde{M} \rightarrow M$ sends two points to one, M is compact if and only if $\tilde{M}$ is—and in the compact case $\chi(M) = \chi(\tilde{M})/2 = 0$. ∎

For example, the only compact surfaces that can be made Lorentz surfaces are the torus and Klein bottle. Also, a sphere S^n admits a Lorentz metric if and only if n is odd ≥ 3.

GEODESICS IN HYPERQUADRICS

For any semi-Riemannian manifold a basic problem is to understand the global behavior of its geodesics. As a simple example we consider hyperquadrics. Proposition 4.28 tells what the geodesics of $S^n_\nu(r)$ are; a closer look at its proof shows where they go. (See Figure 4.3.)

38. Proposition. Let p and q be distinct nonantipodal points of $S^n_\nu(r)$.

(1) If $\langle p, q\rangle > r^2$, then p and q lie on a unique geodesic, which is timelike and one-to-one.

(2) If $\langle p, q\rangle = r^2$, then p and q lie on a unique geodesic, which is also a null geodesic of $\mathbf{R}^{n+1}_\nu$.

(3) If $-r^2 < \langle p, q\rangle < r^2$, then p and q lie on a unique geodesic, which is spacelike and periodic.

(4) If $\langle p, q\rangle \leq -r^2$, there exists no geodesic joining p and q.

Proof. The hypotheses on p and q imply that they lie in a unique plane Π through the origin of $\mathbf{R}^{n+1}_\nu$, and by Proposition 4.28 the only geodesic that can possibly pass through both p and q is a parametrization of a component of the one-dimensional manifold $\Pi \cap S^n_\nu(r)$. Consider now the three cases in the proof of that proposition.

Case 1. Π is positive definite. Then $\Pi \cap S^n_\nu(r)$ is a circle in Π parametrized by a (periodic) spacelike geodesic and $-r^2 < \langle p, q\rangle < r^2$.

Case 2. Π is nondegenerate indefinite. As in Example 2.21, $\Pi \cap S^n_\nu(r) = \{x \in \Pi : \langle x, x\rangle = -r^2\}$ is a hyperbola of two branches. It is easy to see that p and q are on the same branch if and only if $\langle p, q\rangle > r^2$, and are on opposite branches if and only if $\langle p, q\rangle < -r^2$. In the former case p and q lie on a (timelike) geodesic; in the latter on no geodesic.

Case 3. Π is degenerate. Proposition 4.28 shows that $\Pi \cap S^n_\nu(r)$ consists of two parallel null straight lines of $\mathbf{R}^{n+1}_\nu$, with p and q on the same line if and only if $\langle p, q\rangle = r^2$; on opposite lines if and only if $\langle p, q\rangle = -r^2$.

Since the various restrictions on $\langle p, q\rangle$ above are mutually exclusive, the result follows. ∎

The corresponding result for pseudohyperbolic spaces derives as usual from the anti-isometry of Lemma 4.24. These results let us answer reasonable questions about the geodesics of hyperquadrics. For example, it follows, as predicted by the Hopf–Rinow theorem, that spheres and hyperbolic spaces are geodesically connected. By contrast *an indefinite hyperquadric Q is never geodesically connected*, since it always contains points p, q for which $\langle p, q\rangle$ has arbitrarily large positive and negative values. In fact, taking $Q = S^n_\nu(r)$, the points connectable to p by geodesics (that is, the image of $\exp_p$) consist of $-p$ and all points of Q on the same side of the hyperplane $\{x : \langle p, x\rangle = -r^2\} \approx T_{-p}(Q)$ as p itself.

GEODESICS IN SURFACES

A semi-Riemannian surface is either Lorentz or definite—and the latter can be supposed, as usual, to be Riemannian. Even in dimension 2 the geodesic differential equations remain quite complicated, so we consider a special case: surfaces that admit coordinate systems x, y for which E, F, G depend only on one coordinate, say x; hence, $E_y = F_y = G_y = 0$. On the domain of such a coordinate system the geometry is "constant in the y direction"; that is, each coordinate y-translation $(x, y) \to (x, y + c)$ is an isometry.

39. Lemma. Let x, y be surface coordinates with $E_y = F_y = G_y = 0$.

(1) If γ is a geodesic with coordinate functions $x(s)$, $y(s)$ then

$$\langle \gamma', \partial_y \rangle = Fx' + Gy' = C$$

(*the conservation equation*), where C is constant. Hence

$$(EG - F^2)x'^2 = Ge - C^2,$$

where $e = \langle \gamma', \gamma' \rangle = Ex'^2 + 2Fx'y' + Gy'^2$.

(2) The y-coordinate curve, $x = x_0$, is a geodesic if and only if $G_x(x_0) = 0$.

(3) If, furthermore, $F = 0$ then every x-coordinate curve is pregeodesic.

Proof. (0) In view of the identities $\langle \partial_x, D_{\partial_x} \partial_y \rangle = E_y/2$, $\langle \partial_y, D_{\partial_y} \partial_y \rangle = G_y/2$, and $\langle \partial_x, D_{\partial_y} \partial_y \rangle + \langle \partial_y, D_{\partial_x} \partial_y \rangle = F_y$, the hypotheses on E, F, G are equivalent to $\langle V, D_V(\partial_y) \rangle = 0$ for all V. (A more conceptual reason for this identity appears in Chapter 9.)

(1) The derivative of $\langle \gamma', \partial_y \rangle$ is $\langle \gamma'', \partial_y \rangle + \langle \gamma', D_{\gamma'} \partial_y \rangle = 0$. Subtracting $C^2 = (Fx' + Gy')^2$ from the coordinate formula for $Ge = G\langle \gamma', \gamma' \rangle$ gives the differential equation.

(2) This curve β has $\beta' = \partial_y$, hence

$$\langle \beta'', \partial_x \rangle = \langle D_{\partial_y} \partial_y, \partial_x \rangle = -G_x(x_0)/2; \qquad \langle \beta'', \partial_y \rangle = \langle D_{\partial_y} \partial_y, \partial_y \rangle = 0.$$

(3) Since $F = 0$, to show that the x-coordinate curve α is pregeodesic it suffices to observe that $\langle \alpha'', \partial_y \rangle = \langle D_{\partial_x} \partial_x, \partial_y \rangle = F_x - E_y/2 = 0$. ∎

Evidently the lemma remains valid under the reversal $x \leftrightarrow y$ hence $E \leftrightarrow G$.

Since the conservation equation is a first-order differential equation it is generally much easier to deal with than the (second-order) geodesic differential equations. A practical way to search for geodesics, under the hypotheses of the lemma, is to try to find the constant-speed solutions of the conservation equation. This class of curves contains all geodesics—and little else (only linear parametrizations of y-parameter curves).

40. Example. The Poincaré Half-plane P. By Exercise 3.8, P is the region $v > 0$ in $\mathbf{R}^2$ with line element $ds^2 = (du^2 + dv^2)/v^2$ and constant curvature $K = -1$. The preceding lemma applies with $u = y$ and $v = x$.

(1) *Isometries.* For any numbers a and $r > 0$, the mappings $(u, v) \to (\pm u + a, v)$ and $(u, v) \to (ru, rv)$ are isometries.

(2) *Geodesics.* The conservation equation is $\langle \gamma', \partial_u \rangle = u'/v^2 = C$. For $C = 0$ this yields the vertical lines u constant, so suppose $C \neq 0$. Assuming $|\gamma'| = 1$ we get $(u'^2 + v'^2)/v^2 = 1$. Substituting $u' = Cv^2$ yields $v'^2 = v^2(1 - C^2v^2)$. Since u' is never zero,

$$\frac{dv}{du} = \frac{v'}{u'} = \frac{\sqrt{1 - C^2v^2}}{Cv}.$$

Let $r = 1/C$; then an integration gives

$$\sqrt{r^2 - v^2} = u - u_0, \qquad \text{that is,} \quad (u - u_0)^2 + v^2 = r^2.$$

Thus the geodesics of P are the constant speed parametrizations of all vertical lines and all semicircles centered on the u axis.

This surface gives a concrete model for the so-called non-Euclidean geometry of Bolyai and Lobachevski, which satisfies all the axioms of Euclidean plane geometry except the parallel postulate. Indeed, through each point of P not on a geodesic γ there pass infinitely many geodesics that do not meet γ. (It will turn out that P is isometric to the hyperbolic plane $H^2(1)$.)

Lorentz surfaces have two notable features not shared by higher dimensional Lorentz manifolds. First, reversing the metric gives again a Lorentz surface. Thus Lorentz surfaces with $K > 0$ differ from those with $K < 0$ only in causal character.

Second, every null curve is pregeodesic. In fact, $\langle \alpha', \alpha' \rangle = 0$ implies $\langle \alpha'', \alpha' \rangle = 0$, and $\alpha'^{\perp}$ is one-dimensional for a surface, so α'' must be collinear with $\alpha' \neq 0$. Thus α is not turning, and hence (Exercise 3.19) is pregeodesic. In terms of arbitrary coordinates, the null curves are the solutions of

$$\langle \alpha', \alpha' \rangle = Eu'^2 + 2Fu'v' + Gv'^2 = 0.$$

41. Example. The Schwarzschild Half-plane P_I. For constant $\text{M} > 0$ let $h(r) = 1 - (2\text{M}/r)$. Then P_I is the region $r > 2\text{M}$ in the tr-plane, furnished with Lorentz line element

$$ds^2 = -h\, dt^2 + h^{-1}\, dr^2.$$

(1) *Curvature.* In terms of Proposition 3.44(2) take $\varepsilon_1 = -1$, $e = \sqrt{h}$; $\varepsilon_2 = 1$, $g = 1/\sqrt{h}$. Then we compute $K = 2\text{M}/r^3 > 0$.

(2) *Isometries.* The mappings $(t, r) \to (\pm t + b, r)$ are isometries.

(3) *Geodesics.* Lemma 39 applies with $x = r$ and $y = t$. The r-coordinate curves are spacelike pregeodesics. We now find the null geodesics, which obey

$$-h t'^2 + h^{-1} r'^2 = 0.$$

The conservation equation is $\langle \gamma', \partial_t \rangle = -\hbar t' = \text{const}$. Because of (2) it will be enough to find a single null geodesic, so by a convenient choice of constants we reduce the above equation to just $r' = 1$. Hence $r = s + b$. To integrate

$$\left(1 - \frac{2\text{M}}{s + b}\right)\frac{dt}{ds} = 1,$$

choose $b = 2\text{M}$, which yields

$$t = s + 2\text{M} \ln s, \qquad r = s + 2\text{M} \qquad \text{for} \quad s > 0.$$

Applying the isometries in (2) gives explicit parametrizations for all the null geodesics of P_I. Intersecting null geodesics mark out the null cones of P_I (Figure 7).

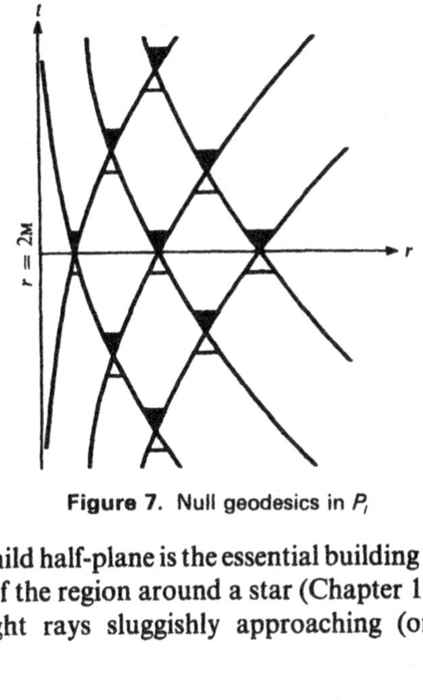

Figure 7. Null geodesics in P_I

The Schwarzschild half-plane is the essential building block in the simplest relativistic model of the region around a star (Chapter 13). Its null geodesics describe radial light rays sluggishly approaching (or departing) radius $r = 2\text{M}$.

42. Definition. A coordinate system u, v in a Lorentz surface is *null* provided its coordinate curves are null. Thus the line element has the form $ds^2 = 2F\,du\,dv$, where as usual $F = \langle \partial_u, \partial_v \rangle$.

The local geometry of the surface can then be traced back to the function F (see Exercise 8). As an example, rotating the natural coordinates t, x of $\boldsymbol{R}_1^2$ by 45° gives null coordinates

$$u = (1/\sqrt{2})(-t + x), \qquad v = (1/\sqrt{2})(t + x)$$

for which $2\,du\,dv = -dt^2 + dx^2$.

COMPLETENESS AND EXTENDIBILITY

For a manifold with indefinite metric, completeness is a more subtle notion than in the Riemannian case, since the Hopf–Rinow theorem (21) has has no satisfactory generalization. In discussing completeness it suffices as usual to consider only geodesics defined on $[0, b)$; left endpoints can be handled similarly. Lemma 8 gives a convenient topological criterion for the inextendibility of a geodesic $\gamma: [0, b) \to M$; namely, *there is a parameter sequence* $\{s_n\} \to b$ *such that* $\{\gamma(s_n)\}$ *does not converge.* (If two such sequences converge to different points, interlacing them gives a nonconvergent sequence.) Thus for example all the geodesics of the Poincaré half-plane are inextendible (at both ends) as are all the null geodesics of the Schwarzschild half-plane.

Sometimes pregeodesics are available but not their geodesic parametrizations. *A spacelike or timelike pregeodesic* $\alpha: [0, b) \to M$ *is complete* (to the right) *if and only if it has infinite length.* This is clear since the unit speed reparametrization of α is a geodesic defined on the interval $[0, L(\alpha))$. In this way it is easy to check that the Poincaré half-plane is complete; for example, a typical "semicircular" pregeodesic $\alpha(s) = (\sin s, \cos s)$, $0 \leq s < \pi/2$, has $\langle \alpha', \alpha' \rangle = \sec^2 s$, hence

$$L(\alpha) = \int_0^{\pi/2} \sec s \, ds = \infty.$$

For null geodesics there is no such simple criterion, but null geodesics are often easier to compute. For example, those of the Schwarzschild half-plane are incomplete as s decreases; since they are inextendible, P_I is incomplete.

On a manifold with indefinite metric, completeness can be separated by causal character into *spacelike completeness* (inextendible spacelike geodesics complete), *null completeness*, and *timelike completeness.* A complete manifold of course satisfies all three, while M minus a point satisfies none, But the conditions are independent [BE]; for instance, here is a Lorentz surface that is null and spacelike complete but timelike incomplete.

43. Example (Geroch). On $\boldsymbol{R}^2$ with coordinates (t, x) it is easy to see that there is a smooth function $f > 0$ satisfying (1) $f = 1$ outside the open strip $|x| < 1$, (2) f is symmetric about the t axis, (3) $\int_0^\infty f(t, 0)\, dt$ is finite. Let M be this plane with line element $f^2(-dt^2 + dx^2)$. Then M is *timelike incomplete but null and spacelike complete.* In fact, by (2), the t-axis can be parametrized as a timelike geodesic, but it is incomplete, since by (3) it has finite length (from origin). Outside the strip the metric is Minkowskian, so inextendible null and spacelike geodesics that avoid the strip are certainly complete. But if such a geodesic β meets the strip it must leave because β is a curve with $|dt/dx| \leq 1$ (since the timecones of M are still Minkowskian), so staying in the strip would produce an endpoint. Having left the strip, β evidently cannot return and hence is complete.

Riemannian geometers are accustomed to assuming their manifolds are complete, but relativity theory necessarily involves Lorentz manifolds that are incomplete. For incomplete manifolds the following notion becomes important.

44. Definition. A connected semi-Riemannian manifold M is *extendible* provided M is (isometric to) an open submanifold of a connected semi-Riemannian manifold $\tilde{M} \neq M$. In general M is *extendible* if one of its connected components is extendible. Otherwise M is *inextendible* (or *maximal*).

Compact manifolds are inextendible. By Exercise 3.7 a complete manifold is inextendible. The converse is false; for example, an ordinary cone (minus vertex) in $\boldsymbol{R}^3$ is inextendible but not complete.

45. Remark. Let M be a connected semi-Riemannian manifold. Suppose that for every inextendible geodesic $\gamma\colon [0, b) \to M$ of given causal character there is a curvature invariant I such that $I(\gamma(s))$ does not approach a finite limit as $s \to b$. (For example the invariant could be $\mathrm{Ric}(\gamma', \gamma')$ or $K(\gamma', V)$ for a parallel vector field V on γ.) Then M is inextendible, for otherwise some γ has an extension past b in $\tilde{M}$ hence $I(\gamma(s))$ has limit $I(\tilde{\gamma}(b))$ as $s \to b$.

Exercises

1. For a Riemannian manifold M, (a) the distance function $d\colon M \times M \to M$ is continuous; (b) if M is complete and C is a closed set of M, then for any $p \in M$ there is a point of C closest to p.

2. In a Lorentz vector space, (a) orthogonal null vectors are collinear, (b) orthogonal nonspacelike vectors are null hence collinear, (c) there exists no two-dimensional subspace on which the scalar product is identically zero.

3. Let V be a Lorentz vector space. (a) Nonspacelike vectors v, w are in the same causal cone if and only if either $\langle v, w\rangle < 0$ or v and w are null with $w = av$, $a > 0$. (b) If u is timelike, $\bar{C}(u) = C(u) \cup$ (one component of Λ) = (closure of $C(u)$) $-$ 0. (c) Causal cones are convex. (d) The components of the set of all nonspacelike vectors in V are the two causal cones in V.

4. If V is a scalar product space of index ν and dimension $\geqslant 2$, the set of all timelike vectors is connected (or empty) if $\nu \neq 1$.

5. In a semi-Riemannian surface, if t, r is a coordinate system with $ds^2 = E(r)\,dt^2 + G(r)\,dr^2$, compute: (a) $H^r(\partial_t, \partial_t) = E_r/2G$, $H^r(\partial_t, \partial_r) = 0$, $H^r(\partial_r, \partial_r) = -G_r/2G$. (b) grad $r = \partial_r/G$. (c) $\Delta r = (1/2G)[(E_r/E) - (G_r/G)]$.

6. If σ is a timelike geodesic in $\mathbf{R}^3_2$ from p to q, then arbitrarily close to σ there are smooth timelike curves from p to q that are (a) longer than σ, (b) shorter than σ.

7. (a) If X and Y are linearly independent vector fields on a neighborhood of a point p in a surface, there is a coordinate system (u, v) at p such that ∂_u and ∂_v are collinear with X and Y, respectively. (Hint: Let (u, y) be a coordinate system such that $Y = \partial/\partial y$.) (b) At each point of a Lorentz surface there is a null coordinate system.

8. For a null coordinate system in a Lorentz surface, (a) $D_{\partial_u} \partial_u = F_u/F\, \partial_u$, $D_{\partial_v} \partial_v = F_v/F\, \partial_v$, $D_{\partial_u} \partial_v = D_{\partial_v} \partial_u = 0$. (b) The Gaussian curvature is $K = (-1/F)(F_u/F)_v = (-1/F)(F_v/F)_u$. (c) For a function f,

$$\operatorname{grad} f = (f_v\,\partial_u + f_u\,\partial_v)/F \qquad \text{and} \qquad \Delta f = 2f_{uv}/F.$$

(d) The geodesic differential equations are

$$u'' + (F_u/F)u'^2 = 0, \qquad v'' + (F_v/F)v'^2 = 0.$$

9. In a Lorentz manifold, if σ is a longest timelike curve joining p and q, then σ is a monotone reparametrization of an (unbroken) timelike geodesic from p to q.

10. Directional derivatives. Let u be a unit vector in $T_p(M)$ and let $f \in \mathfrak{F}(M)$ have $\operatorname{grad}_p f \neq 0$. Prove: (a) If M is Riemannian, then $u(f) = |\operatorname{grad}_p f| \cos \vartheta$ where $\vartheta = \measuredangle(u, \operatorname{grad}_p f)$. (b) If M is Lorentz, and $\operatorname{grad}_p f$ and u are timelike in the same timecone, then $u(f) = -|\operatorname{grad}_p f| \cosh \varphi$, where φ is the Lorentz angle between u and $\operatorname{grad}_p f$ (change sign if opposite timecone). (c) If M is Lorentz and one of u, $\operatorname{grad}_p f$ is null, the other timelike, then $u(f) \neq 0$.

11. *Hyperbolic space.* If $p, q \in H^n(r)$ then (a) there is a unique geodesic $\sigma\colon \mathbf{R} \to H^n(r)$ such that $\sigma(0) = p$ and $\sigma(1) = q$. (b) $d(p, q) = L(\sigma|[0, 1]) = r\varphi$, where φ is the hyperbolic angle between p and q in $\mathbf{R}^{n+1}_1$ (compare Example 20). (c) If also $m \in H^n(r)$, hyperbolic angles satisfy $\measuredangle(p, q) \leq \measuredangle(p, m) + \measuredangle(m, q)$, with equality if and only if $m \in \gamma[0, 1]$.

12. In a complete Riemannian manifold, (a) if each $\sigma_i\colon [0, 1] \to M$ is minimizing and $\{\sigma_i'(0)\}$ converges to $v \in TM$, then $\gamma_v|[0, 1]$ is minimizing. (b) If M is not compact there exists a minimizing ray $\rho\colon [0, \infty) \to M$ starting at $p \in M$ (that is, each subsegment of ρ is minimizing).

13. Call a semi-Riemannian manifold M *Misner-complete* provided no geodesic races to infinity, that is, provided every geodesic $\gamma\colon [0, b) \to M$, $b < \infty$, lies in a compact set. Prove: (a) complete $\Rightarrow$ Misner-complete $\Rightarrow$

inextendible; (b) neither converse holds; (c) a Misner-complete Riemannian manifold is complete. (For (b), see Example 7.16.)

14. If $\alpha\colon [0, B) \to M$, $B \leq \infty$, is an extendible piecewise smooth causal (= nonspacelike) curve in a Lorentz manifold, then α has finite length.

15. Let M be a connected manifold with indefinite metric. Prove: (a) Each point of M has a neighborhood any two points of which can be joined by an at-most-once-broken null geodesic segment. (b) Any two points of M can be joined by a broken null geodesic (hence Riemannian distance on M is identically zero). (c) If M is null complete it is inextendible. (d) Same as (c) with *null* replaced by *timelike* [*spacelike*].

16. Let g_1 and g_2 be scalar products of index 1 on a vector space V. If they have the same nullcone then $g_2 = Cg_1$.

6 SPECIAL RELATIVITY

By the end of the last century it had become clear that there were serious difficulties in classical Newtonian physics, centering around the properties of light. Progress in resolving these difficulties was made by Lorentz, Poincaré, and others, but the first comprehensive solution was given by Einstein in 1905 with the publication of his special theory of relativity. Its mathematical essence was a novel way to change space and time coordinates; in 1908 Minkowski showed that these occur naturally if space $\boldsymbol{R}^3$ and time $\boldsymbol{R}^1$ are merged in a single spacetime $\boldsymbol{R}^4_1$. "Henceforth space by itself, and time by itself," he wrote, "are doomed to fade away into mere shadows, and only a kind of union of the two will preserve an independent reality."

Special relativity has a number of features—bizarre from the Newtonian viewpoint—that will follow rather easily from the geometry of the Minkowski spacetime. To mention a few:

There is no way to determine whether two different events occur at the same time or in the same place.

There is no notion of (absolute) speed for a material particle; indeed there is no way to determine whether a nonaccelerating particle is moving or not.

On the other hand, the speed of light in a vacuum is a constant—independent of the motion of its source.

Moving clocks run slow; moving bodies shorten in the direction of the motion.

NEWTONIAN SPACE AND TIME

For the sake of comparison with relativity we review briefly some basic features of Newtonian motion.

1. Definition. *Newtonian space* is a Euclidean 3-space E, that is, a Riemannian manifold isometric to $\boldsymbol{R}^3$ (with dot product).

Since there are no coordinate axes in nature this definition is better than simply declaring Newtonian space to be $\boldsymbol{R}^3$.

For simplicity, *Newtonian time* will be modeled as the real line $\boldsymbol{R}^1$, but only time differences $s - t$ are significant, not particular times s, t.

Informally, a particle is an object that is negligibly small compared to the typical distances in the problem at hand. Thus an electron is a particle compared to an atom, and a galaxy is a particle compared to the universe at large. Particles have mass, and are capable of motion, described by a function from time to space.

2. Definition. A *Newtonian particle* is a curve $\alpha: I \to E$ in Newtonian space, with I an interval in Newtonian time.

This definition is consistent with the terminology *velocity* $\alpha' = d\alpha/dt$, *speed* $v = |\alpha'|$, *acceleration* $\alpha'' = d^2\alpha/dt^2$, and *arclength* $L(\alpha|[a, b])$.

Fundamentally mass is constant (and positive) in Newtonian physics, but a complex particle, modeling say a rocket expending fuel, can have mass a function of time.

3. Definition. If $\alpha: I \to E$ is a Newtonian particle of mass m, then

(1) The *momentum* of α is the vector field $m\alpha'$ on α; *scalar momentum* is the function $m|\alpha'|$ on I.

(2) The *force* on α is the vector field $d(m\alpha')/dt$ on α.

(3) The *kinetic energy* of α is the function $mv^2/2$ on I, where $v = |\alpha'|$.

When m is constant then (2)—Newton's second law of motion—takes the familiar form: Force equals mass times acceleration.

The definition of Newtonian space shows that it has preferred coordinate systems.

4. Definition. A *Euclidean coordinate system* for E is an isometry $\xi: E \to \boldsymbol{R}^3$.

Since ξ is in particular a diffeomorphism it is indeed a coordinate system for the manifold E. For any coordinate system, $d\xi(\partial/\partial x^i) = \partial/\partial u^i$, hence ξ is Euclidean if and only if $g_{ij} = \delta_{ij}$ for $1 \leq i, j \leq 3$. Euclidean coordinates are highly efficient in dealing with straight line problems. For example, geodesics then have affine coordinates $x^i(\gamma(t)) = a^i t + b^i$, distantly parallel tangent vectors have the same components, and the distance from p to q is given by the usual Pythagorean formula

$$d(p, q) = [\sum (x^i(q) - x^i(p))^2]^{1/2}.$$

In effect, once Euclidean coordinates are introduced Newtonian space turns into $\boldsymbol{R}^3$.

NEWTONIAN SPACE–TIME

Suppose α is a Newtonian particle moving on a line $L \approx \boldsymbol{R}^1$. Then we are all accustomed to drawing its graph $\{(t, \alpha(t)) : t \in I\}$ and interpreting slope as velocity (that is, $\pm$ speed), rate of change of slope as acceleration, and so on. The particle, by definition a curve, has thus become a one-dimensional submanifold of the plane $\boldsymbol{R}^2 =$ (time $\boldsymbol{R}^1$) $\times$ (space $\boldsymbol{R}^1$). This can be done more generally as follows.

5. Definition. *Newtonian space–time* is the Riemannian product manifold $\boldsymbol{R}^1 \times E$ of Newtonian time and Newtonian space.

The definition is superficial since the product metric lacks physical significance; however, it will serve to introduce some ideas that become significant when the correct (Minkowskian) metric tensor is used.

A point (t, x) of $\boldsymbol{R}^1 \times E$ is called an *event*: it is an instantaneous happening at a particular time $t \in \boldsymbol{R}^1$ and position $x \in E$. The natural projections of $\boldsymbol{R}^1 \times E$ on $\boldsymbol{R}^1$ and E are denoted by T and S, respectively. Thus T is the universal Newtonian clock that lets us measure the time interval between any two events.

As in the case above, a Newtonian particle can now be represented not by its *equation of motion* (Definition 2) but by its worldline, roughly speaking its life history in $\boldsymbol{R}^1 \times E$.

6. Definition. A *worldline* in Newtonian space–time is a one-dimensional submanifold W such that $T \mid W$ is a diffeomorphism onto an interval $I \subset \boldsymbol{R}^1$ (see Figure 1).

The transition between particles and worldlines is easy. Given a particle $\alpha: I \to E$, its graph $\{(t, \alpha(t)) : t \in I\}$ is a worldline. Conversely, given a worldline W the corresponding particle is reconstructed as $\alpha = S \circ (T \mid W)^{-1}$.

On a human scale the Newtonian account of motion is quite accurate, but when it is pushed to extremes, difficulties arise:

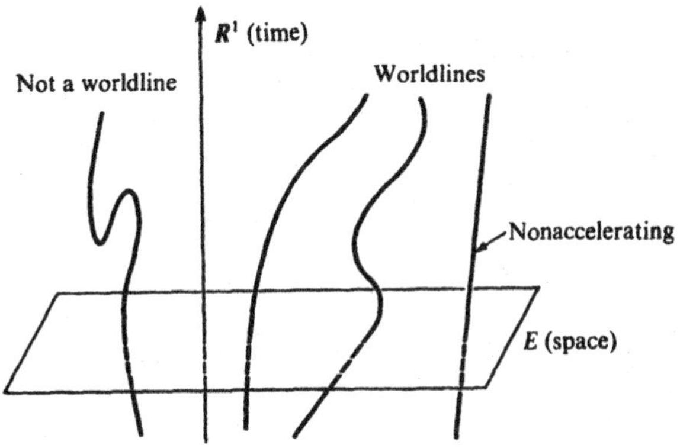

Figure 1

(1) For 300 years it has been known that in a vacuum light travels at a very high but nonetheless finite speed c, and no material object has been observed to travel faster. However, c plays no special role in Newtonian motion; for example, according to Newtonian theory a suitably designed rocket can attain arbitrarily high speeds (Exercise 4).

(2) Suppose that a rocketship is coming directly toward an observer at speed $c/10$. Its searchlight beam darts ahead at speed c relative to the ship and thus by Newtonian addition of velocities reaches the observer at speed $c + c/10$. If the rocketship were traveling directly away from him the light would arrive at speed $c - c/10$. Astronomical observation of double stars should readily reveal such fast and slow light, but in fact the speeds turn out to be the same.

(3) In Newtonian theory a particle is either at rest (the curve is constant) or not—a straightforward dichotomy. Suppose a rocketship ρ is out in space far from any external influence. If ρ is not accelerating how can it be determined whether ρ is at rest or not? A natural approach would be to measure the position of ρ relative to something that *is* at rest. But one by one the candidates failed: earth, sun, "fixed" stars, and the conjectured *ether*. (See, however, *background microwave radiation*, on page 357.) Nor can the crew of ρ perform a distinguishing experiment strictly within their ship. Such experiments deal with measurements relative to ρ and give no information about absolute motion.

In short, Newtonian physics treats light relatively (as in (2)) when it should be treated absolutely (1), and treats motion absolutely when it should be treated relatively (3).

There is a direct way to eliminate the Newtonian difficulty (1) above. The speed of a material particle α at the time t can be read from its worldline

W in terms of the angle between time axis $\boldsymbol{R}^1$ and the tangent line to W at $(t, \alpha(t))$. The tangent directions of light—always at constant speed c—thus determine a cone in every tangent space to $\boldsymbol{R}^1 \times E$. Requiring material particles to have their tangent lines inside the cone thus keeps their speed below c.

Within semi-Riemannian geometry a more natural way to get such cones is to change the sign of the time-coordinate of the metric tensor of $\boldsymbol{R}^1 \times E$, thus producing Minkowski spacetime with its *nullcones*. We shall see that a reasonable attempt to reconstruct Newtonian mechanics in this context eliminates the difficulties (2) and (3) as well, producing special relativity.

7. Remark. Geometric Units. A highly efficient system of physical units is obtained by taking the speed of light c and the gravitational constant G both to be the (dimensionless) number 1. All units then become powers of just one freely chosen unit, e.g., seconds, meters, ... The conversion factors between geometric units and any conventional system of units follow in an obvious way from the values c_{conv} and G_{conv} for that system. For example,

$$c_{\text{cgs}} = 3 \times 10^{10} \quad \text{cm/sec}; \qquad G_{\text{cgs}} = 6.67 \times 10^{-8} \quad \text{cm}^3/\text{g sec}^2.$$

In geometric units, distance and time are measured in the same units, as in the well-known case of distance in (light) years: the time required for light to travel that distance. Conversely time can be measured in units of distance: the distance light travels in the given time. It follows that speed v is dimensionless, with $v = v_{\text{conv}}/c_{\text{conv}}$ for any conventional units.

For example, the earth–sun distance $x_{\text{cgs}} = 1.5 \times 10^{13}$ cm (about 93 million miles) can also be given as $x_{\text{cgs}}/c_{\text{cgs}} = 500$ sec. A rocket of speed $v = 0.01$ has $v_{\text{cgs}} = (0.01)\ c_{\text{cgs}} = 3 \times 10^8$ cm/sec, that is, 3000 km/sec.

In geometric units, mass is also measured in the same units as distance and time. $G_{\text{cgs}}/(c_{\text{cgs}})^2 = 7.42 \times 10^{-29}$ cm/g, hence, for example, the mass of the sun, $\text{M}_{\text{cgs}} \approx 2 \times 10^{33}$ g, becomes 14.8×10^4 cm—about 1.5 km. Using seconds as the geometric unit, the sun's mass is

$$\frac{14.8 \times 10^4 \quad \text{cm}}{3 \times 10^{10} \quad \text{cm/sec}} = 4.9 \times 10^{-6} \quad \text{sec}.$$

Based on the fundamental constants c and G, geometric units have a direct physical significance lacking in more haphazard conventional systems. Certainly speed as $v = v_{\text{conv}}/c_{\text{conv}}$ is more informative than some number of feet per second or kilometers per year. Using geometric units, it also becomes meaningful, for example, to say that the sun's mass is small compared to its radius ($\text{M} = 1.5 \text{ km} \ll r = 7 \times 10^5$ km). As we see in Chapter 13 this fact is decisive for the qualitative character of the sun's gravitational field.

MINKOWSKI SPACETIME

A *spacetime* is a connected time-oriented four-dimensional Lorentz manifold. (Informally, *time-oriented* is often weakened to *time-orientable*.)

8. Definition. A *Minkowski spacetime* M is a spacetime that is isometric to Minkowski 4-space R_1^4.

In this chapter, M will always denote a Minkowski spacetime. As with any (time-oriented) spacetime, the time-orientation of M is called the *future*, and its negative is the *past*. A tangent vector in a future causal cone is said to be *future-pointing* (or *future-directed*). A causal curve is *future-pointing* if all its velocity vectors are future-pointing.

Comparison with Newtonian space–time is helpful in organizing intuitive ideas about Minkowski spacetime; thus points of M are called *events* and particles will be (parametrized) *worldlines*. On the other hand, by contrast with both $R^1 \times E$ and R_1^4, there exists no canonical time function on M. Though there is no Time there are many times:

9. Definition. A *material particle* in M is a timelike future-pointing curve $\alpha: I \to M$ such that $|\alpha'(\tau)| = 1$ for all $\tau \in I$. The parameter τ is called the *proper time* of the particle.

As in the Newtonian case a particle models the life history of an object which in a given context is negligibly small. We imagine that each particle comes equipped with a *clock* (mechanical, atomic, biological, . . .) measuring its proper time. As with Newtonian time, only intervals of proper time are significant. For example if α is a material particle from $\alpha(a) = p$ to $\alpha(b) = q$ then the arclength $b - a$ is its elapsed proper time between the events p and q. For any number d, $\tau \to \alpha(\tau + d)$ can be considered as the same particle with its clock reset, since the time interval between events is unchanged.

10. Definition. A *lightlike particle* is a future-pointing null geodesic $\gamma: I \to M$.

Contemporary physics identifies three types of lightlike particles: *photons* (light itself); *neutrinos* (elementary particles perhaps not perfectly lightlike); and (confidently conjectured) *gravitons* (outside the framework of special relativity).

Any particle $\beta: I \to M$ is a regular curve, and its image $\beta(I)$ is a one-dimensional submanifold of M called the *worldline* of β.

As in the Newtonian case, particles in M have *mass*, positive for material particles but, as we shall see, necessarily zero for lightlike particles.

That light moves geodesically is a fundamental hypothesis of relativity, and since $\langle \gamma', \gamma' \rangle = 0$ for a lightlike particle, parametrization by proper time is out of the question. Being massless it can't carry a clock!

A particle in M that is a geodesic is said to be *freely falling*. In general "freely falling" means moving under the influence of gravity alone. But Chapter 12 will show that the fact that Minkowski space is flat limits special relativity to situations where gravitation is negligible; for example: in elementary particle theory, where electromagnetism dominates; in empty space, far from significant sources of gravity; anywhere, provided times and distances involved are small enough to trivialize gravitational effects.

11. Definition. A *Lorentz* (or *inertial*) coordinate system in M is a time-orientation-preserving isometry $\xi: M \to \mathbf{R}_1^4$.

As in the Newtonian case, it follows immediately that a coordinate system $\xi: M \to \mathbf{R}^4$ is Lorentz if and only if $g_{ij} = \delta_{ij}\varepsilon_j$ (where $\varepsilon = (-1, 1, 1, 1)$) and ∂_0 is future-pointing.

12. Lemma. Given a frame e_0, e_1, e_2, e_3 in $T_p(M)$ such that e_0 is future-pointing, there is a unique Lorentz coordinate system ξ such that $\partial_i|_p = e_i$ for $0 \leq i \leq 3$.

Proof. The normal coordinate system ξ determined by the frame is one such coordinate system, with $\exp_p(\sum x^i(q)e_i) = q$ for all $q \in M$. If η is another, then $\xi^{-1} \circ \eta$ is an isometry of M fixing the given frame. Thus by Proposition 3.62, $\xi^{-1} \circ \eta = \text{id}$; that is, $\eta = \xi$. ∎

In prerelativistic physics the proper role of coordinates was not clear. Lacking a firm notion of *manifold* it was only reasonable to assume that the physics of Newtonian space was linked to the form its laws took in terms of Euclidean coordinates. Not the least of Einstein's contributions was his insistence (*principle of general covariance*) that every physical law has an expression independent of the choice of coordinates. Tensor formalism allows the last phrase to be replaced by "not using coordinates." Indeed, introducing a particular coordinate system in a given context raises the new problem of distinguishing intrinsic properties from coordinate properties. Of course this problem is simplified by using coordinates well adapted to the intrinsic data.

MINKOWSKI GEOMETRY

Since Minkowski spacetime is isometric to $\mathbf{R}_1^4$ we know that (1) for any points $p, q \in M$ there is a unique geodesic σ with $\sigma(0) = p$ and $\sigma(1) = q$, (2) there is a natural linear isometry $T_p(M) \approx T_q(M)$, called distant parallelism, and (3) each exponential map $\exp_p: T_p(M) \to M$ is an isometry. Thus M viewed from p is geometrically the same as $T_p(M)$ viewed from 0.

13. Remark. M is a normal neighborhood of each of its points. Thus for all $p, q \in M$ the displacement vector $\overrightarrow{pq} = \sigma'(0)$ of Chapter 5 is well defined, where σ is the geodesic as in (1) above. Note that $\exp_p(\overrightarrow{pq}) = q$.

In terms of a Lorentz coordinate system ξ, distantly parallel tangent vectors have the same components, and

$$\overrightarrow{pq} = \sum (x^i(q) - x^i(p))\, \partial_i.$$

The preceding remarks let us move causality from the tangent spaces of M to M itself. For an event $p \in M$ the *future timecone* of p is $\{q \in M : \overrightarrow{pq}$ is timelike and future-pointing$\}$. This is a solid cone whose boundary, but for p, is the *future lightcone* of p, namely $\{q \in M : \overrightarrow{pq}$ is null and future-pointing$\}$. The union of these two sets is the *future causal cone* of p. *Past* analogues are defined similarly.

The *lightcone* $\Lambda(p) \subset M$ of p is the union of the past and future lightcones of p (only locally defined in Chapter 5). A point q in neither causal cone of p is *spacelike relative to* p; that is, $\overrightarrow{pq}$ is spacelike.

The appropriateness of the term *causal* now becomes clear. It is natural to say that an event p can *influence* an event q if and only if there is a particle from p to q. By the definition of particles (material and lightlike) it follows from Lemma 5.33 that

(1) The only events that can be influenced by an event p are those in its future causal cone.

(2) The only events that can influence an event p are those in its past causal cone.

Thus "most" events—those that are spacelike relative to p—can neither influence nor be influenced by p. See Figure 2, which obeys the pictorial conventions that the future is upward and (by Lemma 16) light rises at 45°.

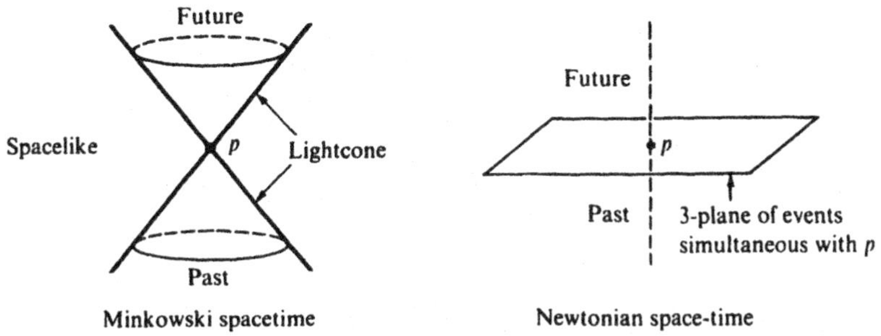

Figure 2

Relativistic causality contrasts sharply with Newtonian causality, where for an event $p = (x_0, t_0)$ the past and future fill the whole space-time except for the 3-plane $t = t_0$ of simultaneous events. Unrestricted by the speed of light, Newtonian rockets can go from x_0 to any distant place x in arbitrarily short time $t - t_0$.

14. Definition. For $p, q \in M$ the number $pq = |\overrightarrow{pq}| \geq 0$ is called the *separation* between p and q.

In terms of a Lorentz coordinate system,

$$pq = \left| -(x^0q - x^0p)^2 + \sum_1^3 (x^jq - x^jq)^2 \right|^{1/2}.$$

Because space and time are merged in M, separation is richer in information than the comparable notion of distance in Euclidean space.

15. Remark. Physical Significance of Separation. Let $p, q \in M$.

(1) If $\overrightarrow{pq}$ is timelike future-pointing, then pq is the elapsed proper time $L(\sigma)$ of the unique freely falling material particle from p to q. (A freely falling spaceship records pq as the time from event p to event q.)

(2) $\overrightarrow{pq}$ is lightlike $\Leftrightarrow pq = 0 \Leftrightarrow$ there is a lightlike particle through p and q.

(3) If $\overrightarrow{pq}$ is spacelike, then $pq \geq 0$ is the *distance from p to q as measured by any freely falling observer* $\perp \overrightarrow{pq}$. (We anticipate some terminology from the next section.)

The k-planes in M are the images under any isometry $\boldsymbol{R}_1^4 \to M$ of the k-planes in $\boldsymbol{R}_1^4$. By Chapter 4 these are totally geodesic and (if nondegenerate) isometric to either $\boldsymbol{R}^k$ or $\boldsymbol{R}_1^k$. If V is a subspace of $T_p(M)$ then

$$P = \exp_p(V) = \{q \in M : \overrightarrow{pq} \in V\}$$

is the unique k-plane in M with $T_p(P) = V$.

Finally we consider some trigonometry in M.

16. Lemma. If $\overrightarrow{op}$ is spacelike and $\overrightarrow{oq}$ is timelike, then any two of the following imply the third:

(1) $\overrightarrow{pq}$ is lightlike,
(2) $\overrightarrow{op} \perp \overrightarrow{oq}$,
(3) $op = oq$.

Proof. Moving the vector $\overrightarrow{pq}$ by parallelism to o gives $\overrightarrow{pq} = \overrightarrow{oq} - \overrightarrow{op}$. Taking scalar products then yields $\pm pq^2 = -oq^2 - 2\langle \overrightarrow{oq}, \overrightarrow{op} \rangle + op^2$, and the result follows. ∎

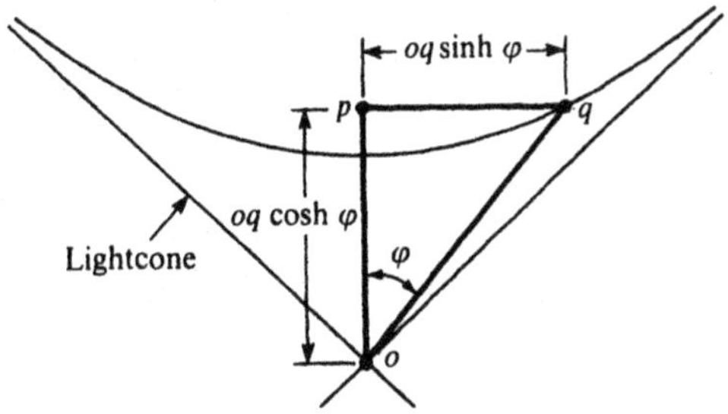

Figure 3. The curved line represents the set of future points at separation *oq* from *o*. (It is a hyperbolic 3-space.)

The notion of hyperbolic angle (5.30) transfers into M in the obvious way: if p and q are in the same timecone of o, then the *hyperbolic angle* $\varphi = \measuredangle\, poq$ is the hyperbolic angle between $\overrightarrow{op}$ and $\overrightarrow{oq}$.

17. Proposition. Let p and q be events in the same timecone of o and such that $op \perp pq$ (Figure 3). Then

(1) $oq^2 = op^2 - pq^2$.

(2) $op = oq \cosh \varphi, \quad pq = oq \sinh \varphi, \quad$ where $\varphi = \measuredangle\, poq$.

Proof. (1) As in the previous lemma, moving the spacelike vector $\overrightarrow{pq}$ to o gives $\overrightarrow{oq} = \overrightarrow{op} + \overrightarrow{pq}$. Then scalar products yield $-oq^2 = -op^2 + pq^2$.

(2) Let u and v be the (timelike) unit vectors in the direction of $\overrightarrow{op}$ and $\overrightarrow{oq}$, respectively. Then $\langle \overrightarrow{op}, \overrightarrow{oq} \rangle = op\, oq \langle u, v \rangle = -op\, oq \cosh \varphi$. But also $\langle \overrightarrow{op}, \overrightarrow{oq} \rangle = \langle \overrightarrow{op}, \overrightarrow{op} + \overrightarrow{pq} \rangle = -op^2$. Thus $op = oq \cosh \varphi$, and hence by (1), $pq^2 = oq^2(\cosh^2 \varphi - 1) = oq^2 \sinh^2 \varphi$. But $\phi \geq 0$, hence $\sinh \varphi \geq 0$, so $pq = oq \sinh \varphi$. ∎

Thus the Pythagorean formula is replaced by (1), and orthogonal projections are given by hyperbolic rather than circular sines and cosines. Note that the timelike projection op is always $\geq oq$ and the same can also hold for the spacelike projection pq.

PARTICLES OBSERVED

An *observer* in M is just a material particle, the terminology suggesting a new role. Let ξ be a Lorentz coordinate system. The x^0 axis of ξ is the worldline of a freely falling observer ω; the natural parametrization of ω has $x^0\omega(t) = t$, so t is proper time for ω. We think of the numbers produced by ξ as measurements taken by the observer ω (see Exercise 3). By Lemma 12, every freely falling observer has many such *associated Lorentz coordinate systems.*

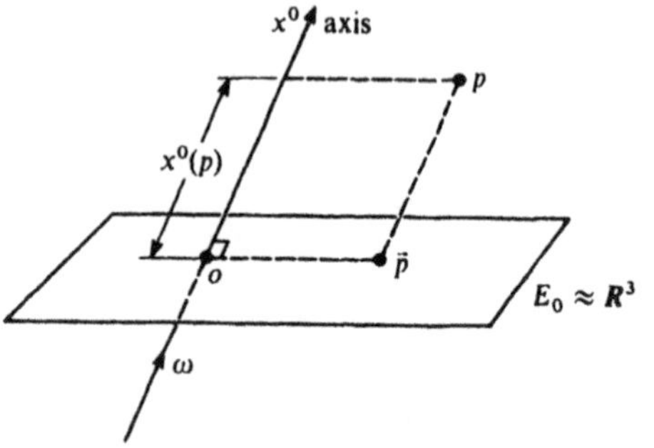

Figure 4

The coordinate slice $x^0 = 0$ of ξ is a Euclidean space E_0 that we identify with $\boldsymbol{R}^3$ by the natural isometry $q \leftrightarrow (x^1(q), x^2(q), x^3(q))$.

In M there is no natural way to define either time or space, but Lorentz coordinates effect an artificial decomposition as follows.

18. Definition. Let ξ be a Lorentz coordinate system in M. For each event $p \in M$ the number $x^0(p)$ is called the *ξ-time* of p and the point $\vec{p} = (x^1(p), x^2(p), x^3(p)) \in \boldsymbol{R}^3$ is called the *ξ-position* of p (see Figure 4).

Now let $\alpha: I \to M$ be a particle, either material or lightlike. For each parameter value $s \in I$ (proper time if α is material), the ξ-time of the event $\alpha(s)$ is $t = x^0(\alpha(s))$ and its ξ-position is $(x^1(\alpha(s)), x^2(\alpha(s)), x^3(\alpha(s)) \in \boldsymbol{R}^3 \approx E_0$. Since α is causal (nonspacelike) and future-pointing,

$$\frac{d(x^0 \circ \alpha)}{ds} = -\langle \alpha', \partial_0 \rangle > 0.$$

Hence $x^0 \circ \alpha$ is a diffeomorphism of I onto some interval $J \subset \boldsymbol{R}^1$. Let $u: J \to I$ be the inverse function. Then at ξ-time $t \in J$, the ξ-position of α is

$$\vec{\alpha}(t) = (x^1 \alpha u(t), x^2 \alpha u(t), x^3 \alpha u(t)).$$

Thus measurements of the particle α in M produce a curve $\vec{\alpha}: J \to \boldsymbol{R}^3 \approx E_0$ called the *ξ-associated Newtonian particle* of α. This is what the observer ω observes of α.

The relationship between α and $\vec{\alpha}$ is a guide for the development of special relativity. Applying Newtonian concepts to $\vec{\alpha}$ suggests how to find their relativistic analogues for α; reinterpreting the relativistic concepts in terms of α shows how the new theory has modified the old.

By convention, the parameters t and s are related by $t = x^0\alpha(s)$ and inversely by $s = u(t)$, and *only* by these functions. Thus it is meaningful to write $dt/ds = d(x^0 \circ \alpha)/ds > 0$ and, by the chain rule,

$$\frac{d\vec{\alpha}}{dt} = \frac{d\vec{\alpha}/ds}{dt/ds}.$$

19. Lemma. Let γ be a lightlike particle in M. For Lorentz coordinate system ξ, the associated Newtonian particle $\vec{\gamma}$ of γ is a straight line in $\boldsymbol{R}^3$ with speed 1.

Proof. Since γ is a geodesic in M, $\xi \circ \gamma$ is a geodesic in $\boldsymbol{R}_1^4$. Thus γ has affine coordinates

$$x^i\gamma(s) = a_i s + b_i \qquad (0 \le i \le 3).$$

Hence the projection $\vec{\gamma}(s) = (x^1\gamma(s), x^2\gamma(s), x^3\gamma(s))$ into $\boldsymbol{R}^3 \approx E_0$ is a straight line, and its reparametrization $\vec{\gamma}(t)$ follows this straight line. Since the vector

$$\frac{d\gamma}{ds} = \sum a_i \, \partial_i = \frac{dt}{ds}\partial_0 + \sum_{j=1}^{3} \frac{d(x^j \circ \gamma)}{ds} \partial_j$$

is null, and dt/ds is positive, it follows that

$$\frac{dt}{ds} = \left|\frac{d\vec{\gamma}}{ds}\right|.$$

Thus the associated particle $\vec{\gamma}$ has speed

$$v = \left|\frac{d\vec{\gamma}}{dt}\right| = \frac{|d\vec{\gamma}/ds|}{dt/ds} = 1. \qquad \blacksquare$$

In particular, light has the same constant speed 1 relative to every freely falling observer.

Now consider the case of a material particle, so the parameter s becomes proper time τ.

20. Proposition. Let ξ be a Lorentz coordinate system in M. If $\vec{\alpha}: J \to \boldsymbol{R}^3$ is the associated Newtonian particle of a material particle $\alpha: I \to M$, then

(1) The *speed* $|d\vec{\alpha}/dt|$ of $\vec{\alpha}$ is $v = \tanh \varphi$ where φ is the Lorentz angle between $\alpha' = d\alpha/d\tau$ and the coordinate vector ∂_0 of ξ. In particular, $0 \le v < 1$.

(2) The time τ of α and its ξ-time t are related by

$$\frac{dt}{d\tau} = \frac{d(x^0 \circ \alpha)}{d\tau} = \cosh \varphi = \frac{1}{\sqrt{1 - v^2}} \ge 1.$$

Here v and φ are, of course, functions of the parameter of α.

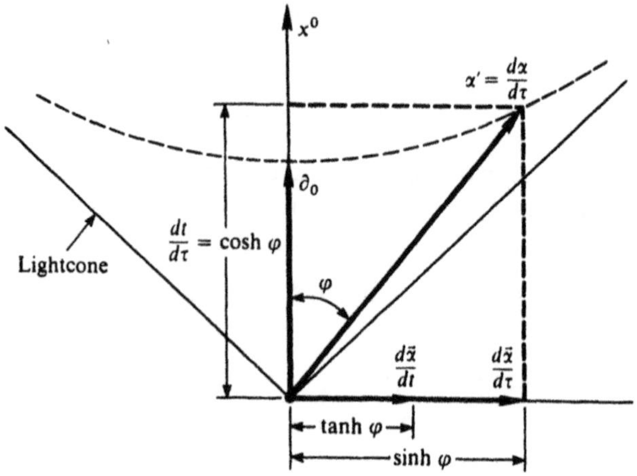

Figure 5. The curved line represents the future-pointing timelike unit vectors (Compare Figure 3.)

Proof. Since both α' and ∂_0 are timelike and future-pointing there is a unique Lorentz angle $\varphi \geq 0$ between them determined by $-\langle \alpha', \partial_0 \rangle = \cosh \varphi \geq 1$. See Figure 5. Since $\alpha' = \sum (d(x^i \circ \alpha)/d\tau)\, \partial_i$, we have

$$\frac{dt}{d\tau} = \frac{d(x^0 \circ \alpha)}{d\tau} = -\langle \alpha', \partial_0 \rangle = \cosh \varphi,$$

and the coordinate expression for $\langle \alpha', \alpha' \rangle = -1$ becomes

$$-\left(\frac{dt}{d\tau}\right)^2 + \left|\frac{d\vec{\alpha}}{d\tau}\right|^2 = -1.$$

Since $\varphi \geq 0$ it follows that

$$\left|\frac{d\vec{\alpha}}{d\tau}\right| = \sqrt{\cosh^2 \varphi - 1} = \sinh \varphi \geq 0.$$

Thus $\vec{\alpha}$ has speed

$$v = \left|\frac{d\vec{\alpha}}{dt}\right| = \frac{|d\vec{\alpha}/d\tau|}{dt/d\tau} = \frac{\sinh \varphi}{\cosh \varphi} = \tanh \varphi.$$

Hyperbolic identities then give $\cosh \varphi = 1/\sqrt{1 - v^2}$. ∎

The following interpretations put more emphasis on the freely falling observer ω, and less on the coordinate system ξ.

(1) *Time.* For any Lorentz coordinate system ξ associated with ω, the coordinate hyperplane E_t given by $x^0 = t$ is readily seen to be the (unique)

3-plane through $\omega(t)$ perpendicular to ω. Thus x^0 is the same for all choices of ξ. In effect, x^0 *imposes* ω's *proper time t on all of M*, with E_t consisting of those events that ω considers to be simultaneous with $\omega(t)$.

(2) *Space.* For the observer himself, the ξ-associated Newtonian particle $\vec{\omega}$ is constant; thus E_0 is called the *restspace* of ω. For any s, t, *orthogonal projection* $E_s \to E_t$ sends $p \in E_s$ to the unique point $q \in E_t$ such that $\vec{pq}$ is parallel to ω. (In terms of any associated Lorentz coordinate system this map merely changes x^0 coordinates.) Consequently the E_ts are canonically isometric Newtonian spaces, and any one will serve equally well as the restspace for ω.

(3) *Speed.* In the preceding proposition, since ω is the x^0 coordinate curve, ω' is always distantly parallel to ∂_0. Thus $\varphi(\tau)$ is the hyperbolic angle between $\alpha'(\tau)$ and ω'. The function $v = |d\vec{\alpha}/dt|$ gives the *speed of* α *relative to* ω, and the function $\varphi = \tanh^{-1} v$, though it measures speed, is traditionally called the *velocity parameter of* α *relative to* ω. (Sometimes $\alpha' = d\alpha/d\tau$ is called the 4-*velocity* of α to distinguish it from the relative notion $d\vec{\alpha}/dt$ in 3-space.)

In short, the restspace of ω is the egocentric Newtonian space in which ω perceives all particles moving as Newtonian particles relative to his rest position.

(4) *Time dilation.* For a particle with proper time τ the equations in (2) of the preceding proposition show that the faster the particle is moving relative to the observer (that is, the larger v is) the slower the particle's clock (τ) runs relative to the observer's clock (t). Thus the slogan: *Moving clocks run slow.*

(5) *Distance.* We can now account for Remark 15(3). That $\vec{pq}$ is orthogonal to a freely falling observer ω means that $x^0(p) = x^0(q)$. Hence p and q are in the same hyperplane E_t, and their separation is ordinary Euclidean distance:

$$pq = \left(\sum_{j=1}^{3} (x^j(p) - x^j(q))^2 \right)^{1/2}.$$

Thus distance between events is meaningful only for observers who consider the events to be simultaneous.

SOME RELATIVISTIC EFFECTS

The preceding section has shown that each freely falling observer ω in M has his own notion of time and space. Many characteristic features of relativity that seem paradoxical arise in comparing the conclusions of two different observers.

For example, if ω_1 and ω_2 are nonparallel they have different restspaces. Thus if $\overrightarrow{pq}$ is orthogonal to ω_1 but not ω_2, the events p and q are simultaneous for ω_1 but not for ω_2. Analogously if $\overrightarrow{pr}$ is parallel to ω_1 then it cannot be parallel to ω_2, and (projecting into restspaces) the events p and r occur at the same position for ω_1 but at different positions for ω_2.

Using distant parallelism the notion of relative speed can be generalized as follows. If α and β are material particles, then the hyperbolic angle $\varphi = \measuredangle(\alpha'(\sigma), \beta'(\tau))$ is their instantaneous velocity parameter and $v = \tanh \varphi$ is the instantaneous *relative speed.* For a freely falling particle β, freely falling observers parallel to β regard it as being at rest, while for others it can have arbitrary constant speeds $0 < v < 1$. In particular, two freely falling observers regard themselves as moving at constant relative speed $\tanh \measuredangle(\alpha', \beta')$.

The difficulties with Newtonian motion mentioned earlier do not arise here: only relative motion is defined, light moves at speed 1 relative to every freely falling observer, and all material particles have relative speeds $v < 1$ (in the empty space modeled by M). The essential dichotomy now is not between rest and motion but between free fall and acceleration—for if β'' is not identically zero no freely falling observer considers β to be at rest.

21. Example. Relativistic Addition of Velocities. A rocketship ρ leaves a space station σ (both freely falling) at relative speed $v_1 > 0$. A spaceman μ is ejected from ρ in the plane of ρ and σ with constant (signed) speed v_2 relative to ρ. Here $v_2 > 0$ means forward, away from σ, and $v_2 < 0$ means backward, toward σ. What is the speed v of μ relative to σ?

The Newtonian answer is, of course, $v = v_1 + v_2$, but Einstein's answer is different.

Figure 6 illustrates the case $v_2 > 0$. Event p is the departure of ρ from σ; event q is the departure of μ from ρ. Thus $v_1 = \tanh \varphi_1$ and $v_2 = \tanh \varphi_2$. By

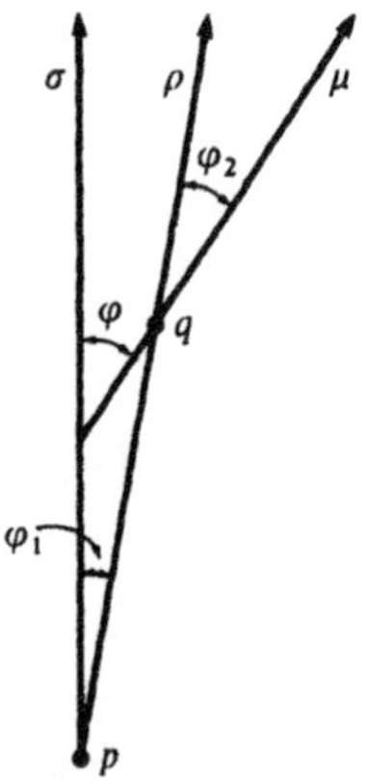

Figure 6

distant parallelism, ρ' is between σ' and μ', hence by the additivity of hyperbolic angles (Exercise 5.11) the angle $\varphi = \measuredangle(\sigma', \mu')$ is $\varphi_1 + \varphi_2$. Thus *addition of velocity parameters replaces Newtonian addition of speed.* Indeed, since

$$v = \tanh \varphi = \tanh(\varphi_1 + \varphi_2) = \frac{\tanh \varphi_1 + \tanh \varphi_2}{1 + \tanh \varphi_1 \tanh \varphi_2},$$

we find

$$v = \frac{v_1 + v_2}{1 + v_1 v_2}.$$

The same formula holds if $v_2 < 0$.

When, as above, several particles lie in a single timelike 2-plane $P \approx \mathbf{R}^2_1$, they are said to be *moving on a line*, though of course the particular Newtonian line and motion depends on the observer.

Einstein dramatized the relativity of time in the following scenario (as formulated in [MTW]).

22. Example. The Twin Paradox. On their 21st birthday Peter leaves his twin Paul behind on their freely falling spaceship and departs at constant relative speed $v = 24/25$ for a free fall of seven years of his proper time. Then he turns and comes back symmetrically in another seven years. Upon his arrival he is thus 35 years old—but Paul is 71.

To compute Paul's age, drop a perpendicular from the turn p to the world-line of the spaceship (Figure 7). By Proposition 17,

$$ox = op \cosh \varphi = \frac{7}{[1 - (24/25)^2]^{1/2}} = 25.$$

Symmetrically, $xq = 25$. Thus Paul's age at Peter's return is $21 + 2(25) = 71$ yr.

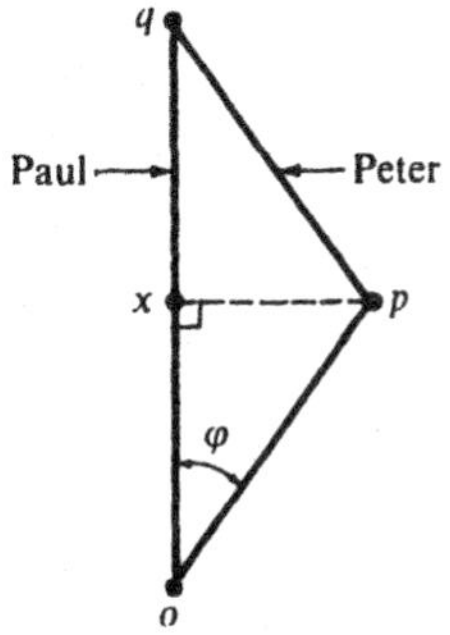

Figure 7

It is useless to object that this phenomenon—no paradox—involves somehow the difficulty of building accurate clocks. "Clock" is merely a name for a time measurer, whether based on the rhythms of atoms, mechanical devices, or biological processes. While they were apart, Paul's heart has beaten 50/14 times more than Peter's.

A more telling objection is that if Peter is to survive the trip the corners at o, p, and q must be smoothed to keep accelerations low. But the phenomenon remains.

23. Corollary. Let $\alpha\colon [a, b] \to M$ be a material particle from p to q in M. Its elapsed proper time $\Delta\tau = b - a$ is at most pq, with equality if and only if α is freely falling.

Since M is a normal neighborhood of each of its points this follows immediately from Proposition 5.34. Thus free fall is the unique slowest way to go from one event to another. Roughly speaking any other way is more nearly lightlike hence quicker.

LORENTZ–FITZGERALD CONTRACTION

Let α and β be parallel freely falling material particles in M. We can consider α and β to be the endpoints of a freely falling rod $[\alpha, \beta]$ in M.

Let ω be a freely falling observer with restspace E_ω through $\omega(0)$. Since α and β are parallel their associated Newtonian particles $\vec{\alpha}$ and $\vec{\beta}$ move along parallel straight lines in E_ω both at constant speed v. Hence ω sees the rod as a line segment $[\vec{\alpha}(t), \vec{\beta}(t)]$ moving in translation, that is, parallel to itself, in E_ω. The *length of the rod as measured by* ω is the constant distance L_ω from $\vec{\alpha}(t)$ to $\vec{\beta}(t)$ in E_ω. For a rider on the rod, say α, the associated Newtonian rod $[\vec{\alpha}, \vec{\beta}]$ is at rest in his restspace where he measures its *restlength* L. This is the length of any vector from α to β orthogonal to both.

24. Proposition. As above let $[\alpha, \beta]$ be a freely falling rod with speed v relative to a freely falling observer ω. In E_ω, (1) If $[\vec{\alpha}, \vec{\beta}]$ moves in a direction orthogonal to its axis, then $L_\omega = L$ (restlength). (2) If $[\vec{\alpha}, \vec{\beta}]$ moves in the direction of its axis, then $L_\omega = L\sqrt{1 - v^2}$.

Proof. Evidently a parallel displacement of ω has no effect on these measurements so we can assume $\omega(0) = \alpha(0)$. (1) Let $A = [\vec{\alpha}(0), \vec{\beta}(0)]$ be the initial position of the Newtonian rod in E_ω. It suffices to show that A is also in the restspace E_α of α through $\alpha(0)$, for then ω and α measure the same length for A. Since $A \subset E_\omega$, $A \perp \omega$. By hypothesis $A \perp \vec{\alpha}$. But α is in the plane

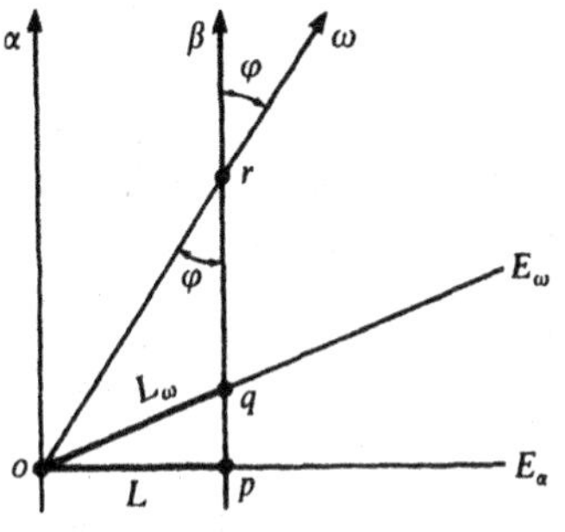

Figure 8

determined by ω and $\vec{\alpha}$, hence $A \perp \alpha$; that is, $A \subset E_\alpha$. (2) This is the case of motion along a line: α, β, and ω are in the same timelike plane, and Figure 8 shows the lines in which the restspaces E_ω and E_α meet that plane.

In the figure, $v = \tanh \varphi$, op is the restlength L of the rod, and oq is L_ω. By Proposition 17, in the right triangle Δopr, $L = ro \sinh \varphi$, and in the right triangle Δroq, $L_\omega = ro \tanh \varphi$. Thus $L = L_\omega \cosh \varphi$, so $L_\omega = L\sqrt{1 - v^2}$. ∎

This phenomenon of shortening in the direction of motion is the celebrated *Lorentz–Fitzgerald contraction* conjectured independently by Fitzgerald and Lorentz some years before Einstein's 1905 organization of special relativity. (For the original papers of Lorentz and Einstein, see [E].)

25. Example. An Einstein Train. A train of restlength 200 m travels a straight stretch of track past a station of restlength 100 m. Thus, when standing at the station, the train is twice as long as the station.

On a particular trip the train passes the station at constant relative speed $v = \sqrt{3}/2 \sim 0.87$. By Proposition 24 the stationmaster, at rest in the station, measures the length of the train as $200\sqrt{1 - v^2} = 100$ m. Hence, as it passes, the train exactly fits the station.

For the conductor, the train is at rest, hence has length 200 m. For him the station is moving at speed v, hence has length $100\sqrt{1 - v^2} = 50$ m. Thus the train is four times as long as the station.

These apparently conflicting measurements coexist harmoniously in the spacetime diagram of Figure 9, which shows the station as $[\alpha, \beta]$ and the train as $[\gamma, \delta]$. The significant times and distances are separations between various pairs of events.

If we suppose the conductor is at the front (δ) of the train and the stationmaster is at the right end (β) of the station, then the event q is their passing each other.

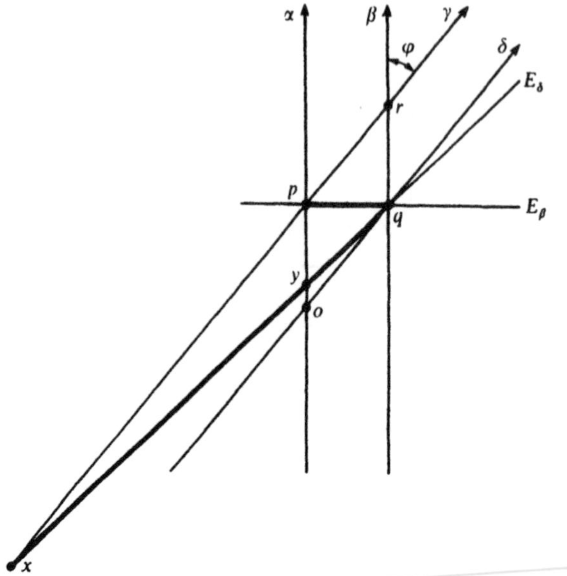

Figure 9. E_β is the restspace of the stationmaster β; E_δ is the restspace of the conductor δ (hence $\delta \perp E_\delta$).

At the station time of q, the stationmaster sees the train as exactly fitting the station—both of them as the 100-m segment $[p, q]$ in his restspace E_β. Orthogonal projection onto E_β would show the train moving to the right through the station.

At the train time of event q, the conductor δ sees the train at rest in E_δ as the 200-m segment $[x, q]$ of which only the first 50-m segment $[y, q]$ is inside the station. Orthogonal projection onto E_δ (along lines parallel to δ) would show the station moving to the left past the resting train.

Event r is the passing of the stationmaster β and the back (γ) of the train. At speed $\sqrt{3}/2$ the 100-m train passes him in elapsed time $op = qr = 200/\sqrt{3}$ m. At the same relative speed the conductor is passed by the 50-m station in elapsed train time $oq = pr = 100/\sqrt{3}$ m.

ENERGY–MOMENTUM

26. Definition. If $\alpha: I \to M$ is a material particle of mass m, its *energy–momentum vector field* is the vector field $P = m\, d\alpha/d\tau$ on α.

To understand this let us consider what a freely falling observer ω makes of it in terms of his notions of time and space. For an associated Lorentz coordinate system ξ the components of P are

$$P^i = m\,\frac{d(x^i \circ \alpha)}{d\tau} \qquad (0 \leq i \leq 3),$$

where as usual τ is the proper time of α. Introducing the proper time t of the observer gives

$$P^i = m\,\frac{d(x^i \circ \alpha)}{dt}\,\frac{dt}{d\tau}, \qquad \text{where} \quad \frac{dt}{d\tau} = \frac{1}{\sqrt{1 - v^2}}.$$

The space components P^1, P^2, P^3 thus describe a vector field

$$\vec{P} = \frac{m}{\sqrt{1 - v^2}}\,\frac{d\vec{\alpha}}{dt}$$

on the associated Newtonian particle $\vec{\alpha}$ in $E_0 \approx \boldsymbol{R}^3$.

This is a reasonable extension of the Newtonian momentum vector field of $\vec{\alpha}$ (Definition 3), since for slow speeds, the time-dilation factor $(1 - v^2)^{-1/2}$ is nearly 1. However the time component of P is something new.

$$P^0 = m\,\frac{d(x^0 \circ \alpha)}{d\tau} = m\,\frac{dt}{d\tau} = \frac{m}{\sqrt{1 - v^2}}.$$

By the binomial theorem (Newton, 1665)

$$P^0 = m + \tfrac{1}{2}mv^2 + O(v^4).$$

The second term here is the Newtonian kinetic energy of $\vec{\alpha}$, and Einstein identified P^0 as the total energy E of the particle as measured by ω, concluding in particular that mass is merely one form of energy. Specifically m is the *rest energy* E_{rest}, since $P^0 = m$ when $v = 0$. Converting to conventional units gives the famous equation $E_{\text{rest}} = mc^2$. To summarize:

27. Definition. Let α be a material particle of mass m in $\boldsymbol{M}$. If ω is a freely falling observer, then the *energy of α relative to ω* is the time component

$$E = m/\sqrt{1 - v^2}$$

of $P = m\,d\alpha/d\tau$, and the *momentum of α relative to ω* is the Euclidean vector field

$$\vec{P} = \frac{m}{\sqrt{1 - v^2}}\,\frac{d\vec{\alpha}}{dt}$$

on $\vec{\alpha}$. Finally the *scalar momentum of α relative to ω* is the function $\mathfrak{p} = |\vec{P}|$.

Just as space and time are merged in special relativity, so are energy and momentum. The unified concept, energy–momentum, can be split up only artificially, relative to a particular observer.

Using distant parallelism, the definition above can be expressed concisely as $P = E\,\partial_0 + \vec{P}$, where

(1) ∂_0 is the timelike coordinate vector field of any Lorentz coordinate system associated with ω; that is, ∂_0 is the (parallel) vector field on α distantly parallel to ω', and

(2) the spacelike vector field $\vec{P}$, orthogonal to ∂_0, has been moved from $\vec{\alpha}$ in the restspace to α itself by distant parallelism.

28. Corollary. Let α be a material particle of mass m in M. Relative to a freely falling observer, the energy E, scalar momentum $\mathfrak{p}$, and speed $v = \tanh\varphi$ of α are related by

(1) $E^2 = m^2 + \mathfrak{p}^2$.

(2) $E = m/\sqrt{1 - v^2} = m\cosh\varphi = -\langle P, \partial_0\rangle$.

(3) $\mathfrak{p} = m\sinh\varphi = mv/\sqrt{1 - v^2}$.

(4) $\mathfrak{p}/E = \tanh\varphi = v$.

Proof. Since $P = m\alpha' = E\,\partial_0 + \vec{P}$, with $P \perp \partial_0$, the assertions follow from Proposition 17, but with the Lorentz trigonometry in each tangent space rather than in M itself. ∎

Let $\{\alpha_i\}$ be a sequence of freely falling material particles whose worldlines approach that of a lightlike particle γ. For any freely falling observer the relative speeds v_i approach 1, hence equations (1) and (4) of the corollary give $E^2 = m^2 + \mathfrak{p}^2$ and $E = \mathfrak{p}$, confirming that lightlike particles have mass zero.

That light carries energy and has momentum is noncontroversial, but since it has no mass a new definition is needed for its energy–momentum.

29. Definition. The *energy–momentum* vector field of a lightlike particle γ is its 4-velocity: $P = \gamma' = d\gamma/ds$.

Then any freely falling observer ω can split P into *energy E relative to* ω and *momentum P relative to* ω by writing $P = E\,\partial_0 + \vec{P}$ with $\vec{P} \perp \partial_0$, just as for a material particle. But since $P = \gamma'$ is lightlike, the energy and scalar momentum are equal: $E = \mathfrak{p} = -\langle\gamma', \partial_0\rangle$. (Compare Lemma 16.) Furthermore, since γ and the observer ω are both geodesics, $E = \mathfrak{p}$ is constant and $\vec{P}$ is parallel.

In geometrical units, energy E and scalar momentum $\mathfrak{p}$ have the same common unit as distance, time, and mass. For a material particle the mass unit for E and $\mathfrak{p}$ is clear from the formulas in Corollary 28. For a lightlike particle γ the affine parameter s is a pure number, hence the formula

$$\mathfrak{p} = E = d(x^0 \circ \gamma)/ds$$

shows the unit is the same as for time.

Conversion to conventional units is accomplished as in Remark 7 using c_{conv} and G_{conv}. For example, in cgs units one gram of energy, the rest energy of a mass of 1 g is

$$E_{\text{cgs}} = (c_{\text{cgs}})^2 \text{ g} = 9 \times 10^{20} \text{ g cm}^2/\text{sec}^2 \quad (=\text{erg}).$$

The wave character of light is measured as follows: A photon of energy E, relative to some observer, has *frequency* $\nu = E/h$, where h is Planck's constant. As usual, frequency times *wavelength* λ is speed c. In geometric units, h is about 1.8×10^{-86} sec^2 and $\lambda\nu = 1$. Since frequency and wavelength derive from energy they too depend on who is doing the observing. Thus visible light for one observer is radio waves for another and x rays for a third.

COLLISIONS

Suppose that a number of particles in M enter a very small spacetime region $\mathcal{O}$ and a (possibly different) number emerge. What takes place in $\mathcal{O}$ may be quite complicated, but if $\mathcal{O}$ is sufficiently small it can be modelled as a single event o.

30. Definition. A *collision* in M is a collection of r *incoming* material or lightlike particles:

$$\alpha_i\colon [a_i, 0] \to M \qquad (1 \le i \le r)$$

and s *outgoing* particles

$$\beta_j\colon [0, b_j] \to M \qquad (1 \le j \le s),$$

such that $\alpha_i(0) = \beta_j(0) = o \in M$ for all i, j. Then o is called the *collision event*.

Such collisions are the stock-in-trade of particle physics, and hundreds of types have been studied experimentally. In all of these the following holds:

31. Law of Conservation of Energy–Momentum. For a collision as above, the total incoming energy–momentum vector equals the total outgoing energy–momentum vector; that is,

$$\sum_{i=1}^{r} P_i = \sum_{j=1}^{s} \bar{P}_j \in T_o(M),$$

where P_i and $\bar{P}_j$ are the energy–momentum vectors of α_i and β_j, respectively, at the collision event o.

32. Example. A Totally Inelastic Collision. Two blobs of putty α_1 and α_2 of mass m_1 and m_2 collide at relative speed $v > 0$ and stick together. Investigate the outgoing blob β.

Conservation of energy–momentum asserts that at the collision event o,

$$m\beta' = m_1\alpha_1' + m_2\alpha_2',$$

where m is the mass of β. Taking scalar products yields

$$-m^2 = -m_1^2 + 2m_1m_2\langle\alpha_1', \alpha_2'\rangle - m_2^2.$$

Here $\langle\alpha_1', \alpha_2'\rangle = -\cosh\varphi$, where $\tanh\varphi = v$. Thus the scalar product is $-(1 - v^2)^{-1/2}$, so that

$$m^2 = (m_1)^2 + [2m_1m_2/\sqrt{1 - v^2}] + (m_2)^2.$$

Thus $m > m_1 + m_2$, so mass is not conserved in the collision. The outgoing blob has drawn additional mass from the energy–momentum of the incoming blobs.

Then β, if freely falling, is completely determined by its 4-velocity

$$\beta' = (m_1/m)\alpha_1' + (m_2/m)\alpha_2'$$

at the collision event o.

A timelike future-pointing unit vector $u \in T_p(M)$ is called an *instantaneous observer* at p. In this role, u is regarded as having that information common to all observers passing through p with 4-velocity u. Specifically, u knows the tangent space $T_p(M)$, and if a particle through p has energy–momentum vector $P \in T_p(M)$, then u can split P into energy E and momentum $\vec{P}$, as usual, by

$$P = Eu + \vec{P}, \qquad \text{with} \quad \vec{P} \perp u.$$

33. Corollary. Relative to an instantaneous observer u at a collision event o, both energy and momentum are preserved in the collision; that is,

$$\sum_{\text{in}} E_i = \sum_{\text{out}} E_j \qquad \text{and} \qquad \sum_{\text{in}} \vec{P}_i = \sum_{\text{out}} \vec{P}_j.$$

Proof. Since the incoming and outgoing total energy–momentum vectors are equal at o, they have the same numerical components in the u direction and the same vector components orthogonal to u. ∎

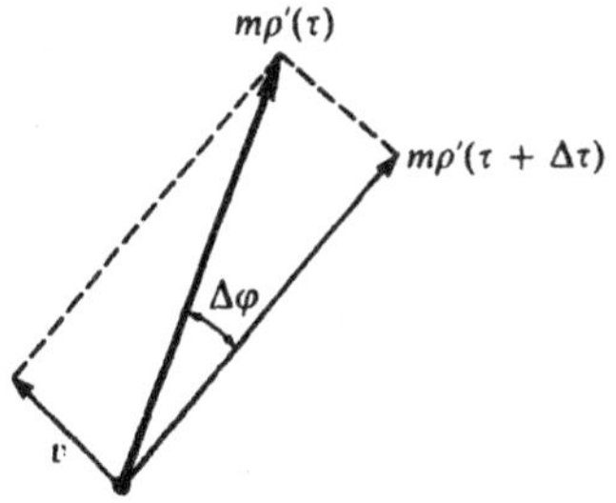

Figure 10

34. Example. Photon Rockets. A rocketship ρ of initial mass m_0 leaves a freely falling space station σ at initial relative speed 0. The exhaust of ρ, aimed toward σ, consists solely of photons. Find the velocity parameter $\varphi(\tau)$ of ρ relative to σ in terms of the mass $m(\tau)$ of $\rho(\tau)$ at its proper time τ since launch.

To set up the necessary calculus we shall construe the events during $(\tau, \tau + \Delta\tau)$ as a collision: the decay of ρ-before-τ into ρ-after-τ and the photons ejected during $\Delta\tau$. Setting $m = m(\tau)$ and $\bar{m} = m(\tau + \Delta\tau)$, conservation of energy–momentum implies that

$$m\rho'(\tau) = \bar{m}\rho'(\tau + \Delta\tau) + v,$$

where v is the energy–momentum of the photons ejected during $\Delta\tau$ (Figure 10). Since v is null, we find

$$-m^2 - \bar{m}^2 - 2m\bar{m}\langle\rho'(\tau), \rho'(\tau + \Delta\tau)\rangle = 0.$$

The scalar product here is $-\cosh \Delta\varphi$, so after writing $\bar{m} = m + \Delta m$ some algebra gives

$$-1 + \cosh \Delta\varphi = (\Delta m)^2/2m\bar{m}.$$

Now take the limit as $\Delta\tau \to 0$. Since $\cosh \Delta\varphi \sim 1 + (\Delta\varphi)^2/2$ and $\bar{m} \to m$, we find $d\varphi = -dm/m$ (minus sign since $\Delta m < 0 < \Delta\varphi$). Integration from 0 to τ then yields

$$\varphi(\tau) = \ln(m_0/m(\tau)).$$

AN ACCELERATING OBSERVER

Let ξ be a Lorentz coordinate system associated with a freely falling observer ω in M, but consider only the timelike 2-plane P: $x^3 = x^4 = 0$ (relativistic motion on a line). Let α be the curve in P with coordinates

$$x^0(\alpha\tau) = g^{-1}\sinh(g\tau), \qquad x^1(\alpha\tau) = g^{-1}\cosh(g\tau) \qquad (g > 0).$$

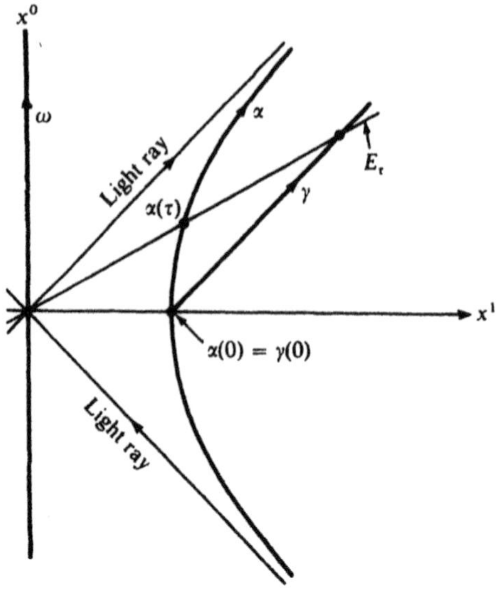

Figure 11. Q is the quadrant between the two light rays.

Differentiation shows that α is an observer with proper time τ but that α is not freely falling; indeed, α has constant scalar acceleration $|\alpha''| = g$.

The observer ω perceives α as a Newtonian particle in his restline $x^0 = 0$, with α moving in from infinity to distance g^{-1}, then retreating symmetrically to infinity. However, α cannot pick up ω at all on his radar screen. In fact, it is clear from Figure 11 that radiation (light) from α will always reach ω but only at ω-proper times $t > 0$—then the reflected radiation can never return to α.

The restline E_τ of α through $\alpha(\tau)$ is the spacelike line through $\alpha(\tau)$ orthogonal to $\alpha'(\tau)$. This line consists of the events that α considers to be simultaneous with $\alpha(\tau)$. It extends only to the origin 0, and such restlines fill the open quadrant Q that α *can* pick up on his radar. In fact, Exercise 7 shows that, using only his clock and radar, α can construct a (non-Lorentz) coordinate system y^0, y^1 on Q such that

$$y^0(q) = \tau \qquad \text{if} \quad q \in E_\tau;$$

$$y^1(q) = 0q - g^{-1} \qquad \text{for all} \quad q \in Q.$$

Here y^0 gives the α-time of each event $q \in Q$, and y^1 gives the α-position. Hence a particle β in Q is perceived by α as the Newtonian particle $\vec{\beta}$ in $\boldsymbol{R}^1$ given by parametrizing $y^1 \circ \beta$ by α's proper time τ. In particular $\vec{\alpha}$ is at rest at 0.

Consider a race between α and the (outgoing) photon γ emitted at $\alpha(0)$ with, say, $\gamma'(0) = \partial_0 + \partial_1$. If α were freely falling the photon would move away from him at speed 1, but since α is accelerating in the direction of γ perhaps he can do better. In fact, he does worse: a straightforward computation shows that γ meets the ray E_τ at $e^{g\tau}\alpha(\tau)$. But $|\alpha(\tau)| = g^{-1}$; hence relative to α,

$$\vec{\gamma}(\tau) = g^{-1}(e^{g\tau} - 1).$$

Thus the photon's speed relative to the observer α is $e^{g\tau}$, which—far from being 1—approaches infinity with increasing τ.

Exercises

1. If $p \in M$ is not on the worldline of a freely falling observer ω, then ω meets the lightcone $\Lambda(p)$ of p in exactly two events: $\omega(\tau^-)$ and $\omega(\tau^+)$. Furthermore, if $q = \omega((\tau^+ + \tau^-)/2)$, then $\vec{pq} \perp \omega$ and $\vec{pq} = q\omega(\tau^-) = q\omega(\tau^+)$.

2. Leaving her twin Jean at rest in their freely falling spaceship, Evelyn departs at constant relative speed $v = 0.4$ for one trip around a circle of radius 2 lightyears in the restspace of their ship. Find their age difference upon Evelyn's return.

3. Radar construction of coordinate systems. Let ω be a freely falling observer in M. (a) Light emitted by ω at $\tau = a$ is reflected off q and received by ω at $\tau = b$. Show q lies in the restspace of ω through $m = \omega((a + b)/2)$ and that $mq = (a + b)/2$. (b) Parallel orthonormal vector fields E_1, E_2, E_3 on ω, $\perp\omega$, represent gyroscopes. Using the spatial direction of the emitted and received light in (a) deduce the coordinates of q for an appropriate Lorentz coordinate system associated with ω.

4. (a) If the exhaust of the rocketship in Example 34 consists of material particles, show that its velocity parameter is $\varphi = v_{ex} \ln(m_0/m)$, where $v_{ex} < 1$ is the (constant) exhaust speed. (b) If the rocket in (a) is Newtonian show that its speed is given by $v = v_{ex} \ln(m_0/m)$. (Thus $v < 1$ in (a) but $v \to \infty$ in (b).)

5. In Example 34, let σ be our solar system and suppose the rocketship ρ is heading directly toward α Centauri. Assume the latter and σ are freely falling and parallel; in their common restspace their distance apart is 4.40 (light)years. If ρ burns fuel at the rate $dm/d\tau = -m/100$ find the approximate time of the trip in terms of (a) σ's proper time t, (b) ρ's proper time τ. (The difference is about 158 days.)

6. (See page 181.) (a) Relative to the freely falling observer ω, find the associated Newtonian particles of accelerating observer α and the photon γ, and verify that the distance between them increases toward $1/g$ as $t \to \infty$. (b) If ω is constantly emitting light at frequency ν_0, show that α measures the frequency of light he receives at proper time τ as $\nu(\tau) = \nu_0 e^{-g\tau}$. (c) If α is constantly emitting light at frequency ν_0, show that ω measures the frequency of the light he receives at proper time $t > 0$ as $\nu(t) = \nu_0/(gt)$.

7. Consider the accelerating observer α on page 181. Light emitted by α at his proper time $\tau = a$ reflects off $q \in Q$ and is received by α at $\tau = b$. Prove: (a) q lies on the ray from 0 through $\alpha((a + b)/2)$; and $0q = g^{-1}e^x$, where $x = \pm g(b - a)/2$, the sign depending on the direction of the emitted light. (b) The functions y^0, y^1 in the text form a coordinate system for Q, with line element

$$ds^2 = -(1 - gy^1)^2 (dy^0)^2 + (dy^1)^2.$$

8. Change of coordinates. Let ξ and η be Lorentz coordinate systems in $\boldsymbol{M}$ such that $x^2 = y^2$ and $x^3 = y^3$. (a) Show that there are unique numbers φ, a, b such that

$$x^0 = y^0 \cosh \varphi + y^1 \sinh \varphi + a, \qquad x^1 = y^0 \sinh \varphi + y^1 \cosh \varphi + b.$$

(b) Explain the meanings of φ, a, b in terms of the freely falling observers constituting the 0-axes of ξ and η. (c) Express $\partial/\partial y^0$, $\partial/\partial y^1$ in terms of $\partial/\partial x^0$, $\partial/\partial x^1$—and vice versa.

This is the Lorentz analogue of a rotation and translation of Euclidean coordinates in $\boldsymbol{R}^2$. (Compare Example 9.4). For coordinate techniques and a variety of interesting problems in special relativity, see [TW].

7 CONSTRUCTIONS

The theory of covering manifolds, whose fundamentals are outlined in Appendix A, is an indispensable tool for dealing with global problems in manifold theory. We establish further properties of covering manifolds that tie them closely to fundamental groups. Furnishing the manifolds with metric tensors brings these techniques to bear in semi-Riemannian geometry, where they provide useful means of constructing new semi-Riemannian manifolds that cover, or are covered by, a given one.

The notion of *vector bundle* is discussed, with the tangent bundle as prototype and the normal bundle of a semi-Riemannian submanifold as a principal application.

A generalization of semi-Riemannian product manifold $M \times N$ is obtained by homothetically distorting the geometry of each fiber $p \times N$ to get a new "warped" metric tensor on the manifold $M \times N$. In particular it turns out that fundamental examples of relativistic spacetimes are such *warped products*. A further generalization, briefly considered, is the notion of *semi-Riemannian submersion.*

DECK TRANSFORMATIONS

1. Definition. A *deck transformation* of a covering $\mathscr{k}: \tilde{M} \to M$ is a diffeomorphism $\phi: \tilde{M} \to \tilde{M}$ such that $\mathscr{k} \circ \phi = \mathscr{k}$.

Thus ϕ merely rearranges the points in each fiber $\mathscr{k}^{-1}(p)$, $p \in M$.

The set $\mathscr{D}$ of all deck transformations of a covering forms a group, with composition of functions as the group operation.

2. Example. The covering map $\exp: \boldsymbol{R}^1 \to S^1$ sending t to $(\cos t, \sin t)$ has periodicity $\exp(t + 2\pi n) = \exp(t)$. Thus any deck transformation must carry each $t \in \boldsymbol{R}$ to a point $t + 2\pi n(t)$, and by continuity the integer $n(t)$ is independent of t. Hence the deck transformation group $\mathscr{D}$ of this covering consists of all translations $\phi_n(t) = t + 2\pi n$. The function $n \to \phi_n$ is an isomorphism from the additive group of integers $\boldsymbol{Z}$ to $\mathscr{D}$, so $\mathscr{D}$ is infinite cyclic.

In the language of Appendix A, a deck transformation ϕ of $\mathscr{k}: \tilde{M} \to M$ is in particular a lift of $\mathscr{k}$ through $\mathscr{k}$. Thus by Proposition A.11, a *deck transformation of a connected covering is determined by its value at a single point*. Explicitly, if $\phi, \psi \in \mathscr{D}$ and $\phi(p) = \psi(p)$ for some $p \in \tilde{M}$, then $\phi = \psi$. In particular, since the identity map of $\tilde{M}$ is a deck transformation, the existence of a single fixed point, $\phi(p) = p$, implies that ϕ is the identity map.

The more symmetrical a covering is, the larger its deck transformation group. A covering $\mathscr{k}: \tilde{M} \to M$ is *normal* (sometimes called *regular*) when its deck transformation group is as large as possible, that is, when $\mathscr{k}(p) = \mathscr{k}(q)$ implies there is a deck transformation ϕ such that $\phi(p) = q$.

3. Examples. (1) Every double covering is normal, with deck-transformation group consisting of the identity map and the map ρ that reverses the two points in each fiber $\mathscr{k}^{-1}(p)$.

(2) Every simply connected covering is normal (by Proposition A.11).

4. Proposition. If $\mathscr{k}: \tilde{M} \to M$ is a simply connected covering, then its deck transformation group $\mathscr{D}$ is isomorphic to the fundamental group $\pi_1(M)$ of M.

Proof. As in Appendix A, let $\pi_1(M)$ be based at $q \in M$. Fix a point $p \in \mathscr{k}^{-1}(q)$. If $\alpha: [0, 1] \to M$ is a loop at q, that is, $\alpha(0) = \alpha(1) = q$, let $\tilde{\alpha}$ be the unique lift of α starting at p.

Any two loops α_1, α_2 in the same homotopy class $a \in \pi_1(M)$ are by definition fixed-endpoint homotopic, hence Corollary A.10 implies $\tilde{\alpha}_1(1) = \tilde{\alpha}_2(1)$. Denote by ϕ_a the unique deck transformation such that $\phi_a(p)$ is the common value of $\tilde{\alpha}(1)$ for all α in a. We assert that the function $a \to \phi_a$ is an isomorphism from $\pi_1(M)$ to $\mathscr{D}$.

First we show that it is a homomorphism, that is, $\phi_a \circ \phi_b = \phi_{ab}$ for all $a, b \in \pi_1(M)$. For α in a and β in b, $\tilde{\alpha}$ runs from p to $\phi_a(p)$ and $\tilde{\beta}$ runs from p to $\phi_b(p)$. Thus $\phi_a \circ \tilde{\beta}$ runs from $\phi_a(p)$ to $\phi_a(\phi_b(p))$. In particular the path product $\tilde{\gamma} = \tilde{\alpha} * (\phi_a \circ \tilde{\beta})$ is well defined. Now $\tilde{\gamma}$ starts at p, and since $\mathscr{k} \circ \phi_a = \mathscr{k}$,

$$\mathscr{k} \circ \tilde{\gamma} = (\mathscr{k} \circ \tilde{\alpha}) * (\mathscr{k} \circ \phi_a \circ \tilde{\beta}) = \alpha * \beta.$$

By definition ab is the element of $\pi_1(M)$ containing $\alpha * \beta$, hence

$$\phi_{ab}(p) = \tilde{\gamma}(1) = \phi_a(\tilde{\beta}(1)) = \phi_a\phi_b(p).$$

Since $\tilde{M}$ is connected, $\phi_{ab} = \phi_a\phi_b$.

The homomorphism above is *onto* $\mathscr{D}$. In fact, if $\psi \in \mathscr{D}$ and $\tilde{\alpha}$ is a curve in $\tilde{M}$ from p to $\psi(p)$, then $\psi = \phi_a$, where a is the homotopy class of the loop $k \circ \tilde{\alpha}$.

To show that the homomorphism is one-to-one, suppose ϕ_a is the identity map. By construction this means that any loop $\alpha \in a$ lifts to a loop at p. Then Corollary A.10 implies that $\tilde{\alpha}$ is fixed-endpoint homotopic to a constant, that is, a is the identity element of $\pi_1(M)$. ∎

This result is often the easiest way to compute fundamental groups. For instance, applied to Example 2, it shows that $\pi_1(S^1)$ is infinite cyclic.

Various situations offer a choice of two objects at each point of a manifold; in favorable cases a covering manifold results, as follows.

5. Remark. Let $k: \Sigma \to M$ be a two-to-one map of a set Σ onto a manifold M. Let $\mathscr{L}$ be a collection of functions λ: $\mathscr{U} \to \Sigma$ ($\mathscr{U}$ an open set of M) such that

(1) $k \circ \lambda = \text{id}$ for all $\lambda \in \mathscr{L}$.

(2) If $\lambda(p) = \mu(p)$ for λ, $\mu \in \mathscr{L}$, then $\lambda = \mu$ on some neighborhood of p in M.

(3) Every point Σ is in the image of some $\lambda \in \mathscr{L}$.

Then there is a unique way to make Σ a manifold so that $k: \Sigma \to M$ is a smooth double covering map and each $\lambda \in \mathscr{L}$ is a smooth local cross section of k.

The proof is a straightforward exercise in the use of Proposition 1.42.

ORBIT MANIFOLDS

Let Γ be a group of diffeomorphisms of a manifold M. For $p \in M$ the set $\{\phi(p): \phi \in \Gamma\}$ is called the *orbit of p under* Γ. The collection of all such orbits is denoted by M/Γ. The *natural map* $k: M \to M/\Gamma$ sends each point to its orbit under Γ. We want conditions on Γ so that M/Γ becomes a manifold and k a covering map.

For example, on the sphere S^n let Γ be the group $\{\pm 1\}$ consisting of the identity map and the antipodal map $p \to -p$. Then the orbits are unordered pairs $\{p, -p\}$ and $S^n/\pm 1$ will turn out to be a projective space.

6. Definition. A group Γ of diffeomorphisms of a manifold M is *properly discontinuous* (and *acts freely*) provided

(PD1) Each point $p \in M$ has a neighborhood $\mathscr{U}$ such that if $\phi(\mathscr{U})$ meets $\mathscr{U}$ for $\phi \in \Gamma$ then $\phi = \text{id}$.

(PD2) Points $p, q \in M$ not in the same orbit have neighborhoods $\mathscr{U}$ and $\mathscr{V}$ such that for every $\phi \in \Gamma$, $\phi(\mathscr{U})$ and $\mathscr{V}$ are disjoint.

It is easy to verify that *the deck transformation group of any covering is properly discontinuous*; conversely,

7. Proposition. Let Γ be a properly discontinuous group of diffeomorphisms of a manifold M. There is a unique way to make M/Γ a manifold so that the natural map $k: M \to M/\Gamma$ is a covering map. If M is connected the deck transformation group is Γ, hence the covering is normal.

Proof. Call a neighborhood $\mathscr{U}$ of p in M *special* if $\mathscr{U}$ has the property in (PD1) and is the domain of a coordinate system ξ. Now $k \mid \mathscr{U}$ is one-to-one hence so is the map $\xi \circ (k \mid \mathscr{U})^{-1}: k(\mathscr{U}) \to \boldsymbol{R}^n$. We apply Proposition 1.42 to show that M/Γ is a manifold with these maps as coordinate systems. Evidently their domains cover M/Γ. For another such map $\eta \circ (k \mid \mathscr{V})^{-1}$, the sets $k(\mathscr{U})$ and $k(\mathscr{V})$ meet if and only if there is a $\phi \in \Gamma$ such that $\phi(\mathscr{U})$ meets $\mathscr{V}$. But then

$$\eta \circ (k \mid \mathscr{V})^{-1} \circ [\xi \circ (k \mid \mathscr{U})^{-1}]^{-1} = \eta \circ \phi \circ \xi^{-1},$$

which is smooth on an open set of $\boldsymbol{R}^n$.

The Hausdorff property for M/Γ follows from (PD2), and M second countable implies M/Γ second countable. Thus M/Γ is a manifold.

If $\mathscr{U}$ is special and connected, then $k(\mathscr{U})$ is evenly covered by k, since the components of $k^{-1}(\mathscr{U})$ are the sets $\phi(\mathscr{U})$ for all $\phi \in \Gamma$. By construction, k carries $\mathscr{U}$ diffeomorphically onto $k(\mathscr{U})$, and similarly for any $\phi(\mathscr{U})$ since $k \circ \phi = k$. Thus k is a covering map, and each ϕ is a deck transformation. If M is connected, then Proposition A.11 applies to show that every deck transformation is in Γ.

The construction above is the only way to make k a local diffeomorphism, hence uniqueness holds. ∎

This result is particularly informative when M is simply connected (hence connected), for then Proposition 4 shows $\pi_1(M/\Gamma) \approx \Gamma$. In the example above, the orbit manifold $P^n = S^n/\pm 1$ is *real projective n-space*. For $n \geq 2$, S^n is simply connected, hence $\pi_1(P^n)$ is the two-element group $\boldsymbol{Z}_2$. (P^1 is diffeomorphic to S^1, hence has infinite cyclic fundamental group.)

ORIENTABILITY

Equivalent to the previous definition of orientability (1.41) is an alternative with greater range of development. Consider first some linear algebra. Two bases $e_1, \ldots, e_n$ and $e'_1, \ldots, e'_n$ for a vector space V have *the same orientation* provided

$$\det A > 0, \qquad \text{where} \quad e'_i = \sum A^j_i e_j \qquad (1 \le i \le n).$$

They have *opposite orientation* if $\det A < 0$. It is easy to check that having the same orientation is an equivalence relation on the set of all bases for V, and that there are just two equivalence classes, called *orientations* of V. The orientation containing $e_1, \ldots, e_n$ will be denoted by $[e_1, \ldots, e_n]$.

If ξ is a coordinate system on $\mathscr{U} \subset M$, let

$$\lambda_\xi(p) = [\partial_1|_p, \ldots, \partial_n|_p].$$

An *orientation* λ of a manifold M assigns to each point $p \in M$ an orientation $\lambda(p)$ of $T_p(M)$ and λ is *smooth* in the sense that for each $p \in M$ there is a coordinate system ξ at p such that $\lambda = \lambda_\xi$ on some neighborhood of p.

A manifold is *orientable* if there exists an orientation of M; to *orient* M is to choose a particular orientation. For example, $\boldsymbol{R}^n$ is orientable in view of its *usual orientation* λ_ξ, where ξ is the natural coordinate system.

Any object that determines a unique orientation of M will also be said to *orient* M. An object in agreement with a selected orientation λ is said to be *positively oriented*. For example, a basis $v_1, \ldots, v_n$ for $T_p(M)$ is positively oriented if $[v_1, \ldots, v_n] = \lambda(p)$, and a coordinate system ξ on $\mathscr{U} \subset M$ is positively oriented if $\lambda_\xi = \lambda$ on $\mathscr{U}$.

8. Lemma. A semi-Riemannian hypersurface M of an orientable manifold $\overline{M}$ is orientable if and only if there exists a smooth unit normal vector field on M.

(For a generalization see Exercise 9(b).)

In particular, any hypersurface $f^{-1}(c)$ as in Proposition 4.17 is orientable if $\overline{M}$ is orientable. Thus the hyperquadrics S^n_ν and H^n_ν are orientable.

If ξ and η are overlapping coordinate systems in M, define

$$J(\xi, \eta) = \det(\partial y^j/\partial x^i) \quad \text{on} \quad \mathscr{U}_\xi \cap \mathscr{U}_\eta.$$

Then

$$\lambda_\xi(p) = \lambda_\eta(p) \Leftrightarrow J(\xi, \eta)(p) > 0.$$

In view of the smoothness condition for orientations, if two orientations of M agree at a point, then they agree on some neighborhood of that point.

If λ is an orientation of M, then so is $-\lambda$, which assigns the opposite orientation at each point. If M is connected, then $\pm\lambda$ are its only orientations, since the sets where an orientation μ agrees with λ and with $-\lambda$ constitute a disjoint open covering of M, so one of them must be all of M. Thus in the connected case, orienting a single tangent space serves to orient all of M.

If $\phi: M \to N$ is a local diffeomorphism, then it is easy to verify that for each point $p \in M$

$$\hat{\phi}[e_1, \ldots, e_n] = [d\phi(e_1), \ldots, d\phi(e_n)]$$

is a well-defined one-to-one function from the orientations of $T_p(M)$ to the orientations of $T_{\phi p}(N)$. If M and N are oriented by λ_M and λ_N, we say that ϕ is

$$\begin{cases} \textit{orientation-preserving} \text{ if } \hat{\phi}(\lambda_M(p)) = \lambda_N(\phi p) \text{ for all } p \in M, \\ \textit{orientation-reversing} \text{ if } \hat{\phi}(\lambda_M(p)) = -\lambda_N(\phi p) \text{ for all } p \in M. \end{cases}$$

If M is connected, then these are the only possibilities; otherwise ϕ might preserve orientation on one component, but reverse it on another.

If $\phi: M \to N$ is a local diffeomorphism and N is orientable, then M is orientable. In fact, if λ is an orientation of N, there is a unique orientation $\phi^*(\lambda)$ of M making ϕ orientation-preserving. (Define $\phi^*(\lambda)$ at $p \in M$ to be the orientation of $T_p(M)$ carried to $\lambda(\phi p)$ by $\hat{\phi}$.)

Orientability has a useful expression in terms of covering manifolds. For a manifold M, let $\hat{M}$ be the set of all orientations of tangent spaces of M. Let $\ell: \hat{M} \to M$ send the two orientations of each $T_p(M)$ to p. Then Remark 5 gives

9. Corollary. For a manifold M there is a unique way to make $\hat{M}$, as above, a manifold so that (1) $\ell: \hat{M} \to M$ is a double covering map, and (2) for each coordinate system ξ on $\mathscr{U} \subset M$, the map $\lambda_\xi: \mathscr{U} \to \hat{M}$ is a smooth local cross section.

This covering $\ell: \hat{M} \to M$ is called the *orientation covering* of M. Evidently an orientation of M as defined earlier is exactly a smooth global section $\lambda: M \to \hat{M}$.

10. Lemma. For any manifold M, (1) its orientation covering manifold $\hat{M}$ is orientable, and (2) M is orientable if and only if $\ell: \hat{M} \to M$ is trivial.

Proof. (1) For each coordinate system ξ in M consider the pulled-back orientation $\ell^*(\lambda_\xi)$ on $\lambda_\xi(\mathscr{U})$. For another coordinate system the intersection $\mathscr{W} = \lambda_\xi(\mathscr{U}) \cap \lambda_\eta(\mathscr{V})$ is nonempty if and only if $\lambda_\xi = \lambda_\eta$ on $\ell(\mathscr{W})$. Thus the local orientations $\ell^*(\lambda_\xi)$ for all ξ combine to give a global orientation of M.

The proof of (2) is analogous. ■

Thus for a nonorientable manifold, orientability can be recovered by passing to the orientation covering manifold. Also, in view of Proposition A.14, *a simply connected manifold is orientable.*

SEMI-RIEMANNIAN COVERINGS

The notion of covering map is made geometric in the following obvious way.

11. Definition. A *semi-Riemannian covering map* $k: \tilde{M} \to M$ is a covering map of semi-Riemannian manifolds that is a local isometry.

For example, the exponential map in Example 2 is a semi-Riemannian covering of the unit circle S^1 by $\boldsymbol{R}^1$. A product of semi-Riemannian coverings is again one; thus, $\exp \times \exp: \boldsymbol{R}^2 \to S^1 \times S^1$ is a semi-Riemannian covering of a flat torus by the plane.

If $k: M \to N$ is a covering map of a smooth manifold M onto a semi-Riemannian manifold N with metric tensor g, then assigning M the pulled-back metric tensor $k^*(g)$ makes k a semi-Riemannian covering map.

If $k: \tilde{M} \to M$ is a semi-Riemannian covering and $\phi: P \to M$ is a local isometry, it is easy to see that any lift $\tilde{\phi}: P \to \tilde{M}$ of ϕ through k is again a local isometry. In particular, any deck transformation ψ of the covering is an isometry (since ψ is a diffeomorphic lift of the identity map). Thus *the deck transformation group is a properly discontinuous group of isometries of* $\tilde{M}$. Conversely,

12. Corollary. If Γ is a properly discontinuous group of isometries of a semi-Riemannian manifold M, then there is a unique way to make M/Γ a semi-Riemannian manifold such that $k: M \to M/\Gamma$ is a semi-Riemannian covering. If M is connected, the deck transformation group is Γ.

Proof. By Proposition 7 it suffices to show there is a unique metric tensor on M/Γ such that k is a local isometry. If $w \in T_q(M/\Gamma)$, then for each $p \in k^{-1}(q)$ there is a unique $v_p \in T_p(M)$ such that $dk(v_p) = w$. If $p, p' \in k^{-1}(q)$, an orbit under Γ, there is an isometry $\phi \in \Gamma$ such that $\phi(p) = p'$, and it follows that $d\phi(v_p) = v_{p'}$. Thus for $z, w \in T_q(M/\Gamma)$ the scalar product $\langle z_p, w_p \rangle$ is independent of p in $k^{-1}(q)$, so we have no choice but to define $g(z, w)$ to be this number. Then g is a smooth metric tensor on M/Γ making k a local isometry, since, if $\lambda: \mathscr{U} \to M$ is a local cross section, $g|\mathscr{U}$ is the pullback $\lambda^*(g_M)$ of the metric tensor of M. ∎

On any hyperquadric the identity map and the antipodal map $p \to -p$ constitute a properly discontinuous group of isometries. Thus we obtain semi-Riemannian orbit manifolds $P_\nu^n(r) = S_\nu^n(r)/\pm 1$ and $H_\nu^n(r)/\pm 1$. The Riemannian manifold $P^n(r) = S^n(r)/\pm 1$ is *projective n-space* of radius r, and $H_0^n(r)/\pm 1$ is just the hyperbolic space $H^n(r)$.

Using covering techniques we can illustrate some distinctive features of indefinite metrics.

A curve segment $\alpha: [a, b] \to M$ is *smoothly closed* provided $\alpha(a) = \alpha(b)$ and $\alpha'(b) = c\alpha'(a) \neq 0$ for some $c > 0$. A smoothly closed geodesic σ (traditionally called merely a *closed geodesic*) thus has a geodesic extension $\tilde{\sigma}$ that, by the uniqueness of geodesics, perpetually traverses the route of σ. However it need not be periodic.

13. Proposition. Let $\tilde{\sigma}$ be the maximal geodesic extension of a smoothly closed geodesic $\sigma: [a, b] \to M$, so $\sigma'(b) = c\sigma'(a) \neq 0$ with $c > 0$. Then

(1) $\tilde{\sigma}$ is complete (domain $\tilde{\sigma} = \boldsymbol{R}$) $\Leftrightarrow \tilde{\sigma}$ is periodic $\Leftrightarrow c = 1$.
(2) If σ is nonnull, then $c = 1$.

Proof. (1a) *If $c \neq 1$, then $\tilde{\sigma}$ is not complete hence not periodic.*

Suppose for definiteness that $c > 1$. In order to extend σ past b we must use the formula

$$\tilde{\sigma}(t) = \sigma(ct - cb + a),$$

since only then is the new segment a geodesic satisfying the extension condition

$$\tilde{\sigma}'(b^+) = c\sigma'(a) = \sigma'(b) = \tilde{\sigma}'(b^-).$$

But $ct - cb + a \in [a, b]$ if and only if $t \in [b, b + (b - a)/c]$. Thus this first extension reaches only to $b + (b - a)/c$. Similarly, a second such extension reaches $b + (b - a)/c + (b - a)/c^2$, and so on. There are no problems in extending σ to infinity in the negative direction when $c > 1$. Thus the largest domain of $\tilde{\sigma}$ in this case is $\{t : t < b^*\}$, where $b^* = (bc - a)/(c - 1) > b$. The case $c < 1$ is analogous.

(1b) *If $c = 1$, then $\tilde{\sigma}$ is periodic, hence complete.*

In fact, the formula $\tilde{\sigma}(t + n(b - a)) = \sigma(t)$ for all $t \in [a, b]$ and all integers n gives $\tilde{\sigma}: \boldsymbol{R} \to M$ as the unique geodesic extension of σ, and $\tilde{\sigma}$ is periodic—of period $b - a$ if $\sigma | [a, b)$ is one-to-one.

(2) Since $c > 0$, the closure condition implies $|\sigma'(b)| = c|\sigma'(a)|$. But σ' is parallel, nonnull, and nonzero, hence $|\sigma'(b)| = |\sigma'(a)| \neq 0$. Thus $c = 1$. ∎

For nonnull geodesics the terms *closed* and *periodic* are often used interchangeably.

14. Example. We describe a closed (null) geodesic that is not periodic. Let M be the right half-plane $\{(u, v) \in \boldsymbol{R}^2 : u > 0\}$ with $ds^2 = 2\,du\,dv$. The map $\phi(u, v) = (u/2, 2v)$ is an isometry of M, and the group $\Gamma = \{\phi^n : n \in \boldsymbol{Z}\}$ it generates is properly discontinuous. The orbit manifold M/Γ is a flat Lorentz surface diffeomorphic to a cylinder $S^1 \times \boldsymbol{R}^1$.

The parametrization $\alpha(t) = (t, 0)$ of the positive u axis is a null geodesic in M. Since $\mathscr{k}: M \to M/\Gamma$ is a local isometry, $\beta = \mathscr{k} \circ \alpha$ is a null geodesic in M/Γ. For all $t > 0$, $\phi(t, 0) = (t/2, 0)$; hence,

$$\beta(t/2) = \mathscr{k}(\alpha(t/2)) = \mathscr{k}\phi\alpha(t) = \mathscr{k}\alpha(t) = \beta(t).$$

Thus β is a closed geodesic, repeatedly traversing a circle C in M/Γ. But obviously γ is not periodic; indeed it makes one circuit of C on each parameter interval $[2^{-k}, 2^{1-k}]$.

The domain $\boldsymbol{R}^+$ of the null geodesic above is maximal; thus the surface M/Γ is not complete. In fact by the preceding proposition:

15. Corollary. If M is either Riemannian or complete, then smoothly closed geodesics in M have periodic extensions.

By the Hopf–Rinow theorem (5.21) a compact Riemannian manifold is complete. However for indefinite metrics compactness does not imply completeness:

16. Example. The Clifton–Pohl Torus [M]. Let M be $\boldsymbol{R}^2 - 0$ with $ds^2 = 2\,du\,dv/(u^2 + v^2)$. Evidently scalar multiplication by any $c \neq 0$ is an isometry of M. Take say $\mu(u, v) = (2u, 2v)$. The group $\Gamma = \{\mu^n\}$ generated by μ is properly discontinuous; thus, $T = M/\Gamma$ is a Lorentz surface. Topologically T is the closed annulus $1 \leq r \leq 2$ with boundary points identified under μ. Thus T is a torus; in particular, it is compact. But T is not complete; for this it suffices to show that M is not complete. By Exercise 5.8 the geodesic differential equations are

$$u'' = \frac{2u}{u^2 + v^2}(u')^2, \qquad v'' = \frac{2v}{u^2 + v^2}(v')^2.$$

Thus the curve $\alpha(t) = (1/(1 - t), 0)$ is a geodesic, defined for $-\infty < t < 1$. Since α fills the positive u axis in M it is inextendible. Thus M is not complete. (For further properties of T, see Exercise 9.12.)

LORENTZ TIME-ORIENTABILITY

The notion of time-orientability of a Lorentz manifold M can be dealt with in much the same way as orientability for a smooth manifold. Let M^T be the set of all timecones in tangent spaces of M, and let $k: M^T \to M$ be the natural two-to-one map. As in the discussion preceding Lemma 5.33, if V is a timelike vector field on $\mathscr{U} \subset M$, then for each $p \in \mathscr{U}$ let $\tau_V(p)$ be the timecone containing V_p. If τ_W is another such *local time-orientation*, then by Lemma 5.30 $\tau_V(p) = \tau_W(p)$ if and only if $\langle V_p, W_p \rangle < 0$. Hence the set of all local time-orientations of M satisfies the conditions in Remark 5, and $k: M^T \to M$ becomes a smooth double covering map. The pulled-back metric tensor on M^T makes this a Lorentz covering, called the *time-orientation covering* of M. Evidently a time-orientation of M as defined in Chapter 5 is just a global section of this covering.

The following analogue of Lemma 10 is proved analogously.

17. Lemma. If M is a Lorentz manifold, then (1) M^T is time-orientable, and (2) M is time-orientable if and only if $k: M^T \to M$ is trivial.

Thus a simply connected Lorentz manifold is time-orientable.

Let $\phi: M \to N$ be a conformal mapping of Lorentz manifolds, so $\langle d\phi v, d\phi w \rangle = h(p)\langle v, w \rangle$ for all $v, w \in T_p(M)$, $p \in M$. Suppose that $h > 0$. Then $d\phi$ preserves the causal character of tangent vectors and hence carries the two timecones of each $T_p(M)$ to the timecones of $T_{\phi p}(N)$. If τ is a time-orientation of N then, just as for ordinary orientation, there is a natural pulled-back time-orientation $\phi^*(\tau)$ of M. Thus N time-orientable implies M time-orientable. If M and N are time-oriented by τ_M and τ_N, then ϕ *preserves time-orientation* provided $\phi^*(\tau_N) = \tau_M$, and *reverses time-orientation* provided $\phi^*(\tau_N) = -\tau_M$. As before, if M is connected, these are the only possibilities.

VOLUME ELEMENTS

Intuitively a volume element on an n-dimensional scalar product space V is a function ω that assigns to n vectors $v_1, \ldots, v_n \in V$ the volume of the parallelepiped with these vectors as sides. (Thus $\omega(v_1, \ldots, v_n) = 0$ if the vectors are linearly dependent, that is, if the parallelepiped collapses.) Equivalently, but more rigorously, ω is multilinear and, if $e_1, \ldots, e_n$ is an orthonormal basis for V, then $\omega(e_1, \ldots, e_n) = \pm 1$. Now we apply this definition to tangent spaces.

18. Definition. A *volume element* on an n-dimensional semi-Riemannian manifold M is a smooth n-form ω such that $\omega(e_1, \ldots, e_n) = \pm 1$ for every frame on M.

Volume elements always exist at least locally.

19. Lemma. On the domain $\mathscr{U}$ of a coordinate system ξ there is a volume element ω_ξ such that $\omega_\xi(\partial_1, \ldots, \partial_n) = |\det(g_{ij})|^{1/2}$.

Proof. For vector fields $V_1, \ldots, V_n$ on $\mathscr{U}$ write $V_j = \sum V^i_j \partial_i$ and define

$$\omega_\xi(V_1, \ldots, V_n) = \det(V^i_j)|\det(g_{ij})|^{1/2}.$$

Properties of determinants show that this uniquely defines ω_ξ as an n-form on $\mathscr{U}$.

If $V_1, \ldots, V_n$ is a frame field, then

$$\delta_{ij}\varepsilon_j = \langle V_i, V_j \rangle = \langle \sum V^r_i \partial_r, \sum V^s_j \partial_s \rangle = \sum V^r_i g_{rs} V^s_j.$$

Taking determinants gives $(-1)^\nu = (\det(V^i_j))^2 \det(g_{ij})$, hence $\omega_\xi(V_1, \ldots, V_n) = \det(V^i_j)|\det(g_{ij})|^{1/2} = \pm 1$. ∎

In the notation of differential forms, $\omega_\xi = |g|^{1/2}\, dx^1 \wedge \cdots \wedge dx^n$, where $|g| = |\det(g_{ij})|$.

20. Lemma. A semi-Riemannian manifold M has a (global) volume element if and only if M is orientable.

Proof. If ω is a volume element, then the bases for $T_p(M)$ such that $\omega(v_1, \ldots, v_n) > 0$ constitute an orientation $\lambda(p)$ of $T_p(M)$, and the function $p \to \lambda(p)$ is smooth.

If M is oriented, then for all positively oriented coordinate systems ξ the local volume elements ω_ξ agree on overlaps, hence give a global volume element. In fact, by the determinant formula, $\omega_\eta = J\omega_\xi$. Then $J > 0$ and $\omega_\eta^2 = \omega_\xi^2$ imply $\omega_\eta = \omega_\xi$ on their common domain. ∎

In Chapter 9 we shall interpret the Lie derivative L_X of Definition 2.16 as a derivative relative to the flow of the vector field X (compare Proposition 1.58). Thus the following result shows that the divergence of X is the logarithmic rate of change of volume under the flow of X.

21. Lemma. If ω is a (local) volume element on M, then $L_X(\omega) = (\operatorname{div} X)\omega$.

Proof. Let $E_1, \ldots, E_n$ be a frame field such that $\omega(E_1, \ldots, E_n) = 1$. Since L_X is a tensor derivation and $L_X(1) = X1 = 0$,

$$(L_X\omega)(E_1, \ldots, E_n) = -\sum \omega(E_1, \ldots, L_X E_i, \ldots, E_n).$$

Write $L_X E_i = [X, E_i]$ as $\sum_j f_{ij} E_j$. Since ω is skew-symmetric, upon substitution the summands other than $f_{ii} E_i$ yield zero. Thus $(L_X \omega)(E_1, \ldots, E_n) = -\sum f_{ii}$.

On the other hand, from Chapter 3,

$$\operatorname{div} X = \sum \varepsilon_i \langle D_{E_i} X, E_i \rangle = -\sum \varepsilon_i \langle [X, E_i], E_i \rangle + \sum \varepsilon_i \langle D_X E_i, E_i \rangle.$$

The latter sum vanishes since $\langle E_i, E_i \rangle$ is constant, leaving $\operatorname{div} X = -\sum f_{ii}$. ∎

The proof of Lemma 20 shows that if M is orientable, it can be oriented by the choice of a volume element—and there are just two, $\pm\omega$, if M is connected. Sometimes we denote a selected volume element on M in classical style by dM.

22. Definition. Let M and N be semi-Riemannian manifolds of the same dimension n, oriented by volume elements dM and dN. If $\phi: M \to N$ is a smooth mapping, the function $J \in \mathfrak{F}(M)$ such that $\phi^*(dN) = J\, dM$ is the *Jacobian function* of ϕ.

Applying the formula to vectors $v_1, \ldots, v_n$ in $T_p(M)$ gives

$$dN(d\phi(v_1), \ldots, d\phi(v_n)) = J(p)\, dM(v_1, \ldots, v_n).$$

Thus the function J is nonvanishing if and only if ϕ is a local diffeomorphism, and then ϕ is orientation-preserving if $J > 0$, orientation-reversing if $J < 0$. In general, if $v_1, \ldots, v_n$ is any basis for $T_p(M)$, the formula

$$|J(p)| = \frac{|dN(d\phi(v_1), \ldots, d\phi(v_n))|}{|dM(v_1, \ldots, v_n)|}$$

exhibits $|J(p)|$ as the rate of change of volume near p under the mapping ϕ.

The following scheme traces back to Gauss. Let M be a hypersurface in $\mathbf{R}^{n+1}_\nu$ oriented by unit normal U. If M has sign ε, that is, if $\langle U, U \rangle = \varepsilon$, let $\mathbf{Q}$ be the unit hyperquadric in $\mathbf{R}^{n+1}_\nu$ with the same sign. Explicitly,

$$\mathbf{Q} = \begin{cases} S^n_\nu(1) & \text{if } \varepsilon = 1, \\ H^n_{\nu-1}(1) & \text{if } \varepsilon = -1. \end{cases}$$

$\mathbf{Q}$ will always be oriented by the position vector field P, which on $\mathbf{Q}$ is a unit normal. For each $p \in M$ the point $\psi(p)$ of $\mathbf{R}^{n+1}_\nu$ canonically corresponding to the vector U_p lies in $\mathbf{Q}$. The resulting smooth mapping $\psi: M \to \mathbf{Q}$ is called the *Gauss map* of $M \subset \mathbf{R}^{n+1}_\nu$.

23. Proposition. Let M be a hypersurface in $\mathbf{R}^{n+1}_\nu$ oriented by U. Then the Jacobian function J of the Gauss map of M is $(-1)^n \det S$, where S is the shape operator of M derived from U.

Proof. By construction U_p and $P_{\psi p}$ canonically correspond to $\psi(p)$ hence to each other. Thus if $v \in T_p(M)$, its canonical correspondent at $\psi(p)$ is in $T_{\psi p}(Q)$.

We assert that the differential map of ψ is essentially just $-S$. If $v \in T_p(M)$, let α be a curve in M with initial velocity v. Writing canonical correspondence as equality gives

$$d\psi(v) = (\psi \circ \alpha)'(0) = (U'_\alpha)(0) = D_v U = -S(v).$$

That M is oriented by U means that a frame $e_1, \ldots, e_n$ in $T_p(M)$ is positively oriented if and only if $e_1, \ldots, e_n, U$ is positively oriented. Then the canonically corresponding frame in $T_{\psi p}(Q)$ is positively oriented (by means of $P \approx U$). For the corresponding volume elements ω,

$$\begin{aligned} J(p) &= \omega(d\psi(e_1), \ldots, d\psi(e_n)) = \omega(-Se_1, \ldots, -Se_n) \\ &= (-1)^n(\det S)\omega(e_1, \ldots, e_n) = (-1)^n \det S. \quad \blacksquare \end{aligned}$$

For an orientable hypersurface $M \subset \mathbf{R}^{n+1}_v$ the function $\det S$ is called the *Gauss–Kronecker curvature*. It is independent of the choice of $\pm U$ in even dimensions n, and for $n = 2$ reduces to Gaussian curvature. The preceding result exhibits $|\det S|$ as the rate of change of volume under the Gauss map.

For the integration of n-forms over (regions in) smooth n-dimensional manifolds, see [W], [Sp], and [BG].

VECTOR BUNDLES

The tangent bundle TM is an instance of the following general notion.

24. Definition. A *k-vector bundle* (E, π) over a manifold M consists of a manifold E and a smooth map $\pi: E \to M$ such that (1) each $\pi^{-1}(p)$, $p \in M$, is a k-dimensional vector space; (2) for each $p \in M$ there is a neighborhood $\mathscr{U}$ of p in M and a diffeomorphism

$$\phi: \mathscr{U} \times \mathbf{R}^k \to \pi^{-1}(\mathscr{U}) \subset E$$

such that for each $q \in \mathscr{U}$, the map $v \to \phi(q, v)$ is a linear isomorphism from $\mathbf{R}^k$ onto $\pi^{-1}(q)$. (See Figure 1.)

Roughly speaking, over small enough regions $\mathscr{U}$ in M, the manifold E is a product $\mathscr{U} \times \mathbf{R}^k$.

Terminology: M is the *base manifold*, E the *total manifold*, π the *projection*, $\pi^{-1}(p)$ the *fiber over p*, $\mathbf{R}^k$ the *standard fiber*, and ϕ a *bundle chart* of the bundle. When the projection π is clear from context we say merely that E is a k-vector bundle over M.

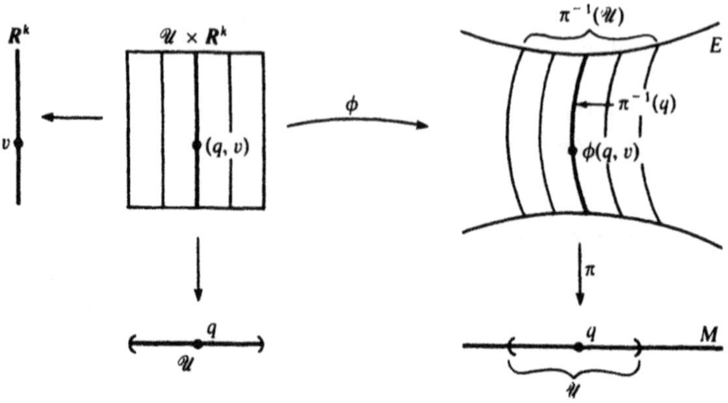

Figure 1

Recall that in the case of the tangent bundle TM each coordinate system ξ on $\mathscr{U} \subset M$ gives rise to a coordinate system $\tilde{\xi}(p, a_1, \ldots, a_n) = \sum a_i \partial_i|_p$ on $\pi^{-1}(\mathscr{U}) \subset TM$. Evidently these functions $\tilde{\xi}$ are bundle charts making TM an n-vector bundle over M^n. A vector field X on M is precisely a section $X: M \to TM$ of the tangent bundle, and this fact is the key to exploiting the analogy between TM and arbitrary vector bundles.

A *local basis* for a k-vector bundle (E, π) is a set of k linearly independent local sections $X_i: \mathscr{U} \to E$ $(1 \le i \le k)$ on an open set $\mathscr{U} \subset M$. There is a natural one-to-one correspondence between bundle charts ϕ and local bases $X_1, \ldots, X_k$ expressed by the formula $\phi(p, a_1, \ldots, a_k) = \sum a_i X_i|_p$.

A k-vector bundle (E, π) over M is *trivial* provided it has k linearly independent global sections, or equivalently a global bundle chart ϕ: $M \times R^k \approx E$. For example, TR^n and TS^1 are trivial, but TS^2 is not, since S^2 does not admit even one nonvanishing tangent vector field.

A vector bundle (E, π) is *orientable* provided there exists a function smoothly assigning to each $p \in B$ an orientation of $\pi^{-1}(p)$. Thus, for example, the tangent bundle (TM, π) is orientable if and only if the manifold M is orientable. (By contrast the *manifold* TM is always orientable.)

If M^n is a semi-Riemannian submanifold of $\bar{M}^{n+k}$, let NM be the set $\bigcup \{T_p(M)^\perp : p \in M\}$ of all normal vectors to M. Let $\pi: NM \to M$ be the map carrying each $T_p(M)^\perp$ to $p \in M$. We shall show that (NM, π) is in a natural way a k-vector bundle over M, called the *normal bundle* of M in $\bar{M}$. For each $p \in M$ there is a normal frame field $E_1, \ldots, E_k$ defined on some neighborhood $\mathscr{U}$ of p in M. The formula

$$\phi(q, a_1, \ldots, a_k) = \sum a_i E_i|_q$$

defines a one-to-one map ϕ of $\mathscr{U} \times R^k$ onto $\pi^{-1}(\mathscr{U}) \subset NM$. As with the tangent bundle, requiring each $\mathscr{U}$ to be a coordinate neighborhood lets us apply Proposition 1.42 to make NM a manifold with maps $(\xi \times \text{id}) \circ \phi^{-1}$ as coordinate systems. Then the maps ϕ are bundle charts making NM a vector bundle over M.

Geometrically the normal bundle NM bears a strong analogy to the tangent bundle TM. Just as vector fields in $\mathfrak{X}(M)$ are the sections of TM, those in $\mathfrak{X}(M)^\perp$ are the sections of NM. Corresponding to the Levi-Civita connection D for TM is the normal connection $D^\perp$ for NP (Definition 4.31). Indeed the theory of vector bundles can be used to unify much of differential geometry by generalizing the notions of tensor field, connection, and metric.

If (E, π) is a vector bundle over M, then the set $\Gamma(E)$ of its sections is in a natural way a module over $\mathfrak{F}(M)$. For tensor fields

$$A: \mathfrak{X}^*(M)^r \times \mathfrak{X}(M)^s \to \mathfrak{F}(M)$$

as defined in Chapter 2, if $\mathfrak{F}(M)$ is replaced by $\Gamma(E)$, then many results, in particular Proposition 2.2, go through as before. For example, the shape tensor of $M \subset \bar{M}$ is an $\mathfrak{F}(M)$-bilinear function

$$II: \quad \mathfrak{X}(M) \times \mathfrak{X}(M) \to \Gamma(NM) = \mathfrak{X}(M)^\perp.$$

Since the fiber of NM at p is $T_p(M)^\perp$ the generalized Proposition 2.2 gives as the value of II at p an $\mathbf{R}$-bilinear function $T_p(M) \times T_p(M) \to T_p(M)^\perp$.

Change notation from $M \subset \bar{M}$ to $P \subset M$. In the trivial case where the submanifold P reduces to a single point $p \in M$, then NP is just the tangent space $T_p(M)$. The exponential map $\exp_p$ generalizes as follows: If M is complete, the *normal exponential map*

$$\exp^\perp: NP \to M$$

sends $v \in NP$ to $\gamma_v(1)$, where γ_v is the M geodesic of initial velocity v. Thus $\exp^\perp$ carries radial lines in $T_p(P)$ to geodesics of M normal to P at p. A differential equations argument as for $\exp: TM \to M$ shows that $\exp^\perp$ is smooth. If M is not complete, then $\exp^\perp$ is defined only on some open domain containing the set Z of all zero vectors in NP. (Henceforth we often write $\exp^\perp$ merely as exp.)

Continuing the analogy, define a neighborhood $\mathscr{U}$ of P in M to be *normal* provided $\mathscr{U}$ is the diffeomorphic image under $\exp^\perp$ of a neighborhood of Z in NP. To prove that normal neighborhoods exist, a preliminary fact is needed.

25. Lemma. If $p \in P \subset M$, then $\exp^\perp$ carries some neighborhood of 0_p in NP diffeomorphically onto a neighborhood of p in M.

Proof. As for any vector bundle, the bundle charts show that the set Z of zeros is a submanifold of NP. Thus exp restricts to a smooth map of Z

onto P. In fact this is a diffeomorphism, since the inverse map is the zero vector field. Thus $d\exp_0$ is one-to-one on $T_0(Z) \subset T_0(NP)$.

Also $N_p = T_p(P)^\perp$ is a submanifold of NP, and as in the one-point case, $d\exp_0$ is the canonical isomorphism $T_0(N_p) \approx N_p$. Since $T_p(M)$ is the direct sum of $T_p(P)$ and N_p it follows that $d\exp_0$ is a linear isomorphism of $T_0(NP)$ onto $T_p(M)$. Then the inverse function theorem gives the result. ∎

26. Proposition. Every semi-Riemannian submanifold $P \subset M$ has a normal neighborhood in M.

Proof. For each $p \in P$ let $\mathcal{N}_p$ be a neighborhood of 0_p in NP on which exp is a diffeomorphism. By definition, P has the induced topology, so by shrinking $\mathcal{N}_p$ we can arrange that, if $\exp(v) \in P$ for $v \in \mathcal{N}_p$, then $v = 0$. By standard point set topology we can further arrange that $\mathcal{N} = \bigcup \mathcal{N}_p$ has this property: given any compact set $K \subset P$, the set $\pi^{-1}(K) \cap \bar{\mathcal{N}}$ is compact, where $\pi: NP \to P$ is the natural projection. It suffices to show that exp is one-to-one on some neighborhood of Z contained in $\mathcal{N}$. Note that by construction, if $v \neq w$ in $\mathcal{N}$ and $\exp(v) = \exp(w)$, then both v and w are nonzero.

There exists a sequence $\mathcal{N} = \mathcal{N}_1 \supset \mathcal{N}_2 \supset \cdots$ of neighborhoods of Z in NP such that $\bigcap \mathcal{N}_j = Z$. We assert that *given any compact set $K \subset P$ there is an index i such that exp is one-to-one on $E_j = (\pi^{-1}K \cap \mathcal{N}_j) \cup Z$*. Assume not. Then for each j there are vectors $v_j \neq w_j$ in E_j such that $\exp(v_j) = \exp(w_j)$. By an above property, neither v_j nor w_j is zero, hence both are in $\pi^{-1}K \cap \mathcal{N}_j \subset \pi^{-1}K \cap \bar{\mathcal{N}}$. Thus, passing to subsequences if necessary, $\{v_j\}$ and $\{w_j\}$ converge, necessarily to zero vectors 0_p and 0_q, respectively. Since exp is continuous, $p = \exp(0_p) = \exp(0_q) = q$. But then there are pairs v_j, w_j (as above) in $\mathcal{N}_p$, a contradiction.

If P is compact, take K to be P and the proof is complete. If P is not compact, then by second countability it contains an increasing sequence of compact sets $K_i \subset \text{int } K_{i+1}$ such that $\bigcup K_i = P$.

For K_1 choose E_i as above. A mild variant of the preceding argument shows that there is an $i' \geq i$ such that exp is one-to-one on $E_i \cup E_{i'}$. Continue by induction. Then $\bigcup E_i$ is the required neighborhood of Z. ∎

This result fails in general for immersed submanifolds. For example, a figure-8 submanifold of $\mathbf{R}^2$ as in Exercise 1.15 evidently has no normal neighborhood.

LOCAL ISOMETRIES

One of the most useful geometric applications of covering methods is this simple consequence of Proposition A.14.

27. Corollary. If $\ell: \tilde{M} \to M$ is a semi-Riemannian covering with $\tilde{M}$ connected and M simply connected, then ℓ is an isometry.

The result fails if ℓ is merely a local isometry onto M; thus it is important to decide which local isometries are covering maps. The following theorem shows that a necessary lift condition (Lemma A.9) is also sufficient.

28. Theorem. Let $\phi: M \to N$ be a local isometry with N connected. Suppose that, given any geodesic $\sigma: [0, 1] \to N$ and point $p \in M$ such that $\phi(p) = \sigma(0)$, there exists a lift $\tilde{\sigma}: [0, 1] \to M$ of σ through ϕ starting at p. Then ϕ is a semi-Riemannian covering map.

Proof. In view of Lemma 3.32 the lift condition implies that ϕ is onto. If $\mathscr{U}$ is a normal neighborhood of q in N, we shall show that $\mathscr{U}$ is evenly covered by ϕ (Definition A.7). The scheme is as follows: $\mathscr{U}$ is filled with radial geodesics; lifting these to start at each $p \in \phi^{-1}(q)$ gives disjoint diffeomorphic copies of $\mathscr{U}$ that fill $\phi^{-1}(\mathscr{U})$.

Let $\tilde{\mathscr{U}}$ be the neighborhood of 0 in $T_q(N)$ mapped diffeomorphically onto $\mathscr{U}$ by $\exp_q$. If $p \in \phi^{-1}(q)$, then $d\phi_p: T_p(M) \to T_q(N)$ is a linear isometry, hence also a diffeomorphism. Thus $\tilde{\mathscr{U}}(p) = (d\phi_p)^{-1}(\tilde{\mathscr{U}})$ is a starshaped neighborhood of 0 in $T_p(M)$.

(1) $\exp_p$ *is defined on* $\tilde{\mathscr{U}}(p)$.

If $v \in \tilde{\mathscr{U}}(p)$, let $\sigma: [0, 1] \to \tilde{\mathscr{U}}$ be the radial geodesic with initial velocity $d\phi(v) \in \tilde{\mathscr{U}}$. By hypothesis, σ has a lift $\tilde{\sigma}: [0, 1] \to M$ starting at p. Since ϕ is locally an isometry, $\tilde{\sigma}$ is a geodesic. Now $d\phi(\tilde{\sigma}'(0)) = \sigma'(0) = d\phi(v)$, hence $v = \tilde{\sigma}'(0)$. But then $\exp_p(v) = \tilde{\sigma}(1)$.

(2) $\mathscr{U}(p) = \exp_p(\tilde{\mathscr{U}}(p))$ *is a normal neighborhood of p in M.*

The proof of (1) shows in fact that, if $v \in \tilde{\mathscr{U}}(p)$, then $\phi(\exp_p(v)) = \exp_q(d\phi(v))$ for all $v \in \tilde{\mathscr{U}}(p)$. Thus ϕ carries $\mathscr{U}(p)$ onto $\mathscr{U}$. Let us agree that these maps are reduced to the sets shown in Figure 2. Then since $d\phi$ and $\exp_q$ are diffeomorphisms, so is $\phi \circ \exp_p$. Thus $\exp_p$ is one-to-one as well as onto, and $d\phi \circ d\exp_p$ is a linear isomorphism at each $v \in \tilde{\mathscr{U}}(p)$. But $d\phi$ is a linear isomorphism, hence so is $(d\exp_p)_v$. It follows that $\mathscr{U}(p)$ is an open submanifold of M and $\exp_p: \tilde{\mathscr{U}}(p) \to \mathscr{U}(p)$ is a diffeomorphism, thus proving (2).

(3) ϕ *carries* $\mathscr{U}(p)$ *diffeomorphically onto* $\mathscr{U}$.

In fact, for reduced maps as above, $\phi = \exp_q \circ d\phi_p \circ \exp_p^{-1}$, a composition of diffeomorphisms.

(4) *If* $p_1 \neq p_2$ *in* $\phi^{-1}(q)$, *then the neighborhoods* $\mathscr{U}(p_1)$ *and* $\mathscr{U}(p_2)$ *are disjoint.*

Assuming that there is a point $m \in \mathscr{U}(p_1) \cap \mathscr{U}(p_2)$, we prove $p_1 = p_2$. For $i = 1, 2$, let $\sigma_i: [0, 1] \to \mathscr{U}(p_i)$ be the radial geodesic—with parametriza-

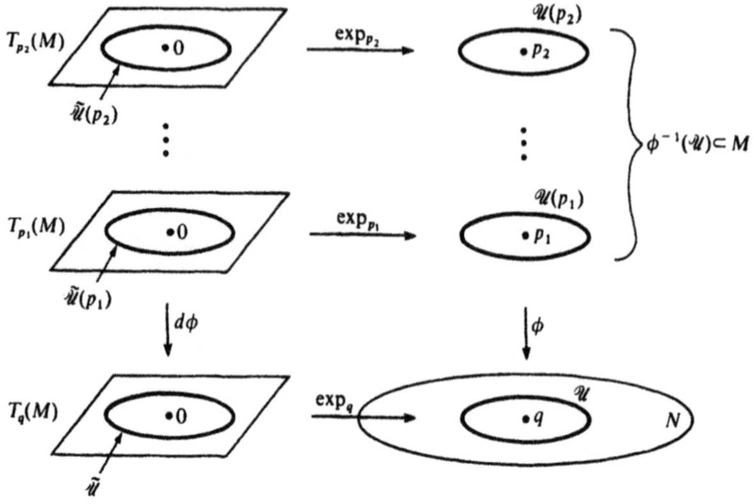

Figure 2

tion reversed—from m to p_i. Then $\phi \circ \sigma_1$ and $\phi \circ \sigma_2$ are reversed radial geodesics in $\mathscr{U}$ from $\phi(m)$ to q, so they are equal. Thus σ_1 and σ_2 are lifts of the same geodesic starting at the same point, hence $\sigma_1 = \sigma_2$. In particular, $p_1 = \sigma_1(1) = \sigma_2(1) = p_2$.

(5) $\phi^{-1}(\mathscr{U}) = \bigcup \{\mathscr{U}(p) : p \in \phi^{-1}(q)\}$.

By (3) the union is contained in $\phi^{-1}(\mathscr{U})$, so we must prove the reverse inclusion.

Let $m \in \phi^{-1}(\mathscr{U})$, and let $\sigma\colon [0, 1] \to \mathscr{U}$ be the reversed radial geodesic from $\phi(m)$ to q. If $\tilde{\sigma}$ is the lift of σ starting at m, let $p = \tilde{\sigma}(1)$. Then $\phi(p) = \sigma(1) = q$, so $p \in \phi^{-1}(q)$. But $\tilde{\sigma}$ must lie in $\tilde{\mathscr{U}}(p)$, hence $m \in \tilde{\mathscr{U}}(p)$. ∎

29. Corollary. Let $\phi\colon M \to N$ be a local isometry, with N connected. Then M is complete if and only if N is complete and ϕ is a semi-Riemannian covering map.

Proof. If M is complete, we assert that the lift condition in the theorem holds. In fact, if $\sigma\colon [0, 1] \to N$ is a geodesic and $\phi(p) = \sigma(0)$, there is a unique vector $v \in T_p(M)$ such that $d\phi(v) = \sigma'(0)$. By completeness, γ_v is defined on $\boldsymbol{R}$, and by the uniqueness of geodesics, $\phi \circ \gamma_v | [0, 1] = \sigma$. Thus the theorem shows that ϕ is a covering map. With the same notation, $\phi \circ \gamma_v$ is a geodesic extension of σ over $\boldsymbol{R}$. Thus N is complete.

Conversely, if $v \in T_p(M)$, then since N is complete, the geodesic with initial velocity $d\pi(v)$ is defined on the whole real line. Since ϕ is a covering map this curve has a lift $\tilde{\gamma}: \boldsymbol{R} \to M$ starting at p and hence having initial velocity v. Since ϕ is a local isometry, $\tilde{\gamma}$ is a geodesic. Thus $\tilde{\gamma}$ is in fact γ_v; hence M is complete. ∎

In this corollary and in the preceding theorem, if N is not connected it suffices to suppose that the image of ϕ meets every component of N.

MATCHED COVERINGS

The goal of this section is to show how a suitable covering $\mathscr{U}^*$ of a manifold M by open sets can be used to construct a covering manifold M^* of M. Suppose M is a surface; we can think of the open sets in $\mathscr{U}^*$ as disks of paper lying on M. If various pairs of overlapping disks are glued together—leaving each in position—then a new surface M^* is constructed together with a mapping $\mathscr{k}: M^* \to M$ sending each point of M^* to the point of M it lies over. In favorable cases $\mathscr{k}$ will be a covering map.

30. Definition. A *matched covering* $(\mathscr{U}^*, \smile)$ of a smooth manifold M is a covering $\mathscr{U}^* = \{\mathscr{U}_a : a \in A\}$ of M by open sets $\mathscr{U}_a$ together with a relation $\smile$ on the index set A such that for all $a, b, c \in A$

(1) $a \smile a$.
(2) If $a \smile b$, then $b \smile a$.
(3) If $a \smile b$, $b \smile c$, and $\mathscr{U}_a \cap \mathscr{U}_b \cap \mathscr{U}_c$ is nonempty, then $a \smile c$.

Then $\smile$ is called the *matching relation*; it tells which sets of $\mathscr{U}^*$ are to be glued together. For such a matched covering let

$$\Sigma = \{(p, a) \in M \times A : p \in \mathscr{U}_a\}.$$

On Σ define $(p, a) \sim (q, b)$ to mean $p = q$ and $a \smile b$. Though $\smile$ need not be an equivalence relation on A, it is easy to check that $\sim$ is an equivalence relation on Σ. Let M^* be the resulting set $\Sigma/\sim$ of equivalence classes.

If $a \in A$, let $\lambda_a: \mathscr{U}_a \to M^*$ be the one-to-one function sending each $p \in \mathscr{U}_a$ to the equivalence class containing (p, a). It follows from a variant of Proposition 1.42 that there is a unique way to make M^* a manifold such that each such λ_a is a diffeomorphism onto an open submanifold of M^*.

A point p^* of M^* is an equivalence class all of whose elements (p, a) have the same first coordinate p. Setting $\mathscr{k}(p^*) = p$ defines the *natural mapping* $\mathscr{k}: M^* \to M$. Then $\mathscr{k}$ is a local diffeomorphism onto M, for it is onto by construction and, if $(p, a) \in p^* \in M^*$, then $\mathscr{k} | \lambda_a(\mathscr{U}_a)$ is the diffeomorphism inverse to $\lambda_a: \mathscr{U}_a \to \lambda_a(\mathscr{U}_a)$.

A further property is required to make $\mathscr{k}$ a covering map.

31. Definition. A matched covering $(\mathcal{U}^*, \backsim)$ of M is *chainable* over a curve $\sigma: [0, 1] \to M$ provided that, given any $a \in A$ such that $\sigma(0) \in \mathcal{U}_a$, there exist numbers $0 = t_0 < t_1 < \cdots < t_k = 1$ and indices $a = a_1 \backsim a_2 \sim \cdots \backsim a_k$ such that

$$\sigma([t_{i-1}, t_i]) \subset \mathcal{U}_{a_i} \qquad \text{for} \quad 1 \leq i \leq k.$$

This is just what is required to guarantee that σ has a smooth lift through $\mathscr{k}: M^* \to M$ starting at any point over $\sigma(0)$. In fact, since $a \backsim b$ implies $\lambda_a = \lambda_b$ on $\mathcal{U}_a \cap \mathcal{U}_b$, the segments $\lambda_{a_i} \circ \sigma | [t_{i-1}, t_i]$ combine to form the required lift.

If $(\mathcal{U}^*, \backsim)$ is chainable over every curve segment, then $\mathscr{k}$ is a covering map; however, we consider only the case where M is semi-Riemannian. As usual the metric $\mathscr{k}^*(g)$ on M^* makes $\mathscr{k}: M^* \to M$ a local isometry, and Theorem 28 then gives

32. Proposition. If a matched covering $(\mathcal{U}^*, \backsim)$ of a semi-Riemannian manifold M is chainable over every geodesic segment $\sigma: [0, 1] \to M$, then $\mathscr{k}: M^* \to M$ is a semi-Riemannian covering map.

WARPED PRODUCTS

On a semi-Riemannian product manifold $B \times F$ the metric tensor is $\pi^*(g_B) + \sigma^*(g_F)$, where π and σ are the projections of $B \times F$ onto B and F, respectively. A rich class of metrics on $B \times F$ will now be obtained by homothetically warping the product metric on each fiber $p \times F$ (see [BO]).

Standard notation is used for the geometry of B, but the metric tensor of F is denoted by $g_F = (,)$ and its Levi-Civita connection by ∇.

33. Definition. Suppose B and F are semi-Riemannian manifolds, and let $f > 0$ be a smooth function on B. The *warped product* $M = B \times_f F$ is the product manifold $B \times F$ furnished with metric tensor

$$g = \pi^*(g_B) + (f \circ \pi)^2 \sigma^*(g_F).$$

Explicitly, if x is tangent to $B \times F$ at (p, q), then

$$\langle x, x \rangle = \langle d\pi(x), d\pi(x) \rangle + f^2(p)(d\sigma(x), d\sigma(x)).$$

The argument proving Lemma 3.5 shows that g is in fact a metric tensor. If $f = 1$, then $B \times_f F$ reduces to a semi-Riemannian product manifold.

B is called the *base* of $M = B \times_f F$, and F the *fiber*. Our goal is to express the geometry of M in terms of *warping function* f and the geometries of B and F.

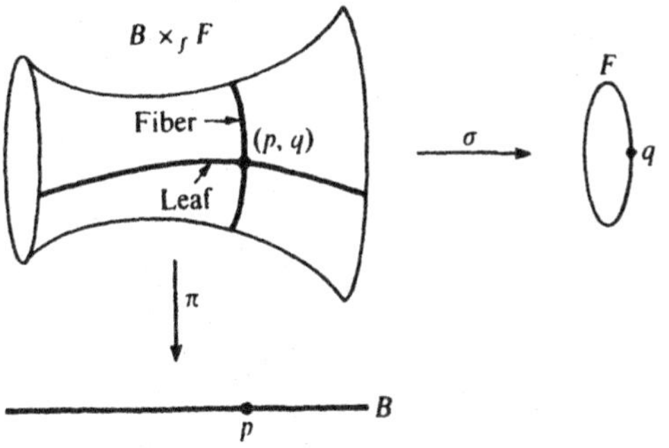

Figure 3

As in the case of a semi-Riemannian product it is easy to see that the *fibers* $p \times F = \pi^{-1}(p)$ and the *leaves* $B \times q = \sigma^{-1}(q)$ are semi-Riemannian submanifolds of M (see Figure 3), and the warped metric is characterized by

(1) For each $q \in F$, the map $\pi|(B \times q)$ is an isometry onto B.

(2) For each $p \in B$, the map $\sigma|(p \times F)$ is a positive homothety onto F, with scale factor $1/f(p)$.

(3) For each $(p, q) \in M$, the leaf $B \times q$ and the fiber $p \times F$ are orthogonal at (p, q).

Vectors tangent to leaves are *horizontal*; vectors tangent to fibers are *vertical*. We denote by $\mathscr{H}$ the orthogonal projection of $T_{(p,q)}(M)$ onto its horizontal subspace $T_{(p,q)}(B \times q)$, and by $\mathscr{V}$ the projection onto the vertical subspace $T_{(p,q)}(p \times F)$.

The horizontal/vertical terminology can be confusing in particular applications, and the following alternative notation for $\mathscr{H}$ and $\mathscr{V}$ is often convenient. It will soon be clear that only the shape tensor of the *fibers* is of interest. Thus, as in Chapter 4, we write *tan* for the projection $\mathscr{V}$ onto $T_{(p,q)}(p \times F)$, and *nor* for the projection $\mathscr{H}$ onto

$$(T_{(p,q)}(p \times F))^{\perp} = T_{(p,q)}(B \times q).$$

Hence for vertical vector fields V, W on M, the formula $II(V, W) = \text{nor}\, D_V W$ gives the shape tensor of all fibers.

Recall from Chapter 1 the notion of *lift* of a vector field on B or F to $B \times F$, the set of all such lifts being denoted as usual by $\mathfrak{L}(B)$ and $\mathfrak{L}(F)$, respectively. Typically we use the same notation for a vector field and for its lift.

The relation of a warped product to the base B is almost as simple as in the special case of a semi-Riemannian product; however, the relation to the fiber F often involves the warping function f.

34. Lemma. If $h \in \mathfrak{F}(B)$, then the gradient of the lift $h \circ \pi$ of h to $M = B \times_f F$ is the lift to M of the gradient of h on B.

Proof. We must show that $\operatorname{grad}(h \circ \pi)$ is horizontal and π-related to $\operatorname{grad} h$ on B.

If v is a vertical tangent vector to M, then $\langle \operatorname{grad}(h \circ \pi), v \rangle = v(h \circ \pi) = d\pi(v)h = 0$, since $d\pi(v) = 0$. Thus $\operatorname{grad}(h \circ \pi)$ is horizontal. If x is horizontal,

$$\begin{aligned}\langle d\pi(\operatorname{grad}(h \circ \pi)), d\pi(x) \rangle &= \langle \operatorname{grad}(h \circ \pi), x \rangle = x(h \circ \pi) \\ &= d\pi(x)h = \langle \operatorname{grad} h, d\pi(x) \rangle.\end{aligned}$$

Hence at each point, $d\pi(\operatorname{grad}(h \circ \pi)) = \operatorname{grad} h$. ∎

Thus there should be no confusion if we simplify the notation by writing h for $h \circ \pi$ and $\operatorname{grad} h$ for $\operatorname{grad}(h \circ \pi)$.

The Levi-Civita connection of M can now be related to those of B and F as follows.

35. Proposition. On $M = B \times_f F$, if X, $Y \in \mathfrak{L}(B)$ and V, $W \in \mathfrak{L}(F)$, then

(1) $D_X Y \in \mathfrak{L}(B)$ is the lift of $D_X Y$ on B.
(2) $D_X V = D_V X = (Xf/f)V$.
(3) $\operatorname{nor} D_V W = II(V, W) = -(\langle V, W \rangle / f) \operatorname{grad} f$.
(4) $\tan D_V W \in \mathfrak{L}(F)$ is the lift of $\nabla_V W$ on F.

Proof. (1) The Koszul formula for $2\langle D_X Y, V \rangle$ reduces to

$$-V\langle X, Y \rangle + \langle V, [X, Y] \rangle$$

since by Corollary 1.44, $[X, V] = [Y, V] = 0$. Because X and Y are lifts from B, $\langle X, Y \rangle$ is constant on fibers. Since V is vertical, $V\langle X, Y \rangle = 0$. But $[X, Y]$ is tangent to leaves, hence $\langle V, [X, Y] \rangle = 0$. Thus $\langle D_X Y, V \rangle = 0$ for all $V \in \mathfrak{L}(F)$, so $D_X Y$ is horizontal. Since each $\pi | (B \times q)$ is an isometry the result follows.

(2) $D_X V = D_V X$ since $[X, V] = 0$. These vector fields are vertical since by (1), $\langle D_X V, Y \rangle = -\langle V, D_X Y \rangle = 0$. All the terms in the Koszul formula for $2\langle D_X V, W \rangle$ vanish except $X\langle V, W \rangle$. By definition of the warped metric tensor, $\langle V, W \rangle(p, q) = f^2(p)(V_q, W_q)$. Writing f for $f \circ \pi$, we have $\langle V, W \rangle = f^2((V, W) \circ \sigma)$. The parenthesized term is constant on leaves, to which X is tangent; hence

$$X\langle V, W \rangle = X[f^2((V, W) \circ \sigma)] = 2fXf((V, W) \circ \sigma) = 2(Xf/f)\langle V, W \rangle.$$

Thus $D_X V = (Xf/f)V$.

(3) By (2),

$$\langle D_V W, X\rangle = -\langle W, D_V X\rangle = -\langle W, (Xf/f)V\rangle = -(Xf/f)\langle V, W\rangle.$$

By Lemma 34, $Xf = \langle \text{grad}\, f, X\rangle$ on M as on B. Thus for all X,

$$\langle D_V W, X\rangle = -\langle ((\langle V, W\rangle/f)\, \text{grad}\, f, X\rangle.$$

(4) Since V and W are tangent to all fibers, Lemma 4.4 asserts that on a fiber, tan $D_V W$ is the fiber covariant derivative applied to the restrictions of V and W to that fiber. Then σ-relatedness follows since homotheties preserve Levi-Civita connections. ∎

36. Corollary. The leaves $B \times q$ of a warped product are totally geodesic; the fibers $p \times F$ are totally umbilic.

Proof. By (1) in the preceding proposition the shape tensor of each leaf is zero. The fiber assertion is immediate from (3). ∎

37. Examples of Warped Products. (1) A surface of revolution is a warped product with leaves the different positions of the rotated curve and fibers the circles of revolution. Explicitly, if M is gotten by revolving a plane curve C about an axis in $\boldsymbol{R}^3$ and $f: C \to \boldsymbol{R}^+$ gives distance to the axis, then M is $C \times_f S^1(1)$.

(2) $\boldsymbol{R}^3 - 0$ as a warped product: In spherical coordinates the line element of $\boldsymbol{R}^3 - 0$ is

$$ds^2 = dr^2 + r^2(d\vartheta^2 + \sin^2 \vartheta\, d\varphi^2).$$

Setting $r = 1$ gives the line element of the unit sphere S^2. Evidently $\boldsymbol{R}^3 - 0$ is diffeomorphic to $\boldsymbol{R}^+ \times S^2$ under the natural map $(t, p) \leftrightarrow tp$. Thus the formula for ds^2 shows that $\boldsymbol{R}^3 - 0$ can be identified with the warped product $\boldsymbol{R}^+ \times_r S^2$. In $\boldsymbol{R}^3 - 0$ the leaves are the rays from the origin and the fibers are the spheres $S^2(r)$, $r > 0$. In general, $\boldsymbol{R}^n - 0$ is naturally isometric to $\boldsymbol{R}^+ \times_r S^{n-1}$.

(3) The standard spacetime models of the universe are warped products (Chapter 12), as are the simplest models of neighborhoods of stars and black holes (Chapter 13).

WARPED PRODUCT GEODESICS

In $B \times_f F$, as in any product manifold, a curve γ can be written as $\gamma(s) = (\alpha(s), \beta(s))$ with α and β the projections of γ into B and F, respectively.

38. Proposition. A curve $\gamma = (\alpha, \beta)$ in $M = B \times_f F$ is a geodesic if and only if

(1) $\alpha'' = (\beta', \beta') f \circ \alpha \operatorname{grad} f$ in B,

(2) $\beta'' = \dfrac{-2}{f \circ \alpha} \dfrac{d(f \circ \alpha)}{ds} \beta'$ in F.

Proof. The result is local in character so it suffices to work in an arbitrarily small interval around say $s = 0$.

Case 1. $\gamma'(0)$ is neither horizontal nor vertical. Then α and β are regular; hence by Exercise 3.12 we can suppose α is an integral curve of X on B and β is an integral curve of V on F. With X and V denoting also the lifts to M, γ is an integral curve of $X + V$. Thus

$$\gamma'' = D_{X+V}(X + V) = D_X X + D_X V + D_V X + D_V V.$$

Evidently $\gamma'' = 0$ if and only if both $\tan \gamma'' = 0$ and $\operatorname{nor} \gamma'' = 0$. Using Proposition 35, the latter become

$$D_X X - \frac{\langle V, V \rangle}{f} \operatorname{grad} f = 0; \qquad 2\frac{Xf}{f} V + \nabla_V V = 0$$

(evaluated on γ). Since $\langle V, V \rangle = f^2(V, V)$ and $\alpha' = X$, $\beta' = V$, the result follows.

Case 2. $\gamma'(0)$ is horizontal. If γ is a geodesic, then since leaves are totally geodesic, γ remains in $B \times \beta(0)$. Hence β is constant and (1) and (2) are trivial. Conversely, if (2) holds, then since $\beta'(0) = 0$, it follows as in the remark below that β is constant. Then (1) implies that α geodesic; hence so is γ.

Case 3. $\gamma'(0)$ is vertical and nonzero. We can suppose $\operatorname{grad} f \neq 0$ at $p = \alpha(0)$, for otherwise the fiber $p \times F$ is totally geodesic and the result follows as in Case 2. If γ is geodesic, then on no interval around 0 does γ remain in the (totally umbilic) fiber $p \times F$. Hence there is a sequence $\{s_i\} \to 0$ such that for all i, $\gamma'(s_i)$ is neither horizontal or vertical. Then (1) and (2) follow by continuity from Case 1. Conversely, (1) shows that $\alpha''(0) \neq 0$, hence there is a sequence $\{s_i\}$ as above and, again by Case 1, γ is geodesic.

∎

39. Remark. Let $\gamma = (\alpha, \beta)$ be a geodesic in $M = B \times_f F$. By Exercise 3.19, equation (2) above implies that *β is a pregeodesic in F*. Also the function $(f \circ \alpha)^4(\beta', \beta')$ is a constant C, since by (2) its derivative is zero. Thus (1) becomes

$$\alpha'' = \frac{C}{(f \circ \alpha)^3} \operatorname{grad} f = -\operatorname{grad} \phi,$$

where $\phi = C/(2f^2)$. By reparametrization it can be assumed that $C/2$ is $-1, 0$, or $+1$ depending on the causal character of β.

In the special case of a semi-Riemannian product the warping function is constant, so the geodesic equations in Proposition 38 reduce to $\alpha'' = 0$, $\beta'' = 0$. This proves Corollary 3.57(1). The second assertion in the corollary, dealing with completeness, has its strongest generalization in the Riemannian case, as follows.

40. Lemma. If B and F are complete Riemannian manifolds, then $M = B \times_f F$ is complete for every warping function f.

Proof. We use the metric completeness criterion from the Hopf–Rinow theorem (5.21). Note first that if v is tangent to M then, since $f > 0$ and F is Riemannian, $\langle v, v \rangle \geq \langle d\pi v, d\pi v \rangle$. Hence $L(\alpha) \geq L(\pi \circ \alpha)$ for any curve segment; hence $d(m, m') \geq d(\pi m, \pi m')$ for all $m, m' \in M$.

This property implies that, if $\{(p_i, q_i)\}$ is a Cauchy sequence in M, then $\{p_i\}$ is Cauchy in B. Since B is complete, $\{p_i\}$ converges to some point $p \in B$. We can assume then that the sequence lies in some compact set K in B; hence $f \geq c > 0$ on K. Then a variant of the argument above shows that $d(m, m') \geq c d(\sigma m, \sigma m')$ for $m, m' \in K \times F$. Now $\{q_i\}$ is Cauchy in F and thus converges; so the original sequence converges and M is complete. ∎

This result fails for indefinite metrics—even if, as in the following example, both B and F have definite metrics.

41. Example (Beem, Buseman). Let $M = \boldsymbol{R}_1^1 \times_{e^t} \boldsymbol{R}^1$. The geodesic equations in Proposition 38 reduce to the following numerical formulas:

$$\alpha'' = -\beta'^2 e^{2\alpha}, \qquad \beta'' = -2\alpha'\beta'.$$

Then $\gamma(s) = (\ln s, 1/s)$ is a geodesic defined for $s > 0$. Evidently γ is inextendible and incomplete, hence M is incomplete. Reversing the metric of M gives the same result for $\boldsymbol{R}^1 \times_{e^t} \boldsymbol{R}_1^1$.

CURVATURE OF WARPED PRODUCTS

The problem is to express the curvature of a warped product $M = B \times_f F$ in terms of its warping function f and the curvatures of B and F.

For a covariant tensor A on B, its *lift* $\tilde{A}$ to M is just its pullback $\pi^*(A)$ under the projection $\pi\colon M \to B$. In the case of a $(1, s)$ tensor $A\colon \mathfrak{X}(B) \times \cdots \times \mathfrak{X}(B) \to \mathfrak{X}(B)$, if $v_1, \ldots, v_s \in T_{(p,q)}(M)$ define $\tilde{A}(v_1, \ldots, v_s)$ to be the horizontal vector at (p, q) that projects to $A(d\pi v_1, \ldots, d\pi v_s)$ in $T_p(B)$. Thus in both cases $\tilde{A}$ is zero on vectors any one of which is vertical. These definitions involve no geometry, hence are correspondingly valid for lifts from F.

Let ${}^B R$ and ${}^F R$ be the lifts to M of the Riemannian curvature tensors of B and F. Since the projection π is an isometry on each leaf, ${}^B R$ gives the Riemannian curvature of each leaf. The corresponding assertion holds for ${}^F R$, since the projection σ is a homothety. Because leaves are totally geodesic, ${}^B R$ agrees with the curvature tensor R of M on horizontal vectors. This time the corresponding assertion fails for ${}^F R$ and R, since fibers are in general only umbilic.

If $h \in \mathfrak{F}(B)$, the lift to M of the Hessian of h is again denoted by H^h. (This agrees with the Hessian of the lift $h \circ \pi$ generally only on horizontal vectors.)

42. Proposition. Let $M = B \times_f F$ be a warped product with Riemannian curvature tensor R. If $X, Y, Z \in \mathfrak{L}(B)$ and $U, V, W \in \mathfrak{L}(F)$, then

(1) $R_{XY}Z \in \mathfrak{L}(B)$ is the lift of ${}^B R_{XY}Z$ on B.
(2) $R_{VX}Y = (H^f(X, Y)/f)V$, where H^f is the Hessian of f.
(3) $R_{XY}V = R_{VW}X = 0$.
(4) $R_{XV}W = (\langle V, W\rangle/f)\, D_X(\operatorname{grad} f)$.
(5) $R_{VW}U = {}^F R_{VW}U - (\langle \operatorname{grad} f, \operatorname{grad} f\rangle/f^2)\{\langle V, U\rangle W - \langle W, U\rangle V\}$.

These are tensor equations, hence are valid as usual for individual tangent vectors.

Proof. (1) Proved above.

(2) Since $[V, X] = 0$, $R_{VX}Y = -D_V D_X Y + D_X D_V Y$. By Proposition 35,

$$\begin{aligned} D_X D_V Y &= D_X(Yf/f\,V) = X(Yf/f)V + Vf/f\,D_X V \\ &= [XYf/f + YfX(1/f)]V + (Yf/f)(Xf/f)V. \end{aligned}$$

But $X(1/f) = -Xf/f^2$, so this expression reduces to $(XYf/f)V$. Because $D_X Y \in \mathfrak{L}(B)$, $D_V(D_X Y) = ((D_X Y)f/f)V$. Thus

$$R_{VX}Y = [(XYf - (D_X Y)f)/f]V = (H^f(X, Y)/f)V.$$

(3) As usual we can assume that $[V, W] = 0$. Thus

$$R_{VW}X = -D_V D_W X + D_W D_V X.$$

Now

$$D_V D_W X = D_V(Xf/f\,W) = V(Xf/f)W + Xf/f\,D_V W.$$

But Xf/f is constant on fibers, so $V(Xf/f) = 0$. Thus

$$R_{VW}X = (Xf/f)(-D_V W + D_W V) = (Xf/f)[W, V] = 0.$$

Then by a symmetry of curvature, $\langle R_{XY} V, W\rangle = \langle R_{VW} X, Y\rangle = 0$. By (1), $\langle R_{XY} V, Z\rangle = -\langle R_{XY} Z, V\rangle = 0$. These equations hold for all $W \in \mathfrak{L}(F)$ and $Z \in \mathfrak{L}(B)$, hence $R_{XY} V = 0$.

(4) Note first that $R_{XV} W$ is horizontal, since $\langle R_{XV} W, U\rangle = \langle R_{WU} X, V\rangle$, which is zero by (3).

Since $R_{VW} X = 0$, it follows from the symmetries of curvature that $R_{XV} W = R_{XW} V$. But using (2),

$$\begin{aligned}\langle R_{XV} W, Y\rangle = \langle R_{VX} Y, W\rangle &= H^f(X, Y)\langle V, W\rangle/f \\ &= (\langle V, W\rangle/f)\langle D_X(\operatorname{grad} f), Y\rangle.\end{aligned}$$

Since $R_{XV} W$ is horizontal and the equation holds for all Y, the result follows.

(5) $R_{VW} U$ is vertical since, by (3), $\langle R_{VW} U, X\rangle = -\langle R_{VW} X, U\rangle = 0$. Because the projection σ is a homothety on fibers, ${}^F R_{VW} U \in \mathfrak{L}(F)$ is the application to V, W, U of the curvature tensor of each fiber. Thus ${}^F R_{VW} U$ and $R_{VW} U$ are related by the Gauss equation (4.5). Since the shape tensor of the fibers is given by $II(V, W) = -(\langle V, W\rangle/f)\operatorname{grad} f$, the result follows. ∎

Now consider the Ricci curvature Ric of a warped product, writing ${}^B\mathrm{Ric}$ for the lift (pullback by π) of the Ricci curvature of B, and similarly for ${}^F\mathrm{Ric}$.

43. Corollary. On a warped product $M = B \times_f F$ with $d = \dim F > 1$, let X, Y be horizontal and V, W vertical. Then

(1) $\mathrm{Ric}(X, Y) = {}^B\mathrm{Ric}(X, Y) - (d/f)H^f(X, Y)$.
(2) $\mathrm{Ric}(X, V) = 0$.
(3) $\mathrm{Ric}(V, W) = {}^F\mathrm{Ric}(V, W) - \langle V, W\rangle f^{\#}$, where

$$f^{\#} = \frac{\Delta f}{f} + (d - 1)\frac{\langle \operatorname{grad} f, \operatorname{grad} f\rangle}{f^2}$$

and $\Delta f = C(H^f)$ is the Laplacian on B.

The proof is an exercise in tensor computation: Apply the Ricci formula in Lemma 3.52 to a frame field on M whose vector fields are in $\mathfrak{L}(B)$ and $\mathfrak{L}(F)$.

SEMI-RIEMANNIAN SUBMERSIONS

The notion of warped product can be generalized as follows.

44. Definition. A *semi-Riemannian submersion* $\pi: M \to B$ is a submersion (1.38) of semi-Riemannian manifolds such that:

(S1) The fibers $\pi^{-1}(b)$, $b \in B$, are semi-Riemannian submanifolds of M.

(S2) $d\pi$ preserves scalar products of vectors normal to fibers.

Since the fibers of a submersion are smooth submanifolds, (S1) is automatically true if M is Riemannian.

Consistent with previous usage, vectors tangent to fibers are *vertical*, those normal to fibers are *horizontal*. At each point $p \in \pi^{-1}(b) \subset M$, $\mathscr{H}$ and $\mathscr{V}$ denote the orthogonal projections of $T_p(M)$ on its subspaces

$$\mathscr{H}_p = T_p(\pi^{-1}b)^{\perp} \qquad \text{and} \qquad \mathscr{V}_p = T_p(\pi^{-1}b),$$

respectively. Because π is a submersion, $d\pi$ gives a linear isomorphism $\mathscr{H}_p \approx T_b(B)$; (S2) asserts that it is a linear isometry. Just as for a product manifold each vector field X on B has a unique *horizontal lift* $\bar{X}$ on M: $\bar{X}$ is horizontal and π-related to X.

45. Lemma. If $X, Y \in \mathfrak{X}(B)$, then: (1) $\langle \bar{X}, \bar{Y} \rangle = \langle X, Y \rangle \circ \pi$; (2) $\mathscr{H}[\bar{X}, \bar{Y}] = [X, Y]^{-}$; (3) $\mathscr{H}\bar{D}_{\bar{X}}\bar{Y} = (D_X Y)^{-}$, where $\bar{D}$ is the Levi-Civita connection of M.

Proof. (1) is immediate from (S2). For (2) note that by Lemma 1.22, $[\bar{X}, \bar{Y}]$ is π-related to $[X, Y]$, hence $\mathscr{H}[\bar{X}, \bar{Y}]$ is also. (3) holds if both sides have the same scalar product with every horizontal vector field—or merely with every horizontal lift $\bar{Z}$. Thus by (1) it suffices to prove $\langle \bar{D}_{\bar{X}}\bar{Y}, \bar{Z} \rangle = \langle D_X Y, Z \rangle \circ \pi$. This follows by expanding both sides in the Koszul formula, since, using (1) and (2),

$$\bar{X}\langle \bar{Y}, \bar{Z} \rangle = \bar{X}\{\langle Y, Z \rangle \circ \pi\} = (d\pi\bar{X})\langle Y, Z \rangle = X\langle Y, Z \rangle \circ \pi,$$
$$\langle \bar{X}, [\bar{Y}, \bar{Z}] \rangle = \langle \bar{X}, [Y, Z]^{-} \rangle = \langle X, [Y, Z] \rangle \circ \pi. \qquad \blacksquare$$

46. Corollary. Under a semi-Riemannian submersion $\pi: M \to B$, horizontal geodesics in M map to geodesics in B.

Proof. If γ is a nonconstant horizontal geodesic, then $\pi \circ \gamma$ is regular, hence is (locally) an integral curve of a vector field X. But γ is an integral curve of the lift $\bar{X}$, and hence by the lemma,

$$(\pi \circ \gamma)'' = D_X X = d\pi(\mathscr{H}\bar{D}_{\bar{X}}\bar{X}) = d\pi(\bar{D}_{\bar{X}}\bar{X}) = d\pi(\gamma'') = 0. \qquad \blacksquare$$

The preceding results are trivial for warped products since leaves are totally geodesic and project isometrically. However, for a semi-Riemannian submersion $\pi: M \to B$ with dim $B \geq 2$, leaves need not exist, even locally. In view of Frobenius' theorem [W], this failure can be measured by the function assigning to each pair of horizontal vector fields X, Y the vertical vector field $\mathscr{V}[X, Y]$. This function is $\mathfrak{F}(M)$-bilinear and in the case of a warped product is identically zero.

47. Theorem. Let $\pi: M \to B$ be a semi-Riemannian submersion. If horizontal vector fields X, Y on M span nondegenerate planes, then

$$K_B(d\pi X, d\pi Y) = K_M(X, Y) + \tfrac{3}{4}\langle \mathscr{V}[X, Y], \mathscr{V}[X, Y]\rangle / Q(X, Y).$$

The proof is a computation much like those of the preceding section. (See Theorem 3.20 of [CE] or for further details [O2].)

Exercises

1. A local diffeomorphism $\phi: M \to N$ is a covering map (a) if each $\phi^{-1}(q)$, $q \in N$, contains the same finite number of points, or (b) if M is compact and N is connected.

2. Let $k: \overline{M} \to M$ be a double covering with M connected. (a) The map ρ reversing the (two) points in each $k^{-1}(m)$ is a deck transformation—in fact, the only nontrivial one, (b) The covering is trivial if and only if $\overline{M}$ is not connected.

3. Let $k: \tilde{M} \to M$ be a normal covering. (a) Show that M is orientable if and only if $\tilde{M}$ has an orientation preserved by all deck transformations. (b) Prove the analogue for time-orientation and a Lorentz covering.

4. Let $k: \tilde{M} \to M$ be a smooth covering. (a) If $\phi: P \to \tilde{M}$ is continuous and $k \circ \phi$ is smooth, then ϕ is smooth. (b) If $\psi: M \to Q$ is continuous and $\psi \circ k$ is smooth, then ψ is smooth. (Corresponding arguments prove the semi-Riemannian analogues, in which *smooth* is replaced by *locally isometric.*)

5. If V is a vector field, then in terms of coordinates,

$$\operatorname{div} V = \frac{1}{\sqrt{|g|}} \sum_i \frac{\partial}{\partial x^i} (\sqrt{|g|}\, V^i), \quad \text{where} \quad |g| = |\det(g_{ij})|.$$

Hence for the corresponding formula for Δf, replace V^i by $\sum_j g^{ij}\, \partial f / \partial x^j$. (Hint: For a local volume element, $\omega(\partial_1, \ldots, \partial_n) = \pm\sqrt{|g|}$.)

6. Let Γ be a group of diffeomorphisms on a manifold M. (a) If (PD1) of Definition 6 holds, then each orbit of p is a closed set of M and each $p \in M$

has a neighborhood $\mathscr{U}$ such that the sets $\phi(\mathscr{U})$, $\phi \in \Gamma$, are mutually disjoint. (b) If (PD1) holds and M is compact, then Γ is finite. (c) If Γ is finite and no element $\neq$ id has a fixed point, then Γ is properly discontinuous.

7. (a) If Γ is a group of isometries of a Riemannian manifold, then (PD1) of Definition 6 implies (PD2). (b) Let $M = \boldsymbol{R}^2 - 0$ with line element $2\,dx\,dy$. The group Γ generated by the isometry $\phi(x, y) = (x/2, 2y)$ satisfies (PD1) but not (PD2).

8. Definitions of orientation. The following conditions on M^n are equivalent: (i) M is orientable (Definition 1.41); (ii) M has an orientation λ; (iii) there is a nonvanishing n-form on M; (iv) every loop in M is orientation-preserving, that is, lifts as a loop into the orientation covering $\hat{M}$.

9. (a) A product manifold $M \times N$ is orientable if and only if both M and N are. (b) Let (E, π) be a vector bundle over M. If any two of $M, E, (E, \pi)$ are orientable, then so is the third. (c) A submanifold P of an orientable manifold M is orientable if and only if its normal bundle is orientable. (d) If M is orientable and q is a regular value of $\phi\colon M \to N$, then $\phi^{-1}(q)$ is orientable.

10. (a) For any manifold M the manifold TM is orientable. (b) A manifold with trivial tangent bundle is orientable. (c) Lie groups are orientable.

11. A product map $\phi \times \psi$ of a warped product $B \times_f F$ is an isometry if and only if $\psi\colon F \to F$ is an isometry and $\phi\colon B \to B$ is an isometry such that $f \circ \phi = f$.

12. (a) $P_\nu^n(r) = S_\nu^n(r)/\pm 1$ has curvature $K = 1/r^2$ and is complete and geodesically connected; it is orientable if and only if n is odd. (b) The same properties hold for $H_\nu^n(r)/\pm 1$, except $K = -1/r^2$. (c) $P_1^n(r)$ is not time-orientable, but $H_1^n/\pm 1$ is.

13. (a) In Corollary 43, $f^\# = \Delta(f^d)/(f^d d)$. (b) The scalar curvature of $B \times_f F$ is

$$S = {}^BS + {}^FS/f^2 - 2d\Delta f/f - d(d-1)\langle \operatorname{grad} f, \operatorname{grad} f\rangle/f^2,$$

where $d = \dim F$.

14. Express S_ν^n as a warped product with fiber $S^{n-\nu}$ and base H^ν with metric reversed.

8 SYMMETRY AND CONSTANT CURVATURE

A natural condition to impose on a semi-Riemannian manifold is that its curvature tensor R be parallel, that is, have vanishing covariant differential, $DR = 0$. Such a manifold is said to be *locally symmetric*. In particular, manifolds of constant curvature turn out to be locally symmetric.

A fundamental property of curvature is its control over the relative behavior of nearby geodesics. Because a normal neighborhood $\mathscr{U}$ is filled with radial geodesics, curvature thereby gives a description of the geometry of $\mathscr{U}$. Considering only the locally symmetric case, we show that this description is so accurate that, if $\mathscr{U}$ and $\mathscr{U}'$ are normal neighborhoods with the same description (and same dimension and index), then $\mathscr{U}$ and $\mathscr{U}'$ are isometric.

Using covering techniques this local result is given a global formulation that as a first consequence provides a list of all complete, simply connected manifolds of constant curvature.

JACOBI FIELDS

A curve can be compared with nearby curves using the following notion.

1. Definition. A *variation* of a curve segment $\alpha: [a, b] \to M$ is a two-parameter mapping

$$\boldsymbol{x}: [a, b] \times (-\delta, \delta) \to M,$$

such that $\alpha(u) = \boldsymbol{x}(u, 0)$ for all $a \leq u \leq b$.

The u-parameter curves of a variation are called *longitudinal* and the v-parameter curves *transverse*. The *base curve* of $\boldsymbol{x}$ is α. Typically we are interested in the longitudinal curves and the number $\delta > 0$ is not important.

The vector field V on α given by $V(u) = \boldsymbol{x}_v(u, 0)$ is called the *variation vector field* of $\boldsymbol{x}$. Each $V(u)$ is the initial velocity of the transverse curve $v \to \boldsymbol{x}(u, v)$; thus, for $\delta > 0$ sufficiently small, the vector field V is an infinitesimal model of the variation $\boldsymbol{x}$.

If every longitudinal curve of $\boldsymbol{x}$ is geodesic, $\boldsymbol{x}$ is called a *geodesic variation* or *one-parameter family of geodesics*.

2. Definition. If γ is a geodesic, a vector field Y on γ that satisfies the *Jacobi differential equation* $Y'' = R_{Y\gamma'}(\gamma')$ is called a *Jacobi vector field.*

3. Lemma. The variation vector field of a geodesic variation is a Jacobi field.

Proof. Since each longitudinal curve is geodesic, $\boldsymbol{x}_{uu} = 0$. Thus by Proposition 4.44,

$$\boldsymbol{x}_{vuu} = \boldsymbol{x}_{uvu} = \boldsymbol{x}_{uuv} + R(\boldsymbol{x}_v, \boldsymbol{x}_u)\boldsymbol{x}_u = R(\boldsymbol{x}_v, \boldsymbol{x}_u)\boldsymbol{x}_u.$$

Hence $\boldsymbol{x}_v$ satisfies the Jacobi equation on every longitudinal curve, in particular on the base curve, where $\boldsymbol{x}_v$ is the variation vector field. ∎

Because of this result the Jacobi equation is also called the equation of *geodesic deviation*. There is a far-reaching heuristic interpretation. If we think of a geodesic variation $\boldsymbol{x}$ of γ as a one-parameter family of freely falling particles, then the variation vector field V gives the position, relative to γ, of arbitrarily nearby particles. Thus the derivative V' gives relative velocity, and V'' relative acceleration. Assigning these particles unit mass we can read the Jacobi equation $V'' = R_{V\gamma'}\gamma'$ as Newton's second law with the curvature vector $R_{V\gamma'}\gamma'$ in the role of force, the so-called *tidal force*. We shall see in Chapter 12 that this is the key to the interpretation of curvature as gravitation in general relativity.

In the following example, attracting tidal forces pull radiating geodesics back together again.

4. Example. In the sphere $S^2(r)$ consider the variation

$$\boldsymbol{x}(u, v) = r(\cos u \cos v, \cos u \sin v, \sin u)$$

for $-\pi/2 \leq u \leq \pi/2$ and $|v|$ small. The curves v constant are indeed longitudinal, since they parametrize semicircles joining the north and south poles $(0, 0, \pm r)$. Thus $\boldsymbol{x}$ is a geodesic variation of $\gamma(u) = r(\cos u, 0, \sin u)$. The variation vector field is $V(u) = \boldsymbol{x}_v(u, 0) = r \cos u\, \partial_y$. Since ∂_y is parallel in $\boldsymbol{R}^3$

and tangent to the sphere it is parallel in the geometry of the sphere. Hence

$$V'' = -r\cos u\,\partial_y = -V.$$

Since $S^2(r)$ has constant curvature $K = 1/r^2$ and $\gamma' = -r\sin u\,\partial_x + r\cos u\,\partial_z$, we compute

$$R_{V\gamma'}(\gamma') = (1/r^2)[\langle\gamma', V\rangle\gamma' - \langle\gamma', \gamma'\rangle V] = -V.$$

Thus, as predicted by the lemma, V is a Jacobi field.

5. Lemma. Let γ be a geodesic with $\gamma(0) = p$, and let $v, w \in T_p(M)$. Then there is a unique Jacobi field Y on γ such that $Y(0) = v$ and $Y'(0) = w$.

Proof. Let $E_1, \ldots, E_n$ be a parallel frame field on γ, and write $Y = \sum y^i E_i$. Let v^i and w^i $(1 \leq i \leq n)$ be the coordinates of v and w relative to $E_1(0), \ldots, E_n(0)$. Then

$$Y(0) = v \Leftrightarrow y^i(0) = v^i; \qquad Y'(0) = w \Leftrightarrow (dy^i/ds)(0) = w^i.$$

Because γ' is parallel, $\gamma' = \sum a^i E_i$ with constant coefficients. Thus the Jacobi equation $Y'' = R_{Y\gamma'}(\gamma')$ is equivalent to the linear system

$$\frac{d^2(y^m)}{ds^2} = \sum_{ijk} R^m_{ijk} a^i y^j a^k \qquad (1 \leq m \leq n),$$

where the smooth coefficient functions are uniquely determined by

$$R_{E_j E_k}(E_i) = \sum R^m_{ijk} E_m \qquad (1 \leq i, j, k \leq n).$$

Such linear systems have smooth solutions (on the entire domain of γ) that are uniquely determined by initial conditions as above. ∎

Because the Jacobi equation is linear, the set of all Jacobi fields on γ forms a real vector space. The lemma shows that its dimension is $2n$, since v and w can be chosen arbitrarily in the n-dimensional space $T_p(M)$.

Exponential maps are defined in terms of radial geodesics; thus the Jacobi influence of curvature on geodesics leads to a Jacobi description of exponential maps.

6. Proposition. Let o be a point of M and let $x \in T_o(M)$. For $v_x \in T_x(T_o M)$,

$$d\exp_o(v_x) = V(1),$$

where V is the unique Jacobi field on the geodesic γ_x such that

$$V(0) = 0 \qquad \text{and} \qquad V'(0) = v \in T_o(M).$$

Proof. As in the proof of the Gauss lemma (5.1) consider the two-parameter map $\tilde{x}(t, s) = t(x + sv)$ in $T_o(M)$, with $0 \leq t \leq 1$ and $|s|$ small. Its exponential image in M,

$$x(t, s) = \exp_o(t(x + sv)) = \gamma_{x+sv}(t),$$

is a geodesic variation of $\gamma_x | [0, 1]$. Thus the variation vector field

$$V(u) = x_s(u, 0) = d \exp_o(\tilde{x}_s(u, 0))$$

is a Jacobi field on γ_x.

Since $\tilde{x}(1, s) = x + sv$, $V(1) = d \exp_o(v_x)$. The curve $s \to x(0, s)$ is constant at o, hence $V(0) = x_s(0, 0) = 0$. By Proposition 4.44

$$V'(0) = x_{st}(0, 0) = x_{ts}(0, 0).$$

Now $s \to x_t(0, s) = x + sv$ is a vector field on the constant curve at o. Hence by Proposition 3.18, $x_{ts}(0, s) = v$ for all s. In particular, $V'(0) = v$. ∎

This result provides a strong link between M and its curvature tensor.

TIDAL FORCES

A vector field Y on a curve $\alpha: I \to M$ is *tangent to* α if $Y = f\alpha'$ for some $f \in \mathfrak{F}(I)$ and *perpendicular to* α if $\langle Y, \alpha' \rangle = 0$. We consider how these notions relate to Jacobi fields.

If $|\alpha'| > 0$, then each tangent space $T_{\alpha(s)}(M)$ has a direct sum decomposition $\mathbf{R}\alpha' + \alpha'^{\perp}$. Hence each vector field Y on α has a unique expression $Y = Y^{\top} + Y^{\perp}$, where $Y^{\top}$ is tangent to α and $Y^{\perp}$ is perpendicular to α.

If γ is a geodesic, then $Y \perp \gamma$ implies $Y' \perp \gamma$, since $d/ds\langle Y, \gamma' \rangle = \langle Y', \gamma' \rangle$; similarly for tangency. Then, if γ is also nonnull, it follows that $(Y')^{\top} = (Y^{\top})'$ and $(Y')^{\perp} = (Y^{\perp})'$.

7. Lemma. Let Y be a vector field on a geodesic γ.

(1) If Y is tangent to γ then it is a Jacobi field $\Leftrightarrow Y'' = 0 \Leftrightarrow Y(s) = (as + b)\gamma'(s)$ for all s.

(2) If Y is a Jacobi field, then $Y \perp \gamma \Leftrightarrow$ there exist $a \neq b$ such that $Y(a) \perp \gamma$, $Y(b) \perp \gamma \Leftrightarrow$ there exists a such that $Y(a) \perp \gamma$, $Y'(a) \perp \gamma$.

(3) If γ is nonnull, then Y is a Jacobi field $\Leftrightarrow$ both $Y^{\top}$ and $Y^{\perp}$ are Jacobi fields.

Proof. (1) Since $R(v, v) = 0$, the Jacobi equation for $f\gamma'$ is equivalent to $d^2f/ds^2 = 0$. (2) Again since $R(v, v) = 0$, the second derivative of $\langle Y, \gamma' \rangle$ is zero, hence $\langle Y(s), \gamma'(s) \rangle = A + sB$, and the result follows. (3) $R(Y^{\top}, \gamma') = 0$, hence $R(Y^{\perp}, \gamma') = R(Y, \gamma')$. Also, since $R(Y, \gamma')$ is skew-adjoint, $R(Y, \gamma')\gamma' \perp \gamma$.

Thus the Jacobi equation $Y'' = R(Y, \gamma')\gamma'$ splits into the two equations $(Y^{\top})'' = 0$ and $(Y^{\perp})'' = R(Y^{\perp}, \gamma')\gamma'$. The result follows using (1). ∎

A Jacobi field tangent to a geodesic γ is of scant importance since it is the infinitesimal model for a family of geodesics that merely reparametrize γ. Thus in considering the Jacobi equation $Y'' = R_{Y\gamma'}\gamma'$ as *relative acceleration* produced by *tidal force* we shall emphasize the case $Y \perp \gamma$.

8. Definition. For a vector $0 \neq v \in T_p(M)$ the *tidal force operator* $F_v\colon v^{\perp} \to v^{\perp}$ is given by $F_v(y) = R_{yv}v$.

9. Lemma. F_v is a self-adjoint linear operator on $v^{\perp}$, and trace $F_v = -\mathrm{Ric}(v, v)$.

Proof. Clearly F_v is linear (and carries $v^{\perp}$ to itself). The pair symmetry of curvature implies that F_v is self-adjoint, but note that $v^{\perp}$ is a degenerate subspace if v is a null vector. If v is nonnull, let $e_2, \ldots, e_n$ be an orthonormal basis for $v^{\perp}$. Then

$$\mathrm{Ric}(v, v) = \sum \varepsilon_i \langle R_{ve_i}v, e_i\rangle = -\sum \varepsilon_i \langle F_v(e_i), e_i\rangle = -\mathrm{trace}\, F_v.$$

If v is null, let w be a null vector such that $\langle v, w\rangle = -1$; these two vectors span a Lorentz plane Π. Then $e_1 = (v + w)/\sqrt{2}$ and $e_2 = (v - w)/\sqrt{2}$ form an orthonormal basis for Π with e_1 timelike and e_2 spacelike. Let $e_3, \ldots, e_n$ be an orthonormal basis for $\Pi^{\perp} \subset v^{\perp}$. Then

$$\mathrm{Ric}(v, v) = -\langle R_{ve_1}v, e_1\rangle + \langle R_{ve_2}v, e_2\rangle + \sum_{j>2} \varepsilon_j \langle R_{ve_j}v, e_j\rangle.$$

The first two terms on the right-hand side cancel since

$$\langle R_{ve_1}v, e_1\rangle = \tfrac{1}{2}\langle R_{vw}v, w\rangle = \langle R_{ve_2}v, e_2\rangle.$$

Now $v, e_3, \ldots, e_n$ is a basis for $v^{\perp}$, and $F_v(v) = 0$. Hence

$$\mathrm{trace}\, F_v = \sum_{j>2} \varepsilon_j \langle F_v(e_j), e_j\rangle = -\sum_{j>2} \varepsilon_j \langle R_{ve_j}v, e_j\rangle. \qquad ∎$$

In the nonnull case tidal forces are often normalized by taking v to be a unit vector. For example, if M has constant curvature C, then Corollary 3.43 gives $F_v(y) = -\varepsilon C y$, where $\varepsilon = \langle v, v\rangle = \pm 1$.

LOCALLY SYMMETRIC MANIFOLDS

10. Proposition. The following conditions on a semi-Riemannian manifold M are equivalent:

(1) $DR = 0$, that is, M is locally symmetric.

(2) If X, Y, Z are parallel vector fields on a curve α, then the vector field $R_{XY}Z$ on α is also parallel.

(3) Sectional curvature is invariant under parallel translation, that is, the sectional curvature of a nondegenerate tangent plane Π remains constant as Π is parallel translated along any curve α.

Proof. For simplicity assume that the curve α is regular; this will be sufficient for later work.

(1) $\Rightarrow$ (2). Fix an arbitrary point on α, say $\alpha(0)$. By Exercise 3.12 there exist vector fields V, $\bar{X}$, $\bar{Y}$, $\bar{Z}$ on a neighborhood of $\alpha(0)$ in M such that $V_{\alpha(t)} = \alpha'(t)$, $\bar{X}_{\alpha(t)} = X(t)$, and similarly for $\bar{Y}$ and $\bar{Z}$. Since $DR = 0$,

$$0 = (D_V R)_{\bar{X}\bar{Y}}\bar{Z} = D_V(R(\bar{X}, \bar{Y})\bar{Z}) - R(D_V \bar{X}, \bar{Y})\bar{Z} \\ - R(\bar{X}, D_V \bar{Y})\bar{Z} - R(\bar{X}, \bar{Y})(D_V \bar{Z}).$$

Now evaluate at 0. There, for example,

$$(D_V \bar{Y})|_{\alpha(0)} = (\bar{Y}_\alpha)'(0) = Y'(0) = 0,$$

since Y is parallel. Thus the equation above reduces to

$$(R_{XY}Z)'(0) = D_{\alpha'(0)}(R_{XY}Z) = 0.$$

(2) $\Rightarrow$ (1). If $v, x, y, z \in T_p(M)$, let X, Y, Z be the parallel vector fields on γ_v gotten by parallel translation of x, y, z, respectively. Then

$$(D_v R)_{xy} z = (R_{XY}Z)'(0) = 0.$$

(2) $\Rightarrow$ (3). Let $\Pi(0)$ be a nondegenerate tangent plane at $\alpha(0)$, and let X, Y be parallel vector fields on α such that $X(0)$, $Y(0)$ is a basis for $\Pi(0)$. Thus $X(t)$, $Y(t)$ is a basis for the parallel translate $\Pi(t)$ of $\Pi(0)$ along α. Then both $\langle R_{XY}X, Y\rangle$ and $Q(X, Y)$ are constant along α, hence the sectional curvature of $\Pi(t)$ is constant.

(3) $\Rightarrow$ (2). Suppose $\alpha: I \to M$ starts at p. By orthonormal expansion it suffices to show that, if X, Y, Z, W are parallel vector fields on α, then $\langle R_{XY}Z, W\rangle$ is constant. Fix $t \in I$ and define a function $A: T_p(M)^4 \to \boldsymbol{R}$ by

$$A(x, y, z, w) = \langle R_{XY}Z, W\rangle(t),$$

where X, Y, Z, W are parallel vector fields on α extending x, y, z, w. Since $Q(X, Y)$ is constant, (3) gives

$$\frac{A(x, y, x, y)}{Q(x, y)} = K(X(t), Y(t)) = K(x, y)$$

for all $x, y \in T_p(M)$ spanning a nondegenerate tangent plane. It is easy to check that the function A is curvaturelike (page 79); hence by Corollary 3.42

$$A(x, y, z, w) = \langle R_{xy}z, w\rangle.$$

Thus $\langle R_{XY}Z, W\rangle(t)$ is independent of t. ∎

11. Corollary. A semi-Riemannian manifold of constant sectional curvature is locally symmetric.

This is obvious from criterion (3) above. Semi-Riemannian products of locally symmetric manifolds are again locally symmetric but (if nonflat) do not have constant curvature.

ISOMETRIES OF NORMAL NEIGHBORHOODS

Let M and $\bar{M}$ be semi-Riemannian manifolds of the same dimension and index and let $o \in M$ and $\bar{o} \in \bar{M}$. Our goal is to find conditions under which a given linear isometry $T_o(M) \to T_{\bar{o}}(\bar{M})$ is the differential map of an isometry defined on some normal neighborhood of o.

12. Definition. Let $\mathrm{L}: T_o(M) \to T_{\bar{o}}(\bar{M})$ be a linear isometry, and let $\mathscr{U}$ be a normal neighborhood of o in M such that $\exp_{\bar{o}}$ is defined on the set $\mathrm{L}(\exp_o^{-1}(\mathscr{U}))$. Then the mapping

$$\phi_{\mathrm{L}} = \exp_{\bar{o}} \circ \mathrm{L} \circ \exp_o^{-1} : \mathscr{U} \to \bar{M}$$

is called the *polar map* of L on $\mathscr{U}$.

In short, ϕ_{L} sends $\exp_o(v)$ to $\exp_{\bar{o}}(\mathrm{L}v)$ for all $v \in \mathscr{U} \subset T_o(M)$.

Polar maps always exist for $\mathscr{U}$ sufficiently small, and the first two properties below show that *if the isometry we seek exists it must be a polar map of* L.

13. Lemma. With notation as above,

(1) ϕ_{L} carries radial geodesics to radial geodesics; explicitly, if $v \in T_o(M)$, then $\phi_{\mathrm{L}} \circ \gamma_v = \gamma_{\mathrm{L}v}$, where both sides are defined.

(2) The differential map of ϕ at o is L.

(3) If $\mathscr{U}$ is sufficiently small, then ϕ_{L} is a diffeomorphism onto a normal neighborhood of o in $\bar{M}$.

(4) If M is complete, ϕ_{L} is defined on every normal neighborhood of o.

Proof. (1) Since $\gamma_v(t) = \exp_o(tv)$,

$$\phi_{\mathrm{L}}(\gamma_v(t)) = \exp_o(\mathrm{L}(tv)) = \exp_o(t\mathrm{L}(v)) = \gamma_{\mathrm{L}v}(t)$$

for all t such that $\gamma_v(t)$ remains in $\mathscr{U}$.

(2) If $v \in T_o(M)$, then by (1)

$$d\phi_{\mathrm{L}}(v) = d\phi_{\mathrm{L}}(\gamma_v'(0)) = (\phi_{\mathrm{L}} \circ \gamma_v)'(0) = \gamma_{\mathrm{L}v}'(0) = \mathrm{L}v.$$

(3) If $\mathscr{U}$ is sufficiently small, $\phi_{\mathrm{L}}(\mathscr{U})$ is contained in a normal neighborhood of $\bar{o}$. Then $\exp_{\bar{o}}$ is a diffeomorphism of $\tilde{\mathscr{V}} = \mathrm{L}(\exp_o^{-1} \mathscr{U})$ onto a (normal) neighborhood $\mathscr{V}$ of $\bar{o}$. Thus $\phi_{\mathrm{L}}: \mathscr{U} \to \mathscr{V}$ is a composition of diffeomorphisms.

Finally, (4) is clear since $\exp_{\bar{o}}$ is defined on all of $T_{\bar{o}}(\bar{M})$. ∎

If L is to be the differential map at o of an isometry, then, as we saw in Chapter 3, it must *preserve curvature* at o; that is,

$$\mathrm{L}(R_{xy}z) = R_{\mathrm{L}x,\mathrm{L}y}(\mathrm{L}z) \qquad \text{for} \quad x, y, z \in T_o(M).$$

By Corollary 3.42 it is equivalent that L preserve sectional curvature: $K(\Pi) = K(\mathrm{L}\Pi)$ for nondegenerate planes in $T_o(M)$.

This necessary condition is sufficient.

14. Theorem. Let M and $\bar{M}$ be locally symmetric semi-Riemannian manifolds, and let $\mathrm{L}: T_o(M) \to T_{\bar{o}}(\bar{M})$ be a linear isometry that preserves curvature. Then

(1) If $\mathscr{U}$ is a sufficiently small normal neighborhood of o, there is a unique isometry ϕ of $\mathscr{U}$ onto a normal neighborhood $\mathscr{V}$ of $\bar{o}$ such that $d\phi_o = \mathrm{L}$.

(2) If $\bar{M}$ is complete then for any normal neighborhood $\mathscr{U}$ of o, there is a unique local isometry $\phi: \mathscr{U} \to \bar{M}$ such that $d\phi_o = \mathrm{L}$.

Proof. In both cases uniqueness follows from Proposition 3.62. By the preceding lemma, existence in both cases follows if every polar map $\phi = \phi_{\mathrm{L}}: \mathscr{U} \to \bar{M}$ is a local isometry. Thus for $v \in T_p(M)$, $p \in \mathscr{U}$, we must prove that $\langle d\phi v, d\phi v\rangle = \langle v, v\rangle$. The idea is to use Proposition 6 and show that corresponding Jacobi fields grow at the same rate in both manifolds.

(1) Let $\tilde{\mathscr{U}}$ be the neighborhood in $T_o(M)$ corresponding to the normal neighborhood $\mathscr{U}$. There is a unique $x \in \tilde{\mathscr{U}}$ and $y_x \in T_x(T_o M)$ such that $d\exp_o(y_x) = v$. By Proposition 6, $\langle v, v\rangle = \langle Y(1), Y(1)\rangle$, where Y is the Jacobi field on γ_x such that $Y(0) = 0$ and $Y'(0) = y \in T_o(M)$.

(2) Now look at the corresponding situation in $\bar{M}$. Since L is linear, $d\mathrm{L}(y_x) = (\mathrm{L}y)_{\mathrm{L}x}$. Hence by the definition of the polar map ϕ,

$$d\phi(v) = d\exp_{\bar{o}}((\mathrm{L}y)_{\mathrm{L}x}).$$

Thus, as above, $\langle d\phi v, d\phi v\rangle = \langle \bar{Y}(1), \bar{Y}(1)\rangle$, where $\bar{Y}$ is the unique Jacobi field on $\gamma_{\mathrm{L}x}$ such that $\bar{Y}(0) = 0$ and $\bar{Y}'(0) = \mathrm{L}y \in T_{\bar{o}}(\bar{M})$.

(3) Let $E_1, \ldots, E_n$ and $\bar{E}_1, \ldots, \bar{E}_n$ be parallel frame fields on γ_x and $\gamma_{\mathrm{L}x}$ respectively, such that $\mathrm{L}(E_i(0)) = \bar{E}_i(0)$ for all i. Since L is a linear isometry, the coordinates of x and y relative to the $E_i(0)$s are the same as the coordinates of Lx and Ly relative to the $\bar{E}_i(0)$s.

If we write $Y = \sum y^i E_i$, then the functions $y^1, \ldots, y^n$ satisfy the system of differential equations in the proof of Lemma 5 as well as the initial conditions corresponding to $Y(0) = 0$ and $Y'(0) = y$. Over in $\bar{M}$, writing $\bar{Y} = \sum \bar{y}^i E_i$ gives the corresponding differential equations

$$\frac{d^2(\bar{y}^m)}{dt^2} = \sum_{ijk} \bar{R}^m_{ijk} a^i \bar{y}^j a^k \qquad (1 \leq m \leq n).$$

Furthermore, by a remark above, the functions $y^1, \ldots, y^n$ and $\bar{y}^1, \ldots, y^n$ satisfy exactly the same initial conditions.

(4) We assert that $\bar{R}^m_{ijk} = R^m_{ijk}$ on common domain I of γ_x and γ_{Lx} (keeping the former in $\mathscr{U}$). In fact, since the linear isometry L preserves curvature, it follows from our choice of corresponding frame fields that $\bar{R}^m_{ijk}(0) = R^m_{ijk}(0)$. Since M is locally symmetric, Proposition 1 shows that the functions $\langle R_{E_j E_k} E_i, E_m \rangle$ and hence R^m_{ijk} are constant on I. Similarly $\bar{R}^m_{ijk}$ is constant, giving the result.

(5) The functions $\bar{y}^i$ and y^i $(1 \leq i \leq n)$ satisfy the same system of differential equations and the same initial conditions. By the uniqueness of such solutions, $\bar{y}^i = y^i$ for all i; hence

$$\langle d\phi v, d\phi v \rangle = \langle \bar{Y}(1), \bar{Y}(1) \rangle = \sum \varepsilon_i (y^i(1))^2 = \langle Y(1), Y(1) \rangle = \langle v, v \rangle. \quad \blacksquare$$

A first consequence of the theorem is that constant curvature determines local geometry:

15. Corollary. Let M and $\bar{M}$ be semi-Riemannian manifolds with the same dimension and index and the same constant curvature C. Then any points $o \in M$ and $\bar{o} \in \bar{M}$ have isometric neighborhoods.

This is clear since a linear isometry from $T_o(M)$ to $T_{\bar{o}}(\bar{M})$ exists (by 2.27) and must preserve curvature.

Another application will explain the term *locally symmetric*. For $p \in M$ let ζ_p be the polar map of the linear isometry $v \to -v$ of $T_p(M)$. If $\mathscr{U}$ is a suitably chosen normal neighborhood, then $\zeta_p \colon \mathscr{U} \to \mathscr{U}$ is a diffeomorphism. Evidently ζ_p reverses geodesics through p: if $\gamma(0) = p$, then $\zeta_p(\gamma(s)) = \gamma(-s)$. In fact this property uniquely determines ζ_p (once $\mathscr{U}$ is specified), hence ζ_p is called the *local geodesic symmetry* of M at p.

16. Corollary. The following conditions on a semi-Riemannian manifold are equivalent:

(1) M is locally symmetric.

(2) If L: $T_p(M) \to T_q(M)$ is a local isometry that preserves curvature, then there is an isometry ϕ of normal neighborhoods of p and q such that $d\phi_p =$ L.

(3) At each point p of M the local geodesic symmetry ζ_p is an isometry.

Proof. (1) ⇒ (2). By the theorem.

(2) ⇒ (3). The linear isometry $-id$ on $T_p(M)$ carries each nondegenerate plane $\Pi \subset T_p(M)$ to itself; thus $-\text{id}$ preserves curvature. If $\phi_{-\text{id}}$ is the isometry given by (2), then $\phi_{-\text{id}}$ reverses geodesics through p, hence is the local geodesic symmetry at p.

(3) ⇒ (1). We show that the tensor field DR vanishes at an arbitrary point p. Since ζ_p is an isometry it preserves curvature and covariant derivatives. The differential map of ζ_p at p is $-\text{id}$, hence for all $v, x, y, z \in T_p(M)$

$$-(D_v R)_{xy} z = (D_{-v} R)_{-x,-y}(-z) = (D_v R)_{xy} z.$$

Hence $DR = 0$ at p. ∎

SYMMETRIC SPACES

The local result in Theorem 14 can be generalized as follows.

17. Theorem. Let M and $\bar{M}$ be complete, connected, locally symmetric semi-Riemannian manifolds, with M simply connected. If $\text{L}: T_o(M) \to T_{\bar{o}}(\bar{M})$ is a linear isometry that preserves curvature, then there is a unique semi-Riemannian covering map $\phi: M \to \bar{M}$ such that $d\phi_o = \text{L}$.

We postpone the proof to look first at some consequences. Roughly speaking, the theorem asserts that a complete connected locally symmetric manifold $\bar{M}$ is determined—up to semi-Riemannian covering—by its curvature at one point. The result is due in principle to E. Cartan (1869–1951), who singlehandedly created the theory of symmetric spaces. (Also see [A] and Theorem 1.36 of [CE].)

18. Definition. A semi-Riemannian *symmetric space* is a connected semi-Riemannian manifold M such that for each $p \in M$ there is a (unique) isometry $\zeta_p: M \to M$ with differential map $-\text{id}$ on $T_p(M)$.

The isometry ζ_p is called the *global symmetry* of M at p. Since it reverses the geodesics through p, ζ_p is the unique extension to all of M of the local geodesic symmetry at p. Thus the latter is an isometry; so by Corollary 16, *symmetric* implies *locally symmetric*.

19. Examples. (1) $\boldsymbol{R}^n$ is symmetric, since for each point p the map $p + x \to p - x$ is an isometry. (2) The sphere S^n is symmetric, since for each p it is symmetric in the usual Euclidean sense about the line in $\boldsymbol{R}^{n+1}$ through p and $-p$. (3) In fact, by Proposition 4.30, *every connected hyperquadric is symmetric* (take the frame $f_1, \ldots, f_n$ to be $-e_1, \ldots, -e_n$).

Chapter 11 will exhibit a variety of symmetric spaces with nonconstant curvature.

A locally symmetric manifold need not be complete, since any open submanifold of one is again one; however,

20. Lemma. A semi-Riemannian symmetric space is complete.

Proof. To show that a geodesic $\gamma\colon [0, b) \to M$ is extendible, choose c near b in the interval, and let ζ be the global symmetry at $\gamma(c)$. Since ζ reverses geodesics through $\gamma(c)$, a reparametrization of $\zeta \circ \gamma$ provides the required extension of γ. ∎

21. Corollary. A complete, simply connected, locally symmetric semi-Riemannian manifold is symmetric.

Proof. At any point $p \in M$ the linear isometry $-\text{id}$ on $T_p(M)$ preserves curvature. Thus we can apply Theorem 17, with $\overline{M} = M$, to obtain a semi-Riemannian covering map $\phi\colon M \to M$ such that $d\phi_p = -\text{id}$. Since M is simply connected, ϕ is an isometry, by Corollary 7.27. ∎

It follows that, if M is a complete, connected, locally symmetric manifold, then its simply connected semi-Riemannian covering manifold $\tilde{M}$ is symmetric. In fact, by Corollary 7.29, $\tilde{M}$ is complete and (since the covering map is in particular a local isometry) $\tilde{M}$ is locally symmetric. Thus the preceding corollary shows that $\tilde{M}$ is symmetric.

Now we return to the proof of Theorem 17. The scheme is as follows. By Theorem 14 there is an isometry on some neighborhood of $o \in M$ whose differential map is L. Any point $p \in M$ can be connected to o by a broken geodesic β, so we can parallel translate L along β to p and there get another isometry on a neighborhood of p. Because M is simply connected, these locally defined isometries fit together to give the required map ϕ.

Some notation is required to cope with broken geodesics. For any $k \geq 1$ let v be a k-tuple $(v_1, \ldots, v_k)$ of vectors in $T_o(M)$. Then the broken geodesic $\beta_v\colon [0, k] \to M$ is defined inductively as follows (using the fact that M is complete). Let $\beta_1 = \gamma_{v_1}|[0, 1]$. If β_j has been defined on $[0, j]$, for $1 \leq j < k$, let $w = P(v_{j+1})$ be the parallel translate of v_{j+1} along β_j to its endpoint $\beta_j(j)$. Then attaching

$$t \to \gamma_w(t - j) \qquad (j \leq t \leq j + 1)$$

to β_j gives β_{j+1}.

(1) For v as above, let $P_v\colon T_o(M) \to T_{\beta_v(k)}(M)$ be parallel translation along β_v. Similarly in $\overline{M}$, for $\text{L}v = (\text{L}v_1, \ldots, \text{L}v_k)$ let $P_{\text{L}v}\colon T_{\bar{o}}(\overline{M}) \to T_{\beta_{\text{L}v}(k)}(\overline{M})$ be parallel translation along $\beta_{\text{L}v}$. Then the parallel translate of L along β_k is

$$\text{L}_v = P_{\text{L}v} \circ \text{L} \circ P_v^{-1} : T_p(M) \to T_q(\overline{M}),$$

where $p = \beta_v(k)$ and $q = \beta_{Lv}(k)$. Since M and $\bar{M}$ are locally symmetric the linear isometries P_v and P_{Lv} preserve curvature, hence so does the linear isometry L_v.

By Theorem 14, there is a neighborhood $\mathscr{U}_v$ of each $\beta_v(k)$ and a local isometry $\phi_v: \mathscr{U}_v \to \bar{M}$ whose differential map at $\beta_v(k)$ is L_v. In view of Lemma 5.10, we can suppose that each $\mathscr{U}_v$ is convex and that if $\mathscr{U}_v$ and $\mathscr{U}_w$ meet, then their intersection is convex, hence connected.

(2) We now construct a matched covering of M. Let A be the set of k-tuples in $T_o(M)$ for all $k \geq 1$. Let $\mathfrak{R}$ be the collection of convex open sets $\mathscr{U}_v$ as above for all $v \in A$. Since M is connected, $\mathfrak{R}$ is a covering of M. If $v, w \in A$, define $v \frown w$ to mean that $\phi_v = \phi_w$ on $\mathscr{U}_v \cap \mathscr{U}_w$. Evidently the relation $\frown$ is reflexive and symmetric, so it remains to prove the quasi-transitive condition (3) of Definition 7.30. Suppose $u \frown v$, $v \frown w$, and $p \in \mathscr{U}_u \cap \mathscr{U}_v \cap \mathscr{U}_w$. It follows from the two relations that the differential maps $d\phi_u$, $d\phi_v$, $d\phi_w$ all agree at the point p. Hence by Proposition 3.62 $\phi_u = \phi_w$ on $\mathscr{U}_u \cap \mathscr{U}_w$. Thus $u \frown w$, and $(\mathfrak{R}, \frown)$ is a matched covering.

(3) The essential step of the proof is to show that this matched covering is chainable over any geodesic segment $\sigma: [0, 1] \to M$, since then Proposition 7.32 applies.

Choose $v = (v_1, \ldots, v_k) \in A$ so that β_v runs from o to $\sigma(0)$. Let v_{k+1} be the parallel translate of $\sigma'(0)$ back to o along β_v. For each $s \in [0, 1]$ let $w(s) = (v_1, \ldots, v_k, sv_{k+1}) \in A$. Thus the last segment of $\beta_{w(s)}$ is a reparametrization of $\sigma | [0, s]$. Then there is a partition $0 = s_0 < s_1 < \cdots < s_m = 1$ such that

$$\sigma([s_{i-1}, s_i]) \subset \mathscr{U}_{w(s_i)} \qquad \text{for} \quad 1 \leq i \leq m.$$

See Figure 1. It remains to check that $w(s_{i-1}) \frown w(s_i)$.

Abbreviate $\phi_{w(s_i)}$ to ϕ_i. Let P be parallel translation along σ from $\sigma(s_{i-1})$ to $\sigma(s_i)$, and correspondingly let $\bar{P}$ be parallel translation along $\phi_i \circ \sigma$ from $\phi_i \sigma(s_{i-1})$ to $\phi_i \sigma(s_i)$. It follows from the definition of L_v that

$$L_{w(s_{i-1})} = \bar{P}^{-1} \circ L_{w(s_i)} \circ P.$$

As a local isometry, ϕ_i preserves parallel translation; hence,

$$(d\phi_i)_{\sigma(s_{i-1})} = \bar{P}^{-1} \circ (d\phi_i)_{\sigma(s_i)} \circ P.$$

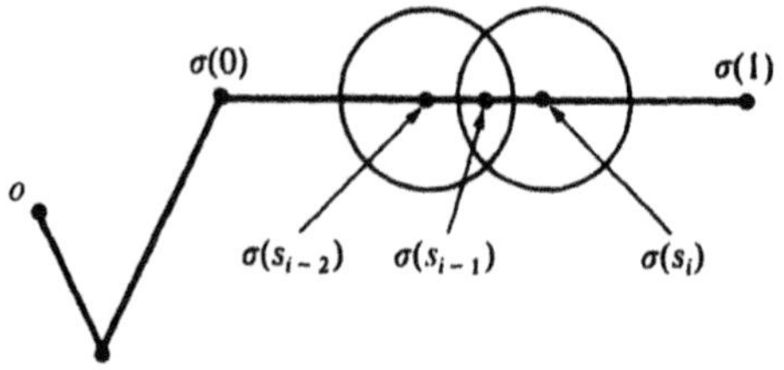

Figure 1

By construction, for each i, $L_{w(s_i)}$ is the differential map of ϕ_i at $\beta_{w(s_i)}(k+1) = \sigma(s_i)$. Thus the two preceding formulas yield

$$(d\phi_i)_{\sigma(s_{i-1})} = \bar{P}^{-1} \circ L_{w(s_i)} \circ P = L_{w(s_{i-1})} = (d\phi_{i-1})_{\sigma(s_{i-1})}.$$

Hence, $\phi_{i-1} = \phi_i$ on $\mathscr{U}_{w(s_{i-1})} \cap \mathscr{U}_{w(s_i)}$; that is, $w(s_{i-1}) \smile w(s_i)$.

(4) By Proposition 7.32, the matched covering $(\mathfrak{K}, \sim)$ gives a semi-Riemannian covering map $\mathscr{k}: M^* \to M$. As with any matched covering, for each $v \in A$ there is local section $\lambda_v: \mathscr{U}_v \to M^*$, and $\lambda_v(\mathscr{U}_v)$ meets $\lambda_w(\mathscr{U}_w)$ if and only if $v \smile w$. It follows that the maps $\phi_v \circ \mathscr{k}: \lambda_v(\mathscr{U}_2) \to \bar{M}$ for all $v \in A$ agree on overlaps, hence define a single map $\phi^*: M^* \to \bar{M}$. This is a local isometry since $\mathscr{k}$ and every ϕ_v are.

Since M is simply connected, Corollary A.14 applies, showing that there is a (locally isometric) global section $\lambda: M \to M^*$ such that $\lambda | \mathscr{U}_0 = \lambda_0$ for $(0) \in A$. Thus $\phi = \phi^* \circ \lambda: M \to \bar{M}$ is a local isometry with $\phi | \mathscr{U}_0 = \phi_0$; hence $d\phi_0 = L$. ∎

SIMPLY CONNECTED SPACE FORMS

22. Definition. A *space form* is a complete connected semi-Riemannian manifold of constant curvature.

Space forms must be regarded as the simplest important class of semi-Riemannian manifolds. In the simply connected case, Theorem 17 is decisive:

23. Proposition. Simply connected space forms are isometric if and only if they have the same dimension, index, and curvature C.

Proof. Obviously the conditions are necessary. For the converse, note that, since constant curvature implies local symmetry, Corollary 21 shows that simply connected space forms are symmetric.

Suppose M and N are simply connected space forms of the same dimension and index. The latter show that linear isometries $T_p(M) \to T_q(N)$ exist. Then Theorem 17 gives a semi-Riemannian covering map $M \to N$ which, since M is simply connected, is an isometry. ∎

Thus up to isometry there is at most one simply connected space form $M(n, v, C)$ of dimension n, index v, and curvature C. Existence for all (n, v, C) is settled for $C = 0$ by the semi-Euclidean spaces $\boldsymbol{R}^n_v$, and, with suitable modifications, for $C > 0$ by pseudospheres and for $C < 0$ by pseudo-hyperbolic spaces.

24. Corollary. For $n \geq 2$,

$$M(n, \nu, C) = \begin{cases} S^n_\nu(r) & \text{if } C = 1/r^2 \text{ and } 0 \leq \nu \leq n-2, \\ \boldsymbol{R}^n_\nu & \text{if } C = 0, \\ H^n_\nu(r) & \text{if } C = -1/r^2 \text{ and } 2 \leq \nu \leq n. \end{cases}$$

In fact, since $S^n_\nu(r) \sim S^{n-\nu} \times \boldsymbol{R}^\nu \sim H^n_{n-\nu}(r)$, these space forms are all simply connected. The remaining cases come in anti-isometric pairs.

One-dimensional semi-Riemannian manifolds are trivially flat; hence by Exercise 8, the only simply connected one-dimensional space forms are $\boldsymbol{R}^1$ and $\boldsymbol{R}^1_1$. For $n \geq 2$ we have

$M(n, n, 1/r^2)$ = one component $cS^n_n(r)$ of $S^n_n(r)$, and by metric reversal,
$M(n, 0, -1/r^2)$ = hyperbolic n-space $H^n(r)$, one component of $H^n_0(r)$.
$M(n, n-1, 1/r^2) = \tilde{S}^n_{n-1}(r)$, the simply connected semi-Riemannian covering manifold of $S^n_{n-1}(r)$, and
$M(n, 1, -1/r^2) = \tilde{H}^n_1(r)$, the simply connected semi-Riemannian covering manifold of $H^n_1(r)$.

Since $\boldsymbol{R}^n$ is the simply connected covering manifold of $S^1 \times \boldsymbol{R}^{n-1}$, these four types are diffeomorphic to $\boldsymbol{R}^n$.

25. Corollary (Hopf). A complete, simply connected, n-dimensional Riemannian manifold of constant curvature C is isometric to

the sphere $S^n(r)$	if $C = 1/r^2$,
Euclidean space $\boldsymbol{R}^n$	if $C = 0$,
hyperbolic space $H^n(r)$	if $C = -1/r^2$.

This result is particularly satisfying from a historical viewpoint, since all three of these constant curvature geometries were well known, at least in low dimensions, before Riemannian geometry was invented (1845) and long before a rigorous proof of this corollary (1926). Euclidean geometry is of course the oldest, and hyperbolic geometry, dating from the early 1800s, the newest.

Writing $\tilde{M}$ for the simply connected semi-Riemannian covering manifold of M, the Lorentz analogue of Hopf's result is

26. Corollary. A complete, simply connected, n-dimensional Lorentz manifold of constant curvature C is isometric to

the Lorentz sphere $S^n_1(r)$	if $C = 1/r^2$ and $n \geq 3$,
$\tilde{S}^2_1(r)$	if $C = 1/r^2$ and $n = 2$;
Minkowski space $\boldsymbol{R}^n_1$	if $C = 0$,
$\tilde{H}^n_1(r)$	if $C = -1/r^2$.

In relativity theory $S_1^4(r)$ is called *de Sitter spacetime*, and $\tilde{H}_1^4(r)$ is *universal anti-de Sitter spacetime*.

27. Example. A model for $\tilde{H}_1^n$. As Figure 2 suggests, H_1^n can be regarded as a hypersurface of revolution in $\boldsymbol{R}_2^{n+1} = \boldsymbol{R}_2^2 \times \boldsymbol{R}^{n-1}$. Unwrapping the circles of revolution gives a covering map $\ell \colon \boldsymbol{R}^1 \times \boldsymbol{R}^{n-1} \to H_1^n$, where

$$\ell(t, x) = (\sqrt{1 + x \cdot x} \cos t, \sqrt{1 + x \cdot x} \sin t, x).$$

Furnished with the pulled-back metric, $\boldsymbol{R}^n$ becomes the simply connected semi-Riemannian covering manifold of H_1^n—and hence can be denoted by $\tilde{H}_1^n$.

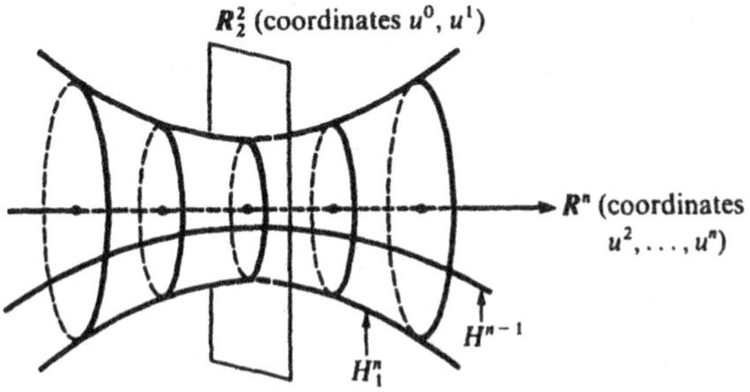

Figure 2. H_1^n in $\boldsymbol{R}_2^{n+1}$.

The geometry of the model can be derived from that of H_1^n as modified by the map ℓ. For example, if the coordinate slice $u^0 = 0$ in $\boldsymbol{R}_2^{n+1}$ is regarded as $\boldsymbol{R}_1^n$, then $\boldsymbol{R}_1^n \cap H_1^n$ is just H_0^{n-1}. The latter's two components H^{n-1} are totally geodesic in H_1^n. By rotational symmetry the same is true for all slices of H_1^n by hyperplanes through $\boldsymbol{R}^{n-1}$.

Thus in $\tilde{H}_1^n$ the slices t constant are totally geodesic spacelike hypersurfaces, isometric to H^{n-1} (or $\boldsymbol{R}^1$ if $n = 2$).

In Riemannian geometry, curvature inequalities such as $K \geq c$ and $a \leq K \leq b$ have been intensively studied. For indefinite metrics we shall now see that such conditions imply constant curvature.

Sectional curvature K is undefined on a degenerate plane Π in $T_p(M)$, but a formula in the proof of Lemma 3.39 shows that $\operatorname{sgn}\langle R_{XY}X, Y\rangle$ is the same for all bases X, Y for Π. This provides a well-defined function $\mathcal{N}$ from the set of degenerate planes in $T_p(M)$ to the set $\{+1, 0, -1\}$.

28. Proposition (Kulkarni *et al.*). Let p be a point of a semi-Riemannian manifold of indefinite metric. The following conditions on $T_p(M)$ are equivalent:

(1) K is constant,

(2) $\mathcal{N} = 0$,

(3) $a \le K$ or $K \le b$, where $a, b \in \mathbf{R}$,
(4) $a \le K \le b$ on indefinite planes,
(5) $a \le K \le b$ on definite planes.

Proof (S. Harris *et al.*). Evidently we can assume $\dim M \ge 3$. Obviously (1) implies (3, 4, 5), and by Corollary 3.43 it implies (2). The reverse implications will use this consequence of $\mathcal{N} = 0$:

(∗) *If X, U, Z are orthonormal vectors and X and U have opposite causal character, then $K(X, Z) = K(U, Z)$.*

To prove this note that $U \pm X$ are null vectors and the planes $(U \pm X, Z)$ are degenerate. Hence

$$\langle R_{X+U,Z}(X + U), Z\rangle = 0 = \langle R_{X-U,Z}(X - U), Z\rangle.$$

Expansion yields first $\langle R_{XZ}U, Z\rangle = 0$, then $\langle R_{XZ}X, Z\rangle + \langle R_{UZ}U, Z\rangle = 0$. Since $Q(X, Z) + Q(U, Z) = 0$, the result follows.

($2 \Rightarrow 1$) *Case I* $\nu = 1$ or $n - 1$. By metric reversal we can assume $\nu = 1$. If U, X, Y is orthonormal with U timelike, then (∗) implies $K(U, X) = K(X, Y)$. It follows that every nondegenerate plane that either contains or is orthogonal to U has the same sectional curvature $k(U)$. If V is an independent timelike vector, the plane spanned by U and V is nondegenerate, hence $k(U) = K(U, V) = k(V)$.

Case II $1 < \nu < n - 1$. Let Π be a positive definite plane, Π' negative definite, and let $X \in \Pi$ and $U \in \Pi'$ be unit vectors. Now $Q(X, U) < 0$, and using Lemma 2.19 we easily check that $\Pi + \mathbf{R}U$ and $\Pi' + \mathbf{R}X$ are nondegenerate. Hence by Case I, $K(\Pi) = K(U, X) = K(\Pi')$. This suffices because, if X, U is an orthonormal basis for any indefinite plane, then extension to an orthonormal basis for $T_p(M)$ will provide planes Π, Π' as above.

($3 \Rightarrow 2$). Suppose $a \le K$ (reversing the metric gives $K \le b$). By Exercise 9 every degenerate plane Π is a limit of definite planes and a limit of indefinite planes. Consider $\Pi(X_i, Y_i) \to \Pi = \Pi(X, Y)$, where $X_i \to X$ and $Y_i \to Y$. Then

$$\frac{\langle R(X_i, Y_i)X_i, Y_i\rangle}{Q(X_i, Y_i)} \ge a.$$

Here the numerator approaches $\langle R(X, Y)X, Y\rangle$ and the denominator approaches zero. Thus, if all $Q(X_i, Y_i) > 0$, then $\langle R(X, Y)X, Y\rangle \ge 0$, hence $\mathcal{N}(\Pi) \ge 0$. Similarly, if all $Q(X_i, Y_i) < 0$, then $\mathcal{N}(\Pi) \le 0$; hence $\mathcal{N}(\Pi) = 0$.

($4 \Rightarrow 2$) and ($5 \Rightarrow 2$). With notation as above, $|\langle R(X_i, Y_i)X_i, Y_i\rangle|/|Q(X_i, Y_i)|$ is bounded, where $Q(X_i, Y_i)$ is negative for (4), positive for (5). Hence $\langle R(X_i, Y_i)X_i, Y_i\rangle$ approaches zero, so $\mathcal{N}(\Pi) = 0$. ∎

If at each point of M, connected and dim ≥ 3, one of the conditions in the proposition holds, then, by Schur's theorem (Exercise 3.21), M has constant curvature.

TRANSVECTIONS

29. Definition. An isometry $\phi: M \to M$ is a *transvection* along a geodesic $\gamma: \boldsymbol{R} \to M$ provided

(1) ϕ translates γ; that is, $\phi(\gamma(s) = \gamma(s + c)$ for all $s \in \boldsymbol{R}$ and some c.

(2) $d\phi$ gives parallel translation along γ; that is, if $x \in T_{\gamma(s)}(M)$, then $d\phi(x) \in T_{\gamma(s+c)}(M)$ is the parallel translate of x along γ.

For example, if ϕ is a rotation of the sphere S^2 about the z-axis in $\boldsymbol{R}^3$, then ϕ is a transvection along the (geodesic) equatorial circle $z = 0$.

In a symmetric space there are transvections along every geodesic.

30. Lemma. If γ is a geodesic in a symmetric space, let ζ_s be the global symmetry of M at $\gamma(s)$. Then for any c, the isometry $\zeta_{c/2}\zeta_0$ is a transvection along γ that translates it by c.

Proof. For all s, $\zeta_0(\gamma(s)) = \gamma(-s)$. But $c/2$ is the midpoint of $[-s, s + c]$, so $\zeta_{c/2}\zeta_0\gamma(s) = \gamma(s + c)$.

If X is a parallel vector field on γ, then for any s, $d\zeta_s(X)$ is a parallel vector field on $\zeta_s \circ \gamma$, which is a reparametrization of γ. (Since parallel translation is independent of parametrization, we can ignore such reparametrizations.)

If $x \in T_{\gamma(s)}M$, then x is parallel along γ to a tangent vector y at $\gamma(0)$. Hence $d\zeta_0(x)$ is parallel to $d\zeta_0(y) = -y$, and also to a vector z at $\gamma(c/2)$. Thus $d\zeta_{c/2}\, d\zeta_0(x)$ is parallel to $-z$, hence to y, hence to x. ∎

31. Corollary. Every nonconstant geodesic in a symmetric space is either one-to-one or simply periodic.

Proof. It suffices to show that, if γ is a geodesic with $\gamma(b) = \gamma(0)$, then $\gamma(s + b) = \gamma(s)$ for all s. For each s let ϕ_s be the transvection that translates γ by s. Then

$$\gamma(s + b) = \phi_s(\gamma(b)) = \phi_s(\gamma(0)) = \gamma(s). \qquad ∎$$

Exercises

1. An isometry ζ of M, connected, is a global symmetry at $p \in M$ if and only if ζ is *involutive* (that is, $\zeta^2 = \text{id}$) and p is an isolated fixed point of ζ.

2. (a) Let M be a semi-Riemannian submanifold of a symmetric space $\bar{M}$. If M is connected, closed in $\bar{M}$, and totally geodesic, then M is symmetric. (b) A semi-Riemannian product $M \times N$ is symmetric if and only if both M and N are symmetric. (c) Let $k: \tilde{M} \to M$ be a simply connected semi-Riemannian covering. If M is symmetric, then $\tilde{M}$ is symmetric (but not conversely).

3. (a) If M is a complete Riemannian manifold, then every element of $\pi_1(M, p)$ contains a geodesic loop. (Hint: Use the Hopf–Rinow theorem in the simply connected covering manifold.) (b) The fundamental group of a Riemannian symmetric space is abelian. (Hint: Show that the geodesic symmetry at p induces the homomorphism $g \to g^{-1}$ on $\pi_1(M, p)$.)

4. A model of $\tilde{S}_1^2$, hence (reversing the metric) $\tilde{H}_1^2$. Consider the smooth map $\boldsymbol{x}: \boldsymbol{R}^2 \to \boldsymbol{R}_1^3$ given by $\boldsymbol{x}(t, \vartheta) = (\sinh t, \cosh t \cos \vartheta, \cosh t \sin \vartheta)$. (a) Prove that $\boldsymbol{x}$ is a covering map of $\boldsymbol{R}^2$ onto the unit pseudosphere S_1^2 in $\boldsymbol{R}_1^3$. (b) Prove that $\boldsymbol{R}^2$ furnished with the pulled-back metric is $\tilde{S}_1^2$ and the line element is $-dt^2 + \cosh^2 t\, d\vartheta^2$. (c) Sketch some null geodesics of this model and determine which points can be reached by geodesics starting at the origin.

5. (a) The model of $\tilde{H}_1^n$ in Example 27 has line element $-S^2\, dt^2 + d\sigma^2$, where $S = (1 + x \cdot x)^{1/2}$ and $d\sigma^2 = dx \cdot dx - ((x \cdot dx)^2/S^2)$ on $\boldsymbol{R}^{n-1}$. (Hint: $S\,dS = x \cdot dx$.) (b) $d\sigma^2$ on $\boldsymbol{R}^{n-1}$ gives H^{n-1}. (For other models of hyperbolic space see [Wo].)

6. Let γ be a geodesic with $\langle \gamma', \gamma' \rangle = \varepsilon = \pm 1$ in a semi-Riemannian surface. If Y is a Jacobi field on γ perpendicular to γ, write $Y = yE$, where E is a parallel unit vector field on α. Show that the Jacobi equation for Y reduces to $y'' + \varepsilon K y = 0$.

7. Let γ be a geodesic in a manifold of constant curvature C, and let Y be a Jacobi field on γ that is $\perp \gamma$. (a) the Jacobi equation for Y is $Y'' + C\langle \gamma', \gamma' \rangle Y = 0$. (b) Let $c = |C\langle \gamma', \gamma' \rangle|^{1/2}$. On γ there exist parallel vector fields $A, B \perp \gamma$ such that

$$
\begin{aligned}
Y(s) &= \cos(cs)\, A(s) + \sin(cs)\, B(s) && \text{if } C\langle \gamma', \gamma' \rangle > 0. \\
Y(s) &= A(s) + sB(s) && \text{if } C\langle \gamma', \gamma' \rangle = 0. \\
Y(s) &= \cosh(cs)\, A(s) + \sinh(cs)\, B(s) && \text{if } C\langle \gamma', \gamma' \rangle < 0.
\end{aligned}
$$

8. (a) Let M be a flat connected semi-Riemannian manifold complete at $o \in M$ (that is, $\exp_o$ is defined on all of $T_o(M)$). Prove that $\exp_o: T_o(M) \to M$ is a semi-Riemannian covering map. (Hint: Use Proposition 6.) (b) Give an example of a connected semi-Riemannian manifold that is complete at one point but not complete.

9. In a vector space with indefinite scalar product, every degenerate 2-plane is a limit of (a) indefinite planes (b) definite planes. (Hint: See proof of Lemma 3.40, but for (b), if $g|\Pi = 0$, consider $\Pi(v + \delta x, w + \delta y)$.)

9 ISOMETRIES

For a semi-Riemannian manifold M, the set $I(M)$ of all isometries $M \to M$ forms a group under composition of mappings. Roughly speaking, the larger $I(M)$ is, the simpler M is. M may have no isometries except the identity map, but many manifolds have isometry groups large enough so that Lie theory, whose rudiments appear in Appendix B, can be applied. In nontrivial cases $I(M)$ is a geometric invariant of M ranking in importance with its curvature and geodesics.

Since each tangent space of M is isometric to $\boldsymbol{R}^n_\nu$, the isometry group $I(\boldsymbol{R}^n_\nu)$ is of fundamental significance in semi-Riemannian geometry. In particular, for manifolds with indefinite metrics it leads to twin notions of time- and space-orientability analogous to ordinary orientability.

By Chapter 7 any connected semi-Riemannian manifold M can be expressed as an orbit manifold $\tilde{M}/\Gamma$, where $\tilde{M}$ is the simply connected covering manifold of M and Γ is a properly discontinuous subgroup of $I(\tilde{M})$. Thus geometric properties of M can be expressed in terms of algebraic properties of Γ, a scheme that is particularly effective when M is a space form.

A Killing vector field on M (Definition 22) is an "infinitesimal isometry" of M. Their tensor properties make Killing vector fields easier to find and study than isometries, and in favorable cases they provide a close link between the geometry of M and the algebra of $I(M)$.

SEMIORTHOGONAL GROUPS

We use the *column vector conventions*, under which an $n \times n$ real matrix g is identified with the linear operator $g: \boldsymbol{R}^n \to \boldsymbol{R}^n$ such that

$$(gx)_i = \sum g_{ij} x_j \qquad \text{for all} \quad 1 \leq i \leq n.$$

Under this identification, composition of functions agrees with matrix multiplication gh. Standing n-tuples x on end as $n \times 1$ matrices ("column vectors") also gives gx by matrix multiplication:

$$\begin{pmatrix} g_{11} & \cdots & g_{1n} \\ & \vdots & \\ g_{n1} & \cdots & g_{nn} \end{pmatrix} \begin{pmatrix} x_1 \\ \vdots \\ x_n \end{pmatrix} = \begin{pmatrix} \sum g_{1j}x_j \\ \vdots \\ \sum g_{nj}x_j \end{pmatrix}.$$

For $0 \leq \nu \leq n$, the *signature matrix* ε is the diagonal matrix $(\delta_{ij}\varepsilon_j)$ whose diagonal entries are $\varepsilon_1 = \cdots = \varepsilon_\nu = -1$ and $\varepsilon_{\nu+1} = \cdots = \varepsilon_n = +1$. Hence $\varepsilon^{-1} = \varepsilon = {}^t\varepsilon$, where ${}^tg$ denotes the transpose of g.

By the identification above, the set of all linear isometries $\boldsymbol{R}^n_\nu \to \boldsymbol{R}^n_\nu$ is the same as the set $O(\nu, n - \nu)$ of all matrices $g \in GL(n, \boldsymbol{R})$ that preserve the scalar product $\langle v, w\rangle = \varepsilon v \cdot w$ of $\boldsymbol{R}^n_\nu$. Evidently $O(\nu, n - \nu)$ is a closed subgroup of $GL(n, \boldsymbol{R})$ and hence is itself a Lie group (see Appendix B). We call it a *semiorthogonal group.*

1. Remark. In this chapter the group $O(\nu, n - \nu)$ will usually be denoted by $O_\nu(n)$. Although the former notation is standard in Lie group theory, the latter has advantages in the present context (compare [Wo]).

2. Lemma. The following conditions on an $n \times n$ matrix are equivalent:

(1) $g \in O_\nu(n)$.

(2) ${}^tg = \varepsilon g^{-1}\varepsilon$.

(3) The columns [rows] of g form an orthonormal basis for $\boldsymbol{R}^n_\nu$ (first ν vectors timelike).

(4) g carries one (hence every) orthonormal basis for $\boldsymbol{R}^n_\nu$ to an orthonormal basis.

Proof. (1) $\Leftrightarrow$ (2). The transpose of an $n \times n$ matrix is its adjoint relative to the dot product. Thus, (1) $\Leftrightarrow \langle gv, gw\rangle = \langle v, w\rangle$ for all $v, w \Leftrightarrow \varepsilon gv \cdot gw = \varepsilon v \cdot w$ for all $v, w \Leftrightarrow {}^tg\varepsilon gv \cdot w = \varepsilon v \cdot w$ for all $v, w \Leftrightarrow {}^tg\varepsilon gv = \varepsilon v$ for all $v \Leftrightarrow {}^tg\varepsilon g = \varepsilon \Leftrightarrow {}^tg = \varepsilon g^{-1}\varepsilon$.

(1) $\Leftrightarrow$ (4). See Lemma 2.27.

The equivalence of (4) and (3, for columns) is clear since the columns of g are just $gu_1, \ldots, gu_n$, where $u_1, \ldots, u_n$ is the natural basis for $\boldsymbol{R}^n$ (orthonormal relative to the scalar product of $\boldsymbol{R}^n_\nu$ for all ν). The equivalence of (4) and (3, for rows) follows since manipulation of (2) shows that $g \in O_\nu(n)$ if and only if ${}^tg \in O_\nu(n)$. ■

In this context, timelike vectors must appear first in orthonormal bases. For example, (0, 1) and (1, 0) are orthogonal unit vectors of $\boldsymbol{R}^2_1$ but $\left(\begin{smallmatrix} 0 & 1 \\ 1 & 0 \end{smallmatrix}\right)$ is not in $O_1(2)$ since (2) fails.

When ν is either 0 or n, the group $O(\nu, n - \nu)$ is the *orthogonal group* $O(n)$ of all linear isometries of Euclidean space $\boldsymbol{R}^n$. For $n \geq 2$, $O_1(n) = O(1, n - 1)$ is the *Lorentz group* of all linear isometries of Minkowski space $\boldsymbol{R}^n_1$. For arbitrary ν, the Lie groups $O_\nu(n)$ and $O_{n-\nu}(n)$ are isomorphic. In fact, conjugation by

$$\sigma = \begin{pmatrix} 0 & I_q \\ I_p & 0 \end{pmatrix}$$

is a Lie group automorphism of $GL(n, \boldsymbol{R})$ that is readily shown to carry $O_\nu(n)$ to $O_{n-\nu}(n)$.

Using the methods of Appendix B, we now compute the matrix Lie algebra of $O_\nu(n)$. Recall that $\mathfrak{gl}(n, \boldsymbol{R})$ is the Lie algebra of all real $n \times n$ matrices, with $[a, b] = ab - ba$, and that the Lie algebra $\mathfrak{o}(n)$ of $O(n)$ consists of all $X \in \mathfrak{gl}(n, \boldsymbol{R})$ that are skew-symmetric: ${}^tX = -X$.

3. Lemma. The Lie algebra $\mathfrak{o}_\nu(n) = \mathfrak{o}(\nu, n - \nu)$ of $O_\nu(n)$ is the subalgebra of $\mathfrak{gl}(n, R)$ consisting of all S for which ${}^tS = -\varepsilon S \varepsilon$. Such S have the form

$$\begin{pmatrix} a & x \\ {}^tx & b \end{pmatrix},$$

where $a \in \mathfrak{o}(\nu)$, $b \in \mathfrak{o}(n - \nu)$, and x is an arbitrary $\nu \times (n - \nu)$ matrix.

Proof. According to Lemma B.14, an $n \times n$ matrix S is in $\mathfrak{o}_\nu(n)$ if and only if $e^{rS} \in O_\nu(n)$ for $|r|$ small. Since $\varepsilon^{-1} = \varepsilon$, Lemma 2 and Lemma B.16 show that $e^{rS} \in O_\nu(n)$ if and only if the matrix $\varepsilon \exp(r{}^tS)\varepsilon^{-1} = \exp(r\varepsilon{}^tS\varepsilon^{-1})$ equals $\exp(-rS)$.

By Lemma B.12, the exponential map is one-to-one near 0, hence $S \in \mathfrak{o}_\nu(n)$ if and only if these exponents are equal for $|r|$ small; that is, $\varepsilon{}^tS = -S\varepsilon$.

To get the matrix description, write S in block form $\left(\begin{smallmatrix} a & x \\ y & b \end{smallmatrix}\right)$, with a $\nu \times \nu$ and b $(n - \nu) \times (n - \nu)$. Since

$$\varepsilon = \begin{pmatrix} -I_\nu & 0 \\ 0 & I_{n-\nu} \end{pmatrix},$$

we find

$${}^tS = \begin{pmatrix} {}^ta & {}^ty \\ {}^tx & {}^tb \end{pmatrix} \quad \text{and} \quad -\varepsilon S \varepsilon = \begin{pmatrix} -a & x \\ y & -b \end{pmatrix}.$$

The result follows. ∎

Because $\dim \mathfrak{o}(\nu) = \nu(\nu - 1)/2$, a count of vector space dimensions shows that $\dim O_\nu(n) = \dim \mathfrak{o}_\nu(n) = n(n - 1)/2$, independent of ν.

The matrix condition ${}^tS = -\varepsilon S\varepsilon$ is equivalent to $\langle Sv, w\rangle = -\langle v, Sw\rangle$ for all $v, w \in \boldsymbol{R}^n_\nu$. Thus we can regard $\mathfrak{o}_\nu(n)$ as consisting of the skew-adjoint linear operators on $\boldsymbol{R}^n_\nu$.

4. Examples. (1) $O(2)$. For each number $\vartheta \in \boldsymbol{R}$, the orthogonal matrix

$$R_\vartheta = \begin{pmatrix} \cos\vartheta & -\sin\vartheta \\ \sin\vartheta & \cos\vartheta \end{pmatrix}$$

is a rotation of $\boldsymbol{R}^2$ through (oriented) angle ϑ. The function $\vartheta \to R_\vartheta$ is a smooth homomorphism from $\boldsymbol{R}$, under addition, into $O(2)$. Its kernel is $2\pi\boldsymbol{Z}$ and its image is the rotation group $O^+(2)$, the component of the identity in $O(2)$. Thus $O^+(2)$ and its other coset $O^-(2)$ are diffeomorphic to circles.

(2) $O_1(2)$. For each $\varphi \in \boldsymbol{R}$, the semiorthogonal matrix

$$B_\varphi = \begin{pmatrix} \cosh\varphi & \sinh\varphi \\ \sinh\varphi & \cosh\varphi \end{pmatrix}$$

is called a *boost* of $\boldsymbol{R}^2_1$ through (oriented) Lorentz angle φ. As above, $\varphi \to B_\varphi$ is a homomorphism, but in this case it is one-to-one. Any $a \in O_1(2)$ must carry each hyperbola $\langle p, p\rangle = 1$ and $\langle p, p\rangle = -1$ into itself but may reverse the branches of each. These two choices split $O_1(2)$ into four disjoint open subsets. The one preserving all branches is exactly the set B of all boosts. B is a subgroup diffeomorphic to $\boldsymbol{R}^1$ and is thus the component of the identity in $O_1(2)$. The other three sets are cosets of B, hence $O_1(2)$ has four components each diffeomorphic to $\boldsymbol{R}^1$. (See Figure 1.)

This example correctly suggests two fundamental differences between the orthogonal group $O(n)$ and the semiorthogonal group $O_\nu(n)$, $0 < \nu < n$:

First, $O(n)$ *is compact*, since (as noted in Appendix B) it is closed and bounded in $\mathrm{gl}(n, \boldsymbol{R}) \approx \boldsymbol{R}^{n^2}$. But $O_\nu(n)$ *is not compact* in the indefinite case,

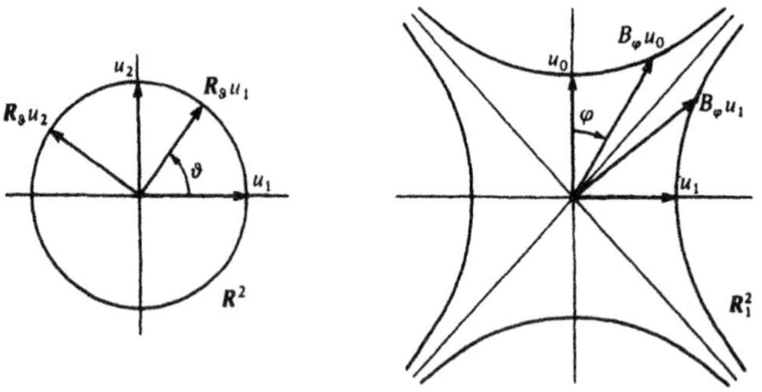

Figure 1

since for example, elements of the form

$$\begin{pmatrix} \cosh \varphi & 0 & \sinh \varphi \\ 0 & I & 0 \\ \sinh \varphi & 0 & \cosh \varphi \end{pmatrix}$$

constitute an unbounded subset of $\boldsymbol{R}^{n^2}$.

Second, $O(n)$ *has two components*, while in the indefinite case $O_\nu(n)$ *has four.*

To prove these connectedness assertions, first assume $0 < \nu < n$ so that $\boldsymbol{R}^n_\nu = \boldsymbol{R}^\nu_\nu \times \boldsymbol{R}^{n-\nu}$. Write $a \in O_\nu(n)$ in block form as

$$\begin{pmatrix} a_T & b \\ c & a_S \end{pmatrix}$$

with a_T $\nu \times \nu$ and a_S $(n - \nu) \times (n - \nu)$. Here the *timelike part*, $a_T: \boldsymbol{R}^\nu_\nu \to \boldsymbol{R}^\nu_\nu$, is $a|\boldsymbol{R}^\nu_\nu$ followed by orthogonal projection on $\boldsymbol{R}^\nu_\nu$, and similarly for the *spacelike part*, $a_S: \boldsymbol{R}^{n-\nu} \to \boldsymbol{R}^{n-\nu}$. Since a is invertible and preserves causal character, it follows readily that both a_T and a_S are invertible.

5. Definition. For $0 < \nu < n$, an element $a \in O_\nu(n)$ *preserves* [*reverses*] *time-orientation* provided $\det a_T > 0$ [<0], and *preserves* [*reverses*] *space-orientation* provided $\det a_S > 0$ [<0].

Thus $O_\nu(n)$ is decomposed into four disjoint sets indexed by the signs of $\det a_T$ and $\det a_S$ (in that order):

$$O^{++}_\nu(n), \quad O^{+-}_\nu(n), \quad O^{-+}_\nu(n), \quad O^{--}_\nu(n).$$

The two determinants are continuous functions of a, so the four sets are open, hence also closed, sets of $O_\nu(n)$.

6. Lemma. $O^{++}_\nu(n)$ is connected for all $0 \le \nu \le$ n.

Proof. Induction on n: For $n = 1$ the result is trivial; for $n = 2$ it is true by Example 4. Assume it is true for some $n \ge 2$. In view of Exercise 1c we can suppose $0 \le \nu \le (n + 1)/2$. (Consider $O^{++}_0(n)$ to be $O^+(n)$.) Given $a \in O^{++}_\nu(n + 1)$, we now construct a curve joining g to the identity element e. Note that there is a natural one-to-one correspondence between $O_\nu(n + 1)$ and the set $F(S^n_\nu)$ of all frames on S^n_ν: A matrix g corresponds to its columns $gu_1, \ldots, gu_{n+1}$; the first n of these vectors, by canonical isomorphism, become a frame tangent to S^n_ν at gu_{n+1}. Clearly every frame in $F(S^n_\nu)$ derives thus from a unique element of $O_\nu(n + 1)$. Since $\nu < n$, S^n_ν is connected. Move the frame corresponding to g to a frame at u_{n+1}, say by parallel translation in the geometry of S^n_ν. Corresponding in $O_\nu(n + 1)$ is a curve α from g to

$\begin{pmatrix} b & 0 \\ 0 & 1 \end{pmatrix} = B$. This curve is continuous (in fact, smooth) and, since the determinant function is continuous, it follows that α remains in $O_\nu^{++}(n+1)$. But then $b \in O_\nu^{++}(n)$. Hence by inductive hypothesis a further curve joins b, hence B, to the identity. ∎

In particular, $O^+(n)$ is connected. This group

$$O^+(n) = SO(n) = \{g \in O(n) : \det g = 1\}$$

is thus the identity component of $O(n)$, and the coset

$$O^-(n) = \{g \in O(n) : \det g = -1\}$$

is the other component.

7. Corollary. Let $0 < \nu < n$. (1) $O^{++} = O_\nu^{++}(n)$ is the identity component of $O_\nu(n)$. (2) The cosets $O^{++}, O^{+-}, O^{-+}, O^{--}$ of O^{++} are the components of $O_\nu(n)$. (3)$O^{++} \cup O^{--}$, $O^{++} \cup O^{+-}$, and $O^{++} \cup O^{-+}$ are subgroups of $O_\nu(n)$.

Proof. (1) Since O^{++} is connected and is both open and closed in $O_\nu(n)$, it is the identity component.

(2) Consider the four *representative matrices* $d(\pm 1, \pm 1) = \mathrm{diag}(\pm 1, 1, \ldots, 1, \pm 1)$ in $O_\nu(n)$. Left multiplication by say $d(-1, +1) \in O^{-+}$ is a diffeomorphism of $O_\nu(n)$ that on each matrix a changes the sign of the first row of a_T without changing a_S. It follows that O^{++} is carried diffeomorphically onto O^{-+}, so the latter is a coset of O^{++} and is connected. The other cases are similar.

(3) One of the representative matrices $d(\pm 1, \pm 1)$ is in each of the four (coset) components. The effect of multiplication and the map $a \to a^{-1}$ on components is thus determined by their effect on the four matrices, hence (3) follows. ∎

Every semiorthogonal matrix has determinant ± 1. In fact, writing the criterion in Lemma 3 as ${}^t a \varepsilon a = \varepsilon$ and taking determinants yields $(\det a)^2 = 1$. Since the determinant function must maintain the same sign on each component of $O_\nu(n)$ the representative matrices show that the *special semi-orthogonal group*

$$SO_\nu(n) = \{g \in O(n) : \det g = 1\}$$

is $O^{++}(n) \cup O^{--}(n)$. Thus the three subgroups in (3) above consist of the linear isometries $\mathbf{R}_\nu^n \to \mathbf{R}_\nu^n$ that preserve orientation, time-orientation, and space-orientation, respectively. (Note that when $\nu = 1$, preservation of time-orientation agrees with the earlier definition requiring timecones to be preserved.) These three choices are not independent; checking signs will show that $a \in O_\nu(n)$ preserves one of the three orientation types if and only if it preserves both or reverses both of the other two types.

SOME ISOMETRY GROUPS

We compute the isometry groups of pseudospheres, pseudohyperbolic spaces, and semi-Euclidean spaces. The pseudoradius r is irrelevant (Exercise 1), so we take $r = 1$, writing S^n_ν and H^n_ν as usual.

8. Proposition. $I(S^n_\nu) = O_\nu(n+1)$ if $\nu < n$; $I(H^n_\nu) = O_{\nu+1}(n+1)$ if $\nu > 0$.

Proof. (1) A linear isometry $a\colon \boldsymbol{R}^{n+1}_\nu \to \boldsymbol{R}^{n+1}_\nu$ carries S^n_ν to itself, and since S^n_ν is a semi-Riemannian submanifold, $a|S^n \in I(S^n)$. Evidently restriction $a \to a|S^n$ is a homomorphism, so once it is shown to be one-to-one and onto we write simply $I(S^n_\nu) = O_\nu(n+1)$.

If $u_1, \ldots, u_{n+1}$ is the natural basis for $\boldsymbol{R}^{n+1}$, its first n vectors correspond to a frame on S^n_ν at the point u_{n+1}. If $\phi \in I(S^n_\nu)$, the proof of Proposition 4.30 shows there is a unique $a \in O_\nu(n+1)$ such that $au_{n+1} = \phi u_{n+1}$ and $da(u_i) = d\phi(u_i)$ for $1 \leq i \leq n$. Since S^n_ν is connected for $\nu < n$, it follows that $\phi = a|S^n$.

(2) For H^n_ν, the proof is the same except that $H^n_\nu \subset \boldsymbol{R}^{n+1}_{\nu+1}$ and u_{n+1} is replaced by u_1. ■

In particular the orthogonal group $O(n+1)$ is the isometry group of the sphere S^n, and the Lorentz group $O_1(n+1)$ is the isometry group of the Lorentz sphere S^n_1.

The proof fails in the nonconnected cases H^n_0 and S^n_n, but what we want are the isometry groups of their components H^n and cS^n_n.

9. Corollary. $I(H^n) = O_1^{++}(n+1) \cup O_1^{+-}(n+1)$;
$I(cS^n_n) = O_n^{++}(n+1) \cup O_n^{-+}(n+1)$.

Proof. Since the hyperbolic space $H^n \subset \boldsymbol{R}^{n+1}_1$ is connected, the preceding proof shows that, if $\phi \in I(H^n)$, there is a unique $a \in O_1(n+1)$ such that $a|H^n = \phi$. Because $\nu = 1$, the matrix a_{T} is just a nonzero number. Since $a(H^n) \subset H^n$, a carries the timecone of $\boldsymbol{R}^{n+1}_1$ containing H^n to itself; thus a preserves time-orientation.

Conversely, if $a \in O_1^{++} \cup O_1^{+-}$, then $a|H^n \in I(H^n)$. The other proof is similar. ■

The linear isometries of $\boldsymbol{R}^n_\nu$ form a subgroup $O_\nu(n)$ of its isometry group $I(\boldsymbol{R}^n_\nu)$. Also if $x \in \boldsymbol{R}^n_\nu$, the translation T_x sending each v to $v + x$ is an isometry. Then $T_x \circ T_y = T_{x+y} = T_y \circ T_x$, and, since T_0 is the identity, $(T_x)^{-1} = T_{-x}$. It follows that the set $\mathbb{R}^n$ of all translations of $\boldsymbol{R}^n_\nu$ is an abelian subgroup of $I(\boldsymbol{R}^n_\nu)$ and that $\mathbb{R}^n$ is isomorphic to $\boldsymbol{R}^n$ (under vector addition) via $T_x \leftrightarrow x$.

10. Proposition. Each isometry of $\mathbf{R}^n_\nu$ has a unique expression as $T_x a$, with $x \in \mathbf{R}^n_\nu$ and $a \in O_\nu(n)$. Furthermore, $T_x a T_y b = T_{x+ay} ab$.

Proof. First we show that, if ϕ is an isometry of $\mathbf{R}^n_\nu$ such that $\phi(0) = 0$, then $\phi \in O_\nu(n)$. In fact, the differential map $d\phi_0$ of ϕ at 0 is a linear isometry, hence it corresponds under the canonical linear isometry $T_0(\mathbf{R}^n_\nu) \approx \mathbf{R}^n_\nu$ to a linear isometry $a: \mathbf{R}^n_\nu \to \mathbf{R}^n_\nu$. But then da_0 is exactly $d\phi_0$, hence $\phi = a$ by Proposition 3.62.

Now if $\phi \in I(\mathbf{R}^n_\nu)$, let $x = \phi(0) \in \mathbf{R}^n_\nu$. Thus $(T_{-x}a)(0) = 0$, so by the preceding remark, $T_{-x}\phi$ equals some $a \in O_\nu(n)$. Hence $\phi = T_x a$.

If $T_x a = T_y b$, then $x = (T_x a)(0) = (T_y b)(0) = y$, hence also $a = b$. Finally, for all v,

$$(aT_y)(v) = a(y + v) = ay + av = (T_{ay}a)(v).$$

Hence $aT_y = T_{ay}a$, and the multiplication rule follows. ∎

The multiplication rule shows that $\mathbb{R}^n$ is a normal subgroup of $I(\mathbf{R}^n)$. The function $(x, a) \to T_x a$ is one-to-one from $\mathbf{R}^n \times O_\nu(n)$ onto $I(\mathbf{R}^n_\nu)$. If $I(\mathbf{R}^n_\nu)$ is made a manifold so that this function is a diffeomorphism, it is easy to check that $I(\mathbf{R}^n_\nu)$ is a Lie group. Then

$$\dim I(\mathbf{R}^n_\nu) = \dim \mathbf{R}^n + \dim O_\nu(n) = n(n + 1)/2.$$

Like $O_\nu(n)$, $I(\mathbf{R}^n_\nu)$ has four components if $0 < \nu < n$, reducing to two for $\nu = 0, n$.

$I(\mathbf{R}^n_\nu)$ is called a *semi-Euclidean group*, $I(\mathbf{R}^n)$ a *Euclidean group*, and, if $n \geq 2$, $I(\mathbf{R}^n_1)$ is the *Poincaré group* or *inhomogeneous Lorentz group*.

TIME-ORIENTABILITY AND SPACE-ORIENTABILITY

For a semi-Riemannian manifold M with indefinite metric we generalize the Lorentz notion of time-orientability and define a complementary notion of space-orientability. The pattern for both is the same as for ordinary orientability; the reason for this is as follows. The manifold description of orientability in Chapter 7 rests solely on the fact that in the full linear group $GL(n, \mathbf{R})$ the matrices with positive determinant form a subgroup of index 2 (one other coset) that is open, hence closed, in $GL(n, \mathbf{R})$—or in the special case of Riemannian manifolds that $SO(n)$ is an open subgroup of index 2 in the orthogonal group $O(n)$. But for $0 < \nu < n$, all three of the subgroups in Corollary 7(3) are open and have index 2. Thus we can give a common treatment, denoting any one of these groups (for arbitrary n) by G. So henceforth

G-orientation should be read as

orientation	if	$G = SO_\nu(n) = O_\nu^{++}(n) \cup O_\nu^{--}(n)$.
time-orientation	if	$G = O_\nu^{++}(n) \cup O_\nu^{+-}(n)$.
space-orientation	if	$G = O_\nu^{++}(n) \cup O_\nu^{-+}(n)$.

Formally the results are valid for all ν, but if $\nu = 0$, time-orientation is undefined and space-orientation becomes orientation (the reverse for $\nu = n$).

Let V be a scalar product space. If $e = (e_1, \ldots, e_n)$ and $f = (f_1, \ldots, f_n)$ are orthonormal bases for V, then $f_j = \sum a_{ij} e_i$ $(1 \leq j \leq n)$ defines a matrix $a = (a_{ij})$ in $O_\nu(n)$. Then e and f are *G-equivalent* provided $a \in G \subset O_\nu(n)$. Because G is a subgroup of index 2, G-equivalence is an equivalence relation with exactly two equivalence classes, called the *G-orientations* of V. To pick one is to *G-orient* V. The G-orientation containing a given orthonormal basis is denoted by $G(e_1, \ldots, e_n)$. If $T: V \to W$ is a linear isometry, then

$$G(e_1, \ldots, e_n) \to G(Te_1, \ldots, Te_n)$$

is a well-defined one-to-one function T_G from the two orientations of V to those of W.

A *G-orientation* of a semi-Riemannian manifold M is a function λ that assigns to each $p \in M$ a G-orientation of $T_p(M)$ and is *smooth* in this sense: For each $p \in M$, there is a coordinate system ξ whose induced local G-orientation $G(\partial_1, \ldots, \partial_n)$ agrees with λ on some neighborhood of p. M is *G-orientable* provided it admits a G-orientation.

Then just as before,

(1) *A connected G-orientable manifold M has exactly two G-orientations* $\pm\lambda$ (these assigning opposite G-orientations to each tangent space).

(2) A local isometry $\phi: M \to N$ is said to *preserve* [*reverse*] G-orientations λ_M and λ_N provided $d\phi_G(\lambda_M(p)) = \lambda_N(\phi p)$ $[= -\lambda_N(\phi p)]$ for all $p \in M$.

(3) *If* $\phi: M \to N$ *is a local isometry and* λ *is a G-orientation of N, there is a unique G-orientation* $\phi^*(\lambda)$ *of M making* ϕ *G-orientation preserving.*

(4) *A semi-Riemannian manifold M has a double covering* $k: M^G \to M$, *called the* G-orientation covering *of M, such that every (local or global) G-orientation of M is a smooth section.*

M^G is always G-orientable, and M is G-orientable if and only if $k: M^G \to M$ is trivial. Since this is a double covering, if M is connected, the latter condition is equivalent to M^G not connected.

If M is connected and G-orientable, it is meaningful to say that an isometry $\phi: M \to M$ preserves (or reverses) G-orientation provided we agree to use the same G-orientation on M as both domain and range of ϕ.

11. Proposition. If $\mathscr{k}: M \to M/\Gamma$ is a connected semi-Riemannian covering, then M/Γ is G-orientable if and only if M is G-orientable and each $\phi \in \Gamma$ preserves G-orientation.

Proof. If λ is a G-orientation of M/Γ, then $\mathscr{k}^*(\lambda)$ is a G-orientation of M. If $\phi \in \Gamma$, then $\phi^*\mathscr{k}^*(\lambda) = (\mathscr{k}\phi)^*(\lambda) = \mathscr{k}^*(\lambda)$, so ϕ preserves G-orientation.

Conversely, let μ be a G-orientation of M. If $q \in M/\Gamma$, we assert that the G-orientation $d\mathscr{k}_G(\mu(p))$ is independent of the choice of p in $\mathscr{k}^{-1}(q)$. In fact any two such points, p and p', are in the same Γ orbit, hence there is a $\phi \in \Gamma$ such that $\phi(p) = p'$. Since ϕ by hypothesis preserves G-orientation,

$$d\mathscr{k}_G(\mu(p')) = d\mathscr{k}_G(d\phi_G(\mu(p))) = d(\mathscr{k}\phi)_G(\mu(p)) = d\mathscr{k}_G(\mu(p)).$$

Thus $\lambda(q) = d\mathscr{k}_G(\mu(p))$ for all $p \in \mathscr{k}^{-1}(q)$, is a valid definition, and using local sections, one can check that λ is smooth, hence is a G-orientation of M/Γ. ∎

Let $\mathscr{k}: M^G \to M$ be the G-orientation covering of M. Since $\mathscr{k}^{-1}(p)$, $p \in M$, consists of the two G-orientations of $T_p(M)$, a lift α of $\tilde{\alpha}$ into M^G amounts to moving a G-orientation of tangent spaces along α. (See Exercises 16 and 24.)

The *natural G-orientation* of $\mathbf{R}^n_\nu$ as a scalar product space is the one containing its natural basis, and the *natural G-orientation* of $\mathbf{R}^n_\nu$ as a semi-Riemannian manifold is the one induced by its natural coordinate system. Thus each canonical isomorphism $T_p(\mathbf{R}^n_\nu) \approx \mathbf{R}^n$ is G-orientation preserving. It follows that $g \in O_\nu(n)$, considered as an isometry of $\mathbf{R}^n_\nu$, preserves G-orientation if and only if $g \in G$.

Pseudospheres and pseudohyperbolic spaces are G-orientable whenever this is meaningful. Furthermore for their isometry groups, as in Proposition 8 and Corollary 9, an isometry $g \mid Q \in I(Q)$ preserves orientation, time-orientation, space-orientation, respectively, if and only if g itself does.

LINEAR ALGEBRA

We consider briefly some basic facts about linear operators T on a scalar product space V.

(1) *Invariant subspaces.* A subspace W of V is *invariant under* T provided $T(W) \subset W$. If W and W' are invariant, so are $W \cap W'$ and $W + W'$. Since the scalars are real, *T has an invariant subspace of dimension* 1 *or* 2. *Proof*: The characteristic polynomial $p(x)$ of T can be factored into a product of linear and quadratic factors. Substituting T for x gives $p(T) = 0$. Hence some factor is singular, and the result follows.

(2) *Self-adjoint and semiorthogonal operators.* Let V be a scalar product space of dimension n and index ν, so $V \approx \boldsymbol{R}^n_\nu$. If S is a self-adjoint operator on V (that is, $\langle Sv, w\rangle = \langle v, Sw\rangle$ for all v, w), then the matrix s of S relative to an orthonormal basis satisfies ${}^t s = \varepsilon s \varepsilon$, where ε is the $(\nu, n - \nu)$ signature matrix. Thus s has the form

$$\begin{pmatrix} a & x \\ -{}^t x & b \end{pmatrix},$$

where a and b are symmetric matrices of size $\nu \times \nu$ and $(n - \nu) \times (n - \nu)$, respectively, while x is an arbitrary $\nu \times (n - \nu)$ matrix.

If A is semiorthogonal, that is, preserves the scalar product, then relative to an orthonormal basis the matrix of A is an element of $O_\nu(n)$. In fact this association is evidently an isomorphism from the group $O(V)$ of all such A to $O_\nu(n)$.

(3) *Perp operation.* Let T be either self-adjoint or semiorthogonal. *If W is invariant under T, then so is $W^\perp$.* For example, if T is semiorthogonal, $v \perp W$ implies $Tv \perp T(W)$. But since T is invertible, $T(W) = W$.

(4) *The definite case.* Let V be an inner product space. A self-adjoint operator S on V is often called *symmetric*. It is well known that S then has an orthonormal basis of eigenvectors, hence the resulting matrix is diagonal.

If A is orthogonal, the facts above lead easily to another standard result: For a suitable orthonormal basis, A has a matrix with ± 1's and elements of $O(2)$ along its main diagonal (zeros elsewhere). In particular, A need not have an eigenvector if n is even. (Every operator has an eigenvector if n is odd.)

(5) *The Lorentz case.* Let V be a Lorentz vector space. If S is self-adjoint, then low dimensions are somewhat irregular, but it turns out that S can be diagonalized outside a subspace of dimension ≤ 3. By contrast with the definite case, a semiorthogonal operator A always has an eigenvector. See Exercises 18–21.

SPACE FORMS†

If M is a space form, then so is its simply connected semi-Riemannian covering manifold $\tilde{M}$, and we saw in Chapter 7 that M can be expressed as $\tilde{M}/\Gamma$, where Γ is a properly discontinuous group of isometries of $\tilde{M}$.

What is important here is not the particular group Γ but its *conjugacy class* in $I(\tilde{M})$. Recall that for a group G, subgroups H and $\bar{H}$ are *conjugate* provided there is a $g \in G$ such that $\bar{H} = gHg^{-1}$. Evidently conjugacy is an equivalence relation.

† This section derives from Chapter 11 of [Wo], which should be consulted for references and further developments.

12. Proposition. If M is a simply connected semi-Riemannian manifold, then orbit manifolds M/Γ and $M/\bar{\Gamma}$ are isometric if and only if Γ and $\bar{\Gamma}$ are conjugate in $I(M)$.

Proof. (1) Suppose $\bar{\Gamma} = \phi\Gamma\phi^{-1}$ for $\phi \in I(M)$. If $p \in M$, then $\phi\Gamma p = \phi\Gamma\phi^{-1}\phi p = \bar{\Gamma}\phi p$; that is, ϕ carries the Γ orbit of p to the $\bar{\Gamma}$ orbit of ϕp. Since these orbits are points of M/Γ and $M/\bar{\Gamma}$, this amounts to a function $\underset{\sim}{\phi}: M/\Gamma \to M/\bar{\Gamma}$ such that $\underset{\sim}{\phi} \circ k = \bar{k} \circ \phi$, where k and $\bar{k}$ are the projections. On any open set $\mathscr{U}$ of M/Γ evenly covered by k, there is a local section $\lambda: \mathscr{U} \to M$, and λ is an isometry onto $\lambda(\mathscr{U})$. But then $\underset{\sim}{\phi}|\mathscr{U} = \bar{k} \circ \phi \circ \lambda$ so $\underset{\sim}{\phi}$ is at least a local isometry. Repeating the above with ϕ replaced by ϕ^{-1} shows that $\underset{\sim}{\phi}$ is an isometry.

(2) Conversely, let $\underset{\sim}{\psi}: M/\Gamma \to M/\bar{\Gamma}$ be an isometry. Since M is simply connected, by Proposition A.11 there is a lift ψ of $\underset{\sim}{\psi} \circ k$ through $\bar{k}$; thus $\bar{k} \circ \psi = \underset{\sim}{\psi} \circ k$. Lifting $\underset{\sim}{\psi}^{-1}$ shows that $\psi \in I(M)$. If $\phi \in \Gamma$, then $\psi\phi\psi^{-1}$ is a deck transformation, since

$$\bar{k}\psi\phi\psi^{-1} = \underset{\sim}{\psi}k\phi\psi^{-1} = \underset{\sim}{\psi}k\psi^{-1} = \bar{k}\psi\psi^{-1} = \bar{k}.$$

Then Corollary 7.12 gives $\psi\phi\psi^{-1} \in \bar{\Gamma}$; that is, $\psi\Gamma\psi^{-1} \subset \bar{\Gamma}$. Symmetrically, $\psi^{-1}\bar{\Gamma}\psi \subset \Gamma$, and it follows that Γ and $\bar{\Gamma}$ are conjugate. ∎

We shall consider only space forms whose fundamental group is finite. Recall that a group is *torsion-free* if the subgroup generated by each element except the identity is infinite. Thus a torsion-free group is finite only if it is trivial, that is, consists solely of the identity; so the following result excludes many space forms from the finite class.

13. Proposition. Let M^n be a space form of curvature C and index ν. Then $\pi_1(M)$ is torsion-free if (a) $C = 0$, (b) $C > 0$ and $\nu = n - 1, n$, or (c) $C < 0$ and $\nu = 0, 1$.

Proof. Write $M = \tilde{M}/\Gamma$ as above. Since $\tilde{M}$ is simply connected, Proposition 7.4 gives $\pi_1(M) \approx \Gamma$. Thus it suffices to show that every finite subgroup F of Γ is trivial. On page 228, we saw that the cases in this proposition are exactly those for which $\tilde{M}$ is diffeomorphic to $\mathbf{R}^n$. A theorem of P. Smith asserts any finite group of diffeomorphisms of $\mathbf{R}^n$ has a common fixed point. Since F consists of deck transformations, it follows that F is trivial. ∎

Geometric proofs for certain of these cases are indicated in Exercise 5.

A space form with curvature C is *positive* if $C > 0$, *negative* if $C < 0$. We shall consider the positive case; reversing the metric gives the corresponding result in the negative case, the index ν changing to $n - \nu$. As before it can be assumed that $C = 1$.

14. Proposition. If M^n, $n \geq 3$, is a positive space form with index $\nu \leq n/2$ (or a negative space form with $\nu \geq n/2$), then the fundamental group of M is finite.

Proof. Since $n \geq 3$ and $2\nu \leq n$ imply $\nu \leq n - 2$, it follows from Corollary 8.24 that the simply connected semi-Riemannian covering manifold $\tilde{M}$ of M is (isometric to) a pseudosphere S^n_ν. Thus we can take M to be S^n_ν/Γ, where Γ is a properly discontinuous subgroup of $I(S^n_\nu) = O_\nu(n + 1)$. For $\nu = 0$, $\pi_1(M) \approx \Gamma$ is finite, since S^n is compact hence Exercise 7.6 applies. But if ν is positive, the benefits of compactness can be gotten indirectly as follows.

Let $W \approx \mathbf{R}^{n+1-\nu}$ be the spacelike coordinate hyperplane in $\mathbf{R}^{n+1}_\nu$. Then $W \cap S^n_\nu$ is a sphere $S^{n-\nu}$ in W. We assert that for every $\phi \in \Gamma$ there is a point $p \in S^{n-\nu}$ such that $\phi(p) \in S^{n-\nu}$. Extended to an element of $O_\nu(n + 1)$, ϕ is linear, hence

$$\dim(W \cap \phi W) + \dim(W + \phi W) = \dim W + \dim \phi(W).$$

It follows that $\dim(W \cap \phi W) \geq n + 1 - 2\nu$, which is positive since $2\nu \leq n$. But any unit vector in $W \cap \phi W$ is a point of $S^{n-\nu}$ that is the ϕ-image of a point of $S^{n-\nu}$.

Assume Γ is infinite. Then there is an infinite sequence $\{\phi_i\}$ of distinct elements of Γ and a sequence $\{p_i\}$ in $S^{n-\nu}$ such that each $\phi_i(p_i)$ is also in $S^{n-\nu}$. Since $S^{n-\nu}$ is compact, by passing if necessary to subsequences we can suppose that $\{p_i\}$ and $\{\phi_i(p_i)\}$ converge to points p and q in $S^{n-\nu}$. This will contradict the proper discontinuity of Γ (Definition 7.6):

Case 1. $q \notin \Gamma p$. The convergent sequences above contradict (PD2).

Case 2. $q \in \Gamma p$. Let $\psi \in \Gamma$ send p to q. The sequence $\{\psi^{-1}\phi_i(p_i)\}$ converges to $\psi^{-1}(q) = p$. But then $\{p_i\} \to p$ contradicts (PD1). ∎

In the omitted dimension 2, the result holds for $\nu = 0$ since S^2 is compact and simply connected, but not for $\nu = 1$ since, for example, S^2_1 has infinite fundamental group.

15. Lemma. Every finite subgroup Γ of $O_\nu(n)$ is conjugate to a subgroup of $O(\nu) \times O(n - \nu)$.

Proof. (*a*) For $v, w \in \mathbf{R}^n$ define

$$b(v, w) = \sum_{g \in \Gamma} gv \cdot gw.$$

Then b is an inner product on $\mathbf{R}^n$ and is preserved by Γ, since, if $h \in \Gamma$, then

$$b(hv, hw) = \sum_g ghv \cdot ghw = \sum_{g'} g'v \cdot g'w = b(v, w).$$

(b) *There is an inner product β on R^n_ν preserved by Γ, and a basis $e_1, \ldots, e_n$ for R^n_ν that is orthonormal relative to both β and the usual scalar product of R^n_ν.*

There is a linear operator S on R^n_ν such that $\langle v, w\rangle = b(Sv, w)$ for all v, w. S is symmetric relative to b, hence R^n_ν is the direct sum of the eigenspaces $\mathcal{N}_r$ of the distinct eigenvalues λ_r of S. These subspaces are mutually orthogonal relative to b, hence to $\langle , \rangle$. Note that $\lambda_r \neq 0$ since S is invertible.

We assert that $g(\mathcal{N}_r) \subset \mathcal{N}_r$ for all $g \in \Gamma$ and all r. If $v \in \mathcal{N}_r$, then $\langle v, w\rangle = \lambda_r b(v, w)$ for all w, hence $\langle gv, gw\rangle = \lambda_r b(gv, gw)$ for all w. Since g is onto, we can take gw to be any vector $x \in \mathcal{N}_s$, $s \neq r$. Then

$$\lambda_s b(gv, x) = \langle gv, x\rangle = \lambda_r b(gv, x).$$

Hence $b(gv, x) = 0$ for all such x, showing that $gv \in \mathcal{N}_r$.

Write v_r for the component of v in $\mathcal{N}_r$. Then $b(v, w) = \sum b(v_r, w_r)$. Define $\beta(v, w) = \sum |\lambda_r| b(v_r, w_r)$. It is easy to verify that β is an inner product on R^n preserved by Γ. Furthermore, if $e_1, \ldots, e_n$ is a $\langle\ ,\ \rangle$-orthonormal basis for R^n adapted to the subspaces $\mathcal{N}_r$, then it is also β-orthonormal, since $\beta(e_i, e_j) = 0$ for $i \neq j$ and

$$\beta(e_j, e_j) = |\lambda_r| b(e_i, e_i) = (|\lambda_r|/\lambda_r)\langle e_i, e_i\rangle = +1.$$

(c) We can now prove the result. Let $a \in O_\nu(n)$ send the natural basis $u_1, \ldots, u_n$ to $e_1, \ldots, e_n$. If $g \in \Gamma \subset O_\nu(n)$, then since $O(n) \cap O_\nu(n) = O(\nu) \times O(n-\nu)$, it suffices to show that $aga^{-1} \in O(n)$. Because g preserves β, its matrix relative to $e_1, \ldots, e_n$ is orthogonal. But this is just the matrix of $a^{-1}ga$ relative to $u_1, \ldots, u_n$, hence $a^{-1}ga \in O(n)$. ∎

By Proposition 12 and Lemma 15, *every positive space form with $\nu \leq n-2$ and finite fundamental group can be expressed as S^n_ν/Γ with $\Gamma \subset O(\nu) \times O(n-\nu+1)$.* Writing $R^{n+1}_\nu = R^\nu_\nu \times R^{n-\nu+1}$ as usual gives

$$S^n_\nu = \{(p, x) : x \cdot x = 1 + p \cdot p\},$$

and $(a, b) \in \Gamma$ acts on S^n_ν by $(a, b)(p, x) = (ap, bx)$. In particular, if $(0, x) = x \in S^{n-\nu} = S^n_\nu \cap R^{n-\nu+1}$, then $(a, b)(0, x) = (0, bx)$.

Let B be the set of all $b \in O(n-\nu+1)$ such that $(a, b) \in \Gamma$ for some $a \in O(\nu)$. *For each b this a is unique*, for if (a, b) and (a', b) are in Γ, then so is $(a, b)^{-1}(a', b) = (a^{-1}a', e)$. But this deck transformation fixes each point $(0, x)$ of $S^{n-\nu} \subset S^n_\nu$, hence is the identity; that is, $a = a'$. Consequently,

(1) B is a subgroup of $O(n-\nu+1)$ isomorphic to Γ under $(a, b) \leftrightarrow b$, and

(2) the map sending each $b \in B$ to the element $a \in O(\nu)$ such that $(a, b) \in \Gamma$ is a homomorphism (whose graph is Γ).

Using these facts we can show that in about half the cases in Proposition 14 the fundamental group is very simple indeed.

16. Proposition. Let M be a positive space form with finite fundamental group. If $n - \nu$ is even, then either M is simply connected or $\pi_1(M) \approx \mathbf{Z}_2$. In the latter case, M is not space-orientable.

Note. The hypotheses hold for all positive space forms with $2\nu \leq n$ and $n - \nu$ even, since the latter eliminates the exceptional case $n = 2$, $\nu = 1$ following Proposition 14.

Proof. By Proposition 13, we can assume $\nu \leq n - 2$; hence $M = S^n/\Gamma$ with $\Gamma \subset O(\nu) \times O(n - \nu + 1)$. The following argument is valid even in the Riemannian case $\nu = 0$, where $\mathbf{R}^\nu$ is trivial and $O(\nu) = \{\pm 1\}$.

Let $(a, b) \in \Gamma$. Since $n - \nu + 1$ is odd, the characteristic polynomial of b has a real root, that is, b has an eigenvector $x \in S^{n-\nu}$. Furthermore $bx = \pm x$, since $|bx| = |x|$.

If $bx = x$, then $(0, x) \in S^n$ is a fixed point of the deck transformation (a, b); hence a and b are identity matrices.

If $bx = -x$, then $b^2x = x$; hence both a^2 and b^2 are identities. Then since a and b are orthogonal they are also symmetric, because $a = a^{-1} = {}^t a$. Thus the inner product spaces $\mathbf{R}^\nu$ and $\mathbf{R}^{n-\nu+1}$ have orthonormal bases composed of eigenvectors of a and b, respectively. As above, all eigenvalues are ± 1. If one of the eigenvalues of b is $+1$, then we have seen that (a, b) has a fixed point and hence is the identity. Thus if (a, b) is not the identity, b is $-e$. Since there is only one a such that $(a, -e) \in \Gamma$, we conclude that, if M is not simply connected, then Γ is a two-element group whose nontrivial element is $(a, -e)$, where a is a matrix with ± 1's on the diagonal, zeros elsewhere. In this case, since $n - \nu + 1$ is odd, $\det(-e) = -1$. Thus $(a, -e)$ reverses space-orientation on $\mathbf{R}^{n+1}_\nu$ and hence also on S^n_ν. Then by Proposition 11, M is not space-orientable. ∎

The proof actually shows more. Conjugation of the $\nu \times \nu$ matrix $a = \text{diag}(\pm 1, \ldots, \pm 1)$ by a suitable permutation matrix will rearrange it so that the -1's (if any) come first. Thus by Proposition 12 there are just $\nu + 1$ cases for M not simply connected, and these are orientable $\Leftrightarrow$ not time-orientable $\Leftrightarrow$ the number of -1's is odd.

17. Corollary. (1) An even-dimensional Riemannian space form with curvature $C = 1$ is isometric to either S^{2k} or $P^{2k} = S^{2k}/\pm 1$, the latter nonorientable.

(2) An odd-dimensional Lorentz space form with curvature $C = 1$ is isometric to one of the following:

S^n_1: orientable, time-orientable, space-orientable.

$S^n_1/\pm 1$: orientable, but not time- or space-orientable.

S^n_1/Γ, where the unique nontrivial element of Γ is $\text{diag}(1, -1, \ldots, -1)$: time-orientable but not orientable or space-orientable.

By contrast with the situation in Proposition 16 there are many positive space forms with finite fundamental group but with $n - \nu$ odd.

18. Example. For any $k \geq 2$, let $R_k \in O(2)$ be a rotation through angle $2\pi/k$. For any n and ν such that $n - \nu$ is odd and ≥ 3, let a be the element of $O(n - \nu + 1)$ with $(n - \nu + 1)/2$ copies of R_k along its main diagonal. Then $(-e, a) \in O_\nu(n + 1)$ generates a cyclic group Γ of order k or $2k$. Then S^n_ν/Γ is a positive space form with fundamental group Γ.

The general problem of classifying all space forms is a difficult one. For Riemannian space forms the positive case has been solved by Wolf [Wo] and much is known about the flat case [Wo, KN].

Every complete Riemannian manifold is geodesically connected, but the space forms S^n_ν and H^n_ν $(0 < \nu < n)$ are not. The following result will show that in the positive Lorentz case, geodesic connectedness is incompatible with time-orientability. Note that for a semi-Riemannian covering $\mathscr{k}$: $M \to M/\Gamma$, the lift and projection properties of geodesics show that M/Γ is geodesically connected if and only if for any p, $q \in M$ there is a geodesic joining p to some point of the orbit Γq.

19. Proposition. A positive Lorentz space form M^n, $n \geq 3$, is geodesically connected if and only if it is not time-orientable.

Proof. For $n \geq 3$, S^n_1 is simply connected, hence time-orientable, but not geodesically connected. Thus we can suppose that $M = S^n_1/\Gamma$, with Γ finite (by Proposition 14) but not trivial.

Suppose first that M is not time-orientable; we must show that given any points $p, q \in S^n_1$ there exists a geodesic from p to Γq. Let $q^* = \sum_{g \in \Gamma} g(q) \in \mathbf{R}^{n+1}_1$. Then q^* is a common fixed point of every $g \in \Gamma$. If q^* is timelike or null, no element of Γ can reverse the causal cones of $\mathbf{R}^{n+1}_1$. This implies that Γ preserves time-orientation: a contradiction. If q^* is spacelike and nonzero, then dividing by its norm gives a point of S^n_1 fixed by the deck transformation group Γ; thus Γ consists only of its identity element: a contradiction. Hence q^* must be 0; so $0 = \langle p, q^* \rangle = \sum_{g \in \Gamma} \langle p, g(q) \rangle$. Thus for some $g \in \Gamma$ we have $\langle p, g(q) \rangle \geq 0$. By Proposition 5.38 there is a geodesic from p to $g(q)$.

Now suppose M is time-orientable. Since Γ is finite, we may suppose $\Gamma \subset O(1) \times O(n)$; by time-orientability $\Gamma \subset 1 \times O(n)$. Thus the spacelike coordinate hyperplane $\mathbf{R}^n$ is invariant under Γ, and $g u_1 = u_1$ for all $g \in \Gamma$. There is a point $p \in S^{n-1} = S^n_\nu \cap \mathbf{R}^n$ that is not in the (finite) orbit Γu_{n+1}. Thus there is an $\varepsilon > 0$ such that $\langle p, g(u_{n+1}) \rangle < 1 - \varepsilon$ for all $g \in \Gamma$. We lift this data homothetically to an arbitrary plane u^1 const, as follows. Writing $S = \sinh t$ and $C = \cosh t$, let

$$p_t = S u_1 + C p, \qquad q_t = S u_1 + C u_{n+1}.$$

For all $g \in \Gamma$, $gu_1 = u_1$ and $u_1 \perp p, g(u_{n+1})$. Hence

$$\begin{aligned}\langle p_t, g(q_t)\rangle &= \langle Su_1 + Cp, Su_1 + Cg(u_{n+1})\rangle \\ &= -S^2 + C^2\langle p, gu_{n+1}\rangle < -S^2 + C^2 - C^2\varepsilon = 1 - C^2\varepsilon.\end{aligned}$$

Thus for t sufficiently large, $\langle p_t, g(u_{n+1})\rangle < -1$. By Proposition 5.38, there exists no geodesic from p_t to Γq_t, showing that M is not geodesically connected. ∎

Let M be a semi-Riemannian manifold with isometry group $I(M)$. If M is simply connected, the isometry group of any M/Γ can be characterized in a purely algebraic way. Recall that, for a subgroup H of a group G, the *normalizer* $N(H)$ in G is the set of all $g \in G$ such that $gHg^{-1} = H$. It is easy to check that $N(H)$ is a subgroup of G containing H; in fact it is the largest subgroup in which H is a normal subgroup, so in particular the quotient group $N(H)/H$ is well defined.

20. Proposition. Let Γ be a properly discontinuous group of isometries of a simply connected semi-Riemannian manifold M. The isometry group $I(M/\Gamma)$ of M/Γ is isomorphic to $N(\Gamma)/\Gamma$, where $N(\Gamma)$ is the normalizer of Γ in $I(M)$.

Proof. That $\phi \in N(H)$ means $\phi\Gamma\phi^{-1} = \Gamma$; hence $\phi\Gamma p = \Gamma\phi p$ for all $p \in M$. Thus there is a unique function $\underset{\sim}{\phi}: M/\Gamma \to M/\Gamma$ such that $\underset{\sim}{\phi} \circ k = k \circ \phi$, where $k: M \to M/\Gamma$ is the natural projection. Local cross sections show that $\underset{\sim}{\phi}$ is smooth, and $\underset{\sim}{\phi}$ is a diffeomorphism since $(\phi^{-1})_\sim$ is its inverse. Since k is a local isometry it follows that $\underset{\sim}{\phi}$ is an isometry.

If $\phi, \psi \in N(H)$, then $\underset{\sim}{\phi}\underset{\sim}{\psi}k = k\phi\psi$. Thus $(\phi\psi)_\sim = \underset{\sim}{\phi}\underset{\sim}{\psi}$; that is, the function $\phi \to \underset{\sim}{\phi}$ is a homomorphism.

If $\phi \in \Gamma$, then obviously $\underset{\sim}{\phi}$ is the identity map of M. Conversely if $\underset{\sim}{\phi}$ is the identity map of M/Γ, then ϕ is a deck transformation; that is, $\phi \in \Gamma$. Thus the kernel of the homomorphism is Γ.

Because M is simply connected, the homomorphism is onto $I(M)$. In fact, if $\mu \in I(M)$, then by Proposition A.11 the map $\mu \circ k: M \to M/\Gamma$ has a lift $\tilde{\mu}: M \to M$ through k. Lifting $\mu^{-1} \circ k$ will show that $\tilde{\mu}$ is an isometry. Since $\mu \circ k = k \circ \tilde{\mu}$, we have $(\tilde{\mu})_\sim = \mu$.

Thus the homomorphism theorem implies $N(\Gamma)/\Gamma \approx I(M/\Gamma)$. ∎

KILLING VECTOR FIELDS

By Definition 2.16, the Lie derivative L_X applied to a vector field Y is $[X, Y]$. Proposition 1.58 interprets this bracket as the rate of change of Y under the flow of X. A similar interpretation holds for L_X applied to any tensor field A; however, for simplicity we take A to be covariant.

21. Proposition. If $X \in \mathfrak{X}(M)$ and $A \in \mathfrak{T}_s^0(M)$, then

$$L_X A = \lim_{t \to 0} \frac{1}{t} [\psi_t^*(A) - A],$$

where $\{\psi_t\}$ is the flow of X. (When the flow is local, the equation holds locally.)

Proof. For simplicity, let $s = 2$. Since L_X is a tensor derivation,

$$(L_X A)(V, W) = XA(V, W) - A([X, V], W) - A(V, [X, W]).$$

Now we work on the right-hand side of the stated formula, abbreviating $\lim_{t\to 0} (1/t)$ to $\mathscr{L}$ and fixing a point p. Then

$$\mathscr{L}(\psi_t^* A - A)(V_p, W_p) = \mathscr{L}\{A(d\psi_t(V_p), d\psi_t(W_p)) - A(V_p, W_p)\}.$$

Adding and subtracting a suitable term turns this into

$$\mathscr{L}\{A(d\psi_t(V_p), d\psi_t(W_p)) - A(V_{\psi_t p}, W_{\psi_t p})\} + \mathscr{L}\{A(V_{\psi_t p}, W_{\psi_t p}) - A(V_p, W_p)\}.$$

Call these two limits I and II. If α is the integral curve of X starting at p, then $\psi_t(p) = \alpha(t)$, and

$$\mathrm{II} = (d/dt)\langle V_\alpha, W_\alpha\rangle|_0 = \alpha'(0)\langle V, W\rangle = X_p\langle V, W\rangle.$$

For I, we use the telescoping identity

$$A(v', w') - A(v, w) = A(v' - v, w') + A(v, w' - w)$$

to get

$$\mathrm{I} = \mathscr{L}\{A(d\psi_t(V_p) - V_{\psi_t p}, d\psi_t(W_p))\} + \mathscr{L}\{A(V_{\psi_t p}, d\psi_t(W_p) - W_{\psi_t(p)})\}.$$

Since A is bilinear and $\psi_t \psi_{-t} = \psi_0 = \mathrm{id}$, the first term can rewritten to use Proposition 1.58 as follows:

$$\begin{aligned}
&\mathscr{L}\{A(d\psi_t(V_p - d\psi_{-t}(V_{\psi_t p})), d\psi_t(W_{\psi_t(p)}))\} \\
&= -A(d\psi_t \mathscr{L}\{d\psi_{-t}(V_{\psi_t p}) - V_p\}, \lim_{t\to 0} d\psi_t(W_{\psi_t(p)})) \\
&= -A([X, V]_p, W_p).
\end{aligned}$$

Similarly, the second term above is $-A(V_p, [X, W]_p)$. Thus, $\mathrm{I} + \mathrm{II} = (L_X A)(V_p, W_p)$ as required. ∎

22. Definition. A *Killing vector field* on a semi-Riemannian manifold is a vector field X for which the Lie derivative of the metric tensor vanishes: $L_X g = 0$.

Thus under the flow of X, the metric tensor does not change; this suggests the following view of a Killing vector field as an *infinitesimal isometry*.

23. Proposition. A vector field X is Killing if and only if the stages ψ_t of all its (local) flows are isometries.

Proof. If each ψ_t is an isometry, then $\psi_t^*(g) = g$. Hence by Proposition 21, $L_X g = 0$.

Conversely, if $L_X g = 0$, let $\{\psi_t\}$ be a local flow of X. If v is a tangent vector at a point in the domain of the flow, then so is $w = d\psi_s(v)$ for s small. By Proposition 21, $\lim_{t\to 0}(1/t)(g(d\psi_t w, d\psi_t w) - g(w, w)) = 0$. Since $\psi_s\psi_t = \psi_{s+t}$,

$$\lim_{t\to 0} \frac{1}{t} \{g(d\psi_{s+t}(v), d\psi_{s+t}(v)) - g(d\psi_s(v), d\psi_s(v))\} = 0.$$

This says that the real-valued function $s \to g(d\psi_s(v), d\psi_s(v))$ has derivative identically zero. Thus it is constant, so $g(d\psi_s(v), d\psi_s(v)) = g(v, v)$ for all v and s. ∎

24. Example. Killing vector fields on the Schwarzschild half-plane P_I (Example 5.41). Since isometries preserve the Gaussian curvature $K = 2\text{M}/r^3$, any isometry must have the form $\phi(t, r) = (f(t, r), r)$. Thus

$$d\phi(\partial_t) = (\partial f/\partial t)\partial_t + \partial_r; \qquad d\phi(\partial_r) = (\partial f/\partial r)\partial_t + \partial_r.$$

Since $ds^2 = -h\, dt^2 + h\, dr^2$ with $h = 1 - (2\text{M}/r)$, computing scalar products shows that ϕ is an isometry if and only if $\partial f/\partial t = 1$ and $\partial f/\partial r = 0$. Thus $f = t + c$ for some number c; that is, the *isometries are t-coordinate translations.* Since the local flows of a Killing vector field consist of isometries, it follows that *every Killing vector field on P_I is a constant multiple of ∂_t.*

Recall that the covariant differential of a vector field is the (1, 1) tensor field DX such that $(DX)(V) = D_V X$ for all $V \in \mathfrak{X}(M)$. Thus at each $p \in M$, $(DX)_p$ is the linear operator on $T_p(M)$ sending v to $D_v X$.

25. Proposition. The following conditions on a vector field X are equivalent:

(1) X is Killing; that is, $L_X g = 0$.
(2) $X\langle V, W\rangle = \langle [X, V], W\rangle + \langle V, [X, W]\rangle$ for all $V, W \in \mathfrak{X}(M)$.
(3) DX is skew-adjoint relative to g; that is, $\langle D_V X, W\rangle + \langle D_W X, V\rangle = 0$ for all $V, W \in \mathfrak{X}(M)$.

Proof. For all V, W the following are equivalent:

$$\langle D_V X, W\rangle + \langle D_W X, V\rangle = 0;$$

$$-\langle [X, V], W\rangle + \langle D_X V, W\rangle - \langle [X, W], V\rangle + \langle D_X W, V\rangle = 0;$$

$$X\langle V, W\rangle = \langle [X, V], W\rangle + \langle V, \langle X, W]\rangle.$$

In view of the product rule the latter is equivalent to $(L_X g)(V, W) = 0$ for all V, W, that is, to $L_X g = 0$. ■

Condition (3) shows that any parallel vector field is Killing.
We call the following highly useful fact the *conservation lemma.*

26. Lemma. Let X be a Killing vector field on M, and let γ be a geodesic in M. Then the restriction X_γ is a Jacobi field and $\langle \gamma', X \rangle$ is constant along γ.

Proof. It suffices to work locally, so near any point of γ let $\{\psi_s\}$ be a local flow of X. Each ψ_s is an isometry, hence the function $(t, s) \to \psi_s(\gamma(t))$ is a geodesic variation of (a segment of) γ. For fixed t, the curve $s \to \psi_s(\alpha(t))$ is an integral curve of X, hence its velocity at $s = 0$ is $X_{\gamma(t)}$. Thus X_γ is the vector field of this geodesic variation, so by Proposition 8.3 it is a Jacobi field. Then, since γ is a geodesic,

$$(d/dt)\langle X_\gamma, \gamma' \rangle = \langle X'_\gamma, \gamma' \rangle = \langle D_{\gamma'} X, \gamma' \rangle.$$

But this last expression is zero by Proposition 25(3); hence $\langle X, \gamma' \rangle$ is constant. ■

27. Lemma. Let X be a Killing vector field on a connected semi-Riemannian manifold M. If $X_p = 0$ and $(DX)_p = 0$ for some one point p of M, then $X = 0$.

Proof. Let A be the set of points of M at which both X and DX vanish. Evidently $M - A$ is open and A is nonempty; thus it suffices to show that A is open. If $o \in A$, let $\mathscr{U}$ be a normal neighborhood of o. If σ is a radial geodesic from o, then by the preceding lemma X_σ is a Jacobi field. But $X_{\sigma(0)} = X_o = 0$ and $(X_\sigma)'(0) = D_{\sigma'(0)} X = (DX)_o(\sigma'(0)) = 0$. By Proposition 8.7, X_σ is identically zero. Hence X is identically zero on $\mathscr{U}$, so DX is also. ■

It will follow that, if X and Y are Killing vector fields such that $X_p = Y_p$ and $DX_p = DY_p$ for some one point, then $X = Y$. Since Killing vector fields are infinitesimal isometries, the lemma is thus an infinitesimal analogue of the uniqueness of local isometries, Proposition 3.62.

THE LIE ALGEBRA *i*(*M*)

Let $i(M)$ be the set of all Killing vector fields on a semi-Riemannian manifold M. By Exercise 2.11,

(1) The Lie derivative L_X is $\mathbf{R}$-linear in X. Hence any linear combination of Killing fields (coefficients constant) is again a Killing field.
(2) $[L_X, L_Y] = L_{[X, Y]}$. Hence a bracket of Killing fields is again Killing.

It follows immediately that $i(M)$ is a real Lie algebra (Definition B.5). In fact, $i(M)$ is a Lie subalgebra of $\mathfrak{X}(M)$, which becomes a Lie algebra under scalar multiplication by numbers rather than functions $f \in \mathfrak{F}(M)$. By contrast with $\mathfrak{X}(M)$, i(M) is always finite-dimensional.

28. Lemma. The Lie algebra $i(M)$ of Killing vector fields on a connected semi-Riemannian manifold M^n has dimension at most $n(n + 1)/2$.

Proof. Fix $p \in M$, and let $\mathfrak{o}(T_pM)$ be the Lie algebra of all skew-adjoint linear operators on $T_p(M)$. (An orthonormal basis converts each such operator into a matrix, thus giving a Lie algebra isomorphism onto $\mathfrak{o}_\nu(n)$.) Let E send each Killing vector field X to $(X_p, (DX)_p)$. Then E is a linear transformation from $i(M)$ to $T_p(M) \times \mathfrak{o}(T_pM)$. Lemma 26 implies that E is one-to-one. Hence

$$\dim i(M) \leq \dim T_p(M) + \dim \mathfrak{o}_\nu(n) = n + n(n - 1)/2 = n(n + 1)/2. \quad \blacksquare$$

29. Example. Killing vector fields on $\boldsymbol{R}^n_\nu$. (1) If $v \in \boldsymbol{R}^n_\nu$, let v^* be the vector field $p \to v_p$. Since v^* is parallel, it is Killing. The flow of v^* consists of the translations $\psi_t(p) = p + tv$, so we call v^* an *infinitesimal translation.*

(2) If S is a skew-adjoint linear operator on $\boldsymbol{R}^n_\nu$, let S^* be the vector field such that $S^*_p = (Sp)_p$ for all $p \in M$. Relative to natural coordinates, $S^* = \sum S^i_j u^j \partial_i$. Thus for any vector field V, $D_V S^* = \sum S^i_j V^j \partial_i = SV$, where we consider S by canonical isomorphism as a (1, 1) tensor field. But S is skew-adjoint, hence S^* is a Killing vector field, called an *infinitesimal linear isometry.*

(3) *The Killing vector fields on $\boldsymbol{R}^n_\nu$ are the vector fields of the form $v^* + S^*$, where $v \in \boldsymbol{R}^n_\nu$ and $S \in \mathfrak{o}_\nu(n)$.* To prove this, it suffices to check that the space of such Killing fields has the maximum dimension $n(n + 1)/2$.

(4) For example, let t, x, y be the natural coordinates on $\boldsymbol{R}^3_1$. The infinitesimal translations on $\boldsymbol{R}^3_1$ have basis $\partial_t, \partial_x, \partial_y$. By Lemma 3 a basis for $\mathfrak{o}_1(3)$ is given by

$$A = \begin{pmatrix} 0 & 1 & 0 \\ 1 & 0 & 0 \\ 0 & 0 & 0 \end{pmatrix}, \qquad B = \begin{pmatrix} 0 & 0 & 1 \\ 0 & 0 & 0 \\ 1 & 0 & 0 \end{pmatrix}, \qquad C = \begin{pmatrix} 0 & 0 & 0 \\ 0 & 0 & -1 \\ 0 & 1 & 0 \end{pmatrix}.$$

As above, these matrices give Killing vector fields

$A^* = x\partial_t + t\partial_x$, infinitesimal boost on the timelike planes, y const;

$B^* = y\partial_t + t\partial_y$, infinitesimal boost on the timelike planes, x const;

$C^* = -y\partial_x + x\partial_y$, infinitesimal rotation on the spacelike planes, t const.

The six Killing fields constitute a basis for $i(\boldsymbol{R}^3_1)$.

Recall that a vector field V is *complete* provided its maximal integral curves are all defined on the whole real line.

30. Proposition. On a complete semi-Riemannian manifold M every Killing vector field V is complete.

Proof. We can assume M is connected. Fix $o \in M$ and $\varepsilon > 0$ such that V has a local flow defined on $\mathscr{U} \times (-\varepsilon, \varepsilon)$, where $\mathscr{U}$ is a neighborhood of o. Let A be the set of all points $p \in M$ such that V has a local flow defined on $\mathscr{V} \times (-\varepsilon, \varepsilon)$, where $\mathscr{V}$ is a neighborhood of p (same ε). It suffices to show that $A = M$, for then every integral curve of V is defined at least on $(-\varepsilon, \varepsilon)$, hence by Lemma 1.52, every maximal integral curve has domain $\boldsymbol{R}$.

We need only show that if a convex open set $\mathscr{C}$ meets A, then $\mathscr{C}$ is contained in A; so suppose that $p \in \mathscr{C} \cap A$ and $r \in \mathscr{C}$.

Let $\sigma: [0, 1] \to \mathrm{M}$ be a geodesic from p to r. Since $p \in A$, the method used in the proof of Lemma 26 gives, for some $\delta > 0$, a geodesic variation $\boldsymbol{x}: [0, \delta] \times (-\varepsilon, \varepsilon) \to \mathrm{M}$ of $\sigma|[0, \delta]$ such that each transverse curve is an integral curve of V. Because M is complete, each longitudinal geodesic of $\boldsymbol{x}$ can be extended to $[0, 1]$, hence we can suppose that $\boldsymbol{x}$ is a geodesic variation of σ. The proof of Lemma 8.3 actually shows that on any longitudinal geodesic $\sigma_s(t) = \boldsymbol{x}(t, s)$ the vector field $\boldsymbol{x}_s$ is a Jacobi field. By Lemma 26, the restriction V_σ is a Jacobi field. By the construction of $\boldsymbol{x}$, the two are equal on $[0, \delta]$ for any s. Thus their initial conditions agree at $t = 0$, so by Lemma 8.5, they are equal on $[0, 1]$. In particular, $\boldsymbol{x}_s(1, s)$ is the value of V at $\beta(s) = \boldsymbol{x}(1, s)$ for all $s \in (-\varepsilon, \varepsilon)$. But this says that $\beta: (-\varepsilon, \varepsilon) \to M$ is an integral curve of V starting at r, so $r \in A$. Hence $\mathscr{C} \subset A$, as required. ∎

On an incomplete manifold, Killing vector fields may or may not be complete. For example, on the unit disk $|p| < 1$ in $\boldsymbol{R}^2$, infinitesimal translations are obviously not complete, but the infinitesimal rotation $-v\partial_u + u\partial_v$ is.

I(M) AS LIE GROUP

The isometry groups of S_ν^n, H_ν^n, and $\boldsymbol{R}_\nu^n$ were, in a natural way, Lie groups. We shall now see that this is true for any semi-Riemannian manifold.

31. Definition. A (*left*) *action* of a Lie group G on a manifold M is a smooth map $G \times M \to M$, denoted by $(g, p) \to gp$, such that

(1) $(gh)p = g(hp)$ for all $g, h \in G$ and $p \in M$.
(2) $ep = p$ for all $p \in M$, where e is the identity element of G.

Here G is also called a *Lie transformation group* on M. The definition makes sense with G an abstract group and M a set, but unless the contrary is mentioned we assume the smooth case.

For a given action, if $g \in G$ is held fixed, then $p \to gp$ is a diffeomorphism with inverse $p \to (g^{-1})p$. An action $G \times M \to M$ is *transitive* provided that for each $p, q \in M$ there is a $g \in G$ such that $gp = q$.

For example, under the column vector conventions, $(g, x) \to gx$ is an action $GL(n, \boldsymbol{R}) \times \boldsymbol{R}^n \to \boldsymbol{R}^n$. This action is not transitive, since $g0 = 0$ for all g—but it is transitive on $\boldsymbol{R}^n - 0$.

If $G \times M \to M$ is an action and o is a point of M, then $H = \{g \in G : go = o\}$ is a closed subgroup of G called the *isotropy subgroup* at o.

Now consider the isometry group $I(M)$ of a semi-Riemannian manifold M. Evaluation of $\phi \in I(M)$ on $p \in M$ gives a natural map $(\phi, p) \to \phi(p)$ from $I(M) \times M$ into M.

32. Theorem. If M is a semi-Riemannian manifold, there is a unique way to make $I(M)$ a manifold such that:

(C1) $I(M)$ is a Lie group.

(C2) The natural action $I(M) \times M \to M$ is smooth.

(C3) A homomorphism $\beta: \boldsymbol{R} \to I(M)$ is smooth if the map $R \times M \to M$ sending (t, p) to $\beta(t)p$ is smooth.

This follows from general results of Palais in Chapter IV of [P].

It can be expected that the Lie algebra $\mathscr{I}(M)$ of the isometry group $I(M)$ is closely related to the Killing vector fields ("infinitesimal isometries") of M. If $X \in \mathscr{I}(M)$, let $t \to \psi_t$ be its one-parameter subgroup. By (C2), the map $\boldsymbol{R} \times M \to M$ sending (t, p) to $\psi_t(p)$ is smooth. For each $p \in M$, let X_p^+ *be* the initial velocity of the curve $t \to \psi_t(p)$. Then X^+ is a smooth vector field on M. Using the identity $\psi_t(\psi_s p) = \psi_{t+s}(p)$, it is easy to show that $\{\psi_t\}$ is the flow of X^+. Since one-parameter groups are defined on the whole real line, X^+ is complete. Since each ψ_t is an isometry, X^+ is Killing.

33. Proposition. Let M be a semi-Riemannian manifold, and let $\mathscr{I}(M)$ be the Lie algebra of its isometry group $I(M)$. Then

(1) The set $ci(M)$ of all complete Killing vector fields on M is a Lie subalgebra of $i(M)$.

(2) The function $X \to X^+$ is a Lie anti-isomorphism $\mathscr{I}(M) \to ci(M)$, that is, a linear isomorphism such that

$$[X^+, Y^+] = -[X, Y]^+ \qquad \text{for all} \quad X, Y \in \mathscr{I}(M).$$

Proof. For each $p \in M$ it follows from (C2) that the map $\pi_p: I(M) \to M$ sending ϕ to $\phi(p)$ is smooth.

(a) *If $X \in \mathscr{I}(M)$, then $d\pi_p(X_e) = X_p^+$ for each $p \in M$.* Let $\alpha(t) = \psi_t$ be the one-parameter subgroup of X. By definition, α is the integral curve of X starting at e in $I(M)$, with $\alpha'(0) = X_e$. Thus $d\pi_p(X_e) = d\pi_p(\alpha'(0)) = (\pi_p \circ \alpha)'(0)$. But $\pi_p \alpha(t) = \alpha(t)(p) = \psi_t p$, so $(\pi_p \circ \alpha)'(0) = X_p^+$.

(b) *The function $X \to X^+$ is a one-to-one linear transformation onto $ci(M)$.* It follows immediately from (a) that the function is $\boldsymbol{R}$-linear. It is one-to-one, for, if $X^+ = 0$, then the integral curves $t \to \psi_t p$ of X are constant for all p. Thus $\psi_t = \text{id}$ for all t, which implies $X = 0$.

We must show that each $Z \in ci(M)$ can be expressed as X^+ for some $X \in \mathscr{I}(M)$. The global flow $\{\psi_t\}$ of Z consists of isometries, so $t \to \psi_t$ is a homomorphism of $\boldsymbol{R}$ into $I(M)$. By (C3) the smoothness of the flow implies $t \to \psi_t$ is smooth. Hence it is a one-parameter subgroup of $I(M)$. If X is the corresponding element of $\mathscr{I}(M)$, then clearly $X^+ = Z$.

(c) $[X^+, Y^+] = -[X, Y]^+$. Let $t \to \psi_t$ be the one-parameter subgroup of X and hence the flow of X^+. Let $q = \psi_t(p)$. Then by (a), $Y_q^+ = d\pi_q(Y_e)$, and furthermore, for all $\phi \in I(M)$,

$$(\psi_{-t}\pi_q)(\phi) = \psi_{-t}\phi\psi_t(p) = (\pi_p R_{\psi_t} L_{\psi_{-t}})(\phi).$$

Since Y is left-invariant it follows that

$$d\psi_{-t}(Y_q^+) = d\pi_p\, dR_{\psi_t}(Y_e).$$

Thus

$$\begin{aligned}
[X^+, Y^+]_p &= \lim_{t \to 0} \frac{1}{t} (d\psi_{-t}(Y_q^+) - Y_p^+) \\
&= \lim_{t \to 0} \frac{1}{t} (d\pi_p\, dR_{\psi_t}(Y_e) - d\pi_p(Y_e)) \\
&= d\pi_p \left\{ \lim_{t \to 0} \frac{1}{t} (dR_{\psi_t}(Y_e) - Y_e) \right\}.
\end{aligned}$$

Since $t \to \psi_t$ is the one-parameter subgroup of X, $t \to \psi_{-t}$ is the one-parameter subgroup of $-X$. Thus by the lemma below, $R_{\psi_{-t}}$ is the flow of $-X$. In view of the signs in Proposition 1.56, the limit above is $[-X, Y]_p$. Hence by (a),

$$[X^+, Y^+]_p = -d\pi_p[X, Y] = -[X, Y]_p^+.$$

Then assertions (1) and (2) are direct consequences of (a), (b), and (c). ∎

34. Lemma. Let $\mathfrak{g}$ be the Lie algebra of a Lie group G. If α is the one-parameter subgroup of $X \in \mathfrak{g}$ then the flow of X is $\{R_{\alpha(t)}\}$.

Proof. That α is the one-parameter subgroup of X means that it is the integral curve of X starting at the identity element e of G. If $g \in G$ then applying the left translation L_g shows that $L_g \circ \alpha$ is the integral curve of $dL_g X = X$ starting at g. But $L_g \alpha(t) = g\alpha(t)$ for all $t \in \boldsymbol{R}$, hence $\{R_{\alpha(t)}\}$ is the flow of X. ∎

HOMOGENEOUS SPACES

One natural way to specify that M has plenty of isometries is as follows.

35. Definition. A semi-Riemannian manifold M is *homogeneous* provided that, given any points $p, q \in M$, there is an isometry ϕ of M such that $\phi(p) = q$.

In short, $I(M)$ is transitive on M, which is also sometimes called a *homogeneous space.* If M is homogeneous, then evidently any geometrical properties at one point of M hold at every point.

To show that a given semi-Riemannian manifold is homogeneous, it suffices to exhibit enough isometries to carry some one fixed point to every point (or vice versa).

36. Lemma. A symmetric semi-Riemannian manifold M is homogeneous.

Proof. Let $\sigma: [0, 1] \to M$ be a geodesic. The global symmetry ζ at $\sigma(\frac{1}{2})$ is an isometry that reverses geodesics, hence carries $\sigma(0)$ to $\sigma(1)$. Since symmetric manifolds are by definition connected, any two points $p, q \in M$ can be joined by a broken geodesic. Thus a finite composition of isometries as above gives an isometry carrying p to q. ∎

37. Remark. Completeness of Homogeneous Spaces. (1) *A Riemannian homogeneous space is complete.* It suffices to show that a unit speed geodesic $\gamma: [0, b)$ has a geodesic extension past b. The existence of normal neighborhoods shows, in the Riemannian case, that for any point $o \in M$ there is an $\varepsilon > 0$ such that every unit speed geodesic starting at o is defined on $[0, \varepsilon)$. If ϕ is an isometry carrying o to $\gamma(b - \varepsilon/2)$, there exists a unit vector $v \in T_o(M)$ such that $d\phi(v) = \gamma'(b - \varepsilon/2)$. Hence the geodesic $\phi \circ \gamma_v$ provides an extension of γ.

(2) *An indefinite homogeneous space need not be complete.* Consider M as in Example 7.14. For each $(a, b) \in M$ the isometry $(u, v) \to (u/a, av)$ carries (a, b) to $(1, ab)$; then the isometry $(u, v) \to (u, v - ab)$ carries this point to $(1, 0)$. Thus M is homogeneous. But M is not complete since clearly the null geodesic $\alpha(t) = (t, 0)$ has maximum domain $\boldsymbol{R}^+$.

Thus a homogeneous space need not be symmetric. Since a homogeneous space has a good supply of isometries, it also has a good supply of Killing vector fields.

38. Corollary. Each tangent vector to a homogeneous semi-Riemannian manifold M extends to a Killing vector field on M.

Proof. For $p \in M$, the projection $g \to gp$ is a submersion $\pi: I(M) \to M$. (This is a general fact about transitive actions; see Proposition 11.13.) Hence, if $v \in T_p(M)$, there is a vector $\tilde{v} \in T_e(IM)$ such that $d\pi(\tilde{v}) = v$. But $\tilde{v}$ extends to a left-invariant vector field V on $I(M)$. Hence by Proposition 33, V^+ is a Killing vector field such that $V_p^+ = v$. ∎

We have now seen that, if a Riemannian manifold is either compact or homogeneous, it is complete, but that neither condition alone suffices for indefinite metrics. However, both together are sufficient.

39. Proposition (Marsden). A compact homogeneous semi-Riemannian manifold is complete.

Proof. To show that a geodesic $\gamma: [0, b) \to M$ is extendible, let $\{s_k\}$ be a sequence in $[0, b)$ that converges to b. Since M is compact, by passing to a subsequence we can suppose that $\{\gamma(s_k)\}$ converges to some point $p \in M$. Let $v_1, \ldots, v_n$ be a basis for $T_p(M)$. By Corollary 38, each v_i extends to a Killing vector field V_i. The conservation lemma (26) says that $\langle \gamma', V_i \rangle$ is a constant c_i for each $i = 1, \ldots, n$. It will follow that *the sequence* $\{\gamma'(s_k)\}$ *in* TM *converges to some vector* $v \in T_p(M)$. We can suppose that the sequence $\{\gamma(s_k)\}$ lies in a neighborhood $\mathscr{U}$ of p in M such that $V_1, \ldots, V_n$ give a basis for each $T_q(M)$, $q \in \mathscr{U}$. As with coordinate vector fields, if $h_{ij} = \langle V_i, V_j \rangle$ then $\det h \neq 0$ and

$$\gamma'(s_k) = \sum (h^{-1})_{ij} c_i V_j \qquad \text{at} \quad \gamma(s_k).$$

Thus the sequence converges as claimed.

By Proposition 3.28, $\gamma': [0, b) \to TM$ is an integral curve of the geodesic vector field G. Hence by the single-sequence criterion in Lemma 1.56, γ' has an extension past b as an integral curve of G. But this extension projects to a geodesic extension of γ. ∎

Putting some dents in, say, the plane $\boldsymbol{R}^2$ will give a manifold whose only isometry is the identity map. At the opposite extreme, a semi-Riemannian manifold M is *frame-homogeneous* provided any frame on M can be carried to any other by the differential map of an isometry of M. Such a manifold M has the largest possible isometry group: M is homogeneous and the isotropy group of $I(M)$ at any point p is the entire group $O(T_pM) \approx O_\nu(n)$ of linear isometries on $T_p(M)$. It follows, using Exercise 14(a), that M has constant curvature. Proposition 4.30 asserts that *hyperquadrics Q are frame-homogeneous.* In the Riemannian case every connected frame-homogeneous manifold ($n \geq 2$) is homothetic to S^n, P^n, $\mathbf{R}^n$, or H^n; for indefinite metrics, the list is longer (see [Wo]).

Exercises

1. (a) If $\psi: M \to N$ is a homothety, show that $\phi \to \psi\phi\psi^{-1}$ is an isomorphism from $I(M)$ to $I(N)$. (b) Prove that, if $-M$ is M with metric reversed, then $I(-M) = I(M)$. (c) Find an explicit matrix formula for a Lie group isomorphism $O_\nu(n) \approx O_{n-\nu}(n)$.

2. Prove: (a) A coordinate vector field ∂_k is Killing if and only if $\partial g_{ij}/\partial x^k = 0$ for all i, j (see 5.39.) (b) For a warped product $M = B \times_f F$, a vector field on F is Killing if and only if its lift to M is Killing.

3. A positive space form M^n with $0 < \nu \leq n/2$ and $n \geq 3$ is noncompact.

4. Let M be a positive space form with finite fundamental group. If $n - \nu$ is odd, then M is space-orientable. (Compare Proposition 16.)

5. (See Proposition 13.) Suppose that $\phi \in I(\tilde{M})$ has $\phi^m = \text{id}$ for some integer $m \neq 0$. Show that ϕ has a fixed point if $\tilde{M}$ is (a) H^n, (b) cS^n_n, (c) $\mathbf{R}^n_\nu$. (Hint: For (a), verify that $p = (1/m)(x + \phi x + \cdots + \phi^{m-1}x)$ is a fixed point in $\mathbf{R}^{n+1}_1$.)

6. Let $S^{n-\nu}/\Gamma$ be a positive Riemannian space form and let $\theta: \Gamma \to O(\nu)$ be a homomorphism. Show that $\Gamma_\theta = \{(\theta b, b): b \in \Gamma\}$ is a finite subgroup of $O_\nu(n+1)$ that is properly discontinuous on S^n_ν. Hence S^n_ν/Γ_θ is a positive space form with fundamental group $\Gamma_\theta \approx \Gamma$.

7. Let P be a semi-Riemannian submanifold of M, and let X be a Killing vector field on M. Prove: (a) If X is tangent to P, then $X|P$ is a Killing vector field on P. (b) If P is totally geodesic, then tan X is Killing on P.

8. If V is an arbitrary vector field and X is Killing, then $D_{[X,V]} = [L_X, D_V]$ and $D_V(DX) = R_{XV}$.

9. If X is a Killing vector field on a semi-Riemannian manifold M, let $f = \frac{1}{2}\langle X, X\rangle$. Prove: (a) grad $f = -D_X X$. (b) $H^f(V, W) = \langle D_V X, D_W X\rangle - \langle R_{XV}X, W\rangle = -\langle D_V(D_X X), W\rangle$, (c) $\Delta f = -\text{trace}(DX \circ DX) - \text{Ric}(X, X)$.

10. (*Continuation*) (a) If an integral curve α of a Killing vector field X starts at a critical point of $\langle X, X\rangle$, then α is a geodesic. (b) Theorem of Bochner: On a compact Riemannian manifold with Ric < 0, every Killing vector field is identically zero. (Hint: At a maximum point of f, H^f is negative semidefinite.)

11. M is said to be *isotropic* at $p \in M$ provided that, if $v, w \in T_p(M)$ have $\langle v, v\rangle = \langle w, w\rangle$, there is a $\phi \in I(M)$ such that $d\phi(v) = w$. M is *isotropic* if it is isotropic at every point. Prove: (a) M isotropic $\Rightarrow$ M complete. (b) M connected and isotropic $\Rightarrow$ M homogeneous. (c) M frame-homogeneous $\Rightarrow$ M isotropic and symmetric. (The converse is false, e.g., CP^n from Chapter 11.) (d) $S^1 \times \boldsymbol{R}^1$ is symmetric but not isotropic.

12. The Clifton–Pohl torus $T = M/\Gamma$ (Example 7.16): (a) Find a group of eight isometries and anti-isometries of M. (b) Show that $s \to (\tan s, 1)$ is a geodesic, and deduce that every null geodesic of M and T is incomplete. (c) $P = u\,\partial_u + v\,\partial_v$ is a Killing vector field on M. (d) If $\gamma(s) = (u(s), v(s))$ is a geodesic, then both $u'v'/r^2$ and $(uv' + vu')/r^2$ are constant, where $r^2 = u^2 + v^2$. (e) T is not geodesically connected. (f) $s \to (s, 1/s)$ is pregeodesic and on $[1, \infty)$ has finite length. (g) T has timelike geodesics that are complete and ones that are not complete; likewise for spacelike.

13. In complex terms the Poincaré plane P of Exercise 3.8 is the region $\operatorname{Im} z > 0$ of $\boldsymbol{C} \approx \boldsymbol{R}^2$ with $ds^2 = dz\, d\bar{z}/(\operatorname{Im} z)^2$. If $A = \left(\begin{smallmatrix} a & b \\ c & d \end{smallmatrix}\right) \in SL(2, \boldsymbol{R})$ let ϕ_A send $z \in P$ to $z' = (az + b)/(cz + d)$. Prove: (a) $\phi_A(P) \subset P$. (b) ϕ_A is an isometry of P. (Hint: Compute $dz'\, d\bar{z}'/(\operatorname{Im} z')^2$.) (c) The map $A \to \phi_A$ is a homomorphism onto the identity component $I_0(P)$ of $I(P)$ with kernel $\pm I$. (d) P is isometric to the hyperbolic plane, hence $I_0(P) \approx O_1^{++}(3) = SO_0(1, 2)$.

14. Prove: (a) If sectional curvature K is constant on the nondegenerate planes of $T_p(M)$ of a given index (0, 1, or 2), then K is constant on all nondegenerate planes in $T_p(M)$. (b) If X is a Killing vector field with $X_p = 0$, then $(DX)_p = -\lim_{t\to 0}(1/t)(d\psi_t - \text{id})$ on $T_p(M)$, where $\{\psi_t\}$ is a local flow of X. (c) If M^n is connected and $\dim i(M) = n(n+1)/2$, then M has constant curvature.

15. If M is complete and connected, the following are equivalent: $\dim I(M) = n(n+1)/2$, $\dim i(M) = n(n+1)/2$, M is frame-homogeneous.

16. *G-orientability* (page 241). Prove: (a) A semi-Riemannian hypersurface M of a G-orientable manifold is G-orientable, where meaningful, if and only if M has a smooth unit normal vector field. (b) An arbitrary semi-Riemannian manifold is G-orientable if and only if every loop in M preserves G-orientation (that is, lifts as a loop into M^G). (c) A connected semi-Riemannian manifold is G-orientable if its fundamental group has no subgroup of order 2.

17. A flat connected frame-homogeneous manifold M is isometric to $\boldsymbol{R}^n_\nu$. To prove this show: (a) $M = \boldsymbol{R}^n_\nu/\Gamma$, where $N(\Gamma) = I(\boldsymbol{R}^n)$; that is, Γ is a normal subgroup. (b) Each element of Γ commutes with each element of the identity component K of $I(\boldsymbol{R}^n_\nu)$. (See Exercise 11.14.) (c) $\Gamma = \{e\}$. (Hint: $K \supset \boldsymbol{R}^n \cup O_\nu^{++}(n)$.)

18. (This exercise requires some advanced linear algebra.) (a) A linear operator S on $V \approx \boldsymbol{R}^n_\nu$ is self-adjoint if and only if V can be expressed as a direct sum of subspaces V_k that are mutually orthogonal (hence nondegenerate) and S-invariant and each $S|V_k$ has matrix of form *either*

$$\begin{pmatrix} \lambda & & & & \\ 1 & \lambda & & 0 & \\ & \ddots & \ddots & & \\ & 0 & 1 & \lambda & \\ & & & 1 & \lambda \end{pmatrix}$$

relative to a basis $v_1, \ldots, v_r$ $(r \geq 1)$ with all scalar products zero except $\langle v_i, v_j \rangle = \varepsilon = \pm 1$ if $i + j = r + 1$, *or*

$$\begin{pmatrix} a & b & & & & & & & \\ -b & a & & & & & & & \\ 1 & 0 & a & b & & & 0 & & \\ 0 & 1 & -b & a & & & & & \\ & & 1 & 0 & a & b & & & \\ & & 0 & 1 & -b & a & & & \\ & & & & & & \ddots & & \\ & 0 & & & & 1 & 0 & a & b \\ & & & & & 0 & 1 & -b & a \end{pmatrix} \qquad (b \neq 0)$$

relative to a basis $u_1, v_1, \ldots, u_m, v_m$ with all scalar products zero except $\langle u_i, u_j \rangle = 1 = -\langle v_i, v_j \rangle$ if $i + j = m + 1$. (Here r, ε, and m depend on k.) (Hint: Complexify, then geometrize the derivation of the Jordan canonical form.) (b) For each type in (a), determine the existence and causal character of eigenvectors, and the index of V_k.

19. (*Continuation*) (a) A self-adjoint linear operator on a Lorentz vector space $V \approx \boldsymbol{R}_1^n$ has a matrix of exactly one of the following four types, where D_k is $k \times k$ diagonal:

Relative to an orthonormal basis, D_n *or*

$$\left(\begin{array}{cc|c} a & b & 0 \\ -b & a & \\ \hline 0 & & D_{n-2} \end{array}\right) \qquad (b \neq 0);$$

relative to a basis $u, v, e_1, \ldots, e_{n-2}$ with all scalar products zero except $\langle u, v\rangle = 1 = \langle e_i, e_i\rangle$ $(1 \le i \le n-2)$,

$$\left(\begin{array}{cc|c} \lambda & 0 & \\ \varepsilon & \lambda & 0 \\ \hline 0 & & D_{n-2} \end{array}\right) (\varepsilon = \pm 1) \quad or \quad \left(\begin{array}{ccc|c} \lambda & 0 & 1 & \\ 0 & \lambda & 0 & 0 \\ 0 & 1 & \lambda & \\ \hline & 0 & & D_{n-3} \end{array}\right).$$

(b) Characterize these cases in terms of eigenvectors (including causal character and multiplicity). (c) For $n = 4$, change tensor type from (1, 1) to (0, 2) and show agreement with the cases in §4.3 of [HE].

20. Let $A \in O_1(n) = O(1, n-1)$. (a) Prove: nonnull eigenvectors of A have eigenvalues ± 1; for two independent null eigenvectors the product of their eigenvalues is $+1$. (b) Analyse eigenvalues and eigenvectors in the case $n = 2$. (c) Prove that if A has a null eigenvalue $\lambda \neq \pm 1$, then it has an (independent) eigenvector with eigenvalue $1/\lambda$.

21. (a) For $A \in O_1(n)$ exactly one of the following is true: (i) A has a timelike eigenvector, (ii) A has a null eigenvector with eigenvalue $\neq \pm 1$, (iii) A has a unique null eigenvector. (b) The first two cases in (a) are equivalent to the existence of an orthonormal basis relative to which the matrix of A is in (i′) $\{\pm 1\} \times O(n-1)$, (ii′) $SO_1(2) \times O(n-2)$. (c) There exists $A \in O_1(3)$ such that A has a unique eigenvector and it is null; hence type (iii) above exists for $n \ge 3$.

22. Semiclassical surface theory. In Exercise 4.7 replace Euclidean space $\boldsymbol{R}^3$ by Minkowski space $\boldsymbol{R}^3_1$. Assume x has sign $\varepsilon = \pm 1$, so $\varepsilon = -1$ if $\mathscr{D}$ is Riemannian, $\varepsilon = 1$ if $\mathscr{D}$ is Lorentz. Prove (a, b, c, d) from Exercise 4.7 with the following modifications: (a) Change a sign in the usual definition of cross product. (c) $K = \varepsilon \det S$, with $\det S$ as before. (d) $H = \varepsilon h U$, with h as before.

23. (*Continuation*) For each of the following immersions in $\boldsymbol{R}^3_1$ compute ε, K, h, and S, and determine the eigenvectors (principal directions) of S: (a) $\mathbf{x}(u, v) = (u, v, (u + v)^2/2)$. (b) $\mathbf{x}(u, v) = (v, u \cos v, u \sin v)$, $|u| < 1$. (c) $\mathbf{x}$ as in (b), but $|u| > 1$. (Hint: All have $h = 0$.)

24. Let M^n have indefinite metric. (a) For a frame field on a loop $\alpha\colon [0, 1] \to M$, consider the matrix $g \in O_\nu(n)$ such that $E_i(1) = \sum g_{ij} E_j(0)$. Express the criterion in Exercise 16(b) in terms of such matrices. (b) Show that if M is G-orientable for any two of the three types (orientable, time-orientable, space-orientable) it is G-orientable for the third. (c) For connected M, express G-orientabilities in terms of a natural homomorphism $\pi_1(M) \to O_\nu(n)/O_\nu^{++}(n) \approx \boldsymbol{Z}_2 \times \boldsymbol{Z}_2$. (d) Find a connected M that is G-orientable for none of the three types.

10 CALCULUS OF VARIATIONS

A basic problem in semi-Riemannian geometry is to measure the change in arc length of a curve segment under small displacements. To formalize the idea let $x: [a,b] \times (-\delta, \delta) \to M$ be a variation of a curve segment α (Definition 8.1), and for each $v \in (-\delta, \delta)$, let $L_x(v)$ be the length of the longitudinal curve $u \to x(u, v)$. Then L_x is a real-valued function with $L_x(0)$ the length of α. Under mild conditions the function $L = L_x$ is smooth, and we shall find useful formulas for the *first* and *second variations of arc length on* x; that is, for

$$L'(0) = \left.\frac{dL}{dv}\right|_{v=0} \quad \text{and} \quad L''(0) = \left.\frac{d^2L}{dv^2}\right|_{v=0},$$

the latter when α is geodesic. The second variation formula involves curvature, and the resulting link between geodesics and curvature has a broad range of applications.

FIRST VARIATION

The class of curves α with $|\alpha'| > 0$ consists of all spacelike regular curves and all timelike (hence regular) curves, the two cases distinguished by the *sign of* α; that is, $\varepsilon = \operatorname{sgn}\langle\alpha', \alpha'\rangle = \pm 1$.

1. Lemma. Let x be a variation of a curve segment $\alpha: [a, b] \to M$ with $|\alpha'| > 0$. If L is the length function of x, then

$$L'(0) = \varepsilon \int_a^b \langle \alpha'/|\alpha'|, V' \rangle \, du,$$

where ε is the sign of α and V is the variation vector field of x.

Proof. $L(v) = \int_a^b |x_u(u, v)|\, du$, and if the v-interval $(-\delta, \delta)$ is small enough, $|x_u|$ is positive, hence differentiable. Then

$$L'(0) = \int_a^b \frac{d}{dv} |x_u|\, du,$$

and furthermore $\operatorname{sgn}\langle x_u, x_u\rangle = \varepsilon$, hence $|x_u| = (\varepsilon\langle x_u, x_u\rangle)^{1/2}$. By Proposition 4.44, $x_{uv} = x_{vu}$, and we compute

$$(d/dv)|x_u| = \tfrac{1}{2}(\varepsilon\langle x_u, x_u\rangle)^{-1/2} 2\varepsilon\langle x_u, x_{uv}\rangle = \varepsilon\langle x_u, x_{vu}\rangle/|x_u|.$$

Setting $v = 0$ gives $x_u(u, 0) = \alpha'(u)$ and, by definition $x_v(u, 0) = V(u)$. Thus $x_{vu}(u, 0) = V'(u)$, and the result follows. ∎

Even simple constructions create piecewise smooth curves by hooking together smooth ones. The preceding formula remains valid in this case, since finitely many discontinuities have no effect on the integral. However, we want to consider an alternative formula in which piecewise smoothness requires care.

A variation x of a piecewise smooth curve $\alpha: [a, b] \to M$ is itself *piecewise smooth* provided x is continuous, and for *breaks* $a < u_1 < \cdots < u_k < b$ the restriction of x to each set $[u_{i-1}, u_i] \times (-\delta, \delta)$ is smooth. There is no loss in generality in assuming that α and x have the same breaks, since we can always add *trivial breaks*—those at which α or x is smooth. The variation vector field V of x is always piecewise smooth. By contrast, the velocity vector field α' will generally have a discontinuity at each break u_i $(1 \leq i \leq k)$. This discontinuity is measured by the tangent vector

$$\Delta\alpha'(u_i) = \alpha'(u_i^+) - \alpha'(u_i^-) \in T_{\alpha(u_i)}(M),$$

where the first term derives from $\alpha | [u_i, u_{i+1}]$ and the second from $\alpha | [u_{i-1}, u_i]$. (We consider $a = u_0$, $b = u_{k+1}$.)

2. Proposition (First Variation Formula). Let $\alpha: [a, b] \to M$ be a piecewise smooth curve segment with constant speed $c > 0$ and sign ε. If x is a variation of α, then

$$L'(0) = -\frac{\varepsilon}{c}\int_a^b \langle \alpha'', V\rangle\, du - \frac{\varepsilon}{c}\sum_{i=1}^{k} \langle \Delta\alpha'(u_i), V(u_i)\rangle + \frac{\varepsilon}{c}\langle \alpha', V\rangle \Big|_a^b,$$

where $u_1 < \cdots < u_k$ are the breaks of α and x.

Proof. Since $|\alpha'| = c$, the integrand in the preceding lemma is $\langle \alpha', V'\rangle/c$. We use integration by parts, a standard device in the calculus of variations.

Since

$$\langle \alpha', V' \rangle = (d/du)\langle \alpha', V \rangle - \langle \alpha'', V \rangle,$$

it follows by the fundamental theorem of calculus that for any subinterval $[u_i, u_{i+1}]$

$$\int_{u_i}^{u_{i+1}} \langle \alpha', V' \rangle \, du = \langle \alpha', V \rangle \Big|_{u_i}^{u_{i+1}} - \int_{u_i}^{u_{i+1}} \langle \alpha'', V \rangle \, du.$$

Taking the sum from $i = 0$ to $i = k$ then gives c times the integral in Lemma 1. In particular, for each $i = 1, 2, \ldots, k$, the contribution of the break at u_i is

$$\langle \alpha'(u_i^-), V(u_i) \rangle - \langle \alpha'(u_i^+, V(u_i) \rangle = -\langle \Delta\alpha'(u_i, V(u_i) \rangle. \qquad \blacksquare$$

Note that the formula depends not on the entire variation $\boldsymbol{x}$ but only on its infinitesimal model, the variation vector field V.

In the summation term, $\Delta\alpha'(u_i)$ is a "one-point acceleration" measuring the change in the velocity of α at the break. If the corner is rounded off, the vanishing of this term will be compensated for by a change in the integral term.

For a *fixed endpoint* variation $\boldsymbol{x}$, the first and last transverse curves are constant, so all longitudinal curves run from $p = \alpha(a)$ to $q = \alpha(b)$. In particular, the variation vector field V vanishes at a and b, and hence so does the last term in the first variation formula.

3. Corollary. A piecewise smooth curve segment α of constant speed $c > 0$ is an (unbroken) geodesic if and only if the first variation of arc length is zero for every fixed endpoint variation of α.

Proof. If α is a geodesic, then $\alpha'' = 0$ and the breaks are all trivial: $\Delta\alpha'(u_i) = 0$. For fixed endpoint variations, $V(a)$ and $V(b)$ are zero. Thus $L'(0) = 0$.

Conversely, suppose $L'(0) = 0$ for every fixed endpoint variation $\boldsymbol{x}$. First we show that each segment $\alpha | [u_i, u_{i+1}]$ is geodesic. It suffices to show that $\alpha''(t) = 0$ for $u_i < t < u_{i+1}$. Let y be any tangent vector to M at $\alpha(t)$, and let f be a bump function on $[a, b]$ with $\operatorname{supp} f \subset [t - \delta, t + \delta] \subset [u_i, u_{i+1}]$. Let Y be the vector field on α obtained by parallel translation of y, and finally let $V = fY$.

Since $V(a)$ and $V(b)$ are both zero, the exponential formula $\boldsymbol{x}(u, v) = \exp_{\alpha(u)}(vV(u))$ produces a fixed endpoint variation of α whose vector field is V. Since $L'(0) = 0$, the formula in Proposition 2 reduces to

$$0 = \int_a^b \langle \alpha'', V \rangle \, du = \int_{t-\delta}^{t+\delta} \langle \alpha'', fY \rangle \, du.$$

This holds for all y, $\delta > 0$, and f. Hence $\langle \alpha''(t), y\rangle = 0$ for all y; hence $\alpha''(t) = 0$.

It remains to show that the breaks are trivial, so that α is an unbroken geodesic. As before, let y be an arbitrary tangent vector at $\alpha(u_i)$, and let f be a bump function at u_i with $\operatorname{supp} f \subset [u_{i-1}, u_{i+1}]$. For a fixed endpoint variation with vector field fY the first variation formula now reduces to

$$0 = L'(0) = -(\varepsilon/c)\langle \Delta\alpha'(u_i), y\rangle \qquad \text{for all} \quad y.$$

Hence $\Delta\alpha'(u_i) = 0$. ∎

SECOND VARIATION

For a variation x of a curve segment α our goal is to compare $L(v)$, v small, with the length $L(0)$ of α. Thus $L''(0)$ is needed only when $L'(0) = 0$. By Corollary 3, it suffices in practice to find a formula for $L''(0)$ in case α is a geodesic.

The vector field $V(u) = x_v(u, 0)$ gives the velocities of the transverse curves of x as they cross the base curve α. Similarly the vector field $A(u) = x_{vv}(u, 0)$ gives their accelerations; we call A the *transverse acceleration vector field* of x.

Recall that if $|\alpha'| > 0$, then any vector field Y on α splits into the sum $Y^{\top} + Y^{\perp}$ of its components tangent to α and perpendicular to α, respectively. If α is a geodesic, then $(Y^{\perp})' = (Y')^{\perp}$ is denoted unambiguously by $\overset{\perp}{Y}{}'$.

4. Theorem (Synge's Formula for Second Variation). Let $\sigma: [a, b] \to M$ be a geodesic segment of speed $c > 0$ and sign ε. If x is a variation of σ, then

$$L''(0) = \frac{\varepsilon}{c}\int_a^b \{\langle \overset{\perp}{V}{}', \overset{\perp}{V}{}'\rangle - \langle R_{V\sigma'}V, \sigma'\rangle\}\, du + \frac{\varepsilon}{c}\langle \sigma', A\rangle\Big|_a^b,$$

where V is the variation vector field, A the transverse acceleration vector field of x.

Proof. Let $h = h(u, v) = |x_u(u, v)|$, so $L(v) = \int_a^b h\, du$, hence $L''(v) = \int_a^b (\partial^2 h/\partial v^2)\, du$. In the proof of Lemma 1, we computed $\partial h/\partial v = (\varepsilon/h)\langle x_u, x_{uv}\rangle$. Thus

$$\frac{\partial^2 h}{\partial v^2} = \frac{\varepsilon}{h^2}\left\{h\frac{\partial}{\partial v}\langle x_u, x_{uv}\rangle - \langle x_u, x_{uv}\rangle\frac{\partial h}{\partial v}\right\}$$

$$= \frac{\varepsilon}{h}\left\{\langle x_{uv}, x_{uv}\rangle + \langle x_u, x_{uvv}\rangle - \frac{\varepsilon}{h^2}\langle x_u, x_{uv}\rangle^2\right\}.$$

The object now is to move u to the right in each term. By Proposition 4.44, $\boldsymbol{x}_{uv} = \boldsymbol{x}_{vu}$ and

$$\boldsymbol{x}_{uvv} = \boldsymbol{x}_{vuv} = R(\boldsymbol{x}_u, \boldsymbol{x}_v)\boldsymbol{x}_v + \boldsymbol{x}_{vvu}.$$

Hence

$$\frac{\partial^2 h}{\partial v^2} = \frac{\varepsilon}{h}\left\{\langle \boldsymbol{x}_{vu}, \boldsymbol{x}_{vu}\rangle + \langle \boldsymbol{x}_u, R(\boldsymbol{x}_u, \boldsymbol{x}_v)\boldsymbol{x}_v\rangle + \langle \boldsymbol{x}_u, \boldsymbol{x}_{vvu}\rangle - \frac{\varepsilon}{h^2}\langle \boldsymbol{x}_u, \boldsymbol{x}_{vu}\rangle^2\right\}.$$

Setting $v = 0$ in this equation produces the following changes: $h \to c$, $\boldsymbol{x}_u \to \sigma'$, $\boldsymbol{x}_v \to V$, $\boldsymbol{x}_{vu} \to V'$, $\boldsymbol{x}_{vv} \to A$, and $\boldsymbol{x}_{vvu} \to A'$. Thus, rearranging the curvature term, we find

$$\left.\frac{\partial^2 h}{\partial v^2}\right|_{v=0} = \frac{\varepsilon}{c}\left\{\langle V', V'\rangle - \langle R_{V\sigma'} V, \sigma'\rangle + \langle \sigma', A'\rangle - \left(\frac{\varepsilon}{c^2}\right)\langle \sigma', V'\rangle^2\right\}.$$

Two simplifications are now possible. First, since σ is a geodesic, $\langle \sigma', A'\rangle = (d/du)\langle \sigma', A\rangle$. Second, since σ'/c is a unit vector field, the tangential component of V' is $\varepsilon\langle V', \sigma'/c\rangle(\sigma'/c)$. Thus

$$V' = (\varepsilon/c^2)\langle V', \sigma'\rangle\sigma' + \overset{\perp}{V}{}';$$

hence

$$\langle V', V'\rangle = (\varepsilon/c^2)\langle V', \sigma'\rangle^2 + \langle \overset{\perp}{V}{}', \overset{\perp}{V}{}'\rangle.$$

Substitution then gives

$$\left.\frac{\partial^2 h}{\partial v^2}\right|_{v=0} = \frac{\varepsilon}{c}\left\{\langle \overset{\perp}{V}{}', \overset{\perp}{V}{}'\rangle - \langle R_{V\sigma'} V, \sigma'\rangle + \frac{d}{du}\langle \sigma', A\rangle\right\}.$$

Finally, integration from a to b yields the required formula. ∎

Comments. (1) The only part of the second variation formula that uses $\boldsymbol{x}$ rather than V is the endpoint term

$$\mathscr{E} = \langle \sigma', A\rangle\Big|_a^b = \langle \sigma'(b), A(b)\rangle - \langle \sigma'(a), A(a)\rangle.$$

Clearly $\mathscr{E}$ measures the contribution to $L''(0)$ of the convexity/concavity of the first and last transverse curves of $\boldsymbol{x}$. For a fixed endpoint variation, $\mathscr{E} = 0$, so $L''(0)$ depends solely on the variation vector field V.

(2) Since $R_{vv} = 0$, the curvature integrand can be written as $\langle R(\overset{\perp}{V}, \sigma')\overset{\perp}{V}, \sigma'\rangle$, so the integral term in the formula depends only on $\overset{\perp}{V}$. This is to be expected, since when $\mathscr{E} = 0$, the tangential component $V^{\top}$ of V amounts essentially to a change of parametrization of σ, which has no effect on arc length.

(3) If $L'(0) = 0$ (as in the fixed endpoint case) and $L''(0) \neq 0$, then obviously the sign of $L''(0)$ tells whether nearby longitudinal curves of x are longer or shorter than the base geodesic σ.

THE INDEX FORM

To examine in more detail the lengths of curves with the same endpoints, fix the following notation: p and q points of M, $[0, b]$ an interval, and $\Omega(p, q)$ the set of all piecewise smooth curve segments $\alpha: [0, b] \to M$ from p to q. *Treating* $\Omega(p, q)$ *as a manifold* establishes a powerful analogy that guides subsequent developments.

If x is a fixed endpoint variation of $\alpha \in \Omega(p, q)$, then each longitudinal curve $u \to x(u, v)$ is a "point" α_v of $\Omega(p, q)$. As a 1-parameter family, $v \to \alpha_v$, the variation is thus a "curve" in $\Omega(p, q)$ starting at α. The initial velocity of a curve is its linear description at its starting point; thus the "initial velocity" of x is its variation vector field V. Since x is a fixed endpoint variation, V is zero at a and b.

The tangent vectors to a manifold at a point p are exactly the initial velocities of all curves starting at p. This suggests

5. Definition. If $\alpha \in \Omega = \Omega(p, q)$ the *tangent space* $T_\alpha(\Omega)$ *to* Ω *at* α consists of all piecewise smooth vector fields V on α such that $V(a) = 0$ and $V(b) = 0$.

As noted earlier, every $V \in T_\alpha(\Omega)$ is the vector field of some fixed endpoint variation of α.

Clearly $T_\alpha(\Omega)$ is a module over the ring of (piecewise smooth) functions on $[0, b]$.

The length function L is a real-valued function on $\Omega(p, q)$. For a function f on a manifold M, if $x \in T_p(M)$, then $x(f) = (d/dt)(f \circ \beta)|_{t=0}$, where β is any curve in M with initial velocity x. By analogy, if $V \in T_\alpha(\Omega)$, we can assign a meaning to $V(L)$. Corresponding to β is a variation x of α with V as its vector field. Consider x as usual as a "curve" $v \to \alpha_v$. Applying L to it gives the length function L_x of x. Thus $V(L)$ must be the first variation $L_x'(0)$.

In manifold theory, $p \in M$ is a *critical point* of $f \in \mathfrak{F}(M)$ provided $x(f) = 0$ for all $x \in T_p(M)$. Thus Corollary 3 is the assertion that the nonnull geodesics in $\Omega(p, q)$ are exactly the nonnull *critical points* of the length function L on $\Omega(p, q)$. (The null case will be dealt with later.)

At a critical point $p \in M$ of $f \in \mathfrak{F}(M)$, the coordinate first derivatives of f vanish and to study f near p we turn to second derivatives, expressed invariantly by the Hessian of f at p. Its properties, developed in Exercise 1.14, show how to define its analogue at a critical point of L.

6. Definition. The *index form* I_σ of a nonnull geodesic $\sigma \in \Omega(p, q)$ is the unique symmetric bilinear form

$$I_\sigma\colon T_\sigma(\Omega) \times T_\sigma(\Omega) \to \mathbf{R},$$

such that if $V \in T_\sigma(\Omega)$, then

$$I_\sigma(V, V) = L''_{\mathbf{x}}(0),$$

where $\mathbf{x}$ is any fixed endpoint variation of σ with variation vector field V.

Recall that in the fixed endpoint case the second variation $L''_{\mathbf{x}}(0)$ depends only on V, not the particular $\mathbf{x}$. The existence of I_σ follows from

7. Corollary. If $\sigma \in \Omega(p, q)$ is a geodesic of speed $c > 0$ and sign ε, then

$$I_\sigma(V, W) = \frac{\varepsilon}{c} \int_0^b \{\langle \overset{\perp}{V}{}', \overset{\perp}{W}{}' \rangle - \langle R_{V\sigma'} W, \sigma' \rangle\}\, du$$

for all $V, W \in T_\sigma(\Omega)$.

Proof. Evidently the formula is bilinear and (by a symmetry of R) symmetric in V and W. When $V = W$, Theorem 4 shows that it gives the required second variation. ∎

In this formula, *tangential vector fields are negligible*. Explicitly, if $V \in T_\sigma(\Omega)$ is tangent to σ, that is, $\overset{\perp}{V} = 0$, then

$$I_\sigma(V, W) = 0 \qquad \text{for all} \quad W \in T_\sigma(\Omega).$$

It follows immediately that

$$I_\sigma(V, W) = I_\sigma(\overset{\perp}{V}, \overset{\perp}{W}) \qquad \text{for all} \quad V, W \in T_\sigma(\Omega).$$

Thus there is no loss of information in restricting the index form I_σ to

$$T_\sigma^\perp(\Omega) = \{V \in T_\sigma(\Omega) : V \perp \sigma\}.$$

We write $I_\sigma^\perp$ for this restriction.

Integration by parts transformed the first variation formula of Lemma 1 to that of Proposition 2; similarly it produces a new version of the formula above.

8. Corollary. Let $\sigma \in \Omega(p, q)$ be a nonnull geodesic. If σ and $V \in T_\sigma(\Omega)$ have breaks $u_1 < \cdots < u_k$, then

$$I_\sigma(V, W) = -\frac{\varepsilon}{c}\int_0^b \langle \overset{\perp}{V}{}'' - R(\overset{\perp}{V}, \sigma')\sigma', \overset{\perp}{W}\rangle\, du - \frac{\varepsilon}{c}\sum_{i=1}^{k} \langle \Delta \overset{\perp}{V}{}', \overset{\perp}{W}\rangle(u_i).$$

Proof. In the preceding corollary, write

$$\langle \overset{\perp}{V}{}', \overset{\perp}{W}{}'\rangle = (d/du)\langle \overset{\perp}{V}{}', \overset{\perp}{W}\rangle - \langle \overset{\perp}{V}{}'', \overset{\perp}{W}\rangle, \qquad \textit{except at breaks}.$$

For the curvature term, substitute

$$\langle R_{V\sigma'} W, \sigma'\rangle = -\langle R_{V\sigma'}\sigma', W\rangle = -\langle R(\overset{\perp}{V}, \sigma')\sigma', \overset{\perp}{W}\rangle.$$

The contributions of $-\langle \overset{\perp}{V}{}'', \overset{\perp}{W}\rangle$ and curvature to the previous integral then constitute the integral in the statement of this corollary. Since the derivative of $\langle \overset{\perp}{V}{}', \overset{\perp}{W}{}'\rangle$ is undefined at breaks, its contribution to the integral is the sum of the terms

$$\int_{u_i}^{u_{i+1}} \frac{d}{du}\langle \overset{\perp}{V}{}', \overset{\perp}{W}\rangle\, du = \langle \overset{\perp}{V}{}', \overset{\perp}{W}\rangle\Big|_{u_i}^{u_{i+1}}.$$

Now W vanishes at 0 and b, and for each break u_i $(1 \le i \le k)$ we obtain

$$\langle \overset{\perp}{V}{}'(u_i^-), \overset{\perp}{W}(u_i)\rangle - \langle \overset{\perp}{V}{}'(u_i^+), \overset{\perp}{W}(u_i)\rangle = -\langle \Delta \overset{\perp}{V}{}', \overset{\perp}{W}\rangle(u_i). \qquad ■$$

This formula for I_σ shows that Jacobi fields are in the offing.

CONJUGATE POINTS

9. Definition. Points $\sigma(a)$ and $\sigma(b)$, $a \neq b$, on a geodesic σ are *conjugate along* σ provided there is a nonzero Jacobi field J on σ such that $J(a) = 0$ and $J(b) = 0$.

We can say that points p and q are conjugate along σ if there is no ambiguity as to the numbers a and b for which $p = \sigma(a)$ and $q = \sigma(b)$. Conjugacy of $p = \sigma(a)$ and $q = \sigma(b)$ along σ is independent of the parametrization of $\sigma|[a, b]$. Hence we often set $a = 0$. Examples will show that points may be conjugate along one geodesic joining them but not along another.

With notation as above, let $\mathscr{J}_{ab}$ be the set of all Jacobi fields on σ that vanish at a and b. Evidently $\mathscr{J}_{ab}$ is a subspace of the n-dimensional space consisting of those vanishing only at a (see Proposition 8.5). The dimension of $\mathscr{J}_{ab}$ is called the *order of conjugacy* of $\sigma(a)$ and $\sigma(b)$ along σ. Either of the following implies that such orders are at most $n - 1$: (1) The tangential Jacobi field $u \to (u - a)\sigma'(u)$ is zero at a but not at b. (2) Jacobi fields in

$\mathscr{J}_{ab}$ vanish twice, hence by Proposition 8.7(2) are everywhere perpendicular to σ.

The following result interprets a conjugate point $\sigma(b)$ of $p = \sigma(0)$ along a geodesic σ as an "almost-meeting point" of geodesics starting from p with initial velocities near $\sigma'(0)$. These neighboring geodesics may, but need not, actually pass through the point $\sigma(b)$.

10. Proposition. Let $\sigma: [0, b] \to M$ be a geodesic starting at p. The following are equivalent:

(1) $\sigma(b)$ is a conjugate point of $p = \sigma(0)$ along σ.

(2) There is a nontrivial variation $\boldsymbol{x}$ of σ through geodesics starting at p such that $\boldsymbol{x}_v(b, 0) = 0$. (Nontrivial means variation vector field not identically zero.)

(3) The exponential map exp: $T_p(M) \to M$ is singular at $b\sigma'(0)$; that is, there is a nonzero tangent vector x to $T_p(M)$ at $b\sigma'(0)$ such that $d\exp_p(x) = 0$.

Proof. (2) $\Rightarrow$ (1). Since each longitudinal curve of $\boldsymbol{x}$ is a geodesic starting at p, the variation vector field J of $\boldsymbol{x}$ is a nonzero Jacobi field vanishing at 0—and, by hypothesis, also at b.

The remainder of the proof is really a corollary of Proposition 8.6. By a reparametrization we can suppose $b = 1$.

(1) $\Rightarrow$ (3). Let J be a nonzero Jacobi field on σ that vanishes at 0 and $b = 1$. Let x be the tangent vector to $T_p(M)$ at $\sigma'(0)$ that canonically corresponds to $J'(0) \in T_p(M)$. Then by Proposition 8.6, $d\exp_p(x) = J(1) = 0$. Since J is nonzero but $J(0) = 0$, we have $J'(0) \neq 0$, hence $x \neq 0$.

(3) $\Rightarrow$ (2). For x as given, let the corresponding vector in $T_p(M)$ also be denoted by x. Then the proof of Proposition 8.6 shows that $\boldsymbol{x}(u, v) = \exp_p(u(\sigma'(0) + vx))$ defines a variation $\boldsymbol{x}$ as required. ∎

There are no conjugate points on sufficiently short geodesics; in fact, by assertion (3) in the proposition, if $\mathscr{N}$ is a normal neighborhood of p, then no point of $\mathscr{N}$ is conjugate to p along a radial geodesic in $\mathscr{N}$.

11. Examples of Conjugate Points. (1) In $\boldsymbol{R}^n_v$ there are no conjugate points, since the geodesics starting at any point are radial lines that obviously do not refocus.

(2) On the sphere $S^n(r)$ all geodesics starting at a point p meet again at the antipodal point $-p$ after arc length πr, so p and $-p$ are conjugate. Also, p is conjugate to itself, since the geodesics meet again at p after arc length $2\pi r$. Then $-p$ is conjugate to p again along geodesics of length $3\pi r$, and so on.

(3) If M has constant curvature C, then on a geodesic σ the Jacobi equation for $Y \perp \sigma$ is $Y'' + C\langle\sigma', \sigma'\rangle Y = 0$. The explicit solutions in Exercise 8.7 readily show the following: if σ is spacelike and $C > 0$, or if σ is timelike and $C < 0$, then points on σ are conjugate if and only if the arc length between them is a multiple of $\pi/\sqrt{|C|}$. (Evidently the conjugacy is of maximum order dim $M - 1$.) Otherwise there are no conjugate points in M. In particular, there are none in flat manifolds or along null geodesics in any constant curvature manifold.

For a nonnull geodesic σ, conjugacy of its endpoints can be read from its index form $I_\sigma^\perp$. In fact, the following shows that their order of conjugacy along σ is precisely the *nullity* of $I_\sigma^\perp$, that is, the (finite) dimension of its *nullspace* (see Exercise 2.12).

12. Corollary. For a nonnull geodesic $\sigma \in \Omega(p, q)$ the nullspace of $I_\sigma^\perp$ is the space $\mathscr{J}_{0b}$ of Jacobi fields on σ that vanish at $p = \sigma(0)$ and $q = \sigma(b)$.

Proof. That $\mathscr{J}_{0b}$ is contained in the nullspace of $I_\sigma^\perp$ is clear from Corollary 8.

The proof of the reverse inclusion is analogous to that of Corollary 3. If V is in the nullspace of $I_\sigma^\perp$, we must assume a priori that it has breaks $u_1 < \cdots < u_k$. First we show that each restriction $V|[u_i, u_{i-1}]$ is Jacobi. For a fixed t inside the interval, let y be an arbitrary tangent vector to M at $\sigma(t)$. Construct $W = fY$ as in the corresponding case for Corollary 3. Then, since $V \perp \sigma$,

$$0 = I_\sigma(V, W) = \int_{t-\delta}^{t+\delta} \langle V'' - R(V, \sigma')\sigma', f\overset{\perp}{Y}\rangle \, du.$$

It follows as before that $V'' - R(V, \sigma')\sigma'$ is zero at t, hence identically zero on $[u_i, u_{i+1}]$, and so V is Jacobi there.

The proof that V is unbroken again follows the same pattern as for Corollary 3. As an unbroken piecewise solution of the Jacobi equation, V is in fact a smooth solution, so $V \in \mathscr{J}_{0b}$. ∎

LOCAL MINIMA AND MAXIMA

For a nonnull geodesic $\sigma \in \Omega(p, q)$, suppose that σ is shorter than its neighboring curves τ in $\Omega(p, q)$—or merely that $L(\sigma) \leq L(\tau)$. If $V \in T_\sigma(\Omega)$, then for any fixed endpoint variation x attached to V, the length function L_x has a local minimum at 0, hence $I_\sigma(V, V) = L_x''(0) \geq 0$. Thus I_σ is positive semidefinite.

In a Riemannian manifold, this situation can certainly occur; for example, if σ is a radial geodesic in a normal ε-neighborhood of $\sigma(0)$. On the other hand, we cannot expect to find Riemannian geodesics with I_σ negative semidefinite, since small ripples in σ will produce nearby longer curves.

13. Lemma. Let σ be a nonnull geodesic of sign ε in a semi-Riemannian manifold M^n of index ν. (1) If I_σ is positive semidefinite, then $\nu = 0$ or n. (2) If I_σ is negative semidefinite, then *either* $\nu = 1$ and $\varepsilon = -1$ *or* $\nu = n - 1$ and $\varepsilon = 1$.

Proof. (1) Assume $0 < \nu < n$. Then there is a unit tangent vector $y \perp \sigma'(0)$ with causal character opposite that of σ. Thus if ε is the sign of σ, $\varepsilon\langle y, y\rangle = -1$. Let Y be gotten by parallel translation of y, and let $\delta > 0$ be such that $\sin(u/\delta)$ is zero at the endpoints of the domain $[a, b]$ of σ. Then $V = \delta \sin(u/\delta) Y \in T_\sigma(\Omega)$. In the second variation formula, take $|\sigma'| = c = 1$ for simplicity, and let $K = K(V, \sigma')$. Then

$$I_\sigma(V, V) = \varepsilon \int_a^b \{\langle V', V'\rangle - K\langle V, V\rangle\varepsilon\}\, du$$

$$= \varepsilon \int_a^b \{\langle y, y\rangle \cos^2(u/\delta) + K\delta^2 \sin^2(u/\delta)\}\, du$$

$$= \int_a^b \{-\cos^2(u/\delta) + \varepsilon K\delta^2 \sin(u/\delta)\}\, du.$$

But K is bounded on $[a, b]$; hence for $\delta > 0$ sufficiently small, $I_\sigma(V, V) < 0$.

(2) If the conclusion is false, there is a vector $y \perp \sigma'(0)$ such that $\varepsilon\langle y, y\rangle = +1$. Then a proof as for (1) shows that I_σ cannot be negative semidefinite. ■

Reversing the metric of M has no effect on the lengths of curves; thus in considering definiteness (or semidefiniteness) of the index form we need only consider

(1) arbitrary geodesics in a Riemannian manifold (where, if I_σ is definite, it can only be *positive* definite);

(2) timelike geodesics in a Lorentz manifold (where, if I_σ is definite, it can only be *negative* definite).

These two cases can be unified as follows.

14. Definition. A geodesic σ in M is *cospacelike* provided the subspace $\sigma'(s)^\perp$ of $T_{\sigma(s)}M$ is spacelike for one (hence every) s.

Then σ is necessarily nonnull, and M is Riemannian or Lorentz depending on the sign of σ.

Our goal now is to relate definiteness of the index form to the existence of conjugate points.

15. Lemma. If V and W are Jacobi fields on a geodesic σ, then $\langle V', W\rangle - \langle V, W'\rangle$ is constant.

Proof. $\langle V', W\rangle' = \langle V'', W\rangle + \langle V', W'\rangle = -\langle R_{V\sigma'}W, \sigma'\rangle + \langle V', W'\rangle$. By a symmetry of curvature this is the same as $\langle V, W'\rangle'$. ∎

16. Lemma. On a geodesic σ let $Y_1, \ldots, Y_k$ be Jacobi fields such that $\langle Y_i', Y_j\rangle = \langle Y_i, Y_j'\rangle$ for all i, j. If $V = \sum f_i Y_i$, then

$$\langle V', V'\rangle - \langle R_{V\sigma'}V, \sigma'\rangle = \langle A, A\rangle + \langle V, B\rangle',$$

where $A = \sum f_i' Y_i$ and $B = \sum f_i Y_i'$.

Proof. Since $V' = A + B$,

$$\begin{aligned}\langle V, B\rangle' = \langle V', B\rangle + \langle V, B'\rangle &= \langle A, B\rangle + \langle B, B\rangle \\ &+ \langle V, \textstyle\sum f_i' Y_i'\rangle + \langle V, \textstyle\sum f_i Y_i''\rangle.\end{aligned}$$

The Jacobi equation $Y_i'' = R_{Y_i\sigma'}\sigma'$ converts the last summand to $-\langle R_{V\sigma'}V, \sigma'\rangle$. Using the hypothesis on the Y_is gives

$$\langle V, \textstyle\sum f_i' Y_i'\rangle = \sum f_j f_i' \langle Y_j, Y_i'\rangle = \sum f_j f_i' \langle Y_j', Y_i\rangle = \langle A, B\rangle.$$

Thus

$$\langle V, B\rangle' = 2\langle A, B\rangle + \langle B, B\rangle - \langle R_{V\sigma'}V, \sigma'\rangle.$$

Since $\langle V', V'\rangle = \langle A + B, A + B\rangle$, the result follows. ∎

17. Theorem. Let $\sigma \in \Omega(p, q)$ be a cospacelike geodesic of sign ε.

(1) If there are no conjugate points of $p = \sigma(0)$ along σ, then the index form $I_\sigma^\perp$ is definite (positive if $\varepsilon = 1$, negative if $\varepsilon = -1$).

(2) If $q = \sigma(b)$ is the only conjugate point of p along σ, then I_σ is semidefinite but not definite.

(3) If there is a conjugate point $\sigma(r)$ of p along σ with $0 < r < b$, then I_σ is not semidefinite.

Proof. (1) We must show that εI_σ is positive definite on $T_\sigma^\perp(\Omega)$. Let $Y_1, \ldots, Y_{n-1}$ be Jacobi fields on σ that vanish at $u = 0$ and have $Y_1'(0), \ldots, Y_{n-1}'(0)$ a basis for $\sigma'(0)^\perp$. These Jacobi fields are then perpendicular to σ by Proposition 8.7. Furthermore, since there are no conjugate points of $\sigma(0)$ along σ, it follows that for each $0 < u \leq b$ the vectors $Y_1, \ldots, Y_{n-1}(u)$ form a basis for $\sigma'(u)^\perp$. Thus if $V \in T_\sigma(\Omega)$, there are (unique) piecewise smooth functions f_i such that $V = \sum f_i Y_i$ on $(0, b]$. Using Exercise 1.17, it is not hard to show that the functions f_i have continuous extensions to $[0, b]$. Since the Y_is vanish at 0, Lemma 15 gives $\langle Y_i', Y_j \rangle = \langle Y_i, Y_j' \rangle$. Thus we can apply Lemma 16 to get

$$\langle V', V' \rangle - \langle R_{V\sigma'} V, \sigma' \rangle = \langle A, A \rangle + \langle V, B \rangle',$$

where $A = \sum f_i' Y_i$ and $B = \sum f_i Y_i'$. Hence

$$\varepsilon I_\sigma(V, V) = \frac{1}{c} \int_0^b \langle A, A \rangle \, du + \frac{1}{c} \langle V, B \rangle \Big|_0^b .$$

The second summand is zero since V vanishes at 0 and b. Since σ is cospacelike and $A \perp \sigma$, we have $\langle A, A \rangle \geq 0$. Hence $\varepsilon I_\sigma(V, V) \geq 0$. Furthermore,

$$I_\sigma(V, V) = 0 \Rightarrow \int_0^b \langle A, A \rangle \, du = 0 \Rightarrow \langle A, A \rangle = 0$$
$$\Rightarrow A = 0 \Rightarrow \text{each } f_i \text{ is constant} \Rightarrow V = 0.$$

(2) Corollary 12 says that $I_\sigma^\perp$ has a nullspace, hence is not definite. A continuity argument will show that I_σ is semidefinite. Briefly, if $V \in T_\sigma(\Omega)$, write $V(u) = (b - u)Z(u)$, and for $b_i = b - (1/i)$, define $V_i(u)$ to be $(b_i - u)Z(u)$ on $[0, b_i]$ and zero thereafter. Then $\{I_\sigma(V_i, V_i)\}$ converges to $I_\sigma(V, V)$. But (1) applies to $\sigma | [0, b_i]$ to show $\varepsilon I_\sigma(V_i, V_i) \geq 0$ for all i.

(3) By hypothesis there is a nonzero Jacobi field J on $\sigma | [0, r]$ that vanishes at 0 and r. Extend J to a vector field Y on σ by defining $Y = 0$ on $[r, b]$. Then $Y'(r^-) = J'(r) \neq 0$ since J is nonzero. But $Y'(r^+) = 0$, so $(\Delta Y')(r) \neq 0$. Choose any $W \in T_\sigma(\Omega)$ such that $W(r) = (\Delta Y')(r)$.

If $\varepsilon = \pm 1$ is the sign of α, it suffices to find a $\delta > 0$ such that $\varepsilon I_\sigma(Y + \delta W, Y + \delta W) < 0$. (By Lemma 13, there is always some $Z \in T_\sigma(\Omega)$ with $\varepsilon I_\sigma(Z, Z) > 0$.) Now

$$\varepsilon I_\sigma(Y + \delta W, Y + \delta W) = \varepsilon\{I_\sigma(Y, Y) + 2\delta I_\sigma(Y, W) + \delta^2 I_\sigma(W, W)\}.$$

By Corollary 8, $I_\sigma(Y, Y) = 0$ since Y is piecewise Jacobi and zero at its only break r. But $\varepsilon I_\sigma(Y, W)$ reduces to

$$-(1/c)\langle \Delta Y', W \rangle(r) = -(1/c)|(\Delta Y'(r)|^2 < 0$$

since σ is cospacelike. Thus the result follows for $\delta > 0$ sufficiently small.

∎

The proof of (3) is a rigorous version of the following heuristic argument. A conjugate point is an almost-meeting point; let us suppose that the meeting actually takes place, so there is another geodesic segment τ near $\sigma|[0,r]$ with the same endpoints and length. (This occurs on spheres, for example.) Then the competitive curve $\tau + \sigma|[r,b]$ has length $L(\sigma)$, but is broken at $\sigma(r)$. In the Riemannian case, rounding off this corner gives a strictly shorter curve $\tilde{\sigma}$. Thus σ is not a local minimum of arc length, and if $V \in T_\sigma(\Omega)$ points from σ to $\tilde{\sigma}$, we expect $I_\sigma(V, V) < 0$. (Inequalities reverse in the Lorentz case.)

The three implications in Theorem 17 involve sets of all-inclusive, mutually exclusive conditions, hence all three converses hold. Thus for example, if $I_\sigma^\perp$ is indefinite there is a conjugate point of $\sigma(0)$ on σ.

18. Remark. In terms of the compact-open topology on $\Omega(p,q)$, the length function L on $\Omega(p,q)$ has a *local minimum* at σ if there is a neighborhood $\mathcal{N}$ of σ in $\Omega(p,q)$ such that $L(\tau) \geqslant L(\sigma)$ for all $\tau \in \mathcal{N}$. A local minimum is *strict* if also $L(\tau) = L(\sigma)$ implies that τ is a reparametrization of σ.

(1) The first assertion in Theorem 17 can be strengthened to: If there are no conjugate points of $\sigma(0)$ along σ, the length function L on $\Omega(p,q)$ has

a strict local minimum if $\varepsilon = 1$ (M Riemannian),

a strict local maximum if $\varepsilon = -1$ (σ timelike in M Lorentz).

The proof uses Proposition 10 and the refinement (2) following Lemma 5.14.

(2) If $\sigma(b)$ is the only conjugate point of $\sigma(0)$ along σ, then no conclusion can be drawn as to whether σ is a local minimum or maximum. For a Riemannian example, let σ be a semicircle of longitude joining the poles p and q of the ordinary sphere S^2, so the only conjugate point of p along σ is q. Being a global minimum for arc length from p to q, σ is certainly a local minimum (but not a strict one). Now change the metric on S^2 symmetrically around σ so that tangent vectors pointing due north or south become slightly shorter—to the fourth order in distance from σ. Then σ is still a geodesic in the new surface Σ and the Gaussian curvature along σ is unchanged, so q is still the only conjugate point of p along σ. But in Σ nearby semicircles of longitude are shorter than σ, so σ is not a local minimum.

(3) If there is a conjugate point $\sigma(r)$ of p along σ with $0<r<b$, then σ is not a local minimum or maximum (This follows from (3) in Theorem 17.)

These results are easiest to picture in the Riemannian case, where we can think of M as a surface in space and a geodesic segment σ as an elastic string under tension and constrained to lie in M with its endpoints fixed. That σ is a geodesic means it is in equilibrium (Corollary 3). If σ is short enough—more precisely, if there are no conjugate points along it—then σ is in *stable* equilibrium: any small displacement makes it longer, thus tending to restore it to its original position. But if σ is long enough to contain a conjugate point, then the equilibrium is unstable: there are arbitrarily nearby positions where the string is shorter, and hence it tends to move farther away from its original position.

SOME GLOBAL CONSEQUENCES

The curvature of a manifold can be linked to its global structure (as a topological space or smooth manifold) by means of conjugate points. Curvature determines conjugate points via the Jacobi equation; conjugate points influence global structure, for example, through exponential maps as in Proposition 10. We consider two extreme cases: one with no conjugate points, the other with many.

Consider the Jacobi equation in terms of tidal forces (8.8) as $Y'' = F_{\sigma'}(Y)$. Then the geometric meaning of conjugate points makes it clear that tidal forces that *attract* tend to cause conjugate points, while those that *repel* tend to prevent them. If the geodesic σ is cospacelike, then for any unit vector $y \perp \sigma$, the y-component of $F_{\sigma'}(y) = R_{y\sigma'}\sigma'$ is $-\langle R_{y\sigma'}y, \sigma'\rangle y$. Thus σ's neighbors in the y direction are attracted if $\langle R_{y\sigma'}y, \sigma'\rangle > 0$, repelled if $\langle R_{y\sigma'}y, \sigma'\rangle < 0$.

19. Lemma. If $\langle R_{v\sigma'}v, \sigma'\rangle \leq 0$ for every vector v perpendicular to the cospacelike geodesic σ, then there are no conjugate points along σ.

Proof. Let $J \neq 0$ be a Jacobi field along σ with $J \perp \sigma$ and $J(0) = 0$. Let $h(s) = \langle J(s), J(s)\rangle$. Then $h' = 2\langle J', J\rangle$, and using the Jacobi equation,

$$\tfrac{1}{2}h'' = \langle J', J'\rangle + \langle J'', J\rangle = \langle J', J'\rangle - \langle R_{J\sigma'}J, \sigma'\rangle.$$

Thus the hypothesis implies $h'' \geq 0$. Evidently $h(0) = h'(0) = 0$. Since $J \neq 0$ we have $J'(0) \neq 0$; hence h is positive near, but not at, 0. Thus $h(t) > 0$ for all $t \neq 0$. ∎

20. Corollary. (1) A Riemannian manifold with sectional curvature $K \leq 0$ has no conjugate points. (2) A Lorentz manifold with $K \geq 0$ on all timelike tangent planes has no conjugate points along any timelike geodesic.

This result can be tested on hyperbolic space $H^n(r)$ and the Lorentz sphere $S_1^n(r)$.

21. Lemma. If M is a complete connected Riemannian manifold with $K \leq 0$, then for each point $p \in M$ the exponential map $\exp_p\colon T_p(M) \to M$ is a covering map.

Proof. By Proposition 10, the absence of conjugate points means that $\exp_p$ is a local diffeomorphism. Assign $T_p(M)$ the induced metric tensor $\exp_p^*(g)$, making $\exp_p$ a local isometry. Since $\exp_p$ carries each ray $t \to tv$ to the geodesic γ_v, it follows that these rays are geodesics. The Riemannian manifold $T_p(M)$ is thus complete at 0; hence by the Hopf–Rinow theorem (5.21) it is complete. The result then follows from Corollary 7.29. ■

Thus Corollary 7.27 gives

22. Theorem (Hadamard). Let H be a complete, simply connected Riemannian manifold with sectional curvature $K \leq 0$. Then for each $p \in H$ the exponential map $\exp_p$: $T_p(H) \to H$ is a diffeomorphism. In particular,

(1) H is diffeomorphic to $\mathbf{R}^n$.
(2) For $p, q \in H$, there is a unique geodesic $\gamma: \mathbf{R} \to M$ such that $\gamma(0) = p$ and $\gamma(1) = q$.

Such manifolds—hyperbolic space, for example—thus have the same manifold structure as Euclidean space and share the best known Euclidean geometric property: "two points determine a line."

To show that conjugate points exist we need only require that *average* tidal forces be sufficiently attractive. In view of Lemma 8.9 this average is given by Ricci curvature; for σ cospacelike,

$$\frac{1}{n-1}\operatorname{Ric}(\sigma', \sigma') = -\frac{1}{n-1}\operatorname{trace} F_{\sigma'} = \frac{1}{n-1}\sum_i \langle R_{e_i\sigma'}e_i, \sigma'\rangle,$$

where $e_2, \ldots, e_n$ is an orthonormal basis for $\sigma'^{\perp}$.

23. Lemma. Let σ be a unit speed cospacelike geodesic in M^n along which $\operatorname{Ric}(\sigma', \sigma') \geq (n-1)C > 0$. If σ has length $L(\sigma) \geq \pi/\sqrt{C}$, there are conjugate points along σ.

Proof. Suppose $\sigma: [0, b] \to M$ with $|\sigma'| = 1$ and $b = L(\sigma) = \pi/\sqrt{C}$. If ε is the sign of σ, then by Theorem 17(1) it suffices to show that $\varepsilon I_\sigma^\perp$ is not positive definite. Thus we want a vector field $0 \neq V \in T_\sigma(\Omega)$ with $\varepsilon I_\sigma(V, V) \leq 0$.

Let $\sigma', E_2, \ldots, E_n$ be a parallel frame field on σ. If f is a smooth function on $[0, b]$ vanishing at the endpoints, then $fE_j \in T_\sigma^\perp(\Omega)$ for $2 \leq j \leq n$. By Corollary 7,

$$\varepsilon I_\sigma(fE_j, fE_j) = \int_0^b [f'^2 - f^2\langle R_{E_j\sigma'}E_j, \sigma'\rangle]\, du.$$

Thus

$$\sum_{j=2}^{n} I_\sigma(fE_j, fE_j) = \int_0^b [(n-1)(f')^2 - \mathrm{Ric}(\sigma', \sigma')f^2]\,du$$
$$\le (n-1)\int_0^b [(f')^2 - Cf^2]\,du.$$

Since $b = \pi/\sqrt{C}$, we are led to choose $f(t) = \sin(\sqrt{C}t)$, for which the preceding integral becomes

$$\int_0^{\pi/\sqrt{C}} C(\cos^2\sqrt{C}u - \sin^2\sqrt{C}u)\,du = 0.$$

Hence $I_\sigma(fE_j, fE_j) \le 0$ for some $2 \le j \le n$. ∎

This lemma has a powerful consequence in the Riemannian case.

24. Theorem (Myers). If M is a complete connected Riemannian manifold with $\mathrm{Ric} \ge (n-1)C > 0$, then

(1) M is compact and has diameter $\le \pi/\sqrt{C}$,
(2) the fundamental group $\pi_1(M)$ is finite.

Proof. (1) As for any metric space, the *diameter* of M is

$$\sup\{d(p,q) : p, q \in M\},$$

where d is the Riemannian metric. If p, $q \in M$, then by Proposition 5.22 there is a shortest geodesic σ from p to q. In particular, σ locally minimizes arclength, so by Theorem 17, q is the only possible conjugate point of p along σ. Thus, by the lemma, $L(\sigma) > \pi/\sqrt{C}$ is impossible, for it forces σ to have an even earlier conjugate point of p. Consequently, $d(p,q) = L(\sigma) \le \pi/\sqrt{C}$. Hence by the Hopf–Rinow theorem, M is compact.

(2) Let $k: \tilde{M} \to M$ be the simply connected Riemannian covering of M. Since $\tilde{M}$ shares the hypothesized properties of M, it is compact. For $m \in M$ the counterimage $k^{-1}(m)$ has no cluster points, since k is locally one-to-one; hence $k^{-1}(m)$ is finite. But Corollary A.10 implies that there is a one-to-one correspondence between $k^{-1}(m)$ and $\pi_1(M)$. ∎

The bound on the diameter of M cannot be reduced, as is shown by $S^n(1/\sqrt{C})$, whose (intrinsic) diameter is exactly $\pi/\sqrt{C}$. The proof of Lemma 23 amounts to a comparison of M with this sphere. Also it is not enough merely to assume $\mathrm{Ric} > 0$ or even $K > 0$, since, for example, a paraboloid of revolution has positive curvature but is not compact.

As an application of the theorems of Hadamard and Myers, consider the product manifold $M = S^2 \times S^1$. The Riemannian product metric has curvature $K \geq 0$, but there can be no metric for which $K \leq 0$, since the simply connected covering manifold of M is $S^2 \times \boldsymbol{R}^1 \neq \boldsymbol{R}^3$. Furthermore, there can be no metric with $K > 0$. In fact, if $K > 0$, then (since M is compact) $K \geq c > 0$, so Myers' theorem applies, showing that $\pi_1(M)$ is finite—a contradiction, since $\pi_1(M) \approx \pi_1(S^1) \approx \boldsymbol{Z}$. It follows in particular that for *every* Riemannian metric on $S^2 \times S^1$ there is a tangent plane Π with $K(\Pi) = 0$. (Whether $S^2 \times S^2$ has this property is not known.)

For Riemannian applications of geodesic variational methods, consult [CE], [Mi], [GKM]; and for some Lorentz analogues and extensive references, see [BE].

THE ENDMANIFOLD CASE

The study of the length of curves joining two points of M can be generalized by replacing one or both of the points by submanifolds. We shall replace only the initial point by such an *endmanifold*.

Fix the notation: P a semi-Riemannian submanifold of M, q any point of M, $[0, b]$ an interval. Then let $\Omega(P, q)$ be the set of all piecewise smooth curves $\alpha: [0, b] \to M$ that run from P to q. Extending the previous analogy, a "curve in $\Omega(P, q)$ starting at α" is a piecewise smooth variation $\boldsymbol{x}$ of α whose longitudinal curves are all in $\Omega(P, q)$. Thus the first transverse curve of $\boldsymbol{x}$ is in P, while the last is constant at q. We call $\boldsymbol{x}$ a (P, q)-*variation* of α. As before, the vector fields of such variations provide the natural notion of tangent vector to $\Omega(P, q)$.

25. Definition. The *tangent space* $T_\alpha(\Omega)$ to $\Omega(P, q)$ at α consists of all piecewise smooth vector fields V on α such that $V(0) \in T_{\alpha(0)}P$ and $V(b) = 0$.

Lemma 49 will show, in particular, that every $V \in T_\alpha(P, q)$ is the variation vector field of some (P, q)-variation of α.

Corollary 3 now generalizes to assert that the nonnull critical points of the length function L on $\Omega(P, q)$ are the geodesics $\sigma \in \Omega(P, q)$ that are *normal to* P: $\sigma'(0) \perp P$.

26. Corollary. Let $\alpha \in \Omega(P, q)$ have $|\alpha'| > 0$. Then $L_{\boldsymbol{x}}'(0) = 0$ for every (P, q)-variation $\boldsymbol{x}$ of α if and only if α is a geodesic normal to P.

Proof. By Proposition 2, $L_{\boldsymbol{x}}'(0)$ is zero for any (P, q)-variation of a normal geodesic. To prove the converse: An argument as for Corollary 3 shows that α is an (unbroken) geodesic. Then for any vector $y \in T_{\alpha(0)}P$, choose $V \in T_\alpha(\Omega)$ so that $V(0) = y$. Let $\boldsymbol{x}$ be a (P, q)-variation of α whose vector field is V. Since $V(b) = 0$,

$$0 = L'_x(0) = (\varepsilon/c)\langle \alpha', V\rangle|_0^b = (\varepsilon/c)\langle \alpha'(0), y\rangle.$$

Hence $\alpha'(0) \perp P$. ∎

For a normal geodesic $\sigma \in \Omega(P, q)$, the second variation of arc length will involve the shape tensor II of the endmanifold P. In fact, for a (P, q)-variation x of σ, the expression $\langle \sigma', A\rangle|_0^b$ in Theorem 4 (with $a = 0$) reduces to $-\langle \sigma'(0), A(0)\rangle$ since the last transverse curve of x is constant. If α is the first transverse curve, then by definition, $A(0) = \alpha''(0)$. Since $\sigma'(0) \perp P$,

$$\langle \sigma'(0), A(0)\rangle = \langle \sigma'(0), \operatorname{nor} \alpha''(0)\rangle = \langle \sigma'(0), II(\alpha'(0), \alpha'(0))\rangle.$$

But $\alpha'(0)$ is just $V(0)$, where V is the variation vector field of x. Thus we conclude that

$$L''_x(0) = \frac{\varepsilon}{c}\int_0^b \{\langle \overset{\perp}{V}{}', \overset{\perp}{V}{}'\rangle - \langle R_{V\sigma'}V, \sigma'\rangle\}\, du - \frac{\varepsilon}{c}\langle \sigma'(0), II(V(0), V(0))\rangle.$$

As before, the index form I_σ of a nonnull normal geodesic $\sigma \in \Omega(P, q)$ is defined to be the unique symmetric $\boldsymbol{R}$-bilinear form on $T_\sigma(\Omega)$ such that $I_\sigma(V, V) = L''_x(0)$ for any P-variation x with vector field $V \in T_\sigma(\Omega)$. Thus the symmetry of the shape tensor II gives the following generalization of Corollary 7.

27. Corollary. If $\sigma \in \Omega(P, q)$ is a normal geodesic of speed $c > 0$ and sign ε, then, for $V, W \in T_\sigma(\Omega)$,

$$I_\sigma(V, W) = \frac{\varepsilon}{c}\int_0^b \{\langle \overset{\perp}{V}{}', \overset{\perp}{W}{}'\rangle - R_{V\sigma'}W, \sigma'\rangle\}\, du - \frac{\varepsilon}{c}\langle \sigma'(0), II(V(0), W(0))\rangle.$$

Again $I_\sigma(V, W) = I_\sigma(\overset{\perp}{V}, \overset{\perp}{W})$, so we need consider I_σ only on $T_\sigma^\perp(\Omega) = \{V \in T_\sigma(\Omega) : V \perp \sigma\}$.

FOCAL POINTS

If σ is a geodesic starting at p, then the infinitesimal model of a variation of σ through geodesics starting at p is a Jacobi field on σ that is zero at p. Now we replace the point p by a semi-Riemannian submanifold P.

28. Proposition. A Jacobi field V on a geodesic σ normal to P is the variation vector field of a variation x of σ through normal geodesics if and only if

$$V(0) \text{ is tangent to } P \qquad \text{and} \qquad \tan V'(0) = \tilde{II}(V(0), \sigma'(0)).$$

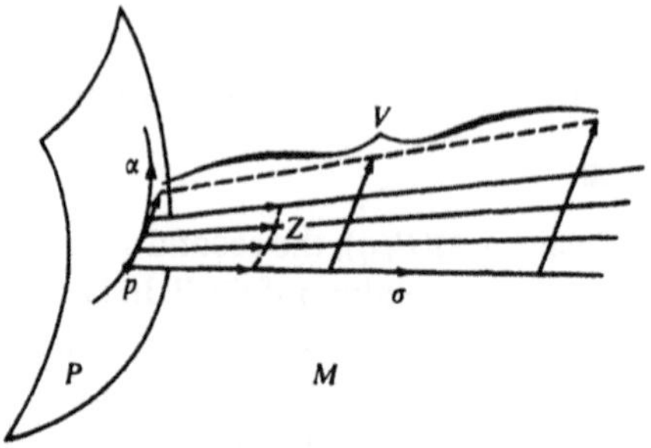

Figure 1

for $\tilde{II}$ as in Remarks 4.39. (A Jacobi field satisfying these conditions is called a *P-Jacobi field* on σ.)

Proof. If V is the variation vector field of such an $\boldsymbol{x}$, then the first transverse curve α of $\boldsymbol{x}$ lies in P, and the vector field $Z(v) = \boldsymbol{x}_u(0, v)$ on α is normal to P. (See Figure 1.) Thus $V(0) = \alpha'(0)$, which is tangent to P, and

$$V'(0) = \boldsymbol{x}_{vu}(0, 0) = \boldsymbol{x}_{uv}(0, 0) = Z'(0).$$

But tan $Z' = \tilde{II}(\alpha', Z)$ and, since $Z(0) = \sigma'(0)$, the second condition follows.

Conversely, suppose V satisfies the conditions and let α be a curve in P with $\alpha'(0) = V(0)$. First we show:

(∗) *There is a normal vector field Z on α such that $Z(0) = \sigma'(0)$ and* $Z'(0) = V'(0)$.

Let A and B be the (normal) vector fields on α gotten by normal parallel translation of $\sigma'(0)$ and nor $V'(0)$ along α. If $Z(v) = A(v) + vB(v)$ for all v, then $Z(0) = \sigma'(0)$. Furthermore, since $\alpha'(0) = V(0)$,

$$Z'(0) = A'(0) + B(0) = \tilde{II}(V(0), \sigma'(0)) + \text{nor } V'(0).$$

By the second hypothesis on V this is just $V'(0)$.

For Z as in (∗), we now define the required variation $\boldsymbol{x}$. Let exp be the exponential map of the normal bundle of P. Then $\boldsymbol{x}(u, v) = \exp(uZ(v))$ defines a variation of σ. The longitudinal curves of $\boldsymbol{x}$ are geodesics with initial velocity $Z(v)$, hence they are normal to P.

If Y is the variation vector field of $\boldsymbol{x}$, then $Y(0) = \alpha'(0) = V(0)$. By construction, $\boldsymbol{x}_u(0, v) = Z(v)$; hence

$$Y'(0) = \boldsymbol{x}_{vu}(0, 0) = \boldsymbol{x}_{uv}(0, 0) = Z'(0) = V'(0).$$

Thus Lemma 8.5 gives $Y = V$. ∎

This generalizes the endpoint case, for when the endmanifold P is a single point p, a P-normal geodesic is a geodesic starting at p, and a P-Jacobi field on σ is a Jacobi field zero at p. Thus it is clear how to generalize the notion of conjugate point.

29. Definition. Let σ be a geodesic of M that is normal to $P \subset M$, that is, $\sigma(0) \in P$, $\sigma'(0) \perp P$. Then $\sigma(r)$, $r \neq 0$, is a *focal point* of P along σ provided there is a nonzero P-Jacobi field J on σ with $J(r) = 0$.

The *focal order* of $\sigma(r)$ is the dimension of the space of P-Jacobi fields on σ that vanish at r. This order is at most $n - 1$, where $n = \dim M$. In fact, since $u \to u\sigma'(u)$ is a P-Jacobi field that vanishes only at $u = 0$, it suffices to show that *the space $\mathscr{J}$ of all P-Jacobi fields on σ has dimension n.* If $x \in T_{\sigma(0)}(P)$ and $z \in T_{\sigma(0)}(P)^{\perp}$, let V be the unique Jacobi field on σ such that

$$V(0) = x, \qquad V'(0) = \tilde{II}(x, \sigma'(0)) + z.$$

Then V is a P-Jacobi field, and the map $x + z \to V$ is a linear isomorphism from $T_{\sigma(0)}M$ to $\mathscr{J}$.

By Lemma 8.7, a *P-Jacobi field vanishing at $r \neq 0$ is everywhere perpendicular to σ* since it is perpendicular to σ at both 0 and r.

The geometrical meaning of focal points is the obvious extension of that of conjugate points; namely, $\sigma(b)$ is a focal point of P along a normal geodesic σ if there is a family of normal geodesics with initial velocities near $\sigma'(0)$ that almost meet at $\sigma(b)$. Formally

30. Proposition. Let $\sigma\colon [0, b] \to M$ be a geodesic normal to P. Then the following are equivalent:

(1) $\sigma(b)$ is a focal point of P along σ.

(2) There is a nontrivial variation $\boldsymbol{x}$ of σ through P-normal geodesics for which $\boldsymbol{x}_v(b, 0) = 0$.

(3) The normal exponential map $\exp\colon NP \to M$ is singular at $b\sigma'(0)$.

Proof. The equivalence of (1) and (2) follows immediately from Proposition 28. By a reparametrization we can suppose $b = 1$.

(3) ⇒ (2). Let x be a nonzero tangent vector to NP at $\sigma'(0)$ for which $d\exp(x) = 0$, and let $p = \sigma(0) \in P$.

Suppose first that x is tangent to the fiber $T_p(P)^{\perp}$ of NP. Since this fiber is a subspace of $T_p(M)$, Proposition 10 applies, showing that $\sigma(1)$ is a conjugate point of p along σ, and hence a focal point of P along σ. If x is not tangent to the fiber, then $d\pi(x) \neq 0$, where π is the projection of NP onto P. Let Z be any curve in NP with initial velocity x. Then $\boldsymbol{x}(u, v) = \exp(uZ(v))$ is the required variation, being nontrivial since $\boldsymbol{x}_v(0, 0) = d\pi(x) \neq 0$.

(2) ⇒ (3). Let $\tilde{x}(u, v) = u x_u(0, v)$. Then $\exp(\tilde{x}(u, v)) = x(u, v)$, hence $d\exp(\tilde{x}_v(1, 0)) = x_v(1, 0) = 0$. $\tilde{x}_v(1, 0)$ is the initial velocity of the curve $v \to x_u(0, v)$ in NP, which projects to $v \to x(0, v)$ in P. If $x_v(0, 0) \neq 0$, then $x_v(1, 0) \neq 0$, hence exp is singular at $\sigma'(0)$. But if $x_v(1, 0) = 0$, then $\sigma(1)$ is a conjugate point of $p = \sigma(0)$, and reversing the argument above, Proposition 10 implies again that exp is singular at $\sigma'(0)$. ∎

31. Example. Let P be the hyperboloid $x^2 + y^2 = 1 + z^2$ in $\boldsymbol{R}^3$ (Figure 2). Consider the x axis as a normal geodesic starting at $p = (1, 0, 0)$. The origin 0 and $q = (2, 0, 0)$ are focal points of P along this geodesic. In fact the normal geodesics along the circle $z = 0$ actually meet at 0, while the normals near p along the hyperbola $y = 0$ almost meet at q.

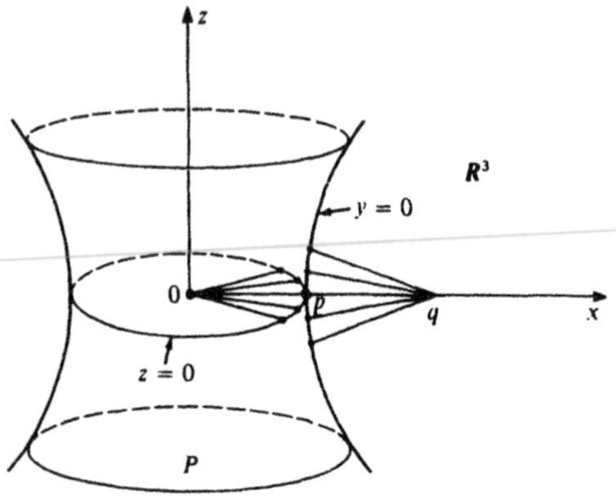

Figure 2

For a submanifold P of semi-Euclidean space, the fact that normal lines near a given normal line σ focus at a point of σ depends only on the shape of P at the foot $\sigma(0)$ of σ. In fact,

32. Proposition. Let σ be a P-normal geodesic in a flat manifold M. The focal points of P along σ are $\sigma(1/k_i)$, where $k_1, \ldots, k_r$ are the distinct nonzero eigenvalues of the shape operator $S_{\sigma'(0)}$ of P.

Proof. Abbreviate $S_{\sigma'(0)}$ to S, and recall from Remarks 4.39 that $S(v) = -\tilde{II}(v, \sigma'(0))$ for all $v \in T_{\sigma(0)}(P)$. First suppose that $S(e) = ke$, where both k and e are nonzero. Let E be the vector field on σ obtained by parallel translation of e. It will suffice to show that $V(u) = (1 - ku)E$ is a P-Jacobi field on σ, since $V(1/k) = 0$.

Evidently V satisfies the flat Jacobi equation $V'' = 0$. By construction, $V(0)$ is tangent to P. Finally, since $V' = -kE$, we have

$$\tan V'(0) = V'(0) = -ke = -S(e) = \tilde{II}(e, \sigma'(0)) = \tilde{II}(V(0), \sigma'(0)).$$

Conversely, let J be a P-Jacobi field on σ for which $J(b) = 0$, $b \neq 0$. Then

$$\tan J'(0) = \tilde{II}(J(0), \sigma'(0)) = -S(J(0)).$$

Since $J'' = 0$, we can write $J(u) = A(u) + uB(u)$, where A and B are gotten by parallel translation of $J(0)$ and $J'(0)$, respectively. Since $J(b) = 0$, it follows that $J(0) + bJ'(0) = 0$. Because J is a P-Jacobi field, $J(0)$ is tangent to P, and hence $J'(0) = -(1/b)J(0)$ is also. Furthermore,

$$S(J(0)) = -\tan J'(0) = -J'(0) = (1/b)J(0). \quad \blacksquare$$

It is easy to check that the focal order of $\sigma(1/k_i)$ above is the multiplicity of the eigenvalue k_i.

In the preceding example, if P is kept the same but $\mathbf{R}^3$ is changed to Minkowski space $\mathbf{R}^3_1$, then the normals to P change—all now go through the origin, which is the only focal point. (This is the idea of the proof of Proposition 4.36.)

A theorem of Sard asserts that, if $\Sigma \subset N$ is the set of singular points of a smooth map $\phi: N \to M$, then $\phi(\Sigma)$ has measure zero in M. Hence, by Proposition 30, the set of focal points of any $P \subset M$ has measure zero in M. By Exercise 8, along any P-normal geodesic the set of focal points is discrete.

33. Lemma. For a nonnull normal geodesic $\sigma \in \Omega(P, q)$ the nullspace of the index form $I_\sigma^\perp$ is the space $\mathscr{J}_b$ of P-Jacobi fields on σ that vanish at $q = \sigma(b)$.

Proof. If $V, W \in T_\sigma^\perp(\Omega)$, then applying integration by parts to the formula in Corollary 27 gives the formula in Corollary 8 but with an additional term

$$-(\varepsilon/c)\langle V'(0) - \tilde{II}(V(0), \sigma'(0)), W(0)\rangle.$$

It follows immediately that if $V \in \mathscr{J}_b$, then $I_\sigma^\perp(V, W) = 0$ for all W. Conversely, if V is in the nullspace of $I_\sigma^\perp$, a standard argument shows $V'' - R_{V\sigma'}\sigma' = 0$, and $V'(0) - \tilde{II}(V(0), \sigma'(0))$ is normal to P. Thus V is a P-Jacobi field zero at b; that is, $V \in \mathscr{J}_b$. $\blacksquare$

For cospacelike geodesics the index form characterizes focal points in the same way as in the special case of conjugate points.

34. Theorem. Let $\sigma \in \Omega(P, q)$ be a cospacelike normal geodesic of sign ε. Then

(1) If there are no focal points of P along σ, then $I_\sigma^\perp$ is definite (positive if $\varepsilon = 1$, negative if $\varepsilon = -1$).

(2) If $q = \sigma(b)$ is the only focal point of P along σ, then $I_\sigma^\perp$ is semidefinite, but not definite.

(3) If there is a focal point $\sigma(s)$, $0 < s < b$, along σ, then I_σ is not semidefinite.

Proof. We proceed as for Theorem 17, with minor adjustments to deal with initial conditions. To prove (1), for example, requires

If V and W are P-Jacobi fields on σ, then $\langle V', W\rangle = \langle V, W'\rangle$.

This will follow from Lemma 15 if $\langle V'(0), W(0)\rangle = \langle V(0), W'(0)\rangle$. But since $W(0)$ is tangent to P,

$$\begin{aligned}\langle V'(0), W(0)\rangle &= \langle \tan V'(0), W(0)\rangle = \langle \tilde{II}(V(0), \sigma'(0)), W(0)\rangle \\ &= -\langle II(V(0), W(0)), \sigma'(0)\rangle,\end{aligned}$$

the last equality by Remarks 4.39. Then symmetry of II gives the result.

Now let $Y_1, \ldots, Y_{n-1}$ be a basis for the P-Jacobi fields on σ perpendicular to σ. As before, if $V \in T_\sigma^\perp(\Omega)$, write $V = \sum f_i Y_i$. Lemma 16 gives

$$\langle V', V'\rangle = \langle R_{V\sigma'} V, \sigma'\rangle = \langle A, A\rangle + \langle V, B\rangle'.$$

By Corollary 27 (assuming $c = 1$),

$$\varepsilon I_\sigma(V, V) = \int_0^b \langle A, A\rangle\, du + \langle V, B\rangle\Big|_0^b - \langle \sigma'(0), II(V(0), V(0))\rangle.$$

The proof is completed as before once we check that the last two terms above cancel. In fact, since $V(b) = 0$ and Y_i is a P-Jacobi field,

$$\begin{aligned}\langle V, B\rangle|_0^b &= -\langle V(0), B(0)\rangle = -\langle V(0), \textstyle\sum f_i(0) Y_i'(0)\rangle \\ &= -\textstyle\sum f_i(0)\langle V(0), \tan Y_i'(0)\rangle \\ &= -\textstyle\sum f_i(0)\langle V(0), \tilde{II}(Y_i(0), \sigma'(0))\rangle = -\langle V(0), \tilde{II}(V(0), \sigma'(0))\rangle \\ &= \langle II(V(0), V(0)), \sigma'(0)\rangle. \quad \blacksquare\end{aligned}$$

APPLICATIONS

A focal point occurs when a P-Jacobi field has a zero; thus to prove that focal points exist, we expect two kinds of assumptions:

(1) on the initial conditions for P-Jacobi fields, that is, on the shape of P (e.g., P is "initially focusing"); and

(2) on the Jacobi equation, that is, on the curvature of M (e.g., tidal forces that cause geodesic convergence).

Here is a basic result of this type.

35. Proposition. Let P be a spacelike submanifold of a Riemannian or Lorentz manifold, and let σ be a P-normal nonnull geodesic. Suppose

(1) $\langle \sigma'(0), II(y, y)\rangle = k > 0$ for some unit vector $y \in T_{\sigma(0)}P$;
(2) $\langle R_{v\sigma'}v, \sigma'\rangle \geq 0$ for all tangent vectors $v \perp \sigma$.

Then there is a focal point $\sigma(r)$ of P along σ with $0 < r \leq 1/k$ provided σ is defined on this interval.

(Note that if $k < 0$, then reversing the orientation of σ would change this sign but not that in (2), hence σ has focal point with $-1/k \leq r < 0$.)

Proof. Assume $|\sigma'| = 1$, so $1/k$ is distance along σ. Since σ is cospacelike, it suffices by Theorem 34 to show that the index form $I_\sigma^\perp$ of σ on $[0, 1/k]$ is indefinite. In view of Lemma 13, if ε is the sign of σ, we want a nonzero $V \in T_\sigma^\perp(\Omega)$ with $\varepsilon I_\sigma(V, V) \leq 0$.

Motivated by the proof of Proposition 32, define $V(u) = (1 - ku)Y(u)$, where $Y(u)$ is the parallel translate of y. Then $V \in T_\sigma^\perp(\Omega)$. Since $V' = -kY$ and $|Y| = 1$,

$$\varepsilon I_\sigma(V, V) = \int_0^{1/k} \{k^2 - \langle R_{V\sigma'}V, \sigma'\rangle\}\, du - \langle II(y, y), \sigma'(0)\rangle.$$

But $\int_0^{1/k} k^2\, du = k$ cancels $-\langle II(y, y), \sigma'(0)\rangle = -k$, leaving $-\int_0^{1/k} \langle R_{V\sigma'}V, \sigma'\rangle\, du$, which by (2) is ≤ 0. ∎

The number k in the proposition can also be expressed as $\langle S_{\sigma'(0)}y, y\rangle$, where S is the shape operator of $P \subset M$. This proposition, and also Proposition 32, suggest that k is an *initial rate of convergence*: Move $\sigma'(0)$ infinitesimally in the y direction to normal vectors z. If $k > 0$, the geodesics γ_z are initial converging toward σ; if $k < 0$, they are initially diverging; if $k = 0$, they are initially parallel. This can be seen directly, as follows. Let Y be any P-Jacobi field on σ such that $Y(0) = y$. Proposition 28 shows that Y is an infinitesimal model for a family of normal geodesics γ_z as above. Thus $|Y|$ measures the distance of neighboring geodesics from σ. But $|Y|'(0) = -k$, since

$$\begin{aligned} |Y|'(0) &= \langle Y(0), Y'(0)\rangle = \langle y, \tan Y'(0)\rangle \\ &= \langle y, \tilde{II}(y, \sigma'(0))\rangle = -\langle \sigma'(0), II(y, y)\rangle = -k. \end{aligned}$$

For a spacelike hypersurface there is an averaged version of the preceding result with sectional curvature replaced by Ricci curvature, as in Myers' theorem, and normal curvature $II(y, y)$ replaced by mean normal curvature H (see page 101).

36. Definition. Let P be a semi-Riemannian submanifold of M with mean curvature vector field H. The *convergence* of P is the real-valued function $\boldsymbol{k}$ on the normal bundle NP such that

$$\boldsymbol{k}(z) = \langle z, H_p \rangle = (1/\dim P) \operatorname{trace} S_z \qquad \text{for} \quad z \in T_p(P)^\perp.$$

For a spacelike hypersurface in M^n,

$$H_p = \frac{1}{n-1} \sum_{i=1}^{n-1} II(e_i, e_i),$$

where $e_1, \ldots, e_{n-1}$ is any orthonormal basis for $T_p(P)$.

37. Proposition. Let P be a spacelike hypersurface in a (necessarily) Riemannian or Lorentz manifold M, and let σ be a geodesic normal to P at $p = \sigma(0)$. Suppose

(1) $\boldsymbol{k}(\sigma'(0)) = \langle \sigma'(0), H_p \rangle > 0$.
(2) $\operatorname{Ric}(\sigma', \sigma') \geq 0$.

Then there is a focal point $\sigma(r)$ of P along σ with $0 < r \leq 1/\boldsymbol{k}(\sigma'(0))$, provided σ is defined on this interval.

Proof. As in the preceding proof suppose $|\sigma'| = 1$ and let $k = \boldsymbol{k}(\sigma'(0))$. If $e_1, \ldots, e_{n-1}$ is an orthonormal basis for $T_p(P)$, parallel translate along σ to obtain $E_1, \ldots, E_{n-1}$. Define $f(u) = 1 - ku$ on $[0, 1/k]$; then $fE_i \in T_\sigma(\Omega)$ and

$$\varepsilon I_\sigma(fE_i, fE_i) = k - \int_0^{1/k} f^2 \langle R_{E_i\sigma'} E_i, \sigma' \rangle \, du - \langle \sigma'(0), II(e_i, e_i) \rangle.$$

Since the E_is are spacelike, adding the expressions above gives

$$(n-1)k - \int_0^{1/k} f^2 \operatorname{Ric}(\sigma', \sigma') \, du - \langle \sigma'(0), (n-1)H_p \rangle.$$

Here the first and last terms cancel, and since $\operatorname{Ric}(\sigma', \sigma') \geq 0$ it follows that $\varepsilon I_\sigma(fE_i, fE_i) \leq 0$ for at least one i. ∎

VARIATION OF *E*

For a curve segment $\alpha\colon [0, b] \to M$ in a semi-Riemannian manifold the integral

$$E(\alpha) = \frac{1}{2} \int_0^b \langle \alpha', \alpha' \rangle \, du$$

(sometimes called *energy* or *action*) lacks the direct geometric significance of arc length, but its variational theory is computationally simpler and, in the following sense, more general. For a piecewise smooth variation x of α let $E_x(v)$ be the value of E on the longitudinal curve $u \to x(u, v)$, so

$$E_x(v) = \frac{1}{2}\int_0^b \langle x_u, x_u \rangle \, du.$$

By contrast with L_x, the function E_x is always smooth without restriction on x. Thus in particular E can be used to study null geodesics.

Formulas for the first and second variations of E are simpler analogues of those for L.

38. Lemma. Let x be a variation of a curve segment α, with V and A the variation and transverse acceleration vector fields of x. If $f = f(u, v) = \langle x_u, x_u \rangle$, then

$$\left.\frac{1}{2}\frac{\partial f}{\partial v}\right|_{v=0} = \langle V', \alpha' \rangle = -\langle V, \alpha'' \rangle + \frac{d}{du}\langle V, \alpha' \rangle,$$

$$\left.\frac{1}{2}\frac{\partial^2 f}{\partial v^2}\right|_{v=0} = \langle V', V' \rangle - \langle R_{V\alpha'} V, \alpha' \rangle + \langle A', \alpha' \rangle$$

$$= -\langle V'' - R_{V\alpha'}\alpha', V \rangle + \langle A', \alpha' \rangle + \frac{d}{du}\langle V', V \rangle.$$

Proof. We readily compute

$$\frac{1}{2}\frac{\partial f}{\partial v} = \langle x_{uv}, x_u \rangle = \langle x_{vu}, x_u \rangle,$$

$$\frac{1}{2}\frac{\partial f^2}{\partial^2 v} = \langle x_{vu}, x_{uv} \rangle + \langle x_{vuv}, x_u \rangle$$

$$= \langle x_{vu}, x_{vu} \rangle - \langle R(x_u, x_v)x_u, x_v \rangle + \langle x_{vvu}, x_u \rangle.$$

Upon evaluation at $v = 0$ the first formula gives $\langle V', \alpha' \rangle$ and the product rule for the derivative of $\langle V, \alpha' \rangle$ gives the alternate version. Computation of the second derivative formulas is similar. ∎

Integration of the formulas in the lemma then gives first and second variation formulas:

39. Proposition. Let x be a variation of $\alpha\colon [0, b] \to M$, with V and A the variation and transverse acceleration vector fields. Then

$$E_x'(0) = \int_0^b \langle V', \alpha' \rangle \, du = -\int_0^b \langle V, \alpha'' \rangle \, du - \sum_{i=1}^{k} \langle V, \Delta\alpha' \rangle(u_i) + \langle V, \alpha' \rangle \Big|_0^b,$$

where $u_1 < \cdots < u_k$ are the breaks of x and α. Furthermore, if α is a geodesic, then

$$E_x''(0) = \int_0^b \{\langle V', V'\rangle - \langle R_{V\alpha'} V, \alpha'\rangle\}\, du + \langle A, \alpha'\rangle \Big|_0^b .$$

As before, if P is a semi-Riemannian submanifold of M and $q \in M$, then E becomes a real-valued function on $\Omega(P, q)$. Using the first variation formula above it is easy to check that *the critical points of E are exactly the normal geodesics from P to q*. If σ is such a geodesic, then strictly analogous to the index form I_σ for L is the *Hessian* H_σ for E. Explicitly, H_σ is the unique $\mathbf{R}$-bilinear form on $T_\sigma(\Omega)$ such that $H_\sigma(V, V) = E_x''(0)$, where x is any variation of σ whose longitudinal curves are in $\Omega(P, q)$ and whose variation vector field is V. By the second variation formula above it follows as in Corollary 27 that

$$H_\sigma(V, W) = \int_0^b \{\langle V', W'\rangle - \langle R_{V\sigma'} W, \sigma'\rangle\}\, du - \langle \sigma'(0), II(V(0), W(0))\rangle,$$

where II is the shape tensor of P.

FOCAL POINTS ALONG NULL GEODESICS

We consider some variational properties of null geodesics in Lorentz manifolds.

40. Corollary. Let σ be a null geodesic normal to a submanifold P of a Lorentz manifold. A P-Jacobi field on σ is the vector field of a variation of σ through null geodesics normal to P if and only if $V \perp \sigma$.

Proof. Suppose x is such a variation. Then $f = \langle x_u, x_u\rangle$ is identically zero, hence Lemma 38 implies $\langle V', \sigma'\rangle = 0$. In particular, $V'(0) \perp \sigma$. But $V(0)$ is tangent to P hence $\perp \sigma$. Thus $V \perp \sigma$ by Lemma 8.7.

For the converse, the proof of Proposition 28 will work provided we can arrange for the vector field Z in assertion (*) to be *null*. Since $V \perp \sigma$, it follows that nor $V'(0) \perp \sigma$; hence by Proposition 5.28, this vector corresponds canonically to a tangent vector to the nullcone $\Lambda \subset T_p(P)^\perp$ at the point $\sigma'(0)$ of Λ. Choose a curve λ in Λ with $\lambda(0) = \sigma'(0)$ and $\lambda'(0) \approx$ nor $V'(0)$. Then on the curve α (see previous proof), let $Z(v)$ be the normal–parallel translate of the vector $\lambda(v)$ along α to $\alpha(v)$. Thus Z is a null vector field on α normal to P, and $Z(0) = \sigma'(0)$. Using a normal–parallel frame field on α, one can check that nor $Z'(0) =$ nor $V'(0)$, hence $Z'(0) = V'(0)$. ∎

A focal point of a submanifold P along a normal geodesic σ is an almost-meeting point of nearby P-normal geodesics *of the same causal character as* σ. If σ is nonnull, this is obvious by continuity (in view of Proposition 30); for σ null, it is immediate from the preceding result, since a P-Jacobi field vanishing at $b \neq 0$ is perpendicular to σ.

If a semi-Riemannian submanifold P admits a normal null geodesic, then $\dim P \leq n - 2$. Also *focal orders along a normal null geodesic are at most* $n - 2$. The idea is that restricting the almost-meeting geodesics to be null reduces the usual maximum $n - 1$ by 1. Formally, the space $\mathscr{J}_b$ of P-Jacobi fields on σ vanishing at $b \neq 0$ is a subspace of the $(n - 1)$-dimensional space $\mathscr{J}^\perp$ of P-Jacobi fields perpendicular to σ. The tangential vector field $u \to u\sigma'(u)$ is never in $\mathscr{J}_b$, but since σ is null it is now in $\mathscr{J}^\perp$.

In particular, in a Lorentz surface there are no focal or conjugate points along null geodesics.

Our goal now is to prove an analogue of Theorem 34 for null geodesics. As before, geometric significance is gained by restricting H_σ to the subspace $T_\sigma^\perp(\Omega) = \{V \in T_\sigma(\Omega) : V \perp \sigma\}$.

41. Proposition. Let P be a spacelike submanifold of a Lorentz manifold. If there are no focal points of P along a normal null geodesic $\sigma \in \Omega(P, q)$, then $H_\sigma^\perp$ is positive semidefinite. Furthermore, if $H_\sigma^\perp(V, V) = 0$, then V is tangent to σ.

Proof. Let $Y_1, \ldots, Y_{n-1}$ be a basis for the space of P-Jacobi fields on σ perpendicular to σ. We can suppose $Y_1(u) = u\sigma'(u)$. Since there are no focal points on σ, if $V \in T_\sigma^\perp(\Omega)$, we can write $V = \sum f_i Y_i$. Lemma 16 gives

$$\langle V', V' \rangle - \langle R_{V\sigma'} V, \sigma' \rangle = \langle A, A \rangle + \langle V, B \rangle'.$$

As in the proof of Theorem 34, $H_\sigma^\perp(V, V)$ then reduces to $\int_0^b \langle A, A \rangle \, du$. Since $A = \sum f_i' Y_i$ is orthogonal to the null vector σ', Lemma 5.28 asserts that $\langle A, A \rangle \geq 0$ and furthermore that $\langle A(u), A(u) \rangle = 0$ if and only if $A(u)$ and $\sigma'(u)$ are collinear. The inequality implies $H_\sigma(V, V) \geq 0$, and it follows that if $H_\sigma(V, V) = 0$, then $A = \sum f_i' Y_i$ is everywhere tangent to σ. Since Y_1 is the only basis vector field ever tangent to σ, we must have $f_i' = 0$ for $i > 1$. But $V(b) = 0$, hence $f_i(b) = 0$, hence $f_i = 0$ for $i > 1$. Thus $V = f_1 Y_1$ which is tangent to σ. ∎

42. Example. Null Focal Points. In $M = \boldsymbol{R}_1^n$, $n \geq 3$, let P be the sphere $S^{n-2}(a)$ in $\boldsymbol{R}^{n-1}$, considered as the hyperplane $u^0 = 0$. Through each point of P run exactly two null normal lines (see Figure 3). In the future directions, the inward-pointing null normals all meet at $(a, 0, \ldots, 0)$, which is evidently a focal point of order $n - 2$ along every such normal.

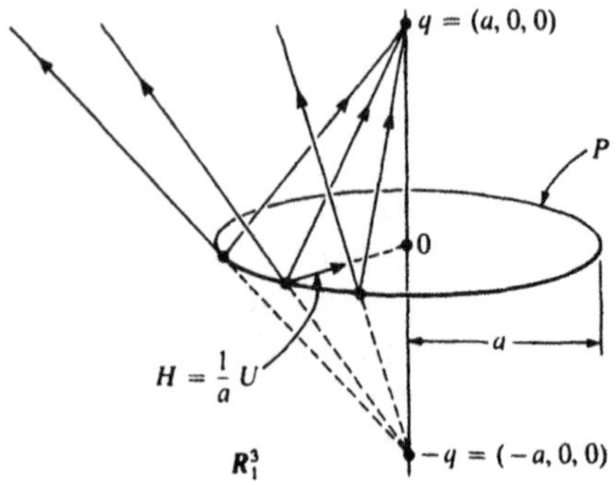

Figure 3

Symmetrically located, $(-a, 0, \ldots, 0)$ is the only other null focal point.

Proposition 37 has the following analogue for null geodesics.

43. Proposition. Let P be a spacelike $(n-2)$-dimensional submanifold of a Lorentz manifold M, with H the mean normal curvature vector field of P. Let σ be a null geodesic normal to P at $p = \sigma(0)$ such that

(1) $k(\sigma'(0)) = \langle \sigma'(0), H_p \rangle > 0$;
(2) $\mathrm{Ric}(\sigma', \sigma') \geq 0$.

Then there is a focal point $\sigma(r)$ of P along σ with $0 < r \leq 1/k$, where $k = k(\sigma'(0))$, provided σ is defined on this interval.

Proof. By (1), $H_p \neq 0$, hence $n = \dim M \geq 3$. Let $e_3, \ldots, e_n$ be an orthonormal basis for $T_p(P)$. Parallel translate these vectors to obtain $E_3, \ldots, E_n$, and as usual let $f(u) = 1 - ku$ on $[0, 1/k]$. Then $fE_i \in T^{\perp}_{\sigma}(\Omega)$, and

$$H^{\perp}_{\sigma}(fE_i, fE_i) = \int_0^{1/k} \{f'^2 - f^2\langle R_{E_i\sigma'}E_i, \sigma'\rangle\}\, du - \langle \sigma'(0), II(e_i, e_i)\rangle$$

$$= k - \int_0^{1/k} f^2\langle R_{E_i\sigma'}E_i, \sigma'\rangle\, du - \langle \sigma'(0), II(e_i, e_i)\rangle.$$

$E_3, \ldots, E_n$ and σ' cover only $n-1$ dimensions of M, but since σ is null, the missing dimension is not involved in the relevant Ricci curvature. In fact, the proof of Lemma 8.9 shows that

$$\mathrm{Ric}(\sigma', \sigma') = \sum_{i=3}^{n} \langle R_{E_i\sigma'}E_i, \sigma'\rangle.$$

Thus adding the formulas above for $i = 3, \ldots, n$ gives

$$\sum H_\sigma^\perp(fE_i, fE_i) = (n-2)k - \int_0^{1/k} f^2 \operatorname{Ric}(\sigma', \sigma')\, du - \langle \sigma'(0), (n-2)H_p \rangle$$

$$= -\int_0^{1/k} f^2 \operatorname{Ric}(\sigma', \sigma')\, du \leq 0.$$

None of the vector fields fE_i is tangent to σ, hence it follows from Proposition 41 that there is a focal point of P on σ. ∎

Here $k = k(\sigma'(0))$ represents the average initial rate of convergence of P-normal *null* geodesics near σ. If $k < 0$ in the proposition, then as before there is a focal point $\sigma(r)$ with $1/k \leq r < 0$.

44. Example. In Example 42, the sphere P is spacelike and $(n-2)$-dimensional. It is totally umbilic in $\boldsymbol{R}^{n-1}$ with mean normal curvature $H = -U/a$, where U is the outward unit normal. Since $\boldsymbol{R}^{n-1}$ is totally geodesic in $M = \boldsymbol{R}_1^n$, H is also the mean normal curvature for P in M.

The two families of normal null geodesics through P can conveniently be parametrized to have future-pointing initial velocities $\pm U + \partial_0$ (see Figure 3). Thus for the inward family, $k = \langle -U + \partial_0, H \rangle = 1/a$, independent of $p \in P$. Then Proposition 41 predicts a focal point with $0 < r \leq 1/k = a$. In fact they occur at $r = a$, that is, at

$$p + a(-p/a + (1, 0, \ldots, 0)) = (a, 0, \ldots, 0).$$

For the outward family, $k = \langle U + \partial_0, H \rangle = -1/a$, and the focal point predicted for $-a = 1/k \leq r < 0$ arrives at $r = -a$, that is, at $(-a, 0, \ldots, 0)$.

Since $\boldsymbol{R}_1^n$ is flat, Proposition 32 would locate these focal points exactly.

A CAUSALITY THEOREM

A fundamental problem in a Lorentz manifold is to determine which pairs of points can be joined by a timelike curve. This section will establish that there are timelike curves from p to q arbitrarily near every causal curve α from p to q—unless α is a null geodesic without conjugate points.

45. Lemma. Let α be a causal curve segment in a Lorentz manifold M, and let $\boldsymbol{x}$ be a variation of α with vector field V. If $\langle V', \alpha' \rangle < 0$, then for all sufficiently small $v > 0$, the longitudinal curve α_v of $\boldsymbol{x}$ is timelike.

Proof. Because α is causal,

$$\langle \boldsymbol{x}_u, \boldsymbol{x}_u \rangle(u, 0) = \langle \alpha'(u), \alpha'(u) \rangle \leq 0 \qquad \text{for all} \quad u.$$

But α is defined on a closed interval $[a, b]$, and

$$\frac{\partial}{\partial v}\langle \boldsymbol{x}_u, \boldsymbol{x}_u\rangle\bigg|_{v=0} = 2\langle V', \alpha'\rangle < 0.$$

Thus if $v > 0$ is sufficiently small, then $\langle \boldsymbol{x}_u, \boldsymbol{x}_u\rangle(u, v) < 0$ for all u; that is, α_v is timelike. ∎

The lemma is valid in the piecewise smooth case, where as usual we understand the hypotheses to hold on unbroken segments. Furthermore it would suffice to assume $\langle V', \alpha'\rangle \leq 0$, with strict inequality only at points where α' is a null vector.

Deforming a causal curve in the direction of its acceleration tends to make it timelike. For example, the acceleration of the null helix $\alpha(t) = (t, \cos t, \sin t)$ always points directly toward the t axis; deforming α in that direction makes it "steeper," hence timelike. The same scheme can be adapted to the instantaneous acceleration $\gamma'(u^+) - \gamma'(u^-)$ at a break of a (piecewise smooth) causal curve. These and similar deformations suffice to prove:

46. Proposition. In a Lorentz manifold M, if α is a causal curve from p to q that is not a null pregeodesic, then there is a timelike curve from p to q arbitrarily close to α.

Proof. We can suppose the domain of α is $[0, 1]$. Consider first two special cases.

Case 1. *$\alpha'(0)$ or $\alpha'(1)$ is timelike.* Assuming the latter, let W be obtained by parallel translation of $\alpha'(1)$ along α. Then W and α' are always in the same causal cone, and since W is timelike, $\langle W, \alpha'\rangle < 0$. By continuity there is a $\delta > 0$ such that $\langle \alpha', \alpha'\rangle < -\delta$ on $[1 - \delta, 1]$. Let f be any smooth function on $[0, 1]$ vanishing at endpoints and with $f' > 0$ on $[0, 1 - \delta]$. Set $V = fW$. Then $\langle V', \sigma'\rangle = f'\langle W, \alpha'\rangle$ is negative on $[0, 1 - \delta]$. Let $\boldsymbol{x}$ be a fixed endpoint variation with vector field V. By a remark above, for $v > 0$ sufficiently small, the longitudinal curve α_v has become timelike on $[0, 1 - \delta]$ and remained timelike on $[1 - \delta, 1]$.

Case 2. *α is a smooth null curve.* Differentiation of $\langle \alpha', \alpha'\rangle = 0$ shows that $\alpha'' \perp \alpha'$. Now α'' cannot always be collinear with α' or, by Exercise 3.19, α could be reparametrized as a null geodesic. Thus the function $\langle \alpha'', \alpha''\rangle \geq 0$ is not identically zero, since orthogonal null vectors are collinear.

Let W be a parallel timelike vector field on α in the same causal cone as α' at each point, so $\langle W, \alpha'\rangle < 0$. Let $V = fW + g\alpha''$, where f and g—vanishing at endpoints—are to be determined so that $\langle V', \alpha'\rangle < 0$.

Since $\langle \alpha'', \alpha'\rangle = 0$ implies $\langle \alpha''', \alpha'\rangle + \langle \alpha'', \alpha''\rangle = 0$, we compute

$$\langle V', \alpha'\rangle = f'\langle W, \alpha'\rangle - g\langle \alpha'', \alpha''\rangle.$$

Because $h = \langle\alpha'', \alpha''\rangle/\langle W, \alpha'\rangle$ is not identically zero, there exists a smooth g, vanishing at endpoints, such that

$$\int_0^1 gh\,du = -1.$$

Let $f(u) = \int_0^u (gh + 1)\,du$. Then f vanishes at endpoints, and $f' = gh + 1 > gh = g\langle\alpha'', \alpha''\rangle/\langle W, \alpha'\rangle$. Consequently $\langle V', \alpha'\rangle < 0$.

To complete the proof, note that if α' is timelike at a nonendpoint s, then Case 1 applies on $[0, s]$ and $[s, 1]$ to give the required result. Thus we are left with the case of a piecewise smooth null curve α. Unless every smooth segment of α can be reparametrized as a null geodesic, then by Case 2 some one can be varied slightly to become timelike—hence another small variation as above gives the result.

There remains only the case of a broken null geodesic α. It suffices to assume there is a single break, $0 < s < 1$. Let W on α be obtained by parallel translation of $\Delta\alpha'(s) = \alpha'(s^+) - \alpha'(s^-)$. Recall that these two velocities are by definition in the same causal cone, so $\langle W, \alpha'\rangle$ is negative on $[0, s^-]$ and positive on $[s^+, 1]$. Now choose a piecewise smooth function f on $[0, 1]$ that vanishes at endpoints and has derivative f' positive on $[0, s^-]$, negative on $[s^+, 1]$. Then for $V = fW$ we have $\langle V', \alpha'\rangle < 0$. ∎

Now the question is: Can a null geodesic segment σ be made timelike by a small fixed endpoint deformation? This should be possible if there is a conjugate point $\sigma(r)$ of $\sigma(0)$ with $0 < r < b$, because, by assuming that some null geodesic τ from $p = \sigma(0)$ that almost reaches $\sigma(r)$ actually reaches $\sigma(r)$, we can apply the preceding proposition to the broken null geodesic $\tau + \sigma|[r, b]$. However, a formal proof of this result requires some work.

47. Lemma. Let $\boldsymbol{x}$ be a variation of a null geodesic σ with variation vector field V perpendicular to σ at its endpoints. If there is a sequence $\{v_i\} \to 0$ such that each longitudinal curve σ_{v_i} is timelike, then V is everywhere perpendicular to σ.

Proof. Since no v_i is zero, by passing to a subsequence we can suppose either all $v_i > 0$ or all $v_i < 0$. Since σ is null and

$$\frac{\partial}{\partial v}\langle \boldsymbol{x}_u, \boldsymbol{x}_u\rangle\Big|_{v=0} = \langle V', \sigma'\rangle,$$

it follows that either $\langle V', \sigma'\rangle \leq 0$ or $\langle V', \sigma'\rangle \geq 0$.

Since σ is a geodesic, $\langle V', \sigma'\rangle = \langle V, \sigma'\rangle'$, hence

$$\int_0^b \langle V', \sigma'\rangle\,du = \langle V, \sigma'\rangle\Big|_0^b = 0.$$

Thus $\langle V', \sigma'\rangle = 0$, so $\langle V, \sigma'\rangle$ is a constant—necessarily zero. ∎

The first derivative criterion in Lemma 45 fails here, since $V \perp \sigma$ implies

$$\frac{\partial}{\partial v}\langle \boldsymbol{x}_u, \boldsymbol{x}_u \rangle \bigg|_{v=0} = 0.$$

Thus we look for variations $\boldsymbol{x}$ with

$$\frac{\partial^2}{\partial v^2}\langle \boldsymbol{x}_u, \boldsymbol{x}_u \rangle \bigg|_{v=0} < 0.$$

It will be no harder to deal with the endmanifold case.

48. Proposition. Let P be a spacelike submanifold of a Lorentz manifold, and let $\sigma \in \Omega(P, q)$ be a normal null geodesic. If there is a focal point of P along σ strictly before q, then there is a timelike curve from P to q arbitrarily near σ.

Proof. Let $\sigma(r)$ be the first focal point of P along σ. By Proposition 46, it suffices to show that for some small $\delta > 0$ there is a small fixed endpoint deformation of $\sigma | [0, r + \delta]$ to a timelike curve segment. Let J be a nonzero P-Jacobi field on σ that vanishes at $\sigma(b)$.

(1) *There is a $\delta > 0$ such that $J = fU$ on $[0, r + \delta]$, where U is a spacelike unit vector field on σ and $f > 0$ on $(0, r)$.*

J is orthogonal to σ, and we assert that it is never tangent to σ on $(0, r)$. In fact, if $J(a) = c\sigma'(a)$ for some $a \in (0, r)$, then $u \to J(u) - (cu/a)\sigma'(u)$ is a nonzero P-Jacobi field on σ that vanishes at a—before the first focal point at r.

Since M is Lorentz, it follows that J is spacelike on $[0, r]$. Now $J(r) = 0$, and whether or not $J(0) = 0$, it follows from Exercise 1.17 that there is a smooth vector field Y on σ such that $J(u) = u(r - u)Y(u)$. Now $J'(r) \neq 0$ (and $J'(0) \neq 0$ if $J(0) = 0$), hence for some $\delta > 0$, Y is never zero on $[0, r + \delta]$. Then $U = Y/|Y|$ and $f(u) = u(r - u)|Y(u)|$ give $J = fU$ as required.

(2) *For some $\delta > 0$, there is a vector field V on σ that vanishes at 0 and $r + \delta$, is perpendicular to σ, and has $\langle V'' - R_{V\sigma'}\sigma', V\rangle > 0$ on $(0, r + \delta)$.*

For $J = fU$ as in (1), let $V = (f + g)U = J + gU$, where g is to be determined. We compute

$$V'' - R_{V\sigma'}\sigma' = g''U + 2g'U' + g[U'' - R_{U\sigma'}\sigma'].$$

Since U is a unit vector field,

(*) $\langle V'' - R_{V\sigma'}\sigma', V\rangle = (f + g)(g'' + gh)$, where $h = \langle U'' - R_{U\sigma'}\sigma', U\rangle$.

Let $-a^2$, with $a > 0$, be a lower bound for h on $[0, r + \delta]$, and define $g(u) = b(e^{au} - 1)$, where $b > 0$ is determined by $g(r + \delta) = -f(r + \delta)$.

Since $g'' = a^2(g + b)$, we get

$$g'' + gh = g(a^2 + h) + a^2 b > 0 \quad \text{on} \quad (0, r + \delta).$$

The function $f + g$ is positive on $(0, r]$ and zero at $r + \delta$. Reducing $\delta > 0$ if necessary, we can suppose that $r + \delta$ is the only zero of $f + g$ in $(0, r + \delta]$. Then $f + g$ and $g'' + gh$ are both positive on $(0, r + \delta)$, hence by (*), V has the required properties.

(3) *There is a fixed endpoint variation* $\mathbf{x}$ *of* $\sigma|[0, r + \delta]$ *whose longitudinal curves* σ_v, *for* v *sufficiently small, are timelike on* $(0, r + \delta)$, *hence causal on* $[0, r + \delta]$.

Let N be a parallel null vector field on σ with $\langle N, \sigma' \rangle = -1$. By the lemma below there is a variation $\mathbf{x}$ of σ with vector field V as in (2) and acceleration vector field $A = \langle V', V \rangle N$. Thus in the second formula for $\partial^2 f/\partial v^2$ in Lemma 38, two terms cancel since

$$\langle A', \sigma' \rangle = \langle V', V \rangle' \langle N, \sigma' \rangle = -\langle V', V \rangle'.$$

The remaining term is negative on $(0, r + \delta)$ by (2), thus (3) follows.

To complete the proof it suffices to apply Proposition 46 to the piecewise smooth curves $\sigma_v + \sigma|[r + \delta, 1]$. ■

49. Lemma. If $\alpha \in \Omega(P, q)$, let $V \in T_\alpha(\Omega)$ and let A be a vector field on α such that nor $A(0) = II(V(0), V(0))$ and $A(b) = 0$. Then there is a (P, q) variation $\mathbf{x}$ of α whose variation and transverse acceleration vector fields are V and A.

Proof. When P is a single point, then $A(0) = V(0) = 0$ and it suffices to define

$$\mathbf{x}(u, v) = \exp_{\sigma(u)}[vV(u) + \tfrac{1}{2}v^2 A(u)].$$

Recall that, at 0, $\exp_p$ preserves velocities and accelerations (up to canonical isomorphism $\approx$).

If $\dim P > 0$, it is easy to find a variation of α of the form $X(u, v) = \exp Z(u, v)$ such that (1) the first transverse curve lies in P and has initial velocity $V(0)$ and initial P-acceleration tan $A(0)$—hence initial M-acceleration $A(0)$—and (2) the last transverse curve is constant at q.

We modify this variation as follows. Denote its variation vector field by $W(u) \approx Z_v(u, 0)$ and its acceleration vector field by $B(u) \approx Z_{vv}(u, 0)$. Both $V - W$ and $A - B$ are zero at endpoints. Hence

$$\exp\{Z(u, v) + v(V(u) - W(u)) + \tfrac{1}{2}v^2(A(u) - B(u))\}$$

defines a variation with the required properties. ■

50. Lemma. If $\sigma \in \Omega(P, q)$ is a null geodesic not normal to P, there is a timelike curve arbitrarily near σ in $\Omega(P, q)$.

Proof. By hypothesis there is a vector $y \in T_{\sigma(0)}P$ such that $\langle y, \sigma'(0)\rangle \neq 0$. Switching if necessary to $-y$, we can suppose $\langle y, \sigma'(0)\rangle > 0$. Define $V(u) = (1 - (u/b))Y(u)$, where $Y(u)$ is the parallel translate of y along σ. Then $V \in T_\sigma(\Omega)$, so there is a (P, q) variation of σ with vector field V. But $V' = -Y/b$, hence $\langle V', \sigma'\rangle = -\langle Y, \sigma'\rangle/b = -\langle y, \sigma'(0)\rangle/b < 0$. Then Lemma 45 completes the proof. ∎

Thus we can summarize as follows.

51. Theorem. Let P be a spacelike submanifold of a Lorentz manifold M. If $\alpha \in \Omega(P, q)$ is a causal curve, there is a timelike curve arbitrarily near α in $\Omega(P, q)$ *unless* α is a P-normal null geodesic along which there are no focal points of P before q.

The exceptional case cannot be eliminated. For example (with P a single point), no timelike curve joins the endpoint of a null geodesic segment in Minkowski space. In Example 42, the P-normal null geodesics in $\Omega(P, q)$ that pass through a focal point $(\pm a, 0, \ldots, 0)$ are exactly those with timelike neighbors in $\Omega(P, q)$.

Exercises

1. In each case, x is a variation of a geodesic segment; compute $L_x''(0)$ by the second variation formula and also directly from L_x. (a) In S^2, $x(u, v) = (\cos v \cos u, \cos v \sin u, \sin v)$, $0 \leq u \leq \pi$. (b) In R^2, $x(u, v) = (u \cosh v, v)$, $-1 \leq u \leq 1$. (c) In R^2, $x(u, v)$ is (u, vu) if $u \in [0, 1]$ and $(u, v(2 - u))$ if $u \in [1, 2]$.

2. Let $\sigma\colon [0, b] \to M$ be a geodesic that is normal to $P \subset M$. If for $0 \leq a < b$ the points $\sigma(a)$ and $\sigma(b)$ are conjugate along σ, then there is a focal point of P along σ. (Hint: Use the index forms—or Hessians—of $\sigma|[a, b]$ and $\sigma \in \Omega(P, \sigma(b))$.)

3. *Two endmanifolds.* Let P and Q be semi-Riemannian submanifolds of M. (a) Establish first and second variation formulas for E on $\Omega(P, Q)$. (b) Prove that the critical points of E are geodesics normal to both P and Q.

4. In the preceding exercise suppose M is a complete connected Riemannian manifold. Prove: (a) If P and Q are compact, there is a shortest geodesic from P to Q and it is normal to both. (b) If P and Q are compact and totally geodesic, and M has $K > 0$, then P and Q meet.

5. Let γ be a cospacelike geodesic in a locally symmetric Riemannian or Lorentz manifold. Let F_v be the tidal force operator of $v = \gamma'(0)$. Prove: (a) $v^\perp$ has an orthonormal basis $e_2, \ldots, e_n$ consisting of eigenvectors: $F_v(e_i) = \lambda_i e_i$. (b) The conjugate points of $\gamma(0)$ along γ are $\gamma(\pi m/\sqrt{-\lambda_j})$ for all negative eigenvalues $\lambda_j < 0$ and integers $m \neq 0$.

6. Let γ be a null geodesic in a locally symmetric Lorentz manifold. (a) Prove an analogue of the preceding exercise by considering the quotient vector space $v^\perp/\boldsymbol{R}v$. (b) In the symmetric space $\boldsymbol{R}^1_1 \times S^2$ find a null geodesic with conjugate points.

7. *Extremal submanifolds.* Let P be a semi-Riemannian submanifold of M. Let Z be a normal vector field on P with compact support (e.g., P itself compact). Then Z can be extended to a neighborhood of P on which its flow ψ_t is defined. (a) If vol(t) is the volume of $\psi_t(P)$, prove that $(\mathrm{vol})'(0) = -(\dim P)\int_P \langle H, Z\rangle\, dP$, where H is the mean curvature vector field of P in M. (Hint: vol(t) is the integral of the Jacobian function of ψ_t.) (b) Deduce that $H = 0$ if and only if P is an *extremal* for volume, that is, $(\mathrm{vol})'(0) = 0$ for all such Z.

8. Let $\sigma(b)$ be an (order m) focal point of P along a P-normal geodesic σ. *If $c \neq b$ is sufficiently near b, then $\sigma(c)$ is not a focal point along σ.* Prove this by showing: (a) There exists a basis $Y_1, \ldots, Y_{n-1}$ for the P-Jacobi fields perpendicular to σ such that $Y_1'(b), \ldots, Y_m'(b), Y_{m+1}(b), \ldots, Y_{n-1}(b)$ is a basis for $\sigma'(b)^\perp$. If $E_1, \ldots, E_{n-1}$ are gotten by parallel translation of the above listed $n - 1$ vectors, let $f = \det\langle Y_i, E_j\rangle$. (b) For $s \neq 0$, $f(s) = 0$ if and only if $\sigma(s)$ is a focal point along σ. (c) Near $u = b$, $f(u) = (u - b)^m g(u)$, where g is a continuous function with $g(b) \neq 0$.

9. *Convergence.* Prove: (a) If z is a vector normal to $P \subset M$, then $\boldsymbol{k}(z)$ is the average of the (possibly complex) eigenvalues of S_z. (b) In the notation of Exercises 4.7 and 9.22, $\boldsymbol{k}(U) = h$. (c) If P is totally umbilic in M, with normal curvature vector field z, then $\boldsymbol{k}(z) = \langle z, z\rangle$.

11 HOMOGENEOUS AND SYMMETRIC SPACES

This chapter will show how symmetric spaces (semi-Riemannian symmetric manifolds) can be constructed, and their geometries described, in terms of Lie groups—in fact, to a large extent, in terms of Lie algebras. Given such a Lie description, curvature and geodesics can often be found by surprisingly simple matrix calculations, and information about isometries and the underlying topology is forthcoming. Initially a broader class of homogeneous spaces, said to be *naturally reductive*, is considered.

Also we consider some complex geometry, and—turning the basic scheme around—use symmetric space information to find properties of several geometrically important Lie groups.

MORE ABOUT LIE GROUPS

We add to Appendix B a few elementary facts needed in this chapter. If G is a Lie group, $\mathfrak{g}$ denotes its Lie algebra and e its identity element.

An *automorphism* of a Lie group G is a map $\phi: G \to G$ that is both a diffeomorphism and a group isomorphism. Since automorphisms preserve the two structures of a Lie group, they must preserve all features of Lie theory on G.

1. Lemma. Let $\phi: G \to G$ be an automorphism. If $X \in \mathfrak{g}$, then the transferred vector field $d\phi(X)$ is in $\mathfrak{g}$, and $d\phi: \mathfrak{g} \to \mathfrak{g}$ is a Lie algebra isomorphism called the *differential* of ϕ.

Proof. (Recall that $(d\phi X)_g = d\phi(X_{\phi^{-1}g})$ for all $g \in G$.) If $a \in G$, let $b = \phi^{-1}(a)$. Then $\phi L_b(g) = \phi(bg) = \phi(b)\phi(g) = a\phi(g)$, hence $\phi L_b = L_a \phi$. Thus if $X \in \mathfrak{g}$,

$$dL_a\, d\phi(X) = d\phi\, dL_b(X) = d\phi(X).$$

Hence $d\phi(X) \in \mathfrak{g}$. The function $d\phi\colon \mathfrak{g} \to \mathfrak{g}$ is certainly linear, and preserves brackets by Proposition 1.22. That it is a linear isomorphism follows since ϕ^{-1} is also an automorphism of G. ■

The differential $d\phi$ contains the same information as the differential map $d\phi\colon TG \to TG$, and is, in fact, completely determined by the single map $d\phi_e\colon T_e(G) \to T_e(G)$. If $a \in G$, let $C_a\colon G \to G$ be the function sending each g to aga^{-1}. C_a is an inner automorphism and, since $C_a = L_a \circ R_{a^{-1}}$, a diffeomorphism. Thus C_a is an automorphism of G.

The differential of C_a is denoted by Ad_a. If $a, b \in G$, then $C_{ab}(g) = abg(ab)^{-1} = a(bgb^{-1})a^{-1}$, hence $C_{ab} = C_a \circ C_b$. Taking differentials gives

$$\mathrm{Ad}_{ab} = \mathrm{Ad}_a \circ \mathrm{Ad}_b.$$

The resulting homomorphism $a \to \mathrm{Ad}_a$ is called the *adjoint representation* of G.

2. Corollary. If $X, Y \in \mathfrak{g}$, then

$$[X, Y] = \lim_{t \to 0} \frac{1}{t} \{\mathrm{Ad}_{\alpha(t)} Y - Y\},$$

where $\alpha(t) = \exp(tX)$ is the one-parameter subgroup of X.

Proof. By Lemma 9.34 the flow of X is $\{R_{\alpha(t)}\}$. Hence by Proposition 1.58,

$$[X, Y] = \lim_{t \to 0} \frac{1}{t} \{dR_{\alpha(-t)} Y - Y\}.$$

Since $\mathrm{Ad}_a = dC_a = dR_{a^{-1}} \circ dL_a$, it follows that Ad_a and $dR_{a^{-1}}$ have the same effect on the (left-invariant) elements of $\mathfrak{g}$. Thus $dR_{\alpha(-t)}$ can be replaced by $\mathrm{Ad}_{\alpha(t)}$ in the preceding formula. ■

This corollary shows that the bracket operation measures failure of commutativity in G. For example, if G is abelian, then $C_a = \mathrm{id}$, hence $\mathrm{Ad}_a = \mathrm{id}$, for all $a \in G$. Thus by the corollary, $[X, Y] = 0$ for all $X, Y \in \mathfrak{g}$; that is, $\mathfrak{g}$ is *abelian*. The converse holds if G is connected (see [KN] or [W]).

Let H be a subgroup of G (perhaps G itself). An object defined on the Lie algebra $\mathfrak{g}$ of G is *$\mathrm{Ad}(H)$-invariant* if it is preserved by $\mathrm{Ad}_h\colon \mathfrak{g} \to \mathfrak{g}$ for all $h \in H$. Let $\mathfrak{h} \subset \mathfrak{g}$ be the Lie algebra of H.

3. Lemma. If a symmetric bilinear form F on $\mathfrak{g}$ is $\mathrm{Ad}(H)$-invariant, then $F([X, W], Y) = F(X, [W, Y])$ for all $X, Y \in \mathfrak{g}$ and $W \in \mathfrak{h}$. The converse holds if H is connected.

Proof. By polarization, the stated formula is equivalent to $F([W, X], X) = 0$ for all X, W. Using the preceding corollary this can be shown equivalent to the constancy of the function $f(s) = F(\mathrm{Ad}_{\alpha(s)} X, \mathrm{Ad}_{\alpha(s)} X)$ on each one-parameter subgroup α in H. (Compare the proof of Proposition 9.23.) The direct assertion follows. For the converse, Lemma B.12 shows that F is preserved by Ad_h for h in some neighborhood of the identity in H. We assume $H = H_0$, and $h \to \mathrm{Ad}_h$ is a homomorphism, so the result follows from Lemma B.13. ∎

If $\mathfrak{g}$ is a Lie algebra and $X \in \mathfrak{g}$, let $\mathrm{ad}_X : \mathfrak{g} \to \mathfrak{g}$ be the map sending each Y to $[X, Y]$. Evidently ad_X is a linear operator, and by the Jacobi identity it is a *Lie derivation*; that is,

$$\mathrm{ad}_Z[X, Y] = [\mathrm{ad}_Z X, Y] + [X, \mathrm{ad}_Z Y].$$

The Jacobi identity also shows that $\mathrm{ad}_{[X,Y]} = [\mathrm{ad}_X, \mathrm{ad}_Y]$. Note that the equation in Lemma 3 asserts that ad_W is skew-adjoint relative to F.

4. Definition. The *Killing form* of a Lie algebra $\mathfrak{g}$ is the function B: $\mathfrak{g} \times \mathfrak{g} \to \mathbf{R}$ given by $B(X, Y) = \mathrm{trace}(\mathrm{ad}_X \mathrm{ad}_Y)$.

5. Lemma. The Killing form B of $\mathfrak{g}$ is a symmetric bilinear form that is invariant under all automorphisms of $\mathfrak{g}$ and satisfies $B([X, Y], Z) = B(X, [Y, Z])$ for $X, Y, Z \in \mathfrak{g}$.

Proof. B is bilinear since the function $X \to \mathrm{ad}_X$ is linear. B is symmetric since trace ST = trace TS. Let μ be an automorphism of $\mathfrak{g}$, that is, a linear isomorphism that preserves brackets. The latter implies $\mathrm{ad}_{\mu X} \circ \mu = \mu \circ \mathrm{ad}_X$, hence $\mathrm{ad}_{\mu X} = \mu \circ \mathrm{ad}_X \circ \mu^{-1}$. Since trace STS^{-1} = trace T, it follows immediately that $B(\mu X, \mu Y) = B(X, Y)$. Finally, since $\mathrm{ad}_{[X,Y]} = [\mathrm{ad}_X, \mathrm{ad}_Y]$, trace properties show that $B([X, Y], Z) = B(X, [Y, Z])$. ∎

If $\mathfrak{g}$ is the Lie algebra of a Lie group G, then the Killing form of $\mathfrak{g}$ is also attributed to G and is in particular $\mathrm{Ad}(G)$-invariant.

Our applications will deal with matrix Lie groups, that is, Lie subgroups of $GL(n, \mathbf{C})$, with their Lie algebras also in matrix form as subalgebras of $\mathfrak{gl}(n, \mathbf{C})$ (Appendix B). In computations the notation $X \cdot Y = \sum X_{ij} Y_{ij}$ is efficient. Note that ${}^tX \cdot {}^tY = X \cdot Y = Y \cdot X$ and $\overline{X \cdot Y} = \bar{X} \cdot \bar{Y}$. On $\mathfrak{gl}(n\, \mathbf{R}) \approx \mathbf{R}^{n^2}$, $X \cdot Y$ is indeed the dot product, while on $\mathfrak{gl}(n, \mathbf{C})$, $X \cdot \bar{Y}$ is the natural Hermitian product.

6. Lemma. Let G be a Lie subgroup of $GL(n, C)$.

(1) $\mathrm{Ad}_a(X) = aXa^{-1}$ for all $a \in G$, $X \in \mathfrak{g} \subset \mathfrak{gl}(n, C)$.
(2) Assume that $X \in \mathfrak{g} \Rightarrow {}^t\bar{X} \in \mathfrak{g}$. Then $B(X, Y) = \mathrm{Re}\,\mathrm{trace}\, XY = \mathrm{Re}\,{}^tX \cdot Y$ is an $\mathrm{Ad}(G)$-invariant scalar product on $\mathfrak{g}$ called the *trace form*.

Proof. (1) The formula $C_a(g) = aga^{-1}$ defining $C_a\colon G \to G$ extends this map to an $\boldsymbol{R}$-linear operator on $\mathfrak{gl}(n, C) \approx \boldsymbol{R}^{2n^2}$. Then dC_a differs only by canonical isomorphism from C_a (Exercise 1.2); hence $\mathrm{Ad}_a = C_a | \mathfrak{g}$.

(2) Clearly B is bilinear and symmetric. The hypothesis on $\mathfrak{g}$ lets us pick Y to be ${}^t\bar{X}$. Then $0 = \mathrm{Re}\, X \cdot \bar{X} = X \cdot \bar{X} = \sum |X_{ij}|^2$ implies $X = 0$, so B is a scalar product. Since $\mathrm{trace}\, STS^{-1} = \mathrm{trace}\, T$, it follows immediately from (1) that $B(\mathrm{Ad}_a X, \mathrm{Ad}_a Y) = B(X, Y)$ for all $a \in G$. ∎

7. Remark. Consider the Lie algebras $\mathfrak{g} = \mathfrak{o}(n)$, $\mathfrak{u}(n)$, $\mathfrak{sp}(n)$ discussed in Appendix B. Then:

(1) $X \in \mathfrak{g} \Rightarrow {}^t\bar{X} \in \mathfrak{g}$;
(2) trace XY is real for all X, $Y \in \mathfrak{g}$. Thus the trace form is nondegenerate in these cases and is given merely by trace XY.
(3) The Killing form B of $\mathfrak{g}$ is proportional to the trace form. In fact, $B(X, Y) = c\,\mathrm{trace}\, XY$, with $c \neq 0$ if $\dim \mathfrak{g} > 1$.

Only the last assertion is not obvious. It can be proved by elementary but tedious computations, but Lie theory produces simplifications [H].

These results are known to hold also for $\mathfrak{g} = \mathfrak{o}(p, q)$, $\mathfrak{u}(p, q)$, and $\mathfrak{sp}(p, q)$. Trace forms are computed for the nonsymplectic cases below and in Example 41.

8. Examples. Trace Forms. (1) *Lie algebra $\mathfrak{o}(n)$ of the orthogonal group $O(n)$.* Since $\mathfrak{o}(n)$ consists of skew-symmetric matrices, trace $XY = -X \cdot Y$.

(2) *Lie algebra $\mathfrak{o}(p, q)$ of the semiorthogonal group $O(p, q)$.* By Lemma 9.3, $\mathfrak{o}(p, q)$ consists of all matrices of the form $X = \begin{pmatrix} a & {}^tx \\ x & b \end{pmatrix}$, where $a \in \mathfrak{o}(p)$, $b \in \mathfrak{o}(q)$, and x is an arbitrary real $q \times p$ matrix. The space of all such x can safely be denoted by $\boldsymbol{R}^{pq}$. (This parametrization of X follows [KN, Volume II], differing trivially from [H].)

If $Y = \begin{pmatrix} c & {}^ty \\ y & d \end{pmatrix} \in \mathfrak{o}(p, q)$, then for X as above,

$$\mathrm{trace}\, XY = X \cdot {}^tY = \begin{pmatrix} a & {}^tx \\ x & b \end{pmatrix} \cdot \begin{pmatrix} -c & {}^ty \\ y & -d \end{pmatrix} = -a \cdot c - b \cdot d + 2x \cdot y.$$

Evidently the vector space $\mathfrak{o}(p, q)$ can be written as an orthogonal direct sum $\mathfrak{o}(p) + \mathfrak{o}(q) + \boldsymbol{R}^{pq}$. The trace form is thus negative definite on $\mathfrak{o}(p) + \mathfrak{o}(q)$ and positive definite on $\boldsymbol{R}^{pq}$.

(3) *Lie algebra* u(n) *of the unitary group* $U(n)$. By Appendix B, u(n) consists of the $n \times n$ complex matrices X that are skew-Hermitian: ${}^t\bar{X} = -X$. Thus trace $XY = X \cdot {}^tY = -X \cdot \bar{Y}$. The trace form is negative definite, since $-X \cdot \bar{X} = -\sum |X_{ij}|^2$. Hence $-$trace XY is an inner product on u(n). (Note that u(n), though constructed of complex numbers, is a real vector space and not a complex one.)

BI-INVARIANT METRICS

Since a Lie group G is in particular a manifold, it can be made semi-Riemannian by furnishing it with a metric tensor. In order to link the resulting geometry of G to the group structure of G, it is customary to use a *left-invariant metric*: one for which left multiplication L_a: G $\to G$ is an isometry for all $a \in G$.

A left-invariant metric on G is virtually the same thing as a scalar product on the Lie algebra g *of G—or a scalar product on* $T_e(G)$. In fact, the last two correspond under the canonical isomorphism $X \to X_e$. If $\langle , \rangle$ is a left-invariant metric tensor on G, then $\langle X, Y\rangle$ *is constant* for X, $Y \in$ g, thereby defining a scalar product on g. Conversely, if $\langle , \rangle$ is a scalar product on $T_e(G)$, the definition

$$\langle x, y\rangle = \langle dL_{a^{-1}}(x), dL_{a^{-1}}(y)\rangle \qquad \text{for } x, y \in T_a(G)$$

gives a left-invariant metric on G.

A metric on G that is both left- and right-invariant (each R_a: $G \to G$ also an isometry) is called *bi-invariant*.

9. Proposition. Let G be a connected Lie group furnished with a left-invariant metric tensor $\langle , \rangle$. Then the following are equivalent:

(1) $\langle , \rangle$ is right-invariant, hence bi-invariant.
(2) $\langle , \rangle$ is Ad(G)-invariant.
(3) The inversion map $g \to g^{-1}$ is an isometry of G.
(4) $\langle X, [Y, Z]\rangle = \langle [X, Y], Z\rangle$ for all X, Y, $Z \in$ g.
(5) $D_X Y = \frac{1}{2}[X, Y]$ for all X, $Y \in$ g.
(6) The geodesics of G starting at e are the one-parameter subgroups of G.

Proof. In fact, conditions (1), (2), and (3) are always equivalent and imply the (equivalent) conditions (4), (5), and (6). The connectedness of G is used only for (4), (5), (6) $\Rightarrow$ (1), (2), (3).

(1) $\Leftrightarrow$ (2) is immediate, since $C_a = L_a \circ R_{a^{-1}}$ and $dL_a | \text{g} = \text{id}$.

(1) ⇔ (3). Let ζ be the inversion map. Then for any one-parameter subgroup, $\zeta\alpha(s) = a(s)^{-1} = \alpha(-s)$; so $d\zeta_e = -\text{id}$. If $a \in G$, then $\zeta = R_{a^{-1}}\zeta L_{a^{-1}}$. Thus the differential map $d\zeta_a\colon T_a(G) \to T_{a^{-1}}(G)$ is $dR_{a^{-1}}\, d\zeta_e\, dL_{a^{-1}}$. Hence (1) ⇒ (3). The converse is clear, since $R_a = \zeta L_{a^{-1}}\zeta$.

(2) ⇔ (4). Immediate from Lemma 3.

(4) ⇔ (5). For elements of $\mathfrak{g}$, scalar products are constant, hence the Koszul formula reduces to $2\langle D_X Y, Z\rangle = -\langle X, [Y, Z]\rangle + \langle Y, [Z, X]\rangle + \langle Z, [X, Y]\rangle$. By (4), the first two summands cancel, yielding (5), Conversely, if (5) holds,

$$\langle X, [Y, Z]\rangle = 2\langle X, D_Y Z\rangle = -2\langle D_Y X, Z\rangle = -\langle [Y, X], Z\rangle = \langle [X, Y], Z\rangle.$$

(5) ⇔ (6). By polarization, (5) is equivalent to $D_X X = 0$ for all $X \in g$. Let α be the one-parameter subgroup of X; hence $\alpha'' = D_X X|_\alpha$. Thus (5) implies that every α is a geodesic; but this is equivalent to (6). Conversely, if α is a geodesic, then $D_X X|_\alpha = 0$. Left multiplications, being isometries, preserve D; hence $D_X X = 0$. ∎

For brevity a Lie group G furnished with a bi-invariant metric will be called a *semi-Riemannian group.* Then G *is a symmetric space* since, by (3) above, the inversion map ζ is the global symmetry at e, and thus $L_a\zeta L_{a^{-1}}$ is the symmetry at $a \in G$. In particular, G is *complete.*

10. Corollary. For a semi-Riemannian group G:

(1) $R_{XY}Z = \frac{1}{4}[[X, Y], Z]$ for $X, Y, Z \in \mathfrak{g}$.
(2) If X and Y span a nondegenerate plane in $\mathfrak{g}$, then

$$K(X, Y) = \frac{1}{4}\frac{\langle [X, Y], [X, Y]\rangle}{\langle X, X\rangle\langle Y, Y\rangle - \langle X, Y\rangle^2}.$$

Proof. (1) Since $D_X Y = \frac{1}{2}[X, Y]$,

$$R_{XY}Z = \tfrac{1}{2}[[X, Y], Z] - \tfrac{1}{4}[X, [Y, Z]] + \tfrac{1}{4}[Y, [X, Z]].$$

But manipulation of the Jacobi identity $\mathfrak{S}[X, [Y, Z]] = 0$ replaces the last two summands by $-\frac{1}{4}[[X, Y], Z]$.

(2) By the shift formula (4) in Proposition 9,

$$\langle R_{XY}X, Y\rangle = \tfrac{1}{4}\langle [[X, Y], X], Y\rangle = \tfrac{1}{4}\langle [X, Y], [X, Y]\rangle. \quad ∎$$

Here $K(X, Y)$ is a number that, for every $g \in G$, gives the sectional curvature of the plane spanned by X_g and Y_g. Some consequences of the curvature formulas are:

(1) If G is abelian, then $K = 0$.
(2) If the metric is Riemannian, then $K \geq 0$.
(3) $\text{Ric}|_\mathfrak{g} = -B/4$, where B is the Killing form of G.

(Proof: $\mathrm{Ric}(X, Y) = \mathrm{trace}\{V \to R_{XV} Y\}$, and $R_{XV} Y = \frac{1}{4}[[X, V], Y] = -1/4(\mathrm{ad}\, Y \,\mathrm{ad}\, X)V$.)

A Lie group is *semisimple* if and only if its Killing form B is nondegenerate. By Lemma 5, the Killing form of G is $\mathrm{Ad}(G)$-invariant. Thus for a semisimple Lie group G, its Killing form provides a bi-invariant metric—and, by (3) above, G is then an Einstein manifold.

To give some further connections between the algebra and the geometry of Lie groups we need: *If G is a compact group of linear operators on a real vector space V, then there is a G-invariant inner product on V* (every $g: V \to V$ is a linear isometry). For a proof, see [W].

11. Proposition. Let B be the Killing form of a connected Lie group G. (1) if G is compact, then B is negative semidefinite; (2) if B is negative definite, then G is compact and has finite fundamental group; and (3) G is compact and semisimple if and only if B is negative definite.

Proof. (1) Since $g \to \mathrm{Ad}_g$ is a continuous homomorphism, $\mathrm{Ad}(G) = \{\mathrm{Ad}_g : g \in G\}$ is a compact group of linear operators on $\mathfrak{g}$. Thus, as mentioned above, there is an $\mathrm{Ad}(G)$-invariant inner product $\langle\, ,\rangle$ on $\mathfrak{g}$. By Lemma 3, ad_X is skew-adjoint relative to $\langle\, ,\rangle$ for each $X \in \mathfrak{g}$. Because $\langle\, ,\rangle$ is positive definite, each ad_X has a skew-symmetric matrix $x_{ij} = -x_{ji}$. But then

$$B(X, X) = \mathrm{trace}(\mathrm{ad}_X)^2 = \sum x_{ij} x_{ji} = -\sum (x_{ij})^2.$$

(2) By the remarks above, $-B$ provides a bi-invariant Riemannian metric on G, making it a complete Einstein manifold of strictly positive scalar curvature. Thus Myers' theorem (10.24) gives the result.

(3) By Exercise 1.12, nondegenerate and semidefinite imply definite. ∎

COSET MANIFOLDS

We describe a simple way to build smooth manifolds out of Lie groups. Geometry is not directly involved, so several proofs will be omitted. However, the construction will be an essential part of the Lie description of semi-Riemannian homogeneous and symmetric spaces.

If H is a closed subgroup (Appendix B) of a Lie group G, let G/H be the set of all left cosets gH of H in G. The *origin* o of G/H is the subgroup H considered as an element of G/H. The *projection* $\pi: G \to G/H$ sends each $g \in G$ to the coset gH containing it. For each $a \in G$ the *translation* $\tau_a: G/H \to G/H$ sends each gH to agH. For $a, b \in G$,

$$\pi \circ L_a = \tau_a \circ \pi, \qquad \text{and} \qquad \tau_{ab} = \tau_a \circ \tau_b.$$

12. Proposition. If H is a closed subgroup of G, there is a unique way to make G/H a manifold so that the projection $\pi: G \to G/H$ is a submersion.

For proofs of this result and of Proposition 13, see [W]. The manifold G/H so constructed is called a *coset manifold.* Since $\pi: G \to G/H$ is a submersion, a map $\phi: G/H \to N$ is smooth if and only if $\phi \circ \pi : G \to N$ is smooth (Exercise 1.10). For example, the identity $\tau_a \circ \pi = \pi \circ L_a$ shows that τ_a is smooth. Hence τ_a is a diffeomorphism since it has inverse map $\tau_{a^{-1}}$.

Recall from Definition 9.31 the notion of action $G \times M \to M$ of a Lie group on a manifold M. For a coset manifold G/H the map $G \times G/H \to G/H$ sending (a, gH) to agH is called the *natural action* of G on G/H. Obviously this action is transitive; we shall now see that every transitive action can be so represented.

If $G \times M \to M$ is an action and $o \in M$, the isotropy subgroup $H = \{g \in G: go = o\}$ is a closed subgroup of G. There is a natural map j from the coset manifold G/H into M that sends each coset aH to the point ao. This map is well defined, since

$$aH = bH \Rightarrow b^{-1}aH = H \Rightarrow b^{-1}a \in H \Rightarrow b^{-1}ao = o \Rightarrow ao = bo.$$

13. Proposition. Let $G \times M \to M$ be a transitive action and let H be its isotropy group at a point $o \in M$. Then the natural map $j: G/H \to M$ (as above) is a diffeomorphism. Hence in particular, the projection $g \to go$ is a submersion $\pi: G \to M$.

14. Examples. Spheres as Coset Manifolds.

(1) $S^n = SO(n+1)/SO(n) = O(n+1)/O(n)$. $SO(n+1)$ acts on the unit sphere S^n in $\boldsymbol{R}^{n+1}$ as a restriction of the usual action of $GL(n, \boldsymbol{R})$ on $\boldsymbol{R}^{n+1}$. This action of $SO(n+1)$ is transitive on S^n; indeed it is already transitive on the set of positively oriented orthonormal bases for $\boldsymbol{R}^{n+1}$.

The isotropy subgroup of $(1, 0, \ldots, 0) \in S^n$ consists of all elements of $SO(n+1)$ of the form $\begin{pmatrix} 1 & 0 \\ 0 & b \end{pmatrix}$, where $b \in SO(n)$. Writing this subgroup as $SO(n)$ and ignoring the natural diffeomorphism j gives $S^n = SO(n+1)/SO(n)$.

Neglecting orientation throughout produces the alternative expression $O(n+1)/O(n)$.

(2) $S^{2n+1} = SU(n+1)/SU(n) = U(n+1)/U(n)$. This is the complex analogue of (1), using the natural Hermitian product on $\boldsymbol{C}^{n+1} = \boldsymbol{R}^{2n+2}$ as in Appendix B.

(3) $S^{4n+3} = Sp(n+1)/Sp(n)$. This is the quaternionic analogue for $\boldsymbol{H}^{n+1} = \boldsymbol{C}^{2n+2} = \boldsymbol{R}^{4n+4}$. See [St]. (Note that $S^0 = O(1)$, $S^1 = U(1)$, and $S^3 = Sp(1)$. It is known that these are the only spheres that can be made Lie groups.)

Ignoring differentiability for the moment, suppose that a Lie group G acts transitively on a set Σ. If H is the isotropy group of an element o of Σ, then we still get the natural one-to-one function j from G/H onto Σ. If the subgroup H is closed, Σ will always be made a manifold by requiring j to be a diffeomorphism. Then the action is smooth, and the projection $g \to go$ is a submersion.

Using this construction we now generalize the preceding example in two slightly different ways. Note that the sphere S^n can be considered as the set of all oriented lines through 0 in $\boldsymbol{R}^{n+1}$. (Such a line meets the unit sphere in two points; the orientation picks one of them.) The unoriented lines give the projective space $P^n = S^n/\pm 1$.

15. Examples. Real Grassmann Manifolds. (1) Unoriented p-subspaces of n-space, where $n = p + q$.

$$G_{pq} = O(p + q)/O(p) \times O(q) = SO(p + q)/S(O(p) \times O(q)).$$

Let G_{pq} be the set of all p-dimensional subspaces of $\boldsymbol{R}^n$, $n = p + q$. $O(n)$ acts in a natural way on G_{pq}; we shall show the action is transitive. Let V_o be the subspace of $\boldsymbol{R}^n$ spanned by the first p vectors in the canonical basis $e_1, \ldots, e_n$. If $V \in G_{pq}$, choose an orthonormal basis $\{e_i'\}$ for $\boldsymbol{R}^n$ whose first p vectors span V. Now let a be the linear operator (matrix) carrying each e_i to e_i'. Then $a \in O(n)$ and $a(V_o) = V$.

If V_0 is invariant under $g \in O(n)$, then so is $V_o^\perp$. Hence the isotropy group of V_o is $O(p) \times O(q)$ considered as all matrices $\left(\begin{smallmatrix} g & 0 \\ 0 & h \end{smallmatrix}\right)$ with $g \in O(p)$, $h \in O(q)$. This gives the first coset expression for G_{pq}.

$SO(n)$ is also transitive on G_{pq}, for by changing one sign (if necessary), we can arrange that the e_i' basis also be positively oriented. The second coset expression follows.

(2) Oriented p-subspaces of n-space.

$$\tilde{G}_{pq} = SO(p + q)/SO(p) \times SO(q).$$

Now let $\tilde{G}_{pq}$ be the set of all *oriented* p-dimensional subspaces of $\boldsymbol{R}^n$. Again $SO(n)$ acts transitively, and for the isotropy group, note that, if $a|V_0$ is orientation-preserving, then so is $a|V_o^\perp$.

In particular, $G_{1n} = P^n$ and $\tilde{G}_{1n} = S^n$.

Since $SO(n + 1)$ is compact and connected, so are both G_{pq} and $\tilde{G}_{pq}$. Since $\dim G/H = \dim G - \dim H$, both G_{pq} and $\tilde{G}_{pq}$ have dimension pq. The map $\tilde{G}_{pq} \to G_{pq}$ that forgets orientation is a smooth two-to-one map, hence necessarily a covering map. It is known that $\pi_1(G_{pq}) \approx \boldsymbol{Z}_2$ (see, for example, [St, p. 134]). Hence $\tilde{G}_{pq}$ is simply connected. The map $W \to W^\perp$ is a diffeomorphism $G_{pq} \approx G_{qp}$; similarly for $\tilde{G}_{pq}$.

16. Lemma. Let $M = G/H$. If the projection $\pi: G \to M$ has a cross section $\lambda: M \to G$, then $\phi(m, h) = \lambda(m)h$ defines a diffeomorphism ϕ: $M \times H \to G$.

Proof. (Such a section amounts to a smooth choice of representative for each coset.) If $g \in G$, let $\psi(g) = (\pi g, (\lambda \pi g)^{-1} g)$. Since $\pi \circ \lambda = \text{id}$, the elements $\lambda \pi g$ and g have the same projection in M. Thus they are in the same coset mod H, so $(\lambda \pi g)^{-1} g \in H$. Hence $\psi(g) \in M \times H$. Both ϕ and ψ are smooth, and it is easily checked that $\phi\psi$ and $\psi\phi$ are identity maps. ∎

The topological properties of H, G, and G/H are closely related. In particular, considerable information about connectedness and fundamental groups of these three manifolds is concentrated in the proposition below. Note first that the set $\pi_0(G)$ of connected components of a Lie group G has a natural group structure. In fact, the identity component G_0 of G is a normal subgroup (Example B.4(3)), and its cosets are the components of G, thus $\pi_0(G)$ can be defined to be the quotient group G/G_0. If $\phi: G \to G'$ is a continuous homomorphism, then $\phi(G_0) \subset G'_0$ and the resulting quotient homomorphism $\phi_0: \pi_0(G) \to \pi_0(G')$ tells how ϕ treats components.

17. Proposition. For a coset manifold $M = G/H$ there is an exact sequence of groups and homomorphisms

$$0 \to \pi_2(M) \xrightarrow{\partial} \pi_1(H) \xrightarrow{i_*} \pi_1(G) \xrightarrow{\pi_*} \pi_1(M) \xrightarrow{\Delta} \pi_0(H) \xrightarrow{i_0} \pi_0(G).$$

Furthermore, i_0 is onto if M is connected.

Here i is the inclusion map $H \subset G$, and *exactness* means that the image of each homomorphism equals the kernel of the next homomorphism.

This sequence derives from the *homotopy sequence* of a *principle bundle* [St], in this case $\pi: G \to M$. Since G is a Lie group, the sequence can be extended to $\pi_0(G)$ ([St, p. 94]), and this final segment can be detached since $\pi_2(G) = 0$.

18. Lemma. If G/H and H are connected [compact], then G is connected [compact].

Proof. The connectness assertion follows from the preceding proposition. For the compactness assertion, by applying Lemma 16 to *local* sections, G can be expressed as a finite union of compact sets $\pi^{-1}K \approx K \times H$. ∎

19. Corollary. (1) The *classical groups* $O(n)$, $SO(n)$, $U(n)$, $SU(n)$, $Sp(n)$ are compact, and all are connected except $O(n)$, which has components $SO(n) = O^+(n)$ and $O^-(n)$.

(2) $SU(n)$ and $Sp(n)$ are simply connected, while $\pi_1 U(n) \approx \boldsymbol{Z}$, and

$$\pi_1 SO(n) \approx \begin{cases} \boldsymbol{Z} & \text{if } n = 2, \\ \boldsymbol{Z}_2 & \text{if } n \geq 3. \end{cases}$$

Proof. The properties of $U(n)$ will follow from the fact that it is diffeomorphic to $SU(n) \times S^1$ (Exercise 3). The proof is by induction on n in Example 14, using Lemma 18 for (1) and Proposition 17 for (2). In the nontrivial case $SO(n)$ in (2), one needs $\pi_2(S^n) = 0$ for $n \geq 3$ [St], and $SO(3)$ diffeomorphic to P^3, hence $\pi_1 SO(3) \approx \boldsymbol{Z}_2$. ∎

20. Example. $G^*_{pq} = O(p, q)/O(p) \times O(q) = O^{++}(p, q)/SO(p) \times SO(q)$.

For this variant of Example 15, let G^*_{pq} be the set of all p-dimensional negative definite subspaces V of $\boldsymbol{R}^n_p$, where $n = p + q$ and $0 < p < n$. (Then $V^\perp$ is q-dimensional and positive definite.) The semiorthogonal group $O(p, q)$ is transitive on G^*_{pq}. If V_o is the subspace of $\boldsymbol{R}^n_p$ spanned by the first p vectors in the canonical basis, then $V_o \in G^*_{pq}$ and the isotropy group of V_o is $O(p) \times O(q)$—since both V_o and $V_o^\perp$ are definite.

By a direct argument as in Example 15 (or by Exercise 6), the identity component $O^{++}(p, q)$ is already transitive on G^*_{pq}. In this case the isotropy group of V_o is evidently $SO(p) \times SO(q)$.

Example 39 will show that G^*_{pq} is diffeomorphic to $\mathbf{R}^{pq}$. Thus it follows from Proposition 17 that

$$\pi_1 O^{++}(p, q) \approx \pi_1 SO(p) \times \pi_1 SO(q).$$

Then the fundamental group of $O^{++}(p, q)$ can be read from Corollary 19(2). ($\pi_1 SO(1)$ is trivial, since $SO(1)$ is a single point.) In particular, $\pi_1 O^{++}(p, q) \approx \boldsymbol{Z}_2 \times \boldsymbol{Z}_2$ for all $p \geq 3, q \geq 3$.

REDUCTIVE HOMOGENEOUS SPACES

If a Lie group G acts on a manifold M, a metric tensor on M is G-*invariant* provided that, for each $g \in G$, the diffeomorphism $p \to gp$ is an isometry. When the action is transitive, such a metric obviously makes M a semi-Riemannian homogeneous space. It is not hard to show that every semi-Riemannian homogeneous space can be expressed as a coset manifold $M = G/H$ with a G-invariant metric. The geometry of M can then be described in Lie terms; to keep this description simple we specialize somewhat.

21. Definition. A coset manifold $M = G/H$ is *reductive* if there is an $\mathrm{Ad}(H)$-invariant subspace $\mathfrak{m}$ of $\mathfrak{g}$ that is complementary to $\mathfrak{h}$ in $\mathfrak{g}$. We call $\mathfrak{m}$ a *Lie subspace* for G/H.

Though $\mathfrak{g}$ is a direct sum $\mathfrak{m} + \mathfrak{h}$ of vector spaces, $\mathfrak{m}$ need not be closed under brackets, as $\mathfrak{h}$ is. By Corollary 2, the invariance of $\mathfrak{m}$ under Ad(H) implies $[\mathfrak{h}, \mathfrak{m}] \subset \mathfrak{m}$.

For $X \in \mathfrak{g}$, denote by $X_{\mathfrak{h}}$ and $X_{\mathfrak{m}}$ the components of X in $\mathfrak{h}$ and $\mathfrak{m}$, respectively.

The *differential map* $d\pi$ *of the projection* $\pi: G \to M = G/H$ *gives a linear isomorphism of* $\mathfrak{m}$ *onto* $T_o(M)$. (See Figure 1.) In fact, identifying $\mathfrak{h} \subset \mathfrak{g}$ as usual with $T_e(H) \subset T_e(G)$, we have $d\pi(\mathfrak{h}) = 0$. Since π is a submersion, $d\pi$ is onto, and counting dimensions shows $d\pi|\mathfrak{m}$ is an isomorphism. In effect, $\mathfrak{m}$ has become the tangent space to M at o.

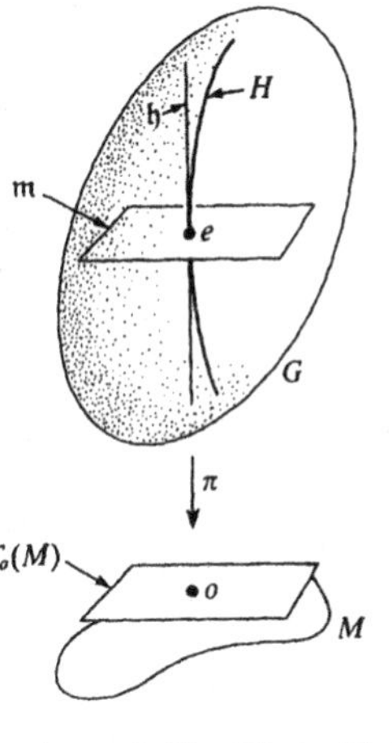

Figure 1. The Lie algebra $\mathfrak{g} = \mathfrak{h} + \mathfrak{m}$ is identified with $T_e(G)$. Then $d\pi: \mathfrak{m} \approx T_o(M)$.

22. Proposition. Let $M = G/H$ be reductive, with Lie subspace $\mathfrak{m}$.

(1) The *linear isotropy group* $\{d\tau_b: b \in H\}$ acting on $T_o(M)$ corresponds under $d\pi$ to Ad(H) on $\mathfrak{m}$.

(2) Requiring $d\pi: \mathfrak{m} \approx T_o(M)$ to be a linear isometry establishes a one-to-one correspondence between Ad(H)-invariant scalar products on $\mathfrak{m}$ and G-invariant metrics on M.

Proof. (1) The assertion is that $d\tau_h \circ d\pi = d\pi \circ \mathrm{Ad}_h$ on $\mathfrak{m}$ for all $h \in H$. This is clear, since $\tau_h \circ \pi = \pi \circ C_h$ for $h \in H$, and $\mathrm{Ad}_h(\mathfrak{m}) \subset \mathfrak{m}$.

(2) Suppose $\langle , \rangle$ is an Ad(H)-invariant scalar product on $\mathfrak{m}$.

Since $d\pi: \mathfrak{m} \approx T_o(M)$ must be a linear isometry, the scalar product $\langle , \rangle_o$ on $T_o(M)$ is determined. But then (1) implies that $d\tau_h: T_o(M) \to T_o(M)$ is a linear isometry for all $h \in H$. This fact will allow $\langle , \rangle_o$ to be extended, in a G-invariant way, over all of M. We assert that if $p = \tau_a(o) = \tau_b(o)$, then the linear isomorphisms $d\tau_{a^{-1}}, d\tau_{b^{-1}}: T_p(M) \to T_o(M)$ pull back $\langle\, , \rangle_o$ to the same scalar product $\langle\, , \rangle_p$ on $T_p(M)$. In fact $\tau_a(o) = \tau_b(o)$ means $aH = bH$, hence $b^{-1}a = h \in H$. Then $\tau_h = \tau_{b^{-1}} \circ \tau_a$; hence for all $x, y \in T_p(M)$,

$$\begin{aligned}\langle d\tau_{a^{-1}}(x), d\tau_{a^{-1}}(y)\rangle &= \langle d\tau_h\, d\tau_{a^{-1}}(x), d\tau_h\, d\tau_{a^{-1}}(y)\rangle \\ &= \langle d\tau_{b^{-1}}(x), d\tau_{b^{-1}}(y)\rangle.\end{aligned}$$

The resulting tensor $\langle\, , \rangle$ on M is easily seen to be G-invariant, and its smoothness can be derived from the existence of local sections (Exercise 1.9).

Reciprocally, if $\langle\, , \rangle$ is a G-invariant metric on M, the linear isotropy group $\{d\tau_h|_o : h \in H\}$ consists of linear isometries. Since $d\pi|\mathfrak{m}$ is required to be a linear isometry, (1) shows that $d\pi$ pulls $\langle\, , \rangle_o$ back to an Ad(H)-invariant scalar product on $\mathfrak{m}$. ∎

Assertion (2) in the above proposition generalizes readily to give a one-to-one correspondence between Ad(H)-invariant (r, s) tensors on the vector space $\mathfrak{m}$ and G-invariant (r, s) tensor fields on M.

The scheme is to treat the geometry of coset manifolds G/H as a generalization of the geometry of Lie groups G (since G/H reduces to G when $H = \{e\}$). From this viewpoint, the isomorphism $\mathfrak{m} \approx T_o(G/H)$ generalizes the canonical isomorphism $\mathfrak{g} \approx T_e(G)$, and a G-invariant metric on G/H generalizes a left-invariant metric on G. The notion of bi-invariant metric on G generalizes as follows.

23. Definition. A *naturally reductive homogeneous space* is a reductive coset manifold $M = G/H$ furnished with a G-invariant metric such that, for the corresponding scalar product on the Lie subspace $\mathfrak{m}$,

$$\langle [X, Y]_{\mathfrak{m}}, Z\rangle = \langle X, [Y, Z]_{\mathfrak{m}}\rangle \qquad \text{for} \quad X, Y, Z \in \mathfrak{m}.$$

In fact, when $H = \{e\}$, hence $\mathfrak{m} = \mathfrak{g}$, this formula is just the shift condition in Proposition 9(4).

To determine the geodesics and curvature of a naturally reductive homogeneous space $M = G/H$ we shall use submersion results from Chapter 7. The Lie subspace $\mathfrak{m}$ has a scalar product; extend this to $\mathfrak{g} = \mathfrak{h} + \mathfrak{m}$ by picking any scalar product on $\mathfrak{h}$ and defining $\mathfrak{h} \perp \mathfrak{m}$. This furnishes G with a left-invariant metric for which elements of $\mathfrak{h}$ are vertical (tangent to fibers) and elements of $\mathfrak{m}$ are horizontal (normal to fibers).

24. Lemma. With notation as above, $\pi: G \to M$ is a semi-Riemannian submersion.

Proof. In Definition 7.44, the condition (S1) holds because $T_e(H) \approx \mathfrak{h}$ is by construction a nondegenerate subspace of $T_e(G) \approx \mathfrak{g}$, and hence left-multiplication by $g \in G$ shows that $T_g(gH)$ is nondegenerate in $T_g(G)$.

To prove (S2), if $g \in G$, let $\mathscr{H}_g$ as usual be the space of horizontal vectors (normal to gH) in $T_g(G)$. Then $dL_{g^{-1}}$ carries $\mathscr{H}_g$ to $\mathscr{H}_e$, and the identity $\tau_g \circ \pi = \pi \circ L_g$ shows that $d\pi\colon \mathscr{H}_g \to T_{\pi g}(M)$ can be expressed as a composition of linear isometries $d\tau_g\, d\pi_e\, dL_{g^{-1}}$. ∎

25. Proposition. If $M = G/H$ is naturally reductive, its geodesics starting at o are given by

$$\gamma_{d\pi X}(t) = \alpha(t)o = \pi\alpha(t) \qquad \text{for all} \quad t \in \boldsymbol{R},$$

where α is the one-parameter subgroup of $X \in \mathfrak{m}$.

Proof. By Corollary 7.46, a submersion carries horizontal geodesics to geodesics, and α is horizontal since it is an integral curve of the horizontal vector field $X \in \mathfrak{m}$. We shall show that if $X, Y \in \mathfrak{m}$, then $D_X Y = \frac{1}{2}[X, Y]$. Thus α is geodesic, since $\alpha'' = D_X X = 0$.

Both $\mathfrak{m}$ and the scalar product on it are Ad(H)-invariant. Hence by a variant of Lemma 3,

$$\langle [X, V], Y\rangle = \langle X, [V, Y]\rangle \qquad \text{if} \quad X, Y \in \mathfrak{m} \quad \text{and} \quad V \in \mathfrak{h}.$$

We call this identity an $\mathfrak{h}$-*shift*, and the one in Definition 23 an $\mathfrak{m}$-*shift*.

If $X, Y \in \mathfrak{m}$, but $W \in \mathfrak{g}$, the Koszul formula gives

$$2\langle D_X Y, W\rangle = -\langle X, [Y, W]\rangle + \langle Y, [W, X]\rangle + \langle W, [X, Y]\rangle.$$

If $W \in \mathfrak{m}$, an $\mathfrak{m}$-shift in the second term cancels the first. If $W \in \mathfrak{h}$, then an $\mathfrak{h}$-shift produces the same result. Hence $D_X Y = [X, Y]/2$. ∎

Naturally reductive homogeneous spaces are complete. In fact, one-parameter subgroups are defined on the whole real line; by the preceding proposition the same is true for inextendible geodesics through o, and, by homogeneity, for all geodesics.

We now find a Lie algebra formula for curvature.

26. Proposition. Let $M = G/H$ be a naturally reductive homogeneous space. If X and Y span a nondegenerate plane in $\mathfrak{m}$, then

$$K(d\pi X, d\pi Y)_o = \frac{\frac{1}{4}\langle [X, Y]_{\mathfrak{m}}, [X, Y]_{\mathfrak{m}}\rangle + \langle [[X, Y]_{\mathfrak{h}}, X], Y\rangle}{Q(X, Y)}.$$

Proof. Continuing in the context of the preceding proof, $\pi\colon G \to M$ is a semi-Riemannian submersion, and for any $Z \in \mathfrak{g}$ the vertical component is $Z_{\mathfrak{h}}$. Hence by Theorem 7.47,

$$K(d\pi X, d\pi Y) = \bar{K}(X, Y) + \frac{\frac{3}{4}\langle [X, Y]_{\mathfrak{h}}, [X, Y]_{\mathfrak{h}}\rangle}{Q(X, Y)}, \tag{1}$$

where $\bar{K}$ is the sectional curvature of G. The previous proof showed that $D_X Y = [X, Y]/2$ for $X, Y \in \mathfrak{m}$, and it follows that

$$\langle \bar{R}_{XY}X, Y\rangle = \langle D_{[X,Y]_{\mathfrak{h}}}X, Y\rangle + \langle D_{[X,Y]_{\mathfrak{m}}}X, Y\rangle - \langle D_X D_Y X, Y\rangle. \quad (2)$$

Now

$$\begin{aligned}\langle D_X D_Y X, Y\rangle &= -\langle D_Y X, D_X Y\rangle = \tfrac{1}{4}\langle [X, Y], [X, Y]\rangle \\ &= \tfrac{1}{4}\langle [X, Y]_{\mathfrak{h}}, [X, Y]_{\mathfrak{h}}\rangle + \tfrac{1}{4}\langle [X, Y]_{\mathfrak{m}}, [X, Y]_{\mathfrak{m}}\rangle, \quad (3)\end{aligned}$$

since $\mathfrak{h} \perp \mathfrak{m}$. Using an $\mathfrak{m}$-shift gives

$$\langle D_{[X,Y]_{\mathfrak{m}}}X, Y\rangle = \tfrac{1}{2}\langle [[X, Y]_{\mathfrak{m}}, X], Y\rangle = \tfrac{1}{2}\langle [X, Y]_{\mathfrak{m}}, [X, Y]\rangle. \quad (4)$$

Abbreviate $[X, Y]_{\mathfrak{h}}$ to V; then by the Koszul formula,

$$2\langle D_V X, Y\rangle = -\langle V, [X, Y]\rangle + \langle X, [Y, V]\rangle + \langle Y, [V, X]\rangle.$$

By an $\mathfrak{h}$-shift the last two terms on the right are equal, hence

$$\langle D_{[X,Y]_{\mathfrak{h}}}X, Y\rangle = -\tfrac{1}{2}\langle [X, Y]_{\mathfrak{h}}, [X, Y]\rangle + \langle [[X, Y]_{\mathfrak{h}}, X], Y\rangle. \quad (5)$$

To the contributions from (3), (4), and (5) add $\tfrac{3}{4}\langle [X, Y]_{\mathfrak{h}}, [X, Y]_{\mathfrak{h}}\rangle$ from (1). Terms of the latter type cancel, leaving the required result. ∎

A point p of a semi-Riemannian manifold is a *pole* provided the exponential map $\exp_p$ is a diffeomorphism. The vertex of a paraboloid of revolution is a pole, and in Hadamard's theorem (10.22) every point of H is a pole. For a homogeneous space, if one point is a pole, every point is.

27. Lemma. If $M = G/H$ is a naturally reductive homogeneous space for which o is a pole, then the map $(X, h) \to (\exp X)h$ is a diffeomorphism of $\mathfrak{m} \times H$ onto G.

Proof. By Proposition 25, if $\exp: \mathfrak{g} \to G$ is the Lie exponential map (Appendix B), then

$$\exp_o \circ d\pi = \pi \circ \exp: \mathfrak{m} \to M.$$

Call this map $E: \mathfrak{m} \to M$. By hypothesis, $\exp_o: T_o(M) \to M$ is a diffeomorphism, and since $d\pi: \mathfrak{m} \to T_o(M)$ is a linear isomorphism, it is also a diffeomorphism. Thus E is a diffeomorphism.

Now $\lambda = \exp \circ E^{-1}: M \to G$ is smooth, and $\pi \circ \lambda = \pi \circ \exp \circ E^{-1} = E \circ E^{-1} = \mathrm{id}$, so λ is a section of $\pi: G \to M$. By Lemma 16, the map $(p, h) \to \lambda(p)h$ is a diffeomorphism of $M \times H$ onto G. Following $E \times \mathrm{id}$ by this diffeomorphism gives the required diffeomorphism $\mathfrak{m} \times H \to G$, since $\lambda(EX) \cdot h = (\exp X)h$. ∎

SYMMETRIC SPACES

First we show how to express a given semi-Riemannian symmetric space M in terms of Lie groups. Since M is homogeneous, $I(M)$ is transitive on M; hence by Exercise 6 the identity component $G = I_0(M)$ is transitive. Thus M can be identified with a coset manifold G/H, where H is the isotropy group of a point o of M. The global symmetry at o then gives further structure.

28. Lemma. With notation as above, if ζ is the global symmetry of $M = G/H$ at o, the map σ sending g to $\zeta g \zeta$ is an involutive automorphism of G. The set $F = \text{Fix}(\sigma) = \{g \in G : \sigma(g) = g\}$ of fixed points of σ is a closed subgroup of G such that $F_0 \subset H \subset F$.

Proof. Since ζ is involutive, $\zeta^{-1} = \zeta$. Thus σ is conjugation by ζ, so σ is an involutive automorphism. Consequently, σ carries $I_0(M)$ to itself, and F is a closed subgroup of G (Example B.4).

If $h \in H$, then the differential map of the isometry $\sigma(h)$ at o is $d\zeta_o\, dh_o\, d\zeta_o$, which is just dh_o, since $d\zeta_o = -\text{id}$. Because M is connected, $\sigma(h) = h$ by Proposition 3.62. Thus $H \subset F$.

To show $F_0 \subset H$, recall that since F_0 is connected, by Appendix B it is generated by the points $\alpha(t)$ of the one-parameter subgroups of F. Thus it suffices to show that $\alpha(t) \in H$. But $\sigma(\alpha(t)) = \alpha(t)$, and hence ζ and $\alpha(t)$ commute. Thus

$$\zeta(\alpha(t)o) = \alpha(t)\zeta(o) = \alpha(t)o \qquad \text{for all } t.$$

Since o is an isolated fixed point of the symmetry ζ, it follows that $\alpha(t)o = o$ for $|t|$ small, hence for all t. Thus $\alpha(t)$ is in the isotropy group H of o. ∎

The preceding makes it clear how to construct symmetric spaces from Lie group data.

29. Theorem. Let H be a closed subgroup of a connected Lie group G. Let σ be an involutive automorphism of G such that $F_0 \subset H \subset F = \text{Fix}(\sigma)$. Then any G-invariant metric tensor on $M = G/H$ makes M a semi-Riemannian symmetric space such that $\zeta \circ \pi = \pi \circ \sigma$, where ζ is the global symmetry of M at o and π is the projection $G \to M$.

Proof. (a) *There is a unique function* $\zeta : M \to M$ *such that* $\zeta \circ \pi = \pi \circ \sigma$.

If $g \in G$, then $\zeta(\pi g) = \pi(\sigma g)$ is a consistent definition, because $\pi g_1 = \pi g_2$ means $g_1 H = g_2 H$, and, since σ fixes H, then $\sigma(g_1)H = \sigma(g_2)H$; that is, $\pi\sigma g_1 = \pi\sigma g_2$.

(b) ζ *is a diffeomorphism.* That ζ is smooth derives as usual from the existence of local sections of the submersion π. Because σ is involutive, it follows that ζ is involutive, hence $\zeta^{-1} = \zeta$.

(c) $d\zeta_o = -\text{id}$. Clearly $\zeta(o) = o$. If $y \in T_o(M)$, we anticipate (2) of the next lemma, which implies that there is a $Y \in \mathfrak{g}$ such that $d\sigma(Y) = -Y$ and $d\pi(Y) = y$. Thus

$$d\zeta(y) = d\zeta(d\pi Y) = d\pi(d\sigma Y) = d\pi(-Y) = -y.$$

(d) $\tau_{\sigma g} = \zeta\tau_g\zeta$ *for* $g \in G$. In fact, for all $a \in G$,

$$\zeta\tau_g \pi a = \zeta\pi(ga) = \pi\sigma(ga) = \pi(\sigma g \cdot \sigma a) = \tau_{\sigma g}\pi(\sigma a) = \tau_{\sigma g}\zeta\pi a.$$

(e) *Relative to any G-invariant metric tensor on M, ζ is an isometry.* If $v \in T_g(M)$, let $v_o = d\tau_{g^{-1}}(v) \in T_o(M)$. Then using (d) and (c),

$$\begin{aligned}\langle d\zeta v, d\zeta v\rangle &= \langle d\zeta\, d\tau_g(v_o), d\zeta\, d\tau_g(v_o)\rangle \\ &= \langle d\tau_{\sigma g} d\zeta(v_o), d\tau_{\sigma g} d\zeta(v_o)\rangle = \langle d\zeta(v_o), d\zeta(v_o)\rangle \\ &= \langle -v_o, -v_o\rangle = \langle v, v\rangle.\end{aligned}$$

The proof of the theorem is completed by observing that if a homogeneous space has a global symmetry ζ at a single point o, it has one at every point $p = \tau(o)$, namely $\tau\zeta\tau^{-1}$. ∎

The existence of the automorphism σ produces striking effects on the Lie algebra $\mathfrak{g}$.

30. Lemma. Let $H \subset G$ and σ be as in the preceding theorem, with $\mathfrak{h} \subset \mathfrak{g}$ the Lie algebras of $H \subset G$. Then

(1) $\mathfrak{h} = \{X \in \mathfrak{g} : d\sigma(X) = X\}$.
(2) $\mathfrak{g}$ is the direct sum of $\mathfrak{h}$ and the subspace $\mathfrak{m} = \{X \in \mathfrak{g} : d\sigma(X) = -X\}$.
(3) $\text{Ad}_h(\mathfrak{m}) \subset \mathfrak{m}$ for all $h \in H$.
(4) $[\mathfrak{h}, \mathfrak{h}] \subset \mathfrak{h}$, $\quad [\mathfrak{h}, \mathfrak{m}] \subset \mathfrak{m}$, $\quad [\mathfrak{m}, \mathfrak{m}] \subset \mathfrak{h}$.

Proof. (1) Since $\sigma|H = \text{id}$, if $X \in \mathfrak{h}$, then $d\sigma(X) = X$. Conversely, suppose $d\sigma(X) = X$. If α is the one-parameter subgroup of X, then α and $\sigma \circ \alpha$ have the same initial velocity. But $\sigma \circ \alpha$ is also a one-parameter subgroup, hence $\sigma \circ \alpha = \alpha$. This means that α lies in F—in fact in its identity component F_0. Since $F_0 \subset H$, we get $X \in \mathfrak{h}$.

(2) For $X \in \mathfrak{g}$ let $X_{\mathfrak{h}} = (X + d\sigma X)/2$ and $X_{\mathfrak{m}} = (X - d\sigma X)/2$. Then $X = X_{\mathfrak{h}} + X_{\mathfrak{m}}$. Because σ is involutive, so is $d\sigma$; hence $d\sigma(X_{\mathfrak{h}}) = X_{\mathfrak{h}}$ and $d\sigma(X_{\mathfrak{m}}) = -X_{\mathfrak{m}}$. Thus $\mathfrak{g} = \mathfrak{h} + \mathfrak{m}$, and the sum is direct, since evidently $\mathfrak{h} \cap \mathfrak{m} = 0$.

(3) If $X \in \mathfrak{m}$ and $h \in H$, we must show that $d\sigma(\text{Ad}_h X) = -\text{Ad}_h X$. Since $\sigma(h) = h$, the automorphisms σ and C_h commute; in fact, $\sigma C_h(a) = \sigma(hah^{-1}) = h\sigma(a)h^{-1}$. Thus

$$d\sigma(\text{Ad}_h X) = d(\sigma C_h)X = d(C_h\sigma)(X) = \text{Ad}_h\, d\sigma X = \text{Ad}_h(-X) = -\text{Ad}_h X.$$

(4) The first inclusion holds since H is a Lie subgroup, the second since $\mathfrak{m}$ is $\mathrm{Ad}(H)$-invariant, but all three are easy consequences of the fact that $\mathfrak{h}$ and $\mathfrak{m}$ are the $+1$ and -1 eigenspaces of $d\sigma$, respectively.

For example, if $X, Y \in \mathfrak{m}$, then

$$d\sigma[X, Y] = [d\sigma X, d\sigma Y] = [-X, -Y] = [X, Y].$$

Hence $[X, Y] \in \mathfrak{h}$. ∎

We call $(G/H, \sigma, B)$ *symmetric data* provided that

(1) H is a closed subgroup of a connected Lie group G.

(2) σ is an involutive automorphism of G such that $F_0 \subset H \subset F = \mathrm{Fix}(\sigma)$.

(3) B is an $\mathrm{Ad}(H)$-invariant scalar product on $\mathfrak{m} = \{X \in \mathfrak{g} \colon d\sigma X = -X\}$.

Theorem 29 shows how such data make G/H a symmetric space under the G-invariant metric corresponding (as in Proposition 22) to B. This construction is presumed when we say that $M = G/H$ is a symmetric space. *G/H is then a naturally reductive homogeneous space with $\mathfrak{m} = \{X \colon d\sigma X = -X\}$ as Lie subspace.* In fact, by Lemma 30, $\mathfrak{m}$ is an $\mathrm{Ad}(H)$-invariant complement to $\mathfrak{h}$, and the shift condition in Definition 23 is trivial since $[\mathfrak{m}, \mathfrak{m}] \subset \mathfrak{h}$.

In this context, $\mathfrak{m}$ will always denote the -1 eigenspace of $d\sigma$, as above.

31. Proposition. Let $M = G/H$ be a semi-Riemannian symmetric space.

(1) The geodesics starting at o are given by

$$\gamma_{d\pi X}(t) = \alpha(t)o = \pi\alpha(t) \qquad \text{for all} \quad t,$$

where α is the one-parameter subgroup of $X \in \mathfrak{m}$.

(2) The curvature tensor at o is given by $R_{xy}z = d\pi[[X, Y], Z]$, where $x, y, z \in T_o(M)$ correspond under $d\pi$ to $X, Y, Z \in \mathfrak{m}$. If x and y span a nondegenerate plane, then

$$K(x, y) = \langle [[X, Y], X], Y \rangle / Q(X, Y).$$

Proof. Since $M = G/H$ is naturally reductive with Lie subspace $\mathfrak{m}$, Proposition 25 gives (1). Proposition 26 gives the sectional curvature formula in (2), since $[X, Y] \in \mathfrak{h}$ for $X, Y \in \mathfrak{m}$. The formula for the curvature operator will follow from Corollary 3.42 once we check that the multilinear function

$$(X, Y, Z, W) \to \langle [[X, Y], Z], W \rangle$$

is curvaturelike on $\mathfrak{m}$. Obviously it is skew-symmetric in X and Y. Cyclic symmetry in X, Y, Z is just the Jacobi identity. Finally, since $[X, Y] \in \mathfrak{h}$, skew-symmetry in Z and W follows from the fact that $\mathfrak{m}$ and the scalar product on it are both $\mathrm{Ad}(H)$-invariant. ∎

To illustrate the theory we now take a well-known symmetric space, but use its geometry only to set up appropriate symmetric data, and from this deduce in particular its geodesics and curvature.

32. Example. $S^n = SO(n+1)/SO(n)$ as in Example 14. As the unit sphere in $\boldsymbol{R}^{n+1}$, S^n is symmetric, with symmetry ζ at $o = (1, 0, \ldots, 0)$ given by $(t_0, t_1, \ldots, t_n) \to (t_0, -t_1, \ldots, -t_n)$.

(1) *The automorphism σ of $SO(n+1)$.* By the column vector conventions, ζ can be regarded as the diagonal matrix with entries $1, -1, \ldots, -1$. Thus by Lemma 28, if $a \in SO(n+1)$,

$$\sigma(a) = \zeta a \zeta = \left[\begin{array}{c|ccc} a_{00} & -a_{01} & \cdots & -a_{0n} \\ \hline -a_{10} & & & \\ \vdots & & (a_{ij}) & \\ -a_{n0} & & & \end{array}\right] \qquad (1 \le i,\ j \le n).$$

So $\mathrm{Fix}(\sigma)$ is $S(O(1) \times O(n))$, and F_0 is the isotropy group $1 \times SO(n) \approx SO(n)$.

(2) *The subspace $\mathfrak{m} = \{X \in \mathfrak{o}(n+1) : d\sigma X = -X\}$.* Since $\zeta = \zeta^{-1}$, σ is conjugation by ζ. So by Lemma 6(1), $d\sigma$ is also conjugation by ζ on the Lie algebra $\mathfrak{o}(n+1)$. It follows that $\mathfrak{m}$ consists of all matrices

$$X = \begin{pmatrix} 0 & -{}^t x \\ x & \tilde{0} \end{pmatrix},$$

where $\tilde{0}$ is the $n \times n$ zero matrix and x is an arbitrary column vector—regarded as usual as an element of $\boldsymbol{R}^n$. Write $X \leftrightarrow x$ for the resulting correspondence between $\boldsymbol{R}^n$ and $\mathfrak{m}$.

(3) *The* Ad(H)*-invariant inner product on* $\mathfrak{m}$. Under $X \leftrightarrow x$, the dot product $x \cdot y$ on $\boldsymbol{R}^n$ corresponds to $B(X, Y) = -\frac{1}{2}\,\mathrm{trace}\, XY = \frac{1}{2} X \cdot Y$ on $\mathfrak{m}$. Here $H = SO(n) \subset SO(n+1)$ and we know that the trace is $\mathrm{Ad}(SO(n+1))$-invariant. It will follow from (5) that the corresponding metric tensor on S^n is its usual one.

(4) *Geodesics.* Let γ be a geodesic of S^n starting at o. By Proposition 31, $\gamma(t) = \exp(tX)o$ for some $X \in \mathfrak{m}$. Direct computation of $\gamma(t)$ using the power series for $\exp(tX) = e^{tX}$ shows that γ is the great circle parametrization $(\cos t)o + (\sin t)x/|x|$, where $X \leftrightarrow x$.

(5) *Identifications.* In (3), $\boldsymbol{R}^n$ is tacitly identified with the last n coordinate space of $\boldsymbol{R}^{n+1}$. Hence canonical isomorphism identifies $T_o(S^n)$ with $\boldsymbol{R}^n$. Then in the notation of (3), $x = \gamma'(0)$. But X is the initial velocity of the one-parameter subgroup projecting to γ. Hence in these terms, the isomorphism $d\pi \colon \mathfrak{m} \approx T_o(S^n)$ is just $X \leftrightarrow x$.

(6) *Linear isotropy.* Apply Proposition 22(1). If $h = \left(\begin{smallmatrix}1 & 0\\ 0 & b\end{smallmatrix}\right) \in H = SO(n)$ and $X \in \mathfrak{m}$, then

$$\mathrm{Ad}_h X = hXh^{-1} = \begin{pmatrix}1 & 0\\ 0 & b\end{pmatrix}\begin{pmatrix}0 & -{}^t x\\ x & 0\end{pmatrix}\begin{pmatrix}1 & 0\\ 0 & b^{-1}\end{pmatrix} = \begin{pmatrix}0 & -{}^t x b^{-1}\\ bx & 0\end{pmatrix}$$

(skew-symmetric since ${}^t b = b^{-1}$). Thus the linear isotropy action of on $T_o(S^n)$ is, via the identifications, just the usual action of $SO(n)$ on $\boldsymbol{R}^n$.

This implies, for example, that S^n is frame-homogeneous.

(7) *Curvature.* In terms of the Lie subspace $\mathfrak{m}$, Proposition 31 asserts that $R_{XY}Z = [[X, Y], Z]$. If, as usual, x, y, z are the corresponding vectors in $\boldsymbol{R}^n \approx T_o(S^n)$, we readily compute first $[X, Y] = \left(\begin{smallmatrix}S & 0\\ 0 & 0\end{smallmatrix}\right)$, where $S = (x_i y_j - x_j y_i)$, and then

$$R_{XY}Z = \begin{pmatrix}0 & -{}^t(Sz)\\ Sz & 0\end{pmatrix}.$$

Thus $R_{XY}Z$ corresponds under identification to Sz, which is just $(x \cdot z)y - (y \cdot z)x$. Hence S^n has constant curvature 1.

33. Remark. Normal Symmetric Spaces. Let G/H and σ be as usual for symmetric data, but let B be a scalar product on $\mathfrak{g}$ that is invariant under both Ad G and $d\sigma$. (If G is semisimple, then by Lemma 5 its Killing form is such a scalar product.) Then

$$\mathfrak{h} \perp \mathfrak{m} \quad \text{relative to} \quad B.$$

In fact, if $X \in \mathfrak{m}$ and $V \in \mathfrak{h}$, then $B(X, V) = B(d\sigma X, d\sigma V) = B(-X, V) = -B(X, V)$.

It follows that $\mathfrak{h}$ and $\mathfrak{m}$ are nondegenerate relative to B. Hence $B|_{\mathfrak{m}}$ is an Ad(H)-invariant scalar product on $\mathfrak{m}$, so the data $(G/H, \sigma, B|_{\mathfrak{m}})$ make $M = G/H$ a symmetric space. Furthermore, because of the shift property in Lemma 3, the curvature formula in Proposition 31 simplifies to

$$K(X, Y) = B([X, Y], [X, Y])/Q(X, Y),$$

where X, Y span a nondegenerate plane in $\mathfrak{m} \approx T_o(M)$.

RIEMANNIAN SYMMETRIC SPACES

In the Riemannian case the study of symmetric spaces is concentrated on the following extreme types:

34. Definition. A Riemannian symmetric space $M = G/H$ is of *compact type* if the Killing form B of G is negative definite, and of *noncompact type* if B is negative definite on $\mathfrak{h}$ and positive definite on $\mathfrak{m}$.

In fact, every simply connected Riemannian symmetric space can be expressed as a product whose factors are compact, noncompact, or Euclidean [H]. The topological and geometrical properties of these types are quite distinctive. (Below, we write, for example, Ric > 0 to mean that the associated quadratic form is positive definite.)

35. Theorem. Let $M = G/H$ be a Riemannian symmetric space.

(1) If M is of compact type, then $K \geq 0$ and Ric > 0, hence M is compact and $\pi_1(M)$ is finite.

(2) If M is of noncompact type, then $K \leq 0$ and Ric < 0, hence M is diffeomorphic to Euclidean space $\boldsymbol{R}^n$ (noncompact, simply connected). Furthermore, G is diffeomorphic to $H \times \boldsymbol{R}^n$.

Proof. Once the curvature assertions have been proved, the topological consequences readily follow:

For (1), Myers' theorem can be applied. In fact, M is complete and, since Ric is positive definite, Ric$(u, u) \geq a > 0$ holds on the unit sphere in some one $T_p(M)$—hence by the homogeneity of M it holds everywhere.

For (2), Proposition 36 will show that M is simply connected. Then by Hadamard's theorem (10.22) M is diffeomorphic to $\boldsymbol{R}^n$ and furthermore has poles, so Lemma 27 gives the assertion about G.

We shall compute curvature only in the commonly occurring special case in which the metric tensor on M derives from an inner product $cB|_{\mathfrak{m}}$, where B is the Killing form of G, and c is a constant with $c > 0$ if M has noncompact type, but $c < 0$ if M has compact type. (The general case uses also some linear algebra like that in the proof of Lemma 9.15; see [H], for example.) Thus by Remark 33, $K(X, Y) = cB([X, Y], [X, Y])/Q(X, Y)$ for $X, Y \in \mathfrak{m}$. Then $[X, Y] \in \mathfrak{h}$, so Definition 34 gives $K \geq 0$ in the compact case, $K \leq 0$ in the noncompact case. Since M is Riemannian, it follows that Ric ≥ 0 and Ric ≤ 0, respectively. Thus it remains to show that Ric$(X, X) = 0$ implies $X = 0$.

Let $X, E_2, \ldots, E_n$ be an orthonormal basis for $\mathfrak{m}$. In neither case does K change sign; hence,

$$\begin{aligned} \mathrm{Ric}(X, X) = 0 &\Rightarrow K(X, E_i) = 0 && \text{for all } i, \\ &\Rightarrow B([X, E_i], [X, E_i]) = 0 && \text{for all } i, \\ &\Rightarrow B([X, Y], [X, Y]) = 0 && \text{for all } Y \in \mathfrak{m}. \end{aligned}$$

Since $cB|_{\mathfrak{m}}$ is an inner product, it follows that $[X, Y] = 0$ for all Y; that is, $ad_X(\mathfrak{m}) = 0$. Since $ad_X \mathrm{ad}_X(\mathfrak{h}) = \mathrm{ad}_X[X, \mathfrak{h}] \subset \mathrm{ad}_X(\mathfrak{m}) = 0$, we conclude that $\mathrm{ad}_X \mathrm{ad}_X = 0$. By the definition of Killing form, $B(X, X) = 0$. Thus $X = 0$. ∎

36. Proposition (S. Kobayashi). A homogeneous Riemannian manifold with $K \leq 0$ and $\text{Ric} < 0$ is simply connected.

The result follows from these three facts about a homogeneous Riemannian manifold M:

(1) Every maximal geodesic of M is either one-to-one or periodic.
(2) If M is not simply connected, it contains a periodic geodesic.
(3) If $K \leq 0$ and $\text{Ric} < 0$, then M contains no periodic geodesics.

Proof of (1). It suffices to show that every geodesic loop is smoothly closed. Suppose $\gamma\colon [0, b] \to M$ has unit speed. By Corollary 9.38 there is a Killing vector field X on M such that $X_p = \gamma'(0)$ at $p = \gamma(0) = \gamma(b)$. Using the conservation lemma (9.26),

$$\begin{aligned}\langle \gamma'(0), \gamma'(b)\rangle &= \langle X_p, \gamma'(b)\rangle = \langle X_p, \gamma'(0)\rangle \\ &= \langle \gamma'(0), \gamma'(0)\rangle = 1.\end{aligned}$$

By the Schwarz inequality the unit vectors $\gamma'(0)$ and $\gamma'(b)$ are equal.

Proof of (2). The simply connected covering $\mathscr{k}\colon \tilde{M} \to M$ is not trivial; that is, there are points $p \neq q$ in $\tilde{M}$ such that $\mathscr{k}(p) = \mathscr{k}(q)$. Being homogeneous Riemannian, M is complete, and hence so is $\tilde{M}$. Thus by the Hopf–Rinow theorem, there is a geodesic β from p to q. Since $\mathscr{k} \circ \beta$ is not one-to-one, the result follows from (1).

Proof of (3). Assume that $\gamma\colon \mathbf{R} \to M$ is a periodic geodesic. Since Ricci curvature is negative definite (and M is Riemannian) there is a tangent vector x at $\gamma(0)$ such that $\langle R_{x\gamma'(0)}x, \gamma'(0)\rangle < 0$. Again there is a Killing vector field X such that $X_{\gamma(0)} = x$.

Let $h(t) = \langle X_{\gamma(t)}, X_{\gamma(t)}\rangle$. Then h is a periodic function. Since X_γ is a Jacobi field on γ, the computation in the proof of (10.19) shows that

$$h'' = 2(|X'|^2 - \langle R_{X\gamma'}X, \gamma'\rangle) \geq 0.$$

But since h is periodic, $h'' = 0$. Thus $\langle R_{X\gamma'}X, \gamma'\rangle = 0$, contradicting the inequality above. ∎

DUALITY

For normal symmetric spaces (Remark 33) there is a remarkable duality that, for our purposes, can be described as follows.

37. Definition. Normal symmetric spaces $M = G/H$ and $M^* = G^*/H^*$ are *dual* provided there exist

(1) a Lie algebra isomorphism $\tilde{\delta}: \mathfrak{h} \to \mathfrak{h}^*$ such that $B^*(\tilde{\delta}V, \tilde{\delta}W) = -B(V, W)$ for all $V, W \in \mathfrak{h}$;

(2) a linear isometry $\delta: \mathfrak{m} \to \mathfrak{m}^*$ such that $[\delta X, \delta Y] = -\tilde{\delta}[X, Y]$ for all $X, Y \in \mathfrak{m}$.

Thus on $\mathfrak{h}$, brackets are preserved and B has signs reversed, while on $\mathfrak{m}$, B is preserved and brackets have signs reversed.

Under the identifications $\mathfrak{m} \approx T_o(M)$ and $\mathfrak{m}^* \approx T_o(M^*)$, δ induces a linear isometry $\delta: T_o(M) \to T_o(M^*)$. Thus duals, M and M^*, have the same dimension and same index.

38. Lemma. Dual symmetric spaces $M = G/H$ and $M^* = G^*/H^*$ have opposite curvatures; that is, $K(\delta\Pi) = -K(\Pi)$ for every nondegenerate plane Π in $T_o(M)$, where $\delta: T_o(M) \to T_o(M^*)$ is the induced linear isometry.

Proof. If X and Y span a nondegenerate plane in $\mathfrak{m} \approx T_o(M)$, then

$$K(\delta X, \delta Y) = \frac{B^*([\delta X, \delta Y], [\delta X, \delta Y])}{Q(\delta X, \delta Y)} = \frac{B^*(\tilde{\delta}[X, Y], \tilde{\delta}[X, Y])}{Q(X, Y)}$$

$$= \frac{-B([X, Y], [X, Y])}{Q(X, Y)} = -K(X, Y). \quad \blacksquare$$

The Grassmann manifolds $\tilde{G}_{pq}$ of Example 15 can be made symmetric spaces by generalizing Example 32 from 1, n to p, q. In a similar way G^*_{pq} from Example 20 becomes a symmetric space dual to $\tilde{G}_{pq}$.

39. Example. Dual Grassmann Manifolds.

$$\begin{aligned} \tilde{G}_{pq} &= SO(p+q)/SO(p) \times SO(q) \\ G^*_{pq} &= O^{++}(p, q)/SO(p) \times SO(q) \end{aligned} \qquad (p \geq 1, \quad q \geq 1).$$

(1) $\tilde{G}_{pq}$. Conjugation by the (p, q) signature matrix ε is an involutive automorphism of $SO(n)$, $n = p + q$. Since $\varepsilon^{-1} = \varepsilon$, $\sigma(g) = \varepsilon g \varepsilon$. Hence

$$\sigma\begin{pmatrix} a & b \\ c & d \end{pmatrix} = \begin{pmatrix} a & -b \\ -c & d \end{pmatrix}, \qquad \text{where} \quad a \text{ is } p \times p, \quad d \text{ is } q \times q.$$

Thus $\mathrm{Fix}(\sigma) = S(O(p) \times O(q))$, whose identity component is $SO(p) \times SO(q)$. By Lemma 6(1), $d\sigma$ on $\mathfrak{o}(n)$ is also conjugation by ε. Thus its -1 eigenspace $\mathfrak{m}$ consists of all matrices

$$X = \begin{pmatrix} 0 & -{}^t x \\ x & 0 \end{pmatrix},$$

where x is an arbitrary $q \times p$ matrix. As in Example 8, we denote the space of all such x by $\boldsymbol{R}^{pq}$. Thus $X \leftrightarrow x$ is a linear isomorphism identifying $\mathfrak{m}$ with $\boldsymbol{R}^{pq}$.

On $\mathfrak{o}(n+1)$ let B be the inner product $B(X, Y) = -\frac{1}{2}$ trace $XY = \frac{1}{2}X \cdot Y$. The factor $\frac{1}{2}$ is introduced so that, as in the case $S^n = \tilde{G}_{1n}$, the restriction $B|_{\mathfrak{m}}$ corresponds under $X \leftrightarrow x$ to the dot product $x \cdot y$ on $\boldsymbol{R}^{pq}$.

Since B is a negative multiple of the Killing form of $\mathfrak{o}(n+1)$, Remark 33 applies, making $\tilde{G}_{pq}$ a symmetric space, in fact a Riemannian symmetric space of compact type. Hence $K \geq 0$ and Ric > 0. As mentioned earlier, $\tilde{G}_{pq}$ is simply connected and has dimension pq. (G_{pq} can be made symmetric, in virtually the same way, so that $\tilde{G}_{pq} \to G_{pq}$ is a Riemannian covering.)

(2) G^*_{pq}. Let σ on $O^{++}(n)$ again be conjugation by the (p, q) signature matrix. Then $\text{Fix}(\sigma) = SO(p) \times SO(q)$, and the -1 eigenspace $\mathfrak{m}^*$ consists of all $X = \begin{pmatrix} 0 & {}^t x \\ x & x \end{pmatrix}$, where x is an arbitrary $p \times q$ matrix.

This time we use $B(X, Y) = \frac{1}{2}$ trace XY, a positive multiple of the Killing form of $\mathfrak{o}(p, q)$. Example 8(2) shows that the restriction of B to $\mathfrak{m}^*$ is an inner product corresponding to the dot product on $\boldsymbol{R}^{pq}$. But B is negative definite on $\mathfrak{h} = \mathfrak{o}(p) + \mathfrak{o}(q)$. Thus G^*_{pq} becomes a Riemannian symmetric space of noncompact type. Hence $K \leq 0$, Ric < 0, and G^*_{pq} is diffeomorphic to $\boldsymbol{R}^{pq}$. Furthermore, $O^{++}(p, q)$ is diffeomorphic to $SO(p) \times SO(q) \times \boldsymbol{R}^{pq}$.

Analogous to the identification $S^n = \tilde{G}_{1n}$ is $H^n = G^*_{1n}$, representing hyperbolic space as the set of timelike lines through 0 in $\boldsymbol{R}^{n+1}_1$.

(3) *Duality*. The subgroup $H = SO(p) \times SO(q)$ is the same for both, so let $\tilde{\delta}$ be the identity map on $\mathfrak{h} = \mathfrak{o}(p) \times \mathfrak{o}(q)$. The scalar products used in (1) and (2) are indeed negatives on $\mathfrak{h}$. Let $\delta\colon m \to m^*$ send $\begin{pmatrix} 0 & -{}^t x \\ x & 0 \end{pmatrix}$ to $\begin{pmatrix} 0 & {}^t x \\ x & 0 \end{pmatrix}$. This map is clearly a linear isometry, and a simple computation shows that $[\delta X, \delta Y] = -[X, Y]$ as required for duality.

Thus $\tilde{G}_{pq}$ and G^*_{pq} are duals, with opposite curvature as specified in Lemma 38. The simplest case is the duality between $S^n = G_{1n}$ and $H^n = G^*_{1n}$.

SOME COMPLEX GEOMETRY

We describe briefly some relations between real and complex geometry, with applications to symmetric spaces.

Let V be a vector space over the complex numbers $\boldsymbol{C}$. If only real scalars are used, then V becomes a real vector space and scalar multiplication by $\sqrt{-1}$ is an $\boldsymbol{R}$-linear operator J on V. Since $J^2 = -\text{id}$, J is nonsingular. If V has complex dimension n, then its real dimension is $2n$. In fact, if $e_1, \ldots, e_n$ is a complex basis, it is easy to check that $e_1, \ldots, e_n, Je_1, \ldots, Je_n$ is a real basis.

A complex line L through 0 in V (that is, a complex one-dimensional subspace) is a real two-dimensional subspace, since if $0 \neq x \in L$, then x, Jx

is a real basis for L. Such *holomorphic sections* are exactly the two-dimensional real subspaces that are invariant under J.

Reciprocally, let W be a real vector space. An $\mathbf{R}$-linear operator J on W such that $J^2 = -\mathrm{id}$ is called a *complex structure* on W. Then the natural definition of complex scalar multiplication as $(a + \sqrt{-1}\,b)x = ax + bJx$ makes W a complex vector space.

Evidently an $\mathbf{R}$-linear operator that *preserves* J (that is, commutes with J) is $\mathbf{C}$-linear.

40. Definition. A *Hermitian scalar product* on a complex vector space V is a function $h: V \times V \to \mathbf{C}$ such that

(1) $h(v, w)$ is $\mathbf{C}$-linear in v;
(2) $h(w, v) = \overline{h(v, w)}$;
(3) h is nondegenerate; that is, $h(v, w) = 0$ for all w implies $v = 0$.

Then $h(v, w)$ is additive in w, but $h(v, cw) = \bar{c}h(v, w)$. Since $h(v, v)$ is always real, terms such as *positive definite* or *semidefinite* are meaningful as before.

On $\mathbf{C}^n$ the analogue of the dot product is the (positive definite) *natural Hermitian product* $(v, w) = v \cdot \overline{w} = \sum v_i \overline{w}_i$.

It is easy to verify that the real part, $\operatorname{Re} h$, of a Hermitian scalar product h on V is a (real) scalar product relative to which J is both orthogonal and skew-adjoint.

41. Example. The Indefinite Unitary Group $U(p, q)$. (1) *Definition.* $U(p, q)$ consists of those matrices in $GL(n, \mathbf{C})$, $n = p + q$, that (as $\mathbf{C}$-linear operators on $\mathbf{C}^n$) preserve the Hermitian scalar product

$$h(x, y) = -x_1\bar{y}_1 - \cdots - x_p\bar{y}_p + x_{p+1}\bar{y}_{p+1} + x_n\bar{y}_n.$$

If ε is the (p, q) signature matrix, then $h(x, y) = \varepsilon x \cdot \bar{y}$, and it follows that $U(p, q) = \{g \in GL(n, \mathbf{C}) : {}^t\bar{g}\varepsilon = \varepsilon g^{-1}\}$. Thus $U(p, q)$ is a closed subgroup of $GL(n, \mathbf{C})$. Note that $U(n, 0) = U(0, n) = U(n)$.

(2) *Lie algebra* $\mathfrak{u}(p, q)$. Recall that $\mathfrak{u}(n)$ consists of skew-Hermitian matrices: ${}^t\overline{X} = -X$. By (1) it follows, as in the analogous real case $\mathfrak{o}(p, q)$, that $\mathfrak{u}(p, q)$ consists of all complex $n \times n$ matrices such that ${}^t\overline{X} = -\varepsilon X \varepsilon$. These have the form

$$X = \begin{pmatrix} a & {}^t\bar{x} \\ x & b \end{pmatrix}, \qquad \text{where} \quad a \in \mathfrak{u}(p), \quad b \in \mathfrak{u}(q), \quad x \in \mathbf{C}^{pq}.$$

As in the real case, $\mathbf{C}^{pq}$ denotes the space of $q \times p$ (or $p \times q$) complex matrices Thus

$$\mathfrak{u}(p, q) = \mathfrak{u}(p) + \mathfrak{u}(q) + \mathbf{C}^{pq} \qquad \text{(direct sum).}$$

(3) *Trace form.* For X as above and, similarly, $Y = \begin{pmatrix} c & {}^t y \\ y & d \end{pmatrix}$,

$$\text{trace } XY = X \cdot {}^t Y = -a \cdot \bar{c} - b \cdot \bar{d} + (x \cdot \bar{y} + \bar{x} \cdot y).$$

We saw in Example 8 that $a \cdot \bar{c}$ is a (real, positive definite) inner product on $\mathfrak{u}(p)$; similarly for $b \cdot \bar{d}$ on $\mathfrak{u}(q)$. The sum in parentheses is just $2 \operatorname{Re} x \cdot \bar{y}$, an inner product on $\boldsymbol{C}^{pq} \approx \boldsymbol{R}^{2pq}$. Thus the trace form on $\mathfrak{u}(p, q)$ is a real scalar product that is negative definite on $\mathfrak{u}(p) + \mathfrak{u}(q)$ and positive definite on $\boldsymbol{C}^{pq} \approx \boldsymbol{R}^{2pq}$. (Like $\mathfrak{u}(n)$, $\mathfrak{u}(p, q)$ is a real but not a complex vector space.)

An *almost complex structure* on a smooth manifold M is a (1, 1) tensor field J such that $J^2 = -\text{id}$. Thus J smoothly assigns a complex structure J_p to each tangent space $T_p(M)$. If a manifold M admits an almost complex structure, it is not hard to show that M is even-dimensional and orientable.

42. Definition. A semi-Riemannian manifold M with almost complex structure J is a *Kähler manifold* provided

(1) J preserves the metric; that is, $\langle JX, JY \rangle = \langle X, Y \rangle$ for all $X, Y \in \mathfrak{X}(M)$;

(2) J is parallel; that is, $D_X(JY) = J(D_X Y)$ for all $X, Y \in \mathfrak{X}(M)$.

The simplest example of a Kähler manifold is Euclidean space $\boldsymbol{R}^{2n}$ with its usual metric and with J derived from $\boldsymbol{R}^{2n} \approx \boldsymbol{C}^n$; explicitly, $J(e_{2k-1}) = e_{2k}$ and $J(e_{2k}) = -e_{2k-1}$.

The *holomorphic curvature* of a Kähler manifold is the restriction of sectional curvature function K to nondegenerate holomorphic sections. It can be seen as assigning to each nonnull tangent vector $x \neq 0$ the sectional curvature $K(x, Jx)$. In fact, the holomorphic plane spanned by x, Jx is nondegenerate since (1) above implies $Jx \perp x$, hence $Q(x, Jx) = \langle x, x \rangle^2 \neq 0$.

43. Proposition. Let $M = G/H$ be a semi-Riemannian symmetric space, and let J_o be an $\operatorname{Ad}(H)$-invariant complex structure on $\mathfrak{m}$ that preserves the scalar product on $\mathfrak{m}$. Then there is a unique G-invariant almost complex structure J on M such that $d\pi\colon \mathfrak{m} \approx T_o(M)$ is J-preserving, and M is Kähler relative to J.

Proof. By the remark following Proposition 22 the (1, 1) tensor J_o on $\mathfrak{m}$ gives rise to a G-invariant (1, 1) tensor field J on M. The scalar product on $\mathfrak{m}$ mentioned in the proposition is, as usual, the one corresponding via $d\pi$ to the scalar product on $T_o(M)$. Since J_o and $J \mid T_o(M)$ correspond via $d\pi$, the latter has $J^2 = -\text{id}$ and preserves the scalar product on $T_o(M)$. These properties then hold at all points of M, since both the metric tensor and J are G-invariant.

To show that J is parallel on M it suffices to show that if Z is a parallel vector field on a geodesic γ, then JZ is also parallel. (Compare proof of Proposition 8.10.) By homogeneity we can suppose that γ starts at $o \in M$ and thus is the projection of a suitable one-parameter subgroup α of G. By Exercise 10, $Z(s) = d\tau_{\alpha(s)}Z(0)$ for all s. Since J is G-invariant,

$$(JZ)(s) = J(Z(s)) = J(d\tau_{\alpha(s)}Z(0)) = d\tau_{\alpha(s)}(JZ(0)).$$

Hence by Exercise 10, JZ is parallel. ∎

The last part of the proof generalizes automatically to the assertion: *A G-invariant tensor field on a symmetric space $M = G/H$ is parallel.*

The dual Grassmann manifolds of Example 39 have complex analogues constructed in a strictly analogous way, as follows.

44. Example.

$$CG_{pq} = U(p+q)/U(p) \times U(q) = SU(p+q)/S(U(p) \times U(q)).$$

(1) *Coset manifold.* Let CG_{pq} be the set of all complex p-dimensional subspaces of $\mathbf{C}^n$, $n = p + q$. $U(n)$ acts naturally on $\mathbf{C}^n$ and thereby on CG_{pq}. Let V_o be the subspace spanned by the first p elements of the natural basis for $\mathbf{C}^n$. The isotropy group of V_o is $U(p) \times U(q)$, since invariance of V_o implies invariance of $V_o^{\perp}$.

(2) *Symmetric space.* Conjugation by the (p, q) signature matrix ε is an involutive automorphism σ of $U(n)$ for which $\mathrm{Fix}(\sigma) = U(p) \times U(q)$.

Since $d\sigma$ is also conjugation by ε, the -1 eigenspace $\mathfrak{m}$ consists of the elements of $u(n)$ of the form $X = \begin{pmatrix} 0 & -{}^t\bar{x} \\ x & 0 \end{pmatrix}$, where x is an arbitrary $q \times p$ complex matrix.

On $\mathfrak{u}(n)$ we know by Example 8 that $B(X, Y) = -\frac{1}{2}\,\mathrm{trace}\, XY = \frac{1}{2}X \cdot \bar{Y}$ is an inner product that is a negative multiple of the Killing form. (As before, the factor $\frac{1}{2}$ means that $B|_{\mathfrak{m}}$ corresponds under $X \leftrightarrow x$ to the natural Hermitian product $x \cdot \bar{y}$ on $\mathbf{C}^{pq}$.) Thus CG_{pq} becomes a Riemannian symmetric space of compact type.

(3) *Properties.* A simple computation shows that the adjoint action of $H = U(p) \times U(q)$ on $\mathfrak{m} \approx \mathbf{C}^{pq}$ corresponds to its action $((a, b), x) \to axb^{-1}$ on $\mathbf{C}^{pq}$.

Scalar multiplication by $\sqrt{-1}$ on $\mathbf{C}^{pq} \approx \mathfrak{m}$ gives a corresponding complex structure J_o on $\mathfrak{m}$. Evidently J_o is $\mathrm{Ad}(H)$-invariant. Hence by Proposition 43, J_o determines an almost complex structure J on CG_{pq} making it a Kähler manifold.

Applying Proposition 17 to the alternate description $CG_{pq} = SU(n)/S(U(p) \times U(q))$ shows that CG_{pq} is simply connected, since $SU(n)$ is simply connected (Corollary 19) and $S(U(p) \times U(q))$ is connected (Exercise 4).

Thus CG_{pq} *is a compact simply connected* $2pq$-dimensional (*Riemannian*) *Kähler symmetric space* with $K \geq 0$ and Ric > 0.

45. Example.

$$CG^*_{pq} = U(p, q)/U(p) \times U(q) = SU(p, q)/S(U(p) \times U(q)).$$

(1) *Coset manifold.* Let CG^*_{pq} be the set of all those complex p-dimensional subspaces of C^n, $n = p + q$, on which the indefinite Hermitian product $\varepsilon x \cdot \bar{y}$ of Example 41 is negative definite. The indefinite unitary group $U(p, q)$ is transitive on CG^*_{pq}, and the isotropy group of the usual subspace V_o is $U(p) \times U(q)$.

(2) *Symmetric space.* As before, conjugation by the (p, q) signature matrix is an involutive automorphism of $U(p, q)$ whose fixed point set is $U(p) \times U(q)$.

In view of the description of $\mathfrak{u}(p, q)$ in Example 41, the -1 eigenspace $\mathfrak{m}^*$ of $d\sigma$ consists of all elements of the form $X = \begin{pmatrix} 0 & {}^t\bar{x} \\ x & 0 \end{pmatrix}$, where x is a $p \times q$ complex matrix.

Let $B^*(X, Y) = \frac{1}{2}\,\text{trace}\, XY = \frac{1}{2}X \cdot \bar{Y}$ on $\mathfrak{u}(p, q)$. Then B^* is a positive multiple of the Killing form of $\mathfrak{u}(p, q)$. Thus it follows from the properties of B^* given by Example 41(3) that it makes CG^*_{pq} a Riemannian symmetric space of noncompact type.

(3) *Properties.* As in the preceding example, the isomorphism $\mathfrak{m}^* \approx C^{pq}$ induces a complex structure on $\mathfrak{m}^*$ that, transmitted to CG^*_{pq}, makes it Kähler. Thus CG^*_{pq} *is a (Riemannian) Kähler symmetric space with* $K \leq 0$ *and* Ric < 0, *and diffeomorphic to* R^{2pq}.

(4) *Duality.* CG_{pq} and CG^*_{pq} are duals. In fact they share the same subalgebra $\mathfrak{h} = \mathfrak{u}(p) + \mathfrak{u}(q)$, on which B and $B^* = -B$ certainly have opposite signs. Let $\delta: \mathfrak{m} \to \mathfrak{m}^*$ send $\begin{pmatrix} 0 & -{}^t\bar{x} \\ x & 0 \end{pmatrix}$ to $\begin{pmatrix} 0 & {}^t\bar{x} \\ x & 0 \end{pmatrix}$. By construction, $B^*(\delta X, \delta Y) = x \cdot y = B(X, Y)$, and, just as in Example 39(3), $[\delta X, \delta Y] = -[X, Y]$.

The simplest real Grassmann manifolds are spheres (or projective spaces) and hyperbolic spaces. The complex Grassmannians above provide complex analogues as follows: *Complex projective space* CP^n is $CG_{1n} = U(n + 1)/U(1) \times U(n)$. Dual to it is *complex hyperbolic space* $CH^n = CG^*_{1n} = U(1, n)/U(1) \times U(n)$. The metrics on the symmetric spaces we have constructed are in effect defined only up to multiplication by a positive constant. Here we choose the constant so that CP^n will have holomorphic curvature 1 (see below). Then duality determines a unique metric on CH^n.

46. Corollary. (1) Complex projective space CP^n is a compact, simply connected, $2n$-dimensional (Riemannian) Kähler symmetric space with constant holomorphic curvature 1, and $\frac{1}{4} \leq K \leq 1$. Each of its geodesics is simply closed of length 2π.

(2) Complex hyperbolic space CH^n is a (Riemannian) Kähler symmetric space of constant holomorphic curvature -1, and $-1 \leq K \leq -\frac{1}{4}$. It is diffeomorphic to $\mathbf{R}^{2n}$, and each of its geodesics in one-to-one.

Proof. (1) (a) The linear isotropy group is transitive on the set of holomorphic planes in $T_o(M)$ and on the unit sphere in $T_o(M)$. In fact, we saw in Example 44 that this action corresponds to the action of $U(1) \times U(n)$ on $\mathbf{C}^n$ by $(e^{i\vartheta}, A)x = e^{-i\vartheta}Ax$. For $\vartheta = 0$, this is already transitive on complex lines, that is, holomorphic planes. But scalar multiplication by all $e^{-i\vartheta} \in U(1)$ is clearly transitive on the unit circle in each holomorphic plane.

(b) It is immediate from (a) that holomorphic curvature is constant on $T_o(M)$ and hence, by homogeneity, is constant everywhere. Multiply B in Example 44 by 4, so $B(X, Y) = -2 \operatorname{trace} XY = 2X \cdot \bar{Y}$. Let $E_1, E_2 \in \mathfrak{m}$ correspond to elements of the natural basis for $\mathbf{C}^n \approx \mathfrak{m}$. In view of (a), an arbitrary tangent plane on CP^n has sectional curvature $K(E_1, Y)$, where $Y = \cos\vartheta\, JE_1 + \sin\vartheta\, E_2$. A simple computation gives $K(E_1, Y) = \frac{1}{4}(1 + 3\cos^2\vartheta)$. Hence $\frac{1}{4} \leq K \leq 1$. Taking $\vartheta = 0$ shows that CP^n has constant holomorphic curvature 1.

(c) *Geodesics.* By (a) and the homogeneity of CP^n, all its geodesics are congruent. Thus it suffices to exhibit a single simply periodic geodesic of length 2π.

The one-parameter subgroup of $E_1 \in \mathfrak{m}$ is

$$\alpha(s) = \left(\begin{array}{cc|c} \cos s & -\sin s & \\ \sin s & \cos s & 0 \\ \hline \multicolumn{2}{c|}{0} & I_{n-2} \end{array}\right).$$

Evidently α is periodic of period 2π, but the geodesic $\gamma = \pi \circ \alpha$ is periodic of period π. In fact, for $s > 0$, the first return of α to $H = U(1) \times U(n)$ is at the first zero of $\sin s$; namely, $s = \pi$. The proof of Proposition 36 shows that $\gamma | [0, \pi]$ is smoothly closed and that γ is simply periodic. Thus by construction, the period of γ is π. Since γ has speed $B(E_1, E_1)^{1/2} = 2$, its length is 2π.

The other stated properties of CP^n are those of CG_{pq} in general.

(2) The curvature properties of CH^n derive from (1) since duality δ reverses curvature signs. Holomorphic curvature is also reversed, since δ commutes with J, hence preserves holomorphic planes.

By the Hadamard theorem (10.22) the geodesics of every CG^*_{pq}, indeed of every Riemannian symmetric space of noncompact type, are one-to-one. ■

If $n = 1$, then in both cases above, K has constant value 1. Since these two surfaces are complete and simply connected, space form classification shows that CP^1 is the unit 2-sphere, and CH^1 is the real hyperbolic plane $H^2(1)$. However, for $n > 1$ the preceding proof shows that K is nonconstant, filling the prescribed interval. Furthermore CP^n has Euler number $n + 1$, hence for $n > 1$ is not even homeomorphic to S^{2n}. (Compare the *Sphere Theorem* in[CE].)

Proposition 8.28 shows that for sectional curvature to fill a finite interval is a distinctively Riemannian phenomenon.

As with the reals and complexes, there are also *quaternionic* projective spaces [H]. More generally, one can define the symmetric space of all vector subspaces of dimension m and index μ in $\boldsymbol{F}^n_\nu$ ($\mu \leq \nu$ and $m - \mu \leq n - \nu$), where $\boldsymbol{F} = \boldsymbol{R}, \boldsymbol{C}$, or $\boldsymbol{H}$. For this and further information about indefinite symmetric spaces, see [Wo].

Exercises

G and H denote Lie groups.

1. Prove: (a) Product structures on $G \times H$ make it a Lie group. (b) If $k: \tilde{G} \to G$ is the simply connected covering of G, then $\tilde{G}$ can be made a Lie group so that k is a homomorphism. (Hint: If μ is the multiplication of G, lift $\mu \circ (k \times k)$ through k.) (c) If K is a closed normal subgroup of G, the quotient group G/K—as a coset manifold—is a Lie group.

2. Let $\phi: G \to G'$ be a smooth homomorphism onto G'. Prove: (a) The kernel K of ϕ is a closed normal subgroup such that the induced group isomorphism $\phi: G/K \to G'$ is a diffeomorphism, hence a Lie group isomorphism. (Hint: G acts on G'.) (b) If ϕ has a smooth cross section, then G is diffeomorphic to $G' \times K$. (Hint: (a) is important because results for $\pi: G \to G/H$ now apply to $\phi: G \to G'$.)

3. Show that $U(n)$ is diffeomorphic to $SU(n) \times S^1$ by considering $\det: U(n) \to U(1) = S^1$.

4. Prove that $S(U(p) \times U(q))$ is diffeomorphic to $SU(p) \times SU(q) \times S^1$. (Hint: Find $a(h)$ so that $(g, h) \to (a(h)g, h)$ is a diffeomorphism of $SU(p) \times U(q)$ onto $S(U(p) \times U(q))$.)

5. If $p, q \geq 1$, prove (a) $U(p, q)$ is diffeomorphic to $SU(p, q) \times U(1)$; (b) $SU(p, q)$ is diffeomorphic to $S(U(p) \times U(q)) \times \boldsymbol{R}^{2pq}$; (c) hence, using Exercise 4, $\pi_1 U(p, q) \approx \boldsymbol{Z} \times \boldsymbol{Z}$.

6. Show that if a Lie group G acts transitively on a connected manifold M, then so does the identity component G_0 of G. (Hint: Since $\pi: G \to M$ is a submersion and G_0 is open, components of G project to open sets of M.)

7. Let H be a closed subgroup of G. Let B be an Ad(G)-invariant scalar product on $\mathfrak{g}$ such that $\mathfrak{h}$ is nondegenerate. Prove: (a) $\mathfrak{m} = \mathfrak{h}^{\perp}$ is a Lie subspace on which $B|_{\mathfrak{m}}$ makes G/H a homogeneous space, said to be *normal.* (b) G/H is naturally reductive. (c) $\langle R_{XY}X, Y\rangle = B([X, Y], [X, Y])$ for $X, Y \in \mathfrak{m} \approx T_o(G/H)$.

8. Prove: (a) If a connected Lie group G admits a bi-invariant Riemannian metric, then every point of G lies on a one-parameter subgroup, that is, $\exp(\mathfrak{g}) = G$. (b) $SL(2, \boldsymbol{R})$ admits a bi-invariant metric but not a bi-invariant Riemannian metric. (Hint: $\begin{pmatrix} -2 & 0 \\ 0 & -1/2 \end{pmatrix}$ has no square root in $SL(2, \boldsymbol{R})$.)

9. Let G consist of all elements of $GL(3, \boldsymbol{R})$ of the form

$$\begin{pmatrix} 1/a & 0 & 0 \\ 0 & a & b \\ 0 & 0 & 1 \end{pmatrix} \quad \text{with} \quad a > 0.$$

Prove (a) G is a closed subgroup of $GL(3, \boldsymbol{R})$; (b) G is nonabelian (hence $\mathfrak{g}$ is nonabelian); (c) G does not admit a bi-invariant semi-Riemannian metric.

10. Let $M = G/H$ be a symmetric space. Prove: (a) If α is the one-parameter subgroup of G associated with $X \in \mathfrak{m}$, then $d\tau_{\alpha(s)}\colon T_oM \to T_{\gamma(s)}M$ is parallel translation along the geodesic $\pi \circ \alpha = \gamma$. (Hint: Use Lemma 8.30). (b) The Levi-Civita connection of M is independent of the choice of G-invariant metric on M.

11. (a) Show that if a curve γ in a symmetric space is reversed by the global symmetry ζ at each of its points, then γ is a geodesic. (b) Deduce Proposition 31(1).

12. Find an almost complex structure on G^*_{2q} (Example 39) making it a Kähler manifold.

13. (a) $GL^+(n, \boldsymbol{R})$ is diffeomorphic to $SL(n, \boldsymbol{R}) \times \boldsymbol{R}^+$. (b) $SL(n, \boldsymbol{R})$ is diffeomorphic to $SO(n) \times \boldsymbol{R}^d$, where $d = (n + 2)(n - 1)/2$.

14. Prove: (a) In Exercise 1(b), $\pi_1(G)$ is isomorphic to the kernel of $\mathscr{k}$. (Hint: See proof of Proposition 7.4.) (b) If N is a discrete normal subgroup of a connected Lie group, then $gn = ng$ for all $n \in N$, $g \in G$. (c) The fundamental group of a Lie group is abelian.

15. Given symmetric data $(G/H, \sigma, B)$ and $(G'/H', \sigma', B')$, let $\phi\colon G \to G'$ be a Lie group isomorphism such that $\phi(H) = H'$, $\phi \circ \sigma = \sigma' \circ \phi$, and $d\phi\colon \mathfrak{m} \to \mathfrak{m}'$ (well defined) preserves the scalar products B, B'. Prove that there is a unique isometry $\mu\colon M \to M'$ of the resulting symmetric spaces such that $\mu \circ \pi = \pi' \circ \phi$.

16. Prove that if $(G/H, \sigma, B)$ is symmetric data, then so is $(G/H_0, \sigma, B)$, and the natural map $G/H_0 \to G/H$ of the resulting symmetric spaces is a semi-Riemannian covering.

17. For $\tilde{G}_{pq}$ as in Example 39, the map $W \to W^\perp$ is an isometry. (Hint: Derive one isometry from Exercise 15 and follow it with a suitable τ_g.)
18. If G is a connected Lie group acting transitively on a set Σ, analyze the effect of a change of origin o on G/H, where the isotropy group H of o is assumed closed. Then show that $\tilde{G}_{pq}$ is unchanged, as a Riemannian manifold, if the origin V_0 is taken to be last p coordinate space instead of first p coordinate space.
19. Represent S_ν^n and H_ν^n as dual symmetric spaces.
20. (a) Establish $\Sigma(p, q) = GL^+(n, \boldsymbol{R})/SO(p, q)$ and $U(p, q)/SO(p, q)$ as dual symmetric spaces. (b) Identify $\Sigma(p, q)$ as the space of all scalar products of index p on $\boldsymbol{R}^n$, $n = p + q$. (c) In the Riemannian case identify $P = \Sigma(n, 0) = \Sigma(0, n)$ with the space of all positive definite symmetric $n \times n$ matrices, and prove P diffeomorphic to $\boldsymbol{R}^{n(n+1)/2}$. (d) Use Exercise 16 to show that in the indefinite case, Σ is not simply connected.

12 GENERAL RELATIVITY; COSMOLOGY

While special relativity remains an entirely satisfactory theory within its range of applicability, there is no way for it to encompass gravity. Newton's law of gravitation, involving space but not time, was lost with the relativistic union of these notions. In the years after 1905, Einstein became convinced that gravitation should be expressed in terms of curvature. By 1915 he had found out how to do it, and in the general theory of relativity flat Minkowski spacetime gives way to spacetimes of arbitrary curvature.

This chapter is a brief account of the fundamentals of general relativity and of the remarkable information it gives about the origin and development of the universe. We leave for the next chapter its equally fruitful application to the neighborhood of a single star.

FOUNDATIONS

General relativity models aspects of the physical universe by *spacetimes*, that is, time-oriented connected four-dimensional Lorentz manifolds. In this section we describe informally some of its fundamentals, starting with with the most obvious.

A. Special Relativity is a Special Case of General Relativity.
In fact special relativity is the general relativity of Minkowski spacetime M, which by Corollary 8.24 is the unique complete, flat, simply connected, Lorentz manifold of dimension 4. Thus any feature of general relativity can be tested in Minkowski spacetime, and, reciprocally, it is natural to carry

over to the general theory those features that do not depend on the distinctive properties of $M \approx \boldsymbol{R}_1^4$. In particular, events, material and lightlike particles, proper time, energy–momentum, and observers are defined just as before.

In sufficiently large regions of the universe, gravity can scarcely be ignored. Thus special relativity is at best a local theory; it is the flatness of M that is physically significant, not its global properties of completeness and simple connectedness. Indeed, by basing itself on arbitrary Lorentz manifolds, the general theory opens the way to the study of global questions.

B. General Relativity Is Locally Approximated by Special Relativity.

If p is an event in a spacetime M, then special relativity makes sense in the tangent space $T_p(M) \approx \boldsymbol{R}_1^4$, and the exponential map $\exp_p$ provides a comparison. We have seen that on sufficiently small neighborhoods—and so long as curvature does not intrude—$T_p(M)$ is a good geometric approximation of M. Historically this approximation, often in the form of normal coordinates, has been essential in attaching physical significance to the geometry of M. In particular, as in special relativity, we call a timelike future-pointing unit vector $u \in T_p(M)$ an *instantaneous observer* at p. The orthogonal decomposition $T_p(M) = Ru + u^\perp$ splits the tangent space into the observer's *time axis* Ru and *restspace* $u^\perp$. As before, if α is a particle with $\alpha(t_0) = p$, then $\alpha'(t_0) = au + x$, $x \in u^\perp$, and correcting x by the time dilation a gives the instantaneous *velocity* x/a of α as measured by u. (Thus the speed $|x|/a$ is 1 for light and less for material particles, as usual.) Similarly, if P is the energy-momentum of α at p, the expression $P = Eu + \vec{P}$, $\vec{P} \in u^\perp$, gives the *energy* E and *momentum* $\vec{P}$ of α at p as measured by u.

As this terminology shows, Newtonian physics also bears comparison with general relativity. Roughly speaking, the scope of the three theories is as given in the accompanying table. When the data meet Newtonian limitations,

	Gravitation	Speeds
General relativity	Arbitrary	Arbitrary
Special relativity	Negligible	Arbitrary
Newtonian physics	Weak	Low

general relativity gives approximately Newtonian results. (The next chapter will provide the classic example.)

C. Gravity Dominates in the Large.

Among nuclear binding forces, electromagnetism, and gravity, the latter is weakest. However, the range of nuclear forces is so small that they are ignored a priori by a theory that models the real world by smooth manifolds. At a somewhat larger scale, the electromagnetic repulsion between two electrons is enormously larger than the gravitational attraction. But charge appears with both plus and minus signs: electromagnetism can attract as well as repel, while gravity can only attract. Thus in aggregates electromagnetic effects can cancel, but gravitational effects accumulate. In fact, at larger scales, gravity is utterly dominant. Although electromagnetism fits into its framework, the essential function of general relativity is to give a spacetime explanation of gravity.

D. Free Fall Is Geodesic; Matter Curves Spacetime.

Free fall is motion solely under the influence of gravity. Newtonian physics distinguishes two cases: accelerating and nonaccelerating. For example, consider two identical spaceships: S_1 coasting in at an angle toward a giant star (idealized infinite radius); S_2 also in free fall but far from the nearest galaxy. According to Newton, S_1, accelerating under the gravitational attraction of the star, follows a curved orbit in space and hence has a curved worldline in Newtonian space-time (6.5). By contrast, S_2 moves in a straight line at constant speed, hence its worldline is straight, that is, geodesic. But if the ships are sealed, neither crew can experimentally determine which ship it is in (*principle of equivalence*). In both, undisturbed objects appear to be at rest and, if pushed, their relative motions do not differ from ship to ship. As in the previous Newtonian dichotomy, rest versus constant velocity, Einstein refused to accept a theoretical distinction between states not experimentally different, and boldly declared that *all free fall is geodesic in spacetime*. The gravitational effect of the star is not to bend the worldline of S_1 but to bend the spacetime in which it is geodesic. (Though the worldline of S_1 is "straight" its orbit in *space* as perceived by observers on the star is curved.)

E. Gravity as Curvature.

If the star in the preceding section is, more realistically, a uniform round ball, then there is a simple experiment enabling the sealed crews to distinguish S_1 from S_2. It suffices to release a few pebbles at rest in each ship. In S_2 they will remain at rest, but in S_1 they will move. Suppose for simplicity that S_1 is falling directly toward the star and that the pebbles are arranged as in Figure 1. Then a and c move inward toward o, while b and d move outward away from o. The Newtonian explanation is that a and c are falling directly toward the center of the star, while by the inverse-square gravitational law, d accelerates toward the star more rapidly than o, and b correspondingly lags.

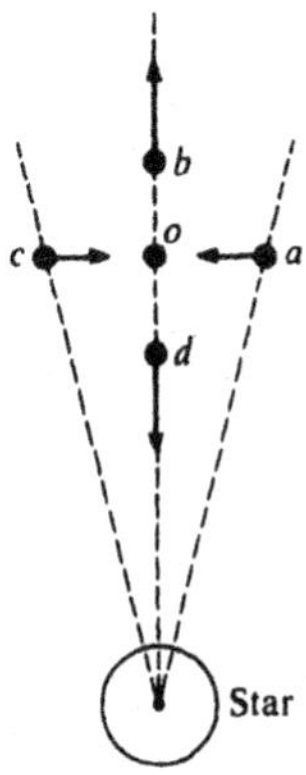

Figure 1. Relative accelerations observed by *o*.

In the relativistic explanation (Chapter 13) the freely falling particles are modeled by timelike geodesics in spacetime. As discussed in Chapter 8, the relative position of neighbors of the pebble γ_o are given by Jacobi fields Y on γ_o. Changes in relative position result from relative acceleration Y'', which by the Jacobi equation is $R_{Y\gamma'}(\gamma')$. In this way curvature, in its role as tidal force, replaces the Newtonian notion of gravitation.

In general, an instantaneous observer $u \in T_p(M)$ measures gravity by the tidal force operator $F_u: u^\perp \to u^\perp$ (Definition 8.8).

F. Sources of Gravity.

"Matter" is an undefined term that we use intuitively to mean all the stuff of the universe. In Newtonian physics the unique source of gravitation is the *mass* of matter. Relativistically, gravitation springs from the *energy–momentum* of matter, to which mass is but one contributor. For a particular form of matter modeled in a spacetime M the energy–momentum content is described infinitesimally by a *stress–energy tensor field* T on M. Lacking a general definition of matter, there can be no general recipe for constructing T, but there are some empirical rules. Let u be an instantaneous observer at $p \in M$. On $u^\perp$ the spatial part of T typically generalizes the classical stress tensor as measured by u; like it, T is a symmetric (0, 2) tensor. The energy density measured by u is $T(u, u)$ and for known forms of matter is non-negative. Finally, conservation of energy–momentum is expressed infinitesimally by $\operatorname{div} T = 0$.

The terms *infinitesimal* and *instantaneous* in the discussions above signal the replacement of global action-at-a-distance Newtonian gravity by a direct-contact differential version.

Two spacetimes on which matter has been modeled are *physically equivalent* provided that there is an isometry of the spacetimes that preserves the matter models.

Geometric units are used (Remark 6.7).

THE EINSTEIN EQUATION

Matter is gravitationally significant only as a carrier of energy–momentum, so for its effect as a source of gravitation (*alias* curvature) we must look to the stress–energy tensor T. But how is T related to the curvature tensor? A brief sketch, in his own words, of Einstein's struggles to answer this question appears in Section 17.7 of [MTW]. Always demanding simplicity, he proposed the formula $G = kT$, where G is some variant of Ricci curvature and k is a constant. Einstein tried several possibilities for G, testing them notably on the problem of the precession of the perihelion of Mercury. The hard work having been done, it is by now easy to pick an obvious candidate. If div T is to vanish, then $G = kT$ implies div $G = 0$. But for Ricci curvature, Proposition 3.54 asserts div Ric $= \frac{1}{2}dS$. Thus subtracting half the scalar curvature $S = C(\text{Ric})$ from Ric produces a good result.

1. Definition. The *Einstein gravitational tensor* of a spacetime M is $G = \text{Ric} - \frac{1}{2}Sg$.

2. Lemma. (1) G is a symmetric (0, 2) tensor field with divergence zero.

(2) $\text{Ric} = G - \frac{1}{2}C(G)g$.

Proof. (1) Both Ric and g are symmetric (0, 2) tensors, hence G is. A direct computation (Exercise 3.10) shows that $\text{div}(Sg) = dS$. Then

$$\text{div}\, G = \text{div}(\text{Ric} - \tfrac{1}{2}Sg) = \tfrac{1}{2}(dS - dS) = 0.$$

(2) Since $C(g) = 4$, $C(G) = C(\text{Ric}) - \frac{1}{2}SC(g) = -S$. Thus the definition of G gives

$$\text{Ric} = G + (\tfrac{1}{2}S)g = G - \tfrac{1}{2}C(G)g. \qquad \blacksquare$$

By (2), the Einstein tensor and the Ricci tensor contain exactly the same information. In fact, since $S = C(\text{Ric})$, each has the same formal expression in terms of the other.

General relativity flows from the following law.

3. The Einstein Equation. If M is a spacetime containing matter with stress–energy tensor T, then

$$G = 8\pi T,$$

where G is the Einstein gravitational tensor.

Many arguments have been advanced to lend plausibility to the Einstein equation, based usually on comparison with Newtonian physics at low speeds and weak gravitation. (The universal constant $k = 8\pi$ is determined from such comparisons; see [HE, Section 3.4] or [MTW, p. 406].) But general relativity cannot be derived; like Newton's gravitational and motion laws, it is Einstein's assertion of how the macroscopic universe works. Tested experimentally in a variety of situations it has given accurate results, notably for cases in which Newtonian physics is inaccurate or inapplicable.

The Einstein equation implies that the stress–energy tensor is a symmetric (0, 2) tensor with divergence zero. As we shall see, the equation $\operatorname{div} T = 0$ gives dynamical laws for the matter that produces T. Roughly speaking,

$G = 8\pi T$ tells how matter determines Ricci curvature;

$\operatorname{div} T = 0$ tells how Ricci curvature moves this matter.

But recall that, as in Section E, the full curvature tensor controls the relative motion of *test particles*—those whose energy–momentum makes negligible contribution to the stress–energy tensor.

If $T = 0$, that is, if M is Ricci flat, then M is said to be a *vacuum* (or *empty*).

PERFECT FLUIDS

The flow of a fluid could be described literally by a vast swarm of particles in a spacetime M. Instead of this discrete model it is easier to deal with a smooth model, where the 4-velocity of the flow is given by a timelike unit vector field U on M. Intuitively, the integral curves of U are the average worldlines of the "molecules" of the fluid.

The classical stress tensor measures internal forces in a body in space by giving at each point m the forces across all surface elements through m. To motivate Definition 4 we apply this scheme heuristically to the spacetime flow of a so-called *perfect fluid*.

Fix $m \in M$. If $v \in T_m(M)$ is a unit vector, then in a hypersurface through m perpendicular to v let $B(v)$ be a small coordinate cube ("box") centered at m.

Let P_B^+ be the total energy–momentum of molecules in $B(v)$ that are crossing from the $-v$ to the $+v$ side of $B(v)$; correspondingly let P_B^- derive from those crossing from $+v$ to $-v$. Then for another unit vector $w \in T_m(M)$, *let* $T(v, w)$ *be the limit as* vol $B(v) \to 0$ *of the* w *component of* $P_B = P_B^+ - P_B^-$.

Now let $u = U_m$ and consider the following choices of v, w.

(1) $T(u, u) = \rho(m)$, *energy density* at m.

The infinitesimal observer u, riding with the flow, can consider $B(u)$ as a local restspace since its tangent space at m is $u^\perp$. Then $P_B^- = 0$, and the usual decomposition $P_B^+ = P_B = E_B u + \vec{P}_B$ gives the energy E_B and momentum $\vec{P}_B$ of the box B as measured by u. Then by the definition above,

$$T(u, u) = \lim_{\text{vol } B \to 0} \frac{E_B}{\text{vol } B}.$$

Clearly this is the energy density of the fluid as measured by u.

(2) If $x, y \in u^\perp$, then $T(x, y) = p(m)\langle x, y\rangle$, where $p(m)$ is the *pressure* at m.

Since $x \perp u$, the box $B(x)$ is a three-dimensional spacetime. Let $B(x) = \Sigma \times I$, where Σ is a spacelike patch of surface through m and I is a time interval of length Δt (see Figure 2). Then if A is the area of Σ,

$$T(x, y) = \lim_{A \to 0} \frac{1}{A} \left\{ \lim_{\Delta t \to 0} \frac{\langle P_B, y\rangle}{\Delta t} \right\}.$$

Here $\vec{P}_B = \vec{P}_B^+ - \vec{P}_B^-$, and a molecule of fluid contributes to $\vec{P}_B^+$ if it crosses Σ from $-x$ to $+x$ during the time interval I. Force is the time rate of change of momentum, hence the time limit above is the force F^+ exerted by the $-x$ side of Σ on the $+x$ side, minus the reverse F^- (but $-F^- = F^+$).

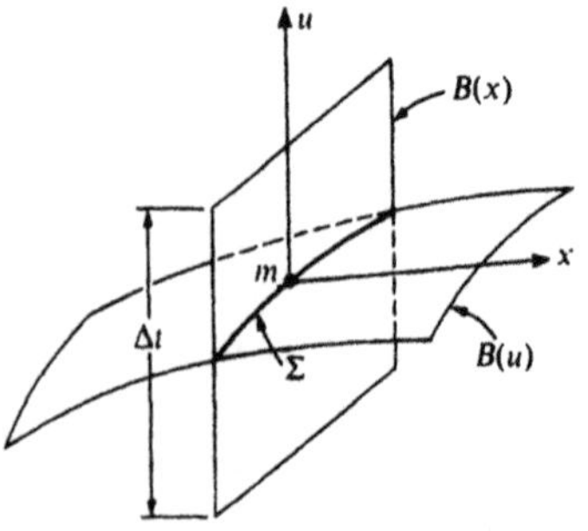

Figure 2. Here dimensions are reduced, since actually dim $\Sigma = 2$ and dim $B(x) = $ dim $B(u) = 3$.

The second limit shows that $T(x, y)$ is the y component of *stress* (force per unit area) across Σ in the x direction. Thus T restricted to $u^\perp$ is the classical stress tensor measured by u in his restspace $u^\perp \approx B(u)$. Since a perfect fluid cannot support shears, the stress above can be written as $\not{p}_x x$. The pressure $\not{p}_x$ is the same in all space directions, hence $T(x, y) = \not{p}(m)\langle x, y\rangle$ (valid for all $x, y \in u^\perp$ since T is bilinear).

(3) If $x \in u^\perp$, then $T(x, u) = T(u, x) = 0$.

Reasoning as in case (2) shows that $T(x, u)$ gives, for u, the energy flow across a patch of spacelike surface perpendicular to x, and $T(u, x)$ measures the density of x-momentum. For a perfect fluid, both are zero.

This discussion can be summarized rigorously as follows.

4. Definition. A *perfect fluid* on a spacetime M is a triple $(U, \rho, \not{p})$ where:

(1) U is a timelike future-pointing unit vector field on M called the *flow vector field.*

(2) $\rho \in \mathfrak{F}(M)$ is the *energy density function*; $\not{p} \in \mathfrak{F}(M)$ is the *pressure function.*

(3) The stress–energy tensor is

$$T = (\rho + \not{p})U^* \otimes U^* + \not{p}g,$$

where U^* is the one-form metrically equivalent to U.

Evidently this formula for T is equivalent to the three equations found above for $X, Y \perp U$, namely:

$$T(U, U) = \rho, \qquad T(X, U) = T(U, X) = 0, \qquad T(X, Y) = \not{p}\langle X, Y\rangle.$$

For a perfect fluid the vanishing of the divergence of the stress–energy tensor has the following consequence.

5. Proposition. If $(U, \rho, \not{p})$ is a perfect fluid,

(1) $U\rho = -(\rho + \not{p}) \operatorname{div} U$ (*energy equation*),

(2) $(\rho + \not{p})D_U U = -\operatorname{grad}_\perp \not{p}$ (*force equation*),

where the *spatial pressure gradient* $\operatorname{grad}_\perp \not{p}$ is the component of $\operatorname{grad} \not{p}$ orthogonal to U.

Proof. If $\bar{T}$ is the (2, 0) tensor field metrically equivalent to T, it is easy to check that $\operatorname{div} T = 0$ is equivalent to $\operatorname{div} \bar{T} = 0$. Writing $\bar{T}$ in terms of coordinates,

$$T^{ij} = (\rho + \not{p})U^i U^j + \not{p}g^{ij}.$$

The divergence is then

$$\sum_j T^{ij}{}_{;j} = \sum_j \{(\rho + \not{p})_{;j} U^i U^j + (\rho + \not{p}) U^i{}_{;j} U^j$$
$$+ (\rho + \not{p}) U^i U^j{}_{;j} + \not{p}_{;j} g^{ij}.$$

Expressed invariantly this is the vector field

$$\operatorname{div} \bar{T} = U(\rho + \not{p})U + (\rho + \not{p}) D_U U + (\rho + \not{p})(\operatorname{div} U)U + \operatorname{grad} \not{p}.$$

But $\operatorname{div} \bar{T} = 0$, and since U is a unit vector field, $D_U U \perp U$. Hence equation (2) is obvious, and $\langle \operatorname{div} \bar{T}, U \rangle = 0$ gives equation (1). ■

Evidently the first equation is a formula for the time rate of change of energy density as measured by U. The second equation is an analogue of Newton's $F = ma$, with force replaced by spatial pressure gradient, and mass replaced by $\rho + \not{p}$, while $D_U U \perp U$ is indeed the spatial acceleration of the molecules of the flow as self-measured. (Compare with classical hydrodynamics.)

The definition of perfect fluid does not tell how to construct a spacetime model of one. Model-building in general relativity is more subtle than in Newtonian physics or special relativity. The latter two theories have fixed universal geometries, $\boldsymbol{R}^1 \times \boldsymbol{R}^3$ and $\boldsymbol{M} \approx \boldsymbol{R}_1^4$, respectively; for a given problem, matter is modeled in that manifold and physical laws then govern its behavior. But in general relativity there can be no universal a priori geometry, since for any spacetime M the Einstein equation already determines the stress–energy tensor; specifically,

$$T = (1/8\pi)(\mathrm{Ric} - \tfrac{1}{2}Sg).$$

Thus a given spacetime can be used to model matter only in the unlikely case that T happens to be a physically realistic stress–energy tensor. Schematically,

$$\text{spacetime} \xrightarrow{\text{geometry}} \begin{matrix}\text{Ricci}\\ \text{curvature}\end{matrix} \xleftrightarrow{\text{Einstein equation}} \begin{matrix}\text{stress–energy}\\ \text{tensor}\end{matrix} \xleftarrow{\text{physics}} \text{matter.}$$

Hence relativistic model-building is a kind of nontrivial matching process.

Since the tidal forces $F_u(y) = R_{yu}u$ in a spacetime M describe gravitation, a natural way to express the empirical fact that *gravity attracts* is $\langle F_u(y), y \rangle \leq 0$; that is, $K(\Pi) \leq 0$ for all timelike tangent planes. Considerably weaker is the *timelike convergence condition,*

$$\mathrm{Ric}(u, u) \geq 0 \qquad \text{for all timelike tangent vectors to } M,$$

which says merely that, *on average, gravity attracts.* (By continuity the inequality holds also for null vectors.) We say the condition holds *strictly* if $\geq$ is replaced by $>$.

In terms of the stress–energy tensor of M the timelike convergence condition becomes

$$T(u, u) \geq \tfrac{1}{2}C(T)\langle u, u\rangle$$

for all timelike (and null) tangent vectors to M. Here $T(u, u)$ is energy density, and in this form the condition is also called the *strong energy condition* on M. For example, see Exercise 10.

ROBERTSON–WALKER SPACETIMES

In the seventeenth century Newton extended the range of physics from the earth to the solar system. Two and a half centuries later, relativity theory, supported by progress in astronomy, began studying the largest physical system: the universe. Astronomical evidence indicates that the universe can be modeled (in smoothed, averaged form) as a spacetime containing a perfect fluid whose "molecules" are the galaxies. At present, the dominant contribution to the energy density of the galactic fluid is the mass of the galaxies, with a much smaller pressure due mostly to radiation.

The decisive fact is that no large asymmetry has been observed in the distribution of the galaxies. They come in clusters, and these clusters in clusters, but at the large scale appropriate to cosmology, the universe—viewed from our galaxy—looks the same in all directions. What evidence there is supports the hypothesis that the same isotropy holds for all galaxies, and it thus becomes possible to build quite simple *cosmological models* whose gross properties have a reasonable chance of being physically realistic. At the very least, such models provide a testing ground for further study.

We start with a smooth manifold $M = I \times S$, where I is a (possibly infinite) open interval in $\boldsymbol{R}^1$ and S is a connected three-dimensional manifold. Let t and σ be the projection onto I and S, respectively. The lines $I \times p$ will be the worldlines of the galactic flow.

Let $U = \partial_t$ be the lift to $I \times S$ of the standard vector field d/dt on $I \subset \boldsymbol{R}^1$, and for each $p \in S$ parametrize $I \times p$ by $\gamma_p(t) = (t, p)$. Since U gives the velocity of each such "galaxy" γ_p, they are its integral curves. Thus the function t will give the common proper time of all galaxies. Holding t constant gives the hypersurface

$$S(t) = t \times S = \{(t, p) : p \in S\}.$$

As usual, the lift $h(t, p) = h(t)$ of a function $h \in \mathfrak{F}(I)$ is again denoted by h, and we write h' for $Uh = dh/dt$.

The geometry of the model will follow from physical assumptions about the galactic flow. Each γ_p is—potentially, at least—a particle with proper time t, hence

(a) $\langle U, U \rangle = -1$.

As might be expected from isotropy, the relative motion of the actual galaxies is negligible on large-scale average. Thus we take each slice $S(t)$ to be a common restspace for their idealizations γ_p, requiring

(b) $U \perp S(t)$ for all $t \in I$.

Hence each such slice becomes a Riemannian (i.e., spacelike) hypersurface.

We formalize the isotropy condition "all spatial directions the same" in local form as follows: Each (t, p) has a neighborhood $\mathcal{N}$ such that, given unit tangent vectors x, y to $S(t)$ at (t, p), there is a *galaxy-preserving* isometry $\phi = \text{id} \times \phi_S$ of $\mathcal{N}$ such that $d\phi(x) = y$.

6. Proposition. Under the conditions above, in $I \times S$

(1) each slice $S(t)$ has constant curvature $C(t)$;

(2) for any $s, t \in I$ the natural map $\mu(s, p) = (t, p)$ from $S(s)$ to $S(t)$ is a homothety.

Proof. (1) In each tangent space to $S(t)$ any plane Π is $x^\perp$ for some unit vector x. Thus for any two such planes there is an isotropy isometry such that $d\phi(\Pi) = \Pi'$. The restriction of $\phi = \text{id} \times \phi_S$ to $\mathcal{N} \cap S(t)$ is again an isometry, hence Π and Π' have the same sectional curvature in the geometry of $S(t)$. Thus Schur's theorem (Exercise 3.21) implies that $S(t)$ has constant curvature.

(2) First we show that *each map μ is conformal*; that is, $|d\mu(x)|$ is the same for all unit vectors x tangent to $S(s)$ at (s, p). Any other such unit vector can be expressed as $d\phi(x)$, where ϕ is an isotropy isometry. Since μ and $\phi = \text{id} \times \phi_S$ commute,

$$|d\mu\, d\phi\, x| = |d\phi\, d\mu\, x| = |d\mu\, x|.$$

Here we have assumed that t is small enough so that (t, p) is in the domain of ϕ, but a finite iteration will give this result for any $t \in I$.

Let $h(s, p, t)$ be the scale factor of $\mu\colon S(s) \to S(t)$ at (s, p), so h is a smooth function on $I \times S \times I$. It suffices to show that $xh = 0$ for any unit vector x tangent to a slice, for then, since S is connected, h depends only on s and t.

Let σ be the geodesic in $S(s)$ with $\sigma(0) = (s, p)$ and $\sigma'(0) = x$. Let ϕ be an isotropy isometry such that $d\phi(x) = -x$. Then $d\phi(\sigma'(u)) = -\sigma'(-u)$, and again since ϕ commutes with μ,

$$h(\sigma(u), t) = |d\mu\, \sigma'(u)| = |d\phi\, d\mu\, \sigma'(u)| = |d\mu\, d\phi\, \sigma'(u)|$$
$$= |d\mu\, \sigma'(-u)| = h(\sigma(-u), t).$$

Hence

$$(xh)(s, p, t) = \frac{d}{du} h(\sigma(u), t)\bigg|_{u=0} = 0. \qquad \blacksquare$$

Since the homothety $\mu = \mu_{st}$ has scale factor $h(s, t)$, Remark 3.65 gives $h(s, t)^2 C(t) = C(s)$. Since μ_{st} is a diffeomorphism, h is never zero. Hence the function C maintains the same sign: $k = -1, 0$, or 1.

Fix $a \in I$, and define $f(a) > 0$ by $C(a)f(a)^2 = k$. Assign S the metric tensor such that the map $j_a(s) = (a, s)$ from S to $S(a)$ is a homothety of scale factor $f(a)$, and define a function $f \in \mathfrak{F}(I)$ by $f(t) = h(a, t)/f(a)$. It follows immediately that

(c) *S has constant curvature k, and every injection j_t: $S \to S(t)$ is a homothety of scale factor $f(t)$.*

In particular, the constant curvature of $S(t)$ is $k/f(t)^2$, and for vectors x, y tangent to $S(t)$

$$\langle x, y\rangle = f^2(t)(d\sigma(x), d\sigma(y)).$$

Thus the conditions (a), (b), and (c) above express $I \times S$ geometrically as a warped product (Definition 7.33) with base I and fiber S.

7. **Definition.** Let S be a connected three-dimensional Riemannian manifold of constant curvature $k = -1, 0$, or 1. Let $f > 0$ be a smooth function on an open interval I in $\boldsymbol{R}^1_1$. Then the warped product

$$M(k, f) = I \times_f S$$

is called a *Robertson–Walker spacetime.* (See Figure 3.)

Explicitly, $M(k, f)$ is the manifold $I \times S$ with line element $-dt^2 + f^2(t)\, d\sigma^2$, where $d\sigma^2$ is the line element of S (lifted to $I \times S$). It is time-oriented by requiring that $U = \partial_t$ be future-pointing. The notation $M(k, f)$ is not completely descriptive, but *sign* k and *scale function* f are the essential ingredients. We say that the interval I is *maximal* provided f cannot be extended to a smooth positive function on an interval strictly larger than I.

The Riemannian manifold S is called the *space* of $M(k, f)$; it is a scale model of each *spacelike slice* $S(t)$. Since S is the fiber of $M(k, f)$, we denote its metric tensor as in Chapter 7 by (,) and its connection by ∇. The *standard choices* for S are the complete simply connected ones: H^3, $\boldsymbol{R}^3$, S^3, with curvatures $-1, 0, +1$, respectively.

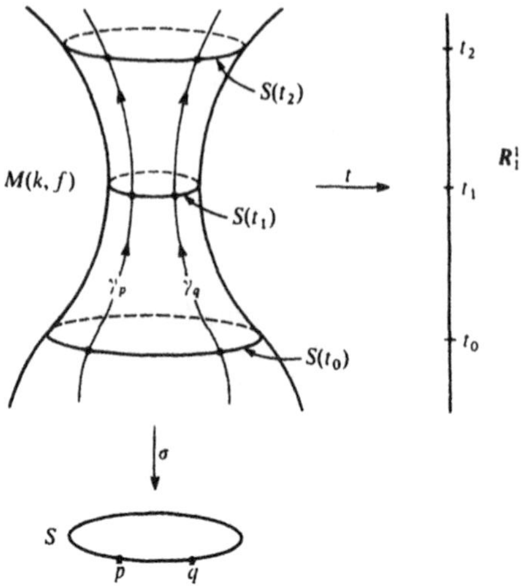

Figure 3

Since S has constant curvature, results from Chapter 8 show that spatially, $M(k, f) = I \times_f S$ is quite homogeneous. If $t \in I$ and $p, q \in S$, there is an isometry ϕ from a neighborhood $\mathscr{U}$ of p to a neighborhood $\mathscr{V}$ of q—in fact, $d\phi_p$ can be preassigned. Then id $\times\, \phi$ is an isometry from $I \times \mathscr{U}$ to $I \times \mathscr{V}$, and if S is one of the standard choices, id $\times\, \phi$ can be defined globally.

An analysis of the connection and curvature of Robertson–Walker spacetimes follows from general results on warped products. As usual let $\mathfrak{L}(S)$ be the set of all lifts to $M(k,f)$ of vector fields on S. Thus $X \in \mathfrak{L}(S)$ if and only if X is everywhere tangent to spacelike slices and is σ-related to a vector field on S. The flow vector field $U = \partial_t$ is a (future-pointing) unit normal to each slice $S(t)$. Thus, as on page 205, nor is orthogonal projection onto the U direction, nor $W = -\langle W, U\rangle U$, and tan projects orthogonally onto the tangent spaces of the slices.

8. Corollary. If $X, Y \in \mathfrak{L}(S)$ on $M(k,f)$, then

(1) $D_U U = 0$,
(2) $D_U X = D_X U = \tilde{II}(X, U) = (f'/f)X$,
(3) nor $D_X Y = II(X, Y) = \langle X, Y\rangle(f'/f)U$,
(4) tan $D_X Y$ is the lift of $\nabla_X Y$ on S.

Proof. Recall that the slices $S(t)$ are fibers not leaves; thus these equations follow from the correspondingly numbered ones in Proposition 7.35, changing notation in the latter by $X, Y \to U$ and $V, W \to X, Y$. (Note that grad $f = -f'U$.) However, a direct proof is quite simple; the Koszul formula is needed only to compute $2\langle D_X U, Y\rangle = U\langle X, Y\rangle$. ∎

By (1), each γ_p is a geodesic (a direct consequence of isotropy), and (3) gives the shape tensor of the totally umbilic fibers.

Similarly, Proposition 7.44 gives this description of Robertson–Walker curvature.

9. Corollary. For vector fields X, Y, Z on $M(k, f)$ tangent to all slices $S(t)$,

(1) $R_{XY}Z = [(f'/f)^2 + (k/f^2)][\langle X, Z\rangle Y - \langle Y, Z\rangle X]$.
(2) $R_{XU}U = (f''/f)X$.
(3) $R_{XY}U = 0$.
(4) $R_{XU}Y = (f''/f)\langle X, Y\rangle U$.

It then follows that every plane containing a vector of U has curvature $K_u = f''/f$, and every plane tangent to a spacelike slice has curvature $K_\sigma = (f'^2 + k)/f^2$ (not to be confused with its curvature k/f^2 in the geometry of $S(t)$). We call these the *principal sectional curvatures* of $M(k, f)$, and they turn up repeatedly in Robertson–Walker geometry and physics.

10. Corollary. For a Robertson–Walker spacetime $M(k, f)$ with flow vector field $U = \partial_t$:

(1) Ricci curvature is given by

$$\mathrm{Ric}(U, U) = -3f''/f, \qquad \mathrm{Ric}(U, X) = 0,$$

$$\mathrm{Ric}(X, Y) = \{2(f'/f)^2 + 2k/f^2 + f''/f\}\langle X, Y\rangle \qquad \text{if} \quad X, Y \perp U.$$

(2) The scalar curvature is

$$S = 6\{(f'/f)^2 + k/f^2 + f''/f\}.$$

These results follow immediately from Corollary 9 and general formulas for Ric and S.

THE ROBERTSON-WALKER FLOW

We show that, for any Robertson–Walker spacetime, the flow given by the vector field $U = \partial_t$ is that of a perfect fluid. Since the stress–energy tensor T of $M(k, f)$ is already determined by the Einstein equation, we must find functions ρ and p such that T has the form required by Definition 4.

11. Theorem. If U is the flow vector field on a Robertson–Walker spacetime $M(k, f)$, then (U, ρ, p) is a perfect fluid with energy density ρ and pressure p given by

$$8\pi\rho/3 = (f'/f)^2 + k/f^2, \qquad -8\pi p = 2f''/f + (f'/f)^2 + k/f^2.$$

Proof. By the Einstein equation, $T = (1/8\pi)[\text{Ric} - \frac{1}{2}Sg]$. The preceding lemma gives $T(U, X) = 0$, and

$$T(X, Y) = \frac{1}{8\pi}\left\{\left[2\left(\frac{f'}{f}\right)^2 + \frac{2k}{f^2} + \frac{f''}{f}\right]\langle X, Y\rangle - 3\left[\left(\frac{f'}{f}\right)^2 + \frac{k}{f^2} + \frac{f''}{f}\right]\langle X, Y\rangle\right\}.$$

This simplifies to $p\langle X, Y\rangle$ for p as above. If $\rho = T(U, U)$, then T has the required form $(\rho + p)U^* \otimes U^* + pg$. Computing $T(U, U)$ from Corollary 10 gives the formula for ρ. ■

In particular, ρ and p are functions of t only—constant on each spacelike slice $S(t)$—as could be seen directly from local isotropy.

In terms of the principal sectional curvatures the equations in the theorem are just

$$K_\sigma = 8\pi\rho/3, \qquad 2K_u + K_\sigma = -8\pi p.$$

Eliminating K_σ gives a basic relation between the scale function f and pressure and density.

12. Corollary. For a Robertson–Walker perfect fluid,

$$3f''/f = -4\pi(\rho + 3p).$$

Now we turn to the dynamics of the flow, that is, to div $T = 0$ as expressed for a perfect fluid by Proposition 5. The force equation is trivial, since $D_U U = 0$ and (p being constant on slices) $\text{grad}_\perp\, p = 0$. The energy equation takes the following form.

13. Corollary. For a Robertson–Walker perfect fluid

$$\rho' = -3(\rho + p)f'/f.$$

Proof. Since $U\rho = \rho'$, it suffices to check that div $U = 3f'/f$. In terms of a frame field with $E_0 = U$,

$$\text{div } U = \sum_{m=0}^{3} \varepsilon_m \langle D_{E_m} U, E_m\rangle = \sum_{j=1}^{3} \langle D_{E_j} U, E_j\rangle$$

since $D_U U = 0$. By Corollary 8 each of the remaining summands is f'/f. ■

This formula for the time rate of change of energy-density can readily be checked intuitively by a Newtonian computation. For any region B in S the galaxies γ_p with $p \in B$ give the history through galactic time of the *comoving* region $B(t) = \{\gamma_p(t): p \in B\}$ in $S(t)$. Since $S(t)$ is just S scaled by the factor $f(t)$, the volume of $B(t)$ is $f^3(t)$ vol B hence its total energy is $\rho(t)f^3(t)$ vol B. Because the fluid is perfect, there is no energy flow across the boundary of $B(t)$. Thus the (algebraic) increase in energy ΔE during a time interval Δt equals the work ΔW done on $B(t)$ by the pressure on its boundary. Taking B for simplicity to be a unit cube, the work done on each pair of opposite faces is

$$\begin{aligned}\text{force} \times \text{distance} &= \text{pressure} \times \text{area} \times \text{distance}\\ &= p(t)f^2(t)(-f'(t)\,\Delta t).\end{aligned}$$

(There is a minus sign, since E increases if $f'(t) < 0$.) Thus

$$(\rho f^3)'(t)\,\Delta t \sim \Delta E = \Delta W \sim -3p(t)(f^3)'(t)\,\Delta t.$$

In the limit, $(\rho f^3)' = -p(f^3)'$, which is equivalent to the formula in the corollary.

ROBERTSON-WALKER COSMOLOGY

A Robertson–Walker spacetime gives a relativistic model of the flow of a perfect fluid; to fit such a model to our universe, data is needed about the physical parameters p, ρ, and f.

14. Remarks. Astronomical Data. (1) According to Hubble (in 1929), all distant galaxies are now moving away from us at a rate proportional to their distance. For galaxies γ_p and γ_q the distance between $\gamma_p(t)$ and $\gamma_q(t)$ in $S(t)$ is $f(t)d(p, q)$, where d is Riemannian distance in the space S. Hubble's discovery, by current estimates, is that

$$H_0 = \frac{f'(t_0)}{f(t_0)} \quad \text{is} \quad \frac{1}{18 \pm 2 \times 10^9 \text{ yr}}.$$

(2) For all known forms of matter, energy density dominates pressure: $\rho > |p|$. At the present time, t_0, the energy density ρ_0 of our universe is estimated to be between 10^{-31} and 5×10^{-29} g/cm^3. The pressure p_0 is positive but very much smaller: $\rho_0 \gg p_0 > 0$.

In particular, (1) shows that currently f has positive derivative: the spaces $S(t)$ are *expanding*. This qualitative fact was already enough to destroy the static models of the universe prevailing in the 1920s, and general relativity deduces the following striking consequence.

15. Proposition. Let $M(k, f) = I \times_f S$. If $H_0 > 0$ for some t_0, and $\rho + 3p > 0$, then I has an initial endpoint t_* with $t_0 - H_0^{-1} < t_* < t_0$, and either (1) $f' > 0$, or (2) f has a maximum point after t_0, and I is a finite interval (t_*, t^*).

Proof. By Corollary 12, $\rho + 3p > 0$ implies $f'' < 0$. Thus the graph of f is—except at t_0—below that of its tangent line at t_0. This line is the graph of

$$F(t) = f(t_0) + H_0 f(t_0)(t - t_0).$$

Thus the Hubble result $H_0 > 0$ shows that as t decreases from t_0, the function $f > 0$ must have a singularity at some t_* before reaching the zero $t_0 - H_0^{-1}$ of F.

Since $f'' < 0$, either f' is always positive on I or f has a maximum point after which $f' < 0$. In the latter case, an argument as before shows that another singularity ensues at $t^* > t_0$. ■

Taking the estimate of *Hubble time* H_0^{-1} above, this says that the universe had a definite beginning some ten to twenty billion years ago and, if it does not continue expanding, must, after contracting for a while, come to an end. The result does not say that the universe begins small ($f \to 0$ as $t \to t_*$) or that in the expanding case ($f' > 0$) it endures forever.

If the energy density ρ approaches infinity as $t \to t_*$ (or t^*), we say that $M(k, f)$ has a *physical singularity* there.

16. Definition. An initial singularity of $M(k, f)$ at t_* is a *big bang* provided $f \to 0$ and $f' \to \infty$ as $t \to t_*$. Similarly, a final singularity is a *big crunch* if $f \to 0$ and $f' \to -\infty$ as $t \to t^*$.

Such singularities are physical, and the converse holds under conditions weaker than those of Remark 14.

17. Theorem. Assume that $M(k, f) = I \times_f S$ has only physical singularities and that I is maximal. If $H_0 > 0$ for some t_0, if $\rho > 0$, and for constants a and A, $-\frac{1}{3} < a \le p/\rho \le A$, then:

(1) The initial singularity is a big bang.
(2) If $k = 0$ or -1, then $I = (t_*, \infty)$ and as $t \to \infty, f \to \infty$ and $\rho \to 0$.
(3) If $k = 1$, then f reaches a maximum followed by a big crunch, hence I is a finite interval (t_*, t^*).

Proof. In particular, $\rho + 3p \ge \varepsilon\rho > 0$ for some $\varepsilon > 0$. Hence the preceding proposition applies, so $f'' < 0$.

To prove (1), note that $f' > 0$ on the interval (t_*, t_0). Since $p \leq A\rho$, Corollary 13 gives

$$\rho' \geq -C\rho f'/f, \qquad \text{where} \quad C = 3(A + 1) > 2.$$

It follows that $(\rho f^C)' \geq 0$. Hence $\rho f^C \leq \rho(t_0)f(t_0)^C$ on (t_*, t_0). By hypothesis $\rho \to \infty$ as $t \to t_*$, hence $f \to 0$.

The inequality $\rho - \varepsilon\rho \geq -3p$ then in a similar way gives

$$\rho' \leq -(2 + \varepsilon)\rho f'/f,$$

and hence $(\rho f^{2+\varepsilon})' \leq 0$ on (t_*, t_0). Thus $\rho f^{2+\varepsilon} \geq \rho(t_0)f(t_0)^{2+\varepsilon}$ on this interval. As $t \to t_*$ we have $f \to 0$ hence $\rho f^2 \to \infty$. Then by the formula for ρ in Theorem 11, $f'^2 + k \to \infty$, and hence $f' \to \infty$. Thus (1) is proved.

Case I. f has a maximum at say t_m.

Since $f'(t_m) = 0$, we get $0 < \rho(t_m) = 3k/(8\pi f^2(t_m))$, hence $k = 1$. Since $f'' < 0$, f' will be negative for $t > t_m$. Then arguments corresponding to those for t_* show that the final singularity at $t^* > t_m$ is a big crunch.

Case II. f has no maximum.

Then $f' > 0$ on the entire interval I, so the results preceding Case I are valid on I. The inequalities $\rho > 0$ and $\rho + 3p > 0$ imply $\rho + p > 0$. Corollary 13 then gives $\rho' < 0$, so there are no physical singularities as t increases. Thus $I = (t_*, \infty)$.

Subcase A. $f \to \infty$ as $t \to \infty$. Since $(\rho f^{2+\varepsilon})' \leq 0$, $\rho f^{2+\varepsilon}$ is bounded for t large. Hence $\rho f^2 \to 0$ as $t \to \infty$. Thus $f'^2 + k \to 0$, hence $k = 0$ or -1.

Subcase B. f is bounded as $t \to \infty$. (It suffices to show that this is impossible.) As $t \to \infty$, $f \to b$ and, since $f'' < 0$, $f' \to 0$. Hence $\rho f^2 \to 3k/8\pi$. Thus $k = 0$ or 1.

As $t \to \infty$, ρf^C is nondecreasing, hence $\rho f^2 \not\to 0$, hence $k \neq 0$.

Finally, if $k = 1$, then $\rho f^2 \to 3/8\pi$. Thus $\rho \geq \delta$ for some $\delta > 0$. Since $f' \to 0$, there is a sequence $\{t_i\} \to \infty$ such that $\{f''(t_i)\} \to 0$. Hence Corollary 12 gives $\{(\rho + 3p)(t_i)\} \to 0$. But this contradicts $\rho + 3p \geq \varepsilon\rho \geq \varepsilon\delta > 0$. ∎

The theorem predicts that our universe begins in a colossal explosion. The limit time t_* is of course not part of the model—both physics and semi-Riemannian geometry fail at t_*—but any fraction of a second later the materials for the construction of the universe are present in the initial fireball.

The following examples show that for a Robertson–Walker model, the big bang origin requires the inequalities in the theorem.

18. Example. (1) $M(0, 1 + t^{2/3}), t > 0$. We compute

$$\rho = 1/(6\pi t^{2/3}(1 + t^{2/3})^2),$$

so 0 is a physical singularity. Also $f' \to \infty$ as $t \to 0$, but obviously $f \to 1$. The upper bound in the theorem fails since $p/\rho = 1/(3t^{2/3})$.

(2) $M(0, \sinh^{-1} t), t > 0$. Here

$$\rho = 3/(8\pi(t^2 + 1)(\sinh^{-1} t)^2),$$

so again 0 is a physical singularity. Also $f \to 0$ as $t \to 0$, but $f' \to 1$. The lower bound in the theorem fails. In fact,

$$\frac{p}{\rho} = \frac{2t \sinh^{-1} t}{3\sqrt{t^2 + 1}} - \frac{1}{3},$$

so $p/\rho > -\frac{1}{3}$ holds but not $p/\rho \geq a > -\frac{1}{3}$.

Under the hypothesis of the theorem, the ultimate fate of the universe depends on the sign k of spatial curvature; this in turn depends on the present energy density ρ_0 and Hubble number $H_0 = f_0'/f_0$.

19. Corollary. Let $\rho_c = 3(H_0)^2/8\pi$, called the *critical energy density*. If $\rho_0 \leq \rho_c$, then $k = 0, -1$ (so the universe expands forever). If $\rho_0 > \rho_c$, then $k = +1$ (so the universe eventually collapses).

Proof. By Theorem 11, $\rho - (3H^2/8\pi) = 3k/8\pi f^2$, so $k = \text{sgn}(\rho_0 - \rho_c)$. ∎

Taking the Hubble number H_0 to be $(18r \times 10^9 \text{ yr})^{-1}$ with $r \sim 1$, the critical density ρ_c is $(3.68/r^2) \times 10^{-22} \text{ yr}^{-2}$. For geometrical units $1 = G = 6.67 \times 10^{-8} \text{ cm}^3/\text{g sec}^2$, and a year contains about 3.16×10^7 sec, giving

$$\rho_c = (5.5/r^2) \times 10^{-30} \quad \text{g/cm}^3.$$

For r near 1 this is within the rather large interval for ρ_0 in Remark 14. The estimation for ρ_0 is difficult. To obtain accuracy, large astronomical regions must be searched, and many elusive forms of matter may contribute to the extremely small average density. Thus even assuming our universe admits an excellent Robertson–Walker model, a firm prediction of its distant future is not yet possible.

FRIEDMANN MODELS

Except in the earliest era of the universe and the final era, if there is one, energy density utterly dominates pressure. Thus Robertson–Walker models with $p = 0$ should be reasonably good for this range, and it is easy to find them all explicitly.

A *dust* is a perfect fluid with $p = 0$ and $\rho > 0$, the terminology suggesting the absence of any nongravitational influence between the "molecules" of the fluid.

20. Lemma. Let $M(k, f)$ be a Robertson–Walker spacetime with f nonconstant. Then the following are equivalent:

(1) The perfect fluid U is a dust.
(2) $\rho f^3 = \text{M}$, a positive constant.
(3) (*Friedmann equation*) $f'^2 + k = A/f$, where $A = 8\pi\text{M}/3 > 0$.

Proof. The equivalence of (2) and (3) is immediate from the formula for ρ in Theorem 11. If (1) holds, so $p = 0$, then Corollary 13 becomes $\rho' f + 3\rho f' = 0$. Hence ρf^3 is a constant, positive since ρ and f are. Conversely, if (2) holds, then it follows from Corollary 13 that $pf' = 0$. The nonconstancy of f (implied by $H_0 > 0$) is needed to prove $p = 0$. Assume p is not identically zero. Then there is a maximal interval $J \subset I$ on which p is never zero. Hence $f' = 0$ on J, so f is constant on J. Thus $J \neq I$. The formula for p in Theorem 11 then shows that p is a nonzero constant on J. Thus p is nonzero on an interval strictly larger than J: a contradiction. ∎

The energy density of a dust derives from mass; thus the equation in (2) asserts the conservation of mass in any comoving region.

A *Friedmann cosmological model* is a Robertson–Walker spacetime such that the galactic fluid is a dust and $H = f'/f$ is positive for some t_0.

21. Remarks. Friedmann Cosmological Models. We find the scale function f in the three cases $k = 0, 1, -1$, putting the big bang at $t_* = 0$.

(1) $k = 0$. The Friedmann equation $ff'^2 = A$ is readily solved for $f = Ct^{2/3}$, where $4C^3 = 9A$. Thus the initial expansion continues forever with $f \to \infty$ and $f' \to 0$.

(2) $k = 1$. Integration of the Friedmann equation gives the parametric solution

$$t = \tfrac{1}{2}A(\vartheta - \sin\vartheta), \qquad f = \tfrac{1}{2}A(1 - \cos\vartheta) \qquad (0 < \vartheta < 2\pi).$$

The graph of f is the cycloid swept out by a point on the rim of a rolling wheel with diameter A (Figure 4). The expansion reaches a maximum $f = A$ at $t = \pi A/2$. Then contraction begins, and f decreases symmetrically toward a final collapse at $t^* = \pi A$.

(3) $k = -1$. The Friedmann equation is now $f'^2 = 1 + (A/f)$. Thus $f' \geq 1$ and the universe expands forever with $f \to \infty$ and $f' \to 1$. An integration as for (2) gives

$$t = \tfrac{1}{2}A(\sinh\eta - \eta), \qquad f = \tfrac{1}{2}A(\cosh\eta - 1) \qquad (\eta > 0).$$

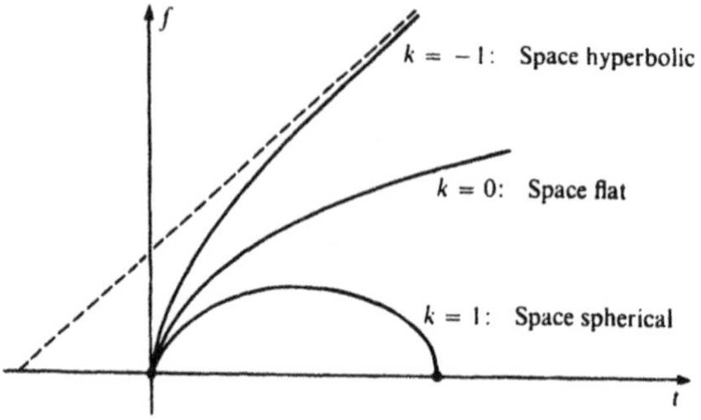

Figure 4. Friedmann scale functions.

These models were proposed by Friedmann several years before the discovery of the expansion of the universe.

The Friedmann equation is the same as the differential equation governing radial motion in Newtonian gravitation. Thus a three-dimensional model of a $k = 1$ Friedmann universe could be made out in space by detonating a ball of dynamite encased in a shell of bricks. The fragments fly radially outward and, provided the charge is insufficient, reach a maximum radius and fall symmetrically back together. With larger charges, escape speed is reached and the fragments continue outward forever. (In the latter case the spacelike splices are topologically wrong, since spheres do not admit metrics of constant curvature $k = 0$ or -1.)

Assuming Hubble time $H_0^{-1} = 18 \times 10^9$ yr, we compute some sample numbers for Friedmann universes, using the standard choices for space S. These form a single family that we parametrize by present energy density ρ_0. Corollary 19 shows that the family splits in two at $\rho_0 = \rho_c$.

(1) The *Einstein–de Sitter cosmological model*, $M(0, t^{2/3}) = \boldsymbol{R}^+ \times_{t^{2/3}} \boldsymbol{R}^3$. The Hubble function $H = f'/f$ is $2/3t$, hence the age of the universe is

$$t_0 = 2/3H_0 = 12 \times 10^9 \quad \text{yr}.$$

Since $k = 0$, $\rho = 3H^2/8\pi$. In particular the present energy density ρ_0 is at the critical value $\rho_c = 5.5 \times 10^{-30}$ g/cm^3.

(2) $M(1, f) = (0, \pi A) \times_f S^3$. The slice $S(t)$ is thus a three-dimensional sphere of radius $f(t)$. By the parametric equations in Remark 21(2), the maximum radius is A and we compute

$$H = f'/f = (2 \sin \vartheta)/A(1 - \cos \vartheta)^2.$$

Then by some trigonometry,

$$\rho = (3/8\pi)(H^2 + (1/f^2)) = 3H^2/4\pi(1 + \cos \vartheta).$$

Since H_0 has been fixed, the present energy density ρ_0 determines the present value ϑ_0 of the angle parameter, and the formula for H then gives A.

In view of Corollary 19, we try ρ_0 at twice the critical density ρ_c. Then $\cos \vartheta_0 = 0$, hence $\vartheta_0 = \pi/2$ and the maximum radius is

$$A = 2/H_0 = 36 \times 10^9 \quad \text{lightyears.}$$

Thus the age of the universe is

$$t_0 = \tfrac{1}{2}A(\tfrac{1}{2}\pi - 1) \approx 10.3 \times 10^9 \quad \text{yr},$$

less than a tenth of its total lifetime πA.

(3) $M(-1, f) = \mathbf{R}^+ \times_{f(t)} H^3$. Much as in the previous example the parametric formulas in Remark 21(3) give

$$H = (2 \sinh \eta)/A(\cosh \eta - 1)^2, \qquad \rho = 3H^2/4\pi(\cosh \eta + 1).$$

Taking ρ_0 to be half the present critical density gives $\cosh \eta_0 = 3$, so $\sinh \eta_0 = 2\sqrt{2}$ and η_0 is about 1.76. Thus $A = \sqrt{2} H_0^{-1}$, and the age of the universe is

$$t_0 = (\sinh \eta_0 - \eta_0)/\sqrt{2} H_0 = 13.5 \times 10^9 \quad \text{yr}.$$

Unless more mass is discovered in our universe, "open" models such as this are probably more realistic than "closed" models as in (2).

The earliest era of the universe and the final one, if it exists, are dominated by radiation. There Friedmann models give way to *radiation models*, for which mass is zero and $\not{p} = \rho/3$ (see Exercise 14).

GEODESICS AND REDSHIFT

Any curve α in $M(k, f) = I \times_f S$ can be written as $(t(s), \beta(s))$, where $t(s)$ is the galactic time of $\alpha(s)$ and β is the projection of α into S. Derivatives with respect to the parameter s are often denoted by a prime; thus in this section we write f_t for $Uf = df/dt$.

22. Proposition. A curve $\alpha = (t, \beta)$ in $M(k,f)$ is a geodesic if and only if

(1) $d^2t/ds^2 + (\beta', \beta')f(t)f_t(t) = 0$.
(2) $\beta'' + 2(f_t(t)/f(t))(dt/ds)\beta' = 0$ (hence β is pregeodesic).

Proof. These equations follow immediately from those in Proposition 7.38, since here $\operatorname{grad} f = -f_t U$ and $(d/ds)(f(t)) = f_t(t) dt/ds$. ∎

23. Corollary. If $\alpha = (t, \beta)$ is a null geodesic in $M(k,f)$, then the function $f(t)\, dt/ds$ is constant.

Proof. Since $\alpha' = (dt/ds)U + \beta'$,

$$0 = \langle \alpha', \alpha' \rangle = -(dt/ds)^2 + (\beta', \beta')f^2.$$

Thus (1) implies

$$0 = f\frac{d^2t}{ds^2} + f_t\left(\frac{dt}{ds}\right)^2 = \frac{d}{ds}\left(f(t)\frac{dt}{ds}\right). \quad \blacksquare$$

This result leads to a relativistic explanation of the physical phenomenon called *cosmological redshift*. When light from a distant galaxy is analyzed in an earth-borne spectroscope, the characteristic pattern of spectral lines is obtained but all wavelengths λ_o are longer than for earth-emitted light—in proportions depending only on the source galaxy. As each wavelength λ_p at emission must be assumed to be the same as on earth, wavelengths have uniformly lengthened during transmission. The fractional increase

$$z = (\lambda_o - \lambda_p)/\lambda_p$$

is called the *redshift parameter* of the source (red having the longest wavelength in the visible band).

Such redshifts occur naturally in any Robertson–Walker model. As in Figure 5, let α be a photon (future-pointing null geodesic) emitted at past galactic time t_p from a galaxy γ_p and received in our galaxy γ_o at the present time t_0. Regard the galaxies as observers, with U giving their 4-velocities. Then, as in special relativity, the decomposition

$$\alpha' = (dt/ds)U + \beta'$$

shows that for each s the galactic observers measure $(dt/ds)(s)$ as the energy $E(s)$ of α and $\beta'(s)$ as its momentum. The relations $E = h\nu$ and $\lambda\nu = 1$ remain valid as before (page 179), where h is Planck's constant.

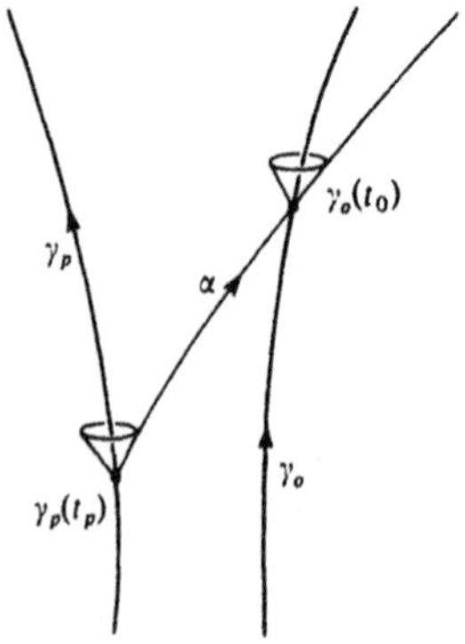

Figure 5

24. Corollary. In $M(k, f)$ a photon emitted (as above) at $\gamma_p(t_p)$ and received at $\gamma_o(t_0)$ has redshift

$$z = (f(t_0)/f(t_p)) - 1.$$

Proof. Since $(dt/ds)(s) = E(s) = h/\lambda(s)$ for all s, the preceding corollary implies that $f(t)/\lambda$ is constant along α. Substituting $\lambda = Cf(t)$ in the definition of z gives the result. ∎

Cosmological redshift is an analogue of the Doppler shifts produced by relative motion. Galaxies generally have negligible relative motion, but their distances apart change as a consequence of the overall expansion (or contraction) of the universe. Since $z > 0$ implies $f(t_0) > f(t_p)$ the fact that in our universe all distant sources have positive redshift is the primary evidence for the expansion of the universe. In any universe that has always been expanding, the redshift z determines the emission time t_p. In fact, z and $f(t_0)$ give $f(t_p)$, and since $f' > 0$, t is uniquely determined.

For example, in the Einstein–de Sitter model, $f(t) = t^{2/3}$ and the estimated age t_0 of the universe is about 12 billion years. If a galaxy has redshift $z = \frac{1}{5}$, then some algebra based on the corollary shows that the light was emitted at $t_p \sim 3t_0/4$, say 3 billion years ago. For a quasar with $z = 2$ the emission time is about $t_0/5$, less than $2\frac{1}{2}$ billion years after the big bang.

In $M(k, f)$ with the standard choices of space S, the times t_p and t_0 determine the distance between the galaxies γ_p and γ_o at one, hence every, galactic time.

25. Corollary. Let $M(k, f)$ have space S^3, $\mathbf{R}^3$, or H^3. If a photon emitted at $\gamma_p(t_p)$ is received at $\gamma_o(t_0)$, then the present distance between the galaxies γ_p and γ_o is

$$f(t_0) \int_{t_p}^{t_0} \frac{dt}{f(t)},$$

provided in the case of S^3 that the integral is $\leq \pi$.

Proof. The projection β of the photon into S is a pregeodesic that, as the proof of Corollary 23 shows, has speed

$$|\beta'|_S = (\beta', \beta')^{1/2} = (1/f(t))(dt/ds).$$

Thus the length of β from p to o is

$$L(\beta) = \int_{s_p}^{s_0} \frac{1}{f(t)} \frac{dt}{ds}\, ds = \int_{t_p}^{t_0} \frac{dt}{f(t)}.$$

With the additional hypothesis for S^3 this length is the Riemannian distance $d(p, o)$ in S, and the result follows. (The extra hypothesis can be avoided; without it, $d(p, o) = \min\{|L(\beta) - 2\pi m| : m \in \mathbf{Z}\}$.) ∎

For example, for the Einstein–de Sitter redshifts mentioned above, the corollary gives a present distance of about 3 billion lightyears for the galaxy with $t_p \sim 3t_0/4$, and present distance of 15 billion lightyears for the quasar with $t_p \sim t_0/5$.

The preceding corollaries lead to an explicit description of the null geodesics in a Robertson–Walker spacetime.

26. Corollary. In $M(k, f) = I \times_f S$ let α be a curve with $t(\alpha(0)) = t_0$. Then α is a null geodesic if and only if

$$\alpha(s) = (t(s), \beta(h(s))),$$

where $t(s)$ is the inverse function of $s = C \int_{t_0}^{t} f(t)\, dt$, β is a unit speed geodesic in S, and $h(s) = \int_{t_0}^{t(s)} dt/f(t)$.

Proof. If α is a null geodesic, the characterization of $t(s)$ is immediate from Corollary 23. The proof of Corollary 25 shows that the pregeodesic $\beta = \sigma \circ \alpha$ has arclength function h as above. Hence α has the required form. Conversely, a null geodesic of this form can be found to realize any given (null) initial velocity. ■

27. Remarks. Consider null geodesics in $M(k, f) = I \times_f S$, with $I = (A, B)$, $-\infty \leq A < B \leq \infty$.

(1) If $\int_{t_0}^{B} f(t)\, dt < \infty$, the preceding corollary shows that no future-pointing null geodesic can be defined on $[0, \infty)$. Hence every null geodesic is future incomplete. Similarly all null geodesics are past incomplete if $\int_{A}^{t_0} f(t)\, dt < \infty$.

(2) Conversely, if both integrals in (1) are infinite, then $t(s)$ and hence $h(s)$ are defined for all $s \in \mathbf{R}$. If S is complete, then β is also defined on $\mathbf{R}$, hence every inextendible null geodesic is complete; that is, $M(k, f)$ is null geodesically complete (hence inextendible).

(3) Again if S is complete, for an inextendible null geodesic the monotone function $t(s)$ traverses the entire interval I. Thus Robertson–Walker photons, unless otherwise destroyed, endure from the initial singularity t_* or $-\infty$ to the final singularity t^* or $+\infty$.

(4) If S is one of the standard choices, then all null geodesics are congruent, mod parametrization. In fact, any two null geodesics can be parametrized to have the same initial values of t and dt/ds. Then there is an isometry ϕ of S such that id $\times\, \phi$ carries $\alpha_1'(0)$ to $\alpha_2'(0)$, hence α_1 to α_2 (*spatial homogeneity*).

28. Examples. Photons in Friedmann Models. We take t_0 in Corollary 26 as the limit value $t_* = 0$; thus the photons start from the big bang.

(1) *Einstein–de Sitter*, $\mathbf{R}^+ \times_{t^{2/3}} \mathbf{R}^3$. Since $s = (3C/5)t^{5/3}$, taking $C = \frac{5}{3}$ gives $t(s) = s^{3/5}$. Then

$$h(s) = \int_0^{t^{3/5}} t^{-2/3}\, dt = 3(t^{3/5})^{1/3} = 3s^{1/5}.$$

Hence a typical photon is

$$\alpha(s) = (s^{3/5}, 3s^{1/5}, 0, 0) \qquad \text{for} \quad s > 0.$$

(2) $M(1, f) = (0, \pi A) \times_f S^3$ (f as in 21(2)). For $\alpha(s) = (t(s), \beta(h(s)),$ let $\vartheta(s) \in (0, 2\pi)$ be the rotation angle corresponding to $t(s) \in (0, \pi A)$. The parametric equations for t and f satisfy $dt = f(\vartheta)\, d\vartheta$, hence

$$h(s) = \int_0^{t(s)} \frac{dt}{f(t)} = \int_0^{\vartheta(s)} d\vartheta = \vartheta(s).$$

Thus as it travels from the big bang to the big crunch, each photon makes exactly one great circle trip around the space S^3.

Although general relativity has made possible the global study of the universe, it also imposes limits on what we can observe. Information from outside our immediate spacetime neighborhood consists of radiation (null geodesics) recorded in detail for only a few decades. Thus the events that have actually been seen by man lie mostly in a thin conical spacetime shell. However, there is good reason to believe that this shell extends to the extremely distant past. The large redshifts of quasars are one indication. More generally, the Robertson–Walker big bang picture of the universe has been reinforced by the 1965 discovery of *microwave background radiation*; this is pervasive, highly isotropic, and typical of the radiation emitted by a hot dense gas. Its properties argue both its origin in an early radiation-dominated era and the isotropy of the universe it has since traversed [HE, Section 10.1]. Thus knowledge of the past is extensive, and an increasingly detailed history of the universe is being constructed, starting from its earliest seconds.

The existence of the big bang, however, limits the spatial range of our observations, since obviously we can only receive radiation that has had time to get here. According to Corollary 25 our galaxy can have received information only from sources whose present distance is at most

$$d_{\max} = f(t_0) \int_{t_*}^{t_0} \frac{dt}{f(t)}.$$

In the Einstein–de Sitter model, for example, this marks out a ball of radius 30 billion lightyears in the infinite space $S(t_0) \approx \mathbf{R}^3$. (See also Exercises 8 and 9.) Thus at vast spatial distances or in the future, the universe could be different from the region we can observe.

OBSERVER FIELDS

An *observer field* on an arbitrary spacetime M is a timelike, future-pointing, unit vector field U. Each integral curve of U is indeed an observer, parametrized by proper time. Thus U describes a family of *U-observers* filling M. The general question is whether they can agree on common notions of space and of time. (Observer fields are called *reference frames* in [SW], from which this section derives.)

Suppose that S is a (necessarily spacelike) hypersurface in M to which an observer field U is normal at every $p \in S$. Then the infinitesimal restspace $U_p^\perp$ (as on page 333) is just $T_p(M)$ for every $p \in S$, hence S is called a *restspace* of U.

29. Example. (1) On a Robertson–Walker spacetime the flow vector field $U = \partial_t$ is an observer field whose U-observers are the galaxies γ_p. The spacelike slices $S(t)$ are restspaces of U.

(2) A single freely falling observer ω in Minkowski spacetime determines a unique observer field U whose observers are all distantly parallel to ω. Each restspace of ω in the special relativity sense is a restspace of U in the above sense.

It is not always possible to integrate the infinitesimal restspaces of U to obtain a restspace.

30. Proposition. For an observer field U the following are equivalent.

(1) There is a restspace of U through each $p \in M$.
(2) If vector fields X and Y are orthogonal to U, then so is $[X, Y]$.
(3) U is *irrotational*; that is, curl U is zero on vector fields $X, Y \perp U$.

Proof. (For properties of curl, see Exercise 3.18.)
(2) $\Leftrightarrow$ (3). If $X, Y \perp U$, then

$$(\text{curl } U)(X, Y) = \langle D_X U, Y\rangle - \langle D_Y U, X\rangle$$
$$= -\langle U, D_X Y\rangle + \langle U, D_Y X\rangle = -\langle U, [X, Y]\rangle.$$

(1) $\Rightarrow$ (2). Let S be a restspace of U through $p \in M$. If $X, Y \perp U$ then X and Y are tangent to S. By Proposition 1.32, $[X, Y]$ is also tangent to S, hence orthogonal to U.

That (2) implies (1) is a special case of Frobenius' theorem [W]. ■

A vector field V_t on $\mathbf{R}^3$ that depends on a time parameter t is *irrotational* provided curl$(V_t) = 0$ for each t. Thus the terminology in (3) above gives the natural relativistic analogue.

An observer field U is *geodesic* if each of its observers is geodesic, that is, if $D_U U = 0$.

31. Corollary. An observer field U is geodesic and irrotational if and only if curl $U = 0$.

Proof. Since curl U is skew-symmetric it suffices by the preceding proposition to show that $D_U U = 0$ is equivalent to (curl U)$(U, Z) = 0$ for all Z. But (curl U)$(U, Z) = \langle D_U U, Z\rangle$, since $\langle D_Z U, U\rangle = \frac{1}{2}Z\langle U, U\rangle = 0$. ∎

In Example 29 both observer fields are geodesic and irrotational.

Now we turn from space-agreement to time-agreement.

32. Definition. An observer field U on M is *proper time synchronizable* provided there exists a function $t \in \mathfrak{F}(M)$ such that $U = -\text{grad } t$. Then t is called a *proper time function* on M.

In fact, t is a common proper time for all U-observers, since

$$d(t \circ \alpha)/d\tau = Ut = -\langle U, \text{grad } t\rangle = -\langle U, U\rangle = 1.$$

Thus $t \circ \alpha$ differs from the proper time τ of α only by an additive "clock-setting" constant.

In Example 29 both observer fields are proper time synchronizable, with t galactic time in (1), and the usual extension of ω's proper time in (2).

33. Corollary. If an observer field U on M is proper time synchronizable, then U is geodesic and irrotational. The converse holds if M is simply connected (hence always holds locally).

Proof. The first assertion is immediate from the corollary above since the curl of a gradient is zero. For the converse, if curl $U = 0$ and M is simply connected, then U is a gradient [BG], hence U is proper time synchronizable. ∎

Suppose U is proper time synchronizable and let α be a U-observer. Any other material particle (or observer) from $\alpha(\tau_1)$ to $\alpha(\tau_2)$ takes less time than α's $\tau_2 - \tau_1$. The proof is a mild variant of that of Proposition 5.34. Thus, for example, the twin phenomenon (6.22) holds for a traveler leaving and eventually returning to a Robertson–Walker galaxy.

Proper time synchronizability is a very strong condition and can be usefully weakened as follows.

34. Definition. An observer field U on M is *synchronizable* provided there are smooth functions $h > 0$ and t on M such that $U = -h \text{ grad } t$.

Here t is a kind of average time agreed on by all U-observers. A synchronizable observer field is irrotational, since U is normal to the level hypersurfaces of t, which are thus restspaces. If τ is the proper time of an U-observer α, the time-dilation $d(t \circ \alpha)/d\tau$ is $1/(h \circ \alpha)$, so the elapsed proper time between two restspaces will generally be different for different U-observers.

STATIC SPACETIMES

35. Definition. A spacetime M is *static* relative to an observer field U provided U is irrotational and there is a smooth function $g > 0$ on M such that gU is a Killing vector field.

Any local flow $\{\psi_t\}$ of gU consists of isometries, and each ψ_t preserves U-observers (though generally distorting their proper time parametrizations). Being irrotational, U has restspaces S, and $\psi_t(S)$ is again a restspace, since $d\psi_t U$ and U are collinear. Thus, locally at least, the spatial universe always looks the same to a U-observer.

We now construct static spacetimes with a given restspace.

36. Definition. Let S be a three-dimensional Riemannian manifold, I an open interval, and $g > 0$ a smooth function on S. Let t and σ as usual be the projections of $I \times S$ onto I and S. The *standard static spacetime* $I\,{}_g\!\times S$ is the manifold $I \times S$ with line element

$$-g(\sigma)^2\, dt^2 + ds^2,$$

where ds^2 is the lift of the line element of S.

This is just the warped product $S \times_g I$ with time coordinate written first. Thus by contrast with a Robertson–Walker spacetime, space remains the same and time is warped. The following result verifies that $I\,{}_g\!\times S$ is static relative to ∂_t/g.

37. Lemma. For $I\,{}_g\!\times S$,

(1) ∂_t is a Killing vector field with global flow isometries given by $\psi_t(s, p) = (s + t, p)$.

(2) The observer field $U = \partial_t/g$ is synchronizable, with $U = -g \operatorname{grad} t$, hence U is irrotational.

(3) The restspaces $t \times S$ of U are isometric under the flow isometries ψ_t, and all are isometric under σ to S.

Proof. (1) Evidently each ψ_t is an isometry of $I\,{}_g\!\times S$. (2) Both U and $\operatorname{grad} t$ are orthogonal to all slices $t \times S$, which are thus restspaces. Since $\langle \operatorname{grad} t, \partial_t \rangle = 1$, it follows that $\operatorname{grad} t = -\partial_t/g^2$, hence $U = -g \operatorname{grad} t$. (3) is obvious. ∎

Every static spacetime is locally standard:

38. Proposition. A spacetime M is static relative to an observer field U if and only if for each $p \in M$ there is a U-preserving isometry of a standard static spacetime onto a neighborhood of p.

Proof. If such isometries exist, then the lemma shows that M is static relative to U. Conversely, since U is irrotational there is a restspace S through p. Let $\{\psi_t : t \in I\}$ be a local flow at p of the Killing vector field $X = gU$. Shrinking S and I if necessary, the map $\psi: I \times S \to \psi_t q$ is a diffeomorphism onto a neighborhood of p in M. Furthermore, $d\psi(\partial_t) = X$, and each ψ_t is an isometry such that $d\psi_t(X) = X$. Thus

(1) $X \perp \psi_t(S)$ for all $t \in I$ (since this is true for $t = 0$).
(2) $g = |X| > 0$ is constant on the curves $t \to \psi_t(q)$, $q \in S$.
(3) $\psi_t | S$ is an isometry of the Riemannian hypersurface S onto $\psi_t(S) = \psi(t \times S)$.

If $0 \times S$ is identified with S, then (2) implies $g \circ \psi = (g|S) \circ \sigma$, where σ is projection on S. Thus the three conditions show that the pullback by ψ^* of the metric tensor of M has the form specified in Definition 36. ■

Thus locally, any static spacetime has the properties given in the lemma; in particular its observer field U is locally synchronizable.

Exercises

1. (a) Let U be a future-pointing timelike unit vector field on a spacetime M. Prove that U is the flow vector field of a perfect fluid if and only if $\mathrm{Ric}(X, Y) = \mathrm{Ric}(X, U) = 0$ whenever X, Y, and U are mutually orthogonal. (b) If (U, ρ, p) is a perfect fluid and $X, Y \perp U$, then $\mathrm{Ric}(U, U) = 4\pi(\rho + 3p)$, $\mathrm{Ric}(U, X) = 0$, and $\mathrm{Ric}(X, Y) = 4\pi(\rho - p)\langle X, Y\rangle$. Hence $S = 8\pi(\rho - 3p)$ and $\Sigma R^{ij}R_{ij} = 64\pi^2(\rho^2 + 3p^2)$.
2. If a spacetime M is an Einstein manifold, so $\mathrm{Ric} = cg$, show that any observer field U on M is the flow vector field of a perfect fluid with $\rho = -p = c/8\pi$.
3. Any Robertson–Walker spacetime $I \times_f S$ is conformally diffeomorphic to a semi-Riemannian product $J \times S$, $J \subset \mathbf{R}^1_1$.
4. (a) Give direct proofs of Corollaries 8 and 9. (b) Deduce Corollary 13 directly from Theorem 11. (c) Given k, deduce the formulas in Theorem 11 from Corollaries 12 and 13.

5. Let (U, ρ, p) be a perfect fluid. An instantaneous observer measures the speed of the fluid as $\tanh \varphi$; find the observer's measurements of density and pressure.

6. Let $U = \partial_t$ on $M(k,f)$. Prove: (a) Corollary 13 is equivalent to $\operatorname{div}(\rho U) = -p \operatorname{div} U$ and also to $L_U(\rho\omega) = -pL_U(\omega)$, where ω can denote either the volume element of $M(k, f)$ or of its spacelike slices. (b) On any spacelike slice $\operatorname{div} U = -\operatorname{trace} S_U$, where S_U is the shape operator of the slice. (c) If $u = U_{(t,p)}$, the tidal force operator F_u is scalar multiplication by f''/f, and $\operatorname{trace} F_u = -\operatorname{Ric}(u, u) = -4\pi(\rho + 3p)$.

7. Prove: (a) Let U be a geodesic observer field on a spacetime M, and let $g \in \mathfrak{F}(M)$. Then $gU \neq 0$ is Killing if and only if g is constant and U Killing. (b) A Robertson–Walker spacetime $M(k,f)$ is static relative to $U = \partial_t$ if and only if f is constant.

8. Consider the Friedmann model $(0, \pi A) \times_f S^3$ of Example 28(2), with $H_0^{-1} = 18 \times 10^9$ yr. (a) For $\rho_0 = 2\rho_c$ as on page 253, show that $f = \frac{1}{2}$, where f is the fraction of all galaxies from which we can have received information. (b) Less realistically, suppose $\rho_0 = 10\rho_c$. Find the maximum radius A, the age of the universe, and f.

9. Consider the Friedmann $k = -1$ model of Example 28(3), with $H_0^{-1} = 18 \times 10^9$ yr. (a) If $\rho_0 = \rho_c/2$ as on page 253, find $d_{\max}$, the largest present distance of galaxies from which we can have received information. (b) If $\rho_0 = \rho_c/10$, find the age of the universe and $d_{\max}$.

10. (a) A perfect fluid satisfies the strong energy condition if and only if $\rho + p \geq 0$ and $\rho + 3p \geq 0$. (b) A Robertson–Walker perfect fluid satisfies the strong energy condition if and only if $K_\sigma \geq K_u \leq 0$.

11. In $M(k,f) = I \times_f S$ prove: (a) If I is not all of $\boldsymbol{R}_1^1$, then every timelike geodesic is incomplete. (Hint: Consider $dt/d\tau$.) (b) If S is complete and $f(t) = (1 - t^2)^{-1}$ on $I = (-1, 1)$, then every null geodesic is complete but no timelike one is.

12. Let M be a spacetime and let $\mathcal{O}$ be an open set of instantaneous observers in $T_p(M)$. Prove: (a) A symmetric (0, 2) tensor T at p is completely determined by its values $T(u, u)$ for $u \in \mathcal{O}$. (b) The curvature tensor R at p is completely determined by g and $F_u(y) = R_{yu}u$ for all $y \perp u \in \mathcal{O}$. (Hint: See Exercise 3.22.)

13. Prove: (a) A Ricci flat Robertson–Walker spacetime is flat, and there are just two types, represented by $\boldsymbol{R}_1^4 = \boldsymbol{R}_1^1 \times \boldsymbol{R}^3$ and $M(-1, \pm t)$. (b) $M(-1, t) = \boldsymbol{R}^+ \times_t H^3$ is physically equivalent to a timecone in $\boldsymbol{R}_1^4$ with the galaxies being the rays from 0.

14. *Radiation models.* (a) If $M(k, f)$ has f nonconstant, show that the following are equivalent: (i) $3p = \rho > 0$; (ii) $\rho f^4 = C > 0$; (iii) $f'^2 + k$

$= a^2/f^2$, where $a^2 = 8\pi C/3$. (b) Prove that the scale function f can be written as $(2at - t^2)^{1/2}$ if $k = 1$, $(2at)^{1/2}$ if $k = 0$, $(2at + t^2)^{1/2}$ if $k = -1$. (c) Identify these as conic sections, and sketch graphs assuming $a = 1$.

15. Let K_u and K_σ be the principal sectional curvatures of $M(k, f)$ at (t, p), and let $u = U_{(t,p)}$. Prove: (a) Every timelike plane Π at (t, p) has a basis x, $(\cosh \varphi)u + (\sinh \varphi)y$, where $\{x, y, u\}$ is orthonormal. Then $K(\Pi) = K_u \cosh^2 \varphi - K_\sigma \sinh^2 \varphi$. (b) A similar formula holds for spacelike planes. (c) At (t, p) and hence on the entire slice $S(t)$, K is constant $\Leftrightarrow K_u(t) = K_\sigma(t)$.

16. If $M(k,f)$ does not have constant curvature, show that every self isometry ϕ is a product as in Exercise 7.11. (Hint: $K_u(t) \neq K_\sigma(t)$ for some t; use the preceding exercise to show that $U \perp \phi(S(t))$.)

17. Prove: (a) S_1^4 is (isometric to) a Robertson–Walker spacetime. (b) For a suitable f, $(0, \pi) \times_f H^3$ is isometric to half of $H_1^4(r)$. (c) $H_1^4(r)$ is not a Robertson–Walker spacetime.

18. A spacetime M is static relative to an observer field U if and only if at each $p \in M$ there is a coordinate system with $U = \partial_0/\sqrt{-g_{00}}$, $g_{0k} = 0$ for $1 \leq k \leq 3$ and $\partial g_{ij}/\partial x^0 = 0$ for all i, j.

13 SCHWARZSCHILD GEOMETRY

Schwarzschild spacetime is the simplest relativistic model of a universe containing a single star. The star is assumed to be static and spherically symmetric—and to be the only source of gravitation for the spacetime. The resulting model can thus be applied to regions around any astronomical object that approximately fulfills these conditions. For example, in the case of the sun it gives a model for the solar system even better than the highly accurate Newtonian model.

Schwarzschild found the spacetime late in 1915, soon after the appearance of general relativity. Intially only half of it, the exterior, seemed to be physically significant. However the neglected half, suitably joined to the exterior, now provides the simplest model of a *black hole*.

BUILDING THE MODEL

Schwarzschild spacetime will emerge naturally from the physical conditions given above.

(1) *Static.* The spacetime is to be static relative to observers comparable to Newtonian observers at rest in Euclidean 3-space. Using Definition 12.36 as a guide, take the restspace to be $\boldsymbol{R}^3$—but with line element $\boldsymbol{q}$ yet to be determined—and let the spacetime be the manifold $\boldsymbol{R}^1 \times \boldsymbol{R}^3$ with line element of the form

$$A(x)\, dt^2 + \boldsymbol{q} \qquad (x \in \boldsymbol{R}^3),$$

where $\boldsymbol{q}$ is lifted from $\boldsymbol{R}^3$. The projection $t\colon \boldsymbol{R}^1 \times \boldsymbol{R}^3 \to \boldsymbol{R}^1$ is called *Schwarzschild time.*

By Lemma 12.37 the lift ∂_t of d/dt from $\boldsymbol{R}^1$ is the Killing vector field required by the definition of static.

(2) *Spherical symmetry.* Since the star and hence the resulting spacetime are to be spherically symmetric, for each $\phi \in O(3)$ the map $(t, x) \to (t, \phi x)$ must be an isometry. Thus it is natural to give a spherical description of $\boldsymbol{R}^3$ (minus the origin) as $\boldsymbol{R}^+ \times S^2$, where $\boldsymbol{R}^+ = \{\rho \in \boldsymbol{R} : \rho > 0\}$ and S^2 is the unit 2-sphere. As in Example 7.37(2), spherical symmetry implies that the line element $\boldsymbol{q}$ on $\boldsymbol{R}^+ \times S^2 \approx \boldsymbol{R}^3 - 0$ can be written as $B(\rho)\, d\rho^2 + C(\rho)\, d\sigma^2$, where $d\sigma^2$ is the line element standard on the unit sphere. For every $\phi \in O(3)$ the differential map of id $\times\ \phi$ carries ∂_t to ∂_t, hence the coefficient function $A(x)$ of dt^2 actually depends only on ρ. Thus the line element on $\boldsymbol{R}^1 \times \boldsymbol{R}^+ \times S^2 \approx \boldsymbol{R}^1 \times (\boldsymbol{R}^3 - 0)$ becomes

$$A(\rho)\, dt^2 + B(\rho)\, d\rho^2 + C(\rho)\, d\sigma^2.$$

(3) *Normalization.* A change of variable in $\boldsymbol{R}^+$ replaces $C(\rho)$ by r^2, so the line element is now

$$E(r)\, dt^2 + G(r)\, dr^2 + r^2\, d\sigma^2.$$

The projection $r: \boldsymbol{R}^1 \times \boldsymbol{R}^+ \times S^2 \to \boldsymbol{R}^+$ is called the *Schwarzschild radius function.*

By this normalization, in each restspace t constant, the surface r constant has line element $r^2\, d\sigma^2$ and is thus the standard 2-sphere $S^2(r)$, with Gaussian curvature $1/r^2$ and area $4\pi r^2$ (see Figure 1).

The line element above exhibits the spacetime as the warped product $P \times_r S^2$ where $P = \boldsymbol{R}^1 \times \boldsymbol{R}^+$ is the half-plane $r > 0$ in the tr-plane, furnished with line element $E(r)\, dt^2 + G(r)\, dr^2$. The projections of $P \times_r S^2$ on P and the unit sphere S^2 are denoted as usual by π and σ.

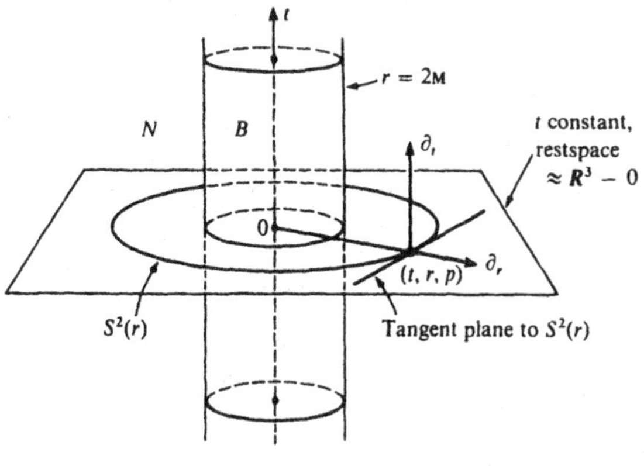

Figure 1. Schwarzschild exterior *N*, black hole *B*.

(4) *Vacuum and Minkowski at infinity.* The only source of gravitation in the Schwarzschild universe is the star itself—which we do not model. Thus the spacetime must be a *vacuum*, that is, Ricci flat.

Sufficiently far away from a source of gravitation, its influence becomes arbitrarily small. Hence we require that, as r approaches infinity, the Schwarzschild metric tensor approaches the Minkowski metric of empty spacetime, which in spherical terms is

$$-dt^2 + dr^2 + r^2\, d\sigma^2.$$

Thus $E(r) \to -1$ and $G(r) \to +1$ as $r \to \infty$.

The conditions above determine the functions E and G and thereby the Schwarzschild metric.

1. Lemma. $P \times_r S^2$ is Ricci flat and Minkowski at infinity if and only if $E = -\mathscr{h}$ and $G = \mathscr{h}^{-1}$, where $\mathscr{h}(r) = 1 - (2\mathrm{M}/r)$, with M an arbitrary constant.

Proof. Applying Corollary 7.43, with $d = \dim S^2 = 2$, the Ricci flat condition is equivalent to

(1) ${}^P\mathrm{Ric}(X, Y) = 2H^r(X, Y)/r$ for $X, Y \in \mathfrak{L}(P)$,
(2) ${}^F\mathrm{Ric}(V, W) = \langle V, W\rangle r^{\#}$ for $V, W \in \mathfrak{L}(S^2)$,

where $r^{\#} = \Delta r/r + \langle \mathrm{grad}\, r, \mathrm{grad}\, r\rangle/r^2$.

We need consider equation (1) only on P itself, since leaves are totally geodesic and project isometrically. Because $\dim P = 2$, ${}^P\mathrm{Ric}(X, Y) = K_P\langle X, Y\rangle$. Choose X, Y from ∂_t, ∂_r; then by Exercise 5.5(a), equation (1) is equivalent to

$$K_P E = E'/2G; \qquad K_P G = -G'/2G.$$

Thus $E'/E = -G'/G$, so EG is constant, and the limit conditions on E and G imply $EG = -1$.

In equation (2), ${}^F\mathrm{Ric}$ is the lift through the projection σ (a homothety on fibers) of the Ricci curvature $\mathrm{Ric} = g = (\ ,\ )$ of the unit sphere. Thus

$${}^F\mathrm{Ric}(V, W) = (d\sigma V, d\sigma W) = \langle V, W\rangle/r^2,$$

so (2) is equivalent to $r^{\#} = 1/r^2$. Using Exercise 5.5 again gives

$$r^{\#} = \frac{1}{2rG}\left[\frac{E'}{E} - \frac{G'}{G}\right] + \frac{1}{r^2 G} = \frac{1}{r^2}.$$

Replacing E'/E by $-G'/G$ yields

$$\frac{-G'}{rG^2} + \frac{1}{r^2 G} = \frac{1}{r^2}.$$

Consequently,

$$\left(\frac{r}{G}\right)' = \frac{G - G'r}{G^2} = 1.$$

Thus r/G is r plus a constant, which we denote by -2M, so

$$G = \frac{r}{r - 2\text{M}} = \left(1 - \frac{2\text{M}}{r}\right)^{-1}, \qquad E = -G^{-1} = -\left(1 - \frac{2\text{M}}{r}\right). \qquad \blacksquare$$

Anticipating the identification of the constant M as the mass of the star, we require $\text{M} > 0$. Thus the *Schwarzschild function* $h(r) = 1 - (2\text{M}/r)$ rises from limit $-\infty$ at $r = 0$ toward limit 1 at $r = \infty$, passing through 0 at $r = 2\text{M}$. Hence the warped product line element

$$-h\, dt^2 + h^{-1}\, dr^2 + r^2\, d\sigma^2$$

fails at $r = 2\text{M}$. We have found not one but two spacetimes.

2. Definition. For $\text{M} > 0$ let P_{I} and P_{II} be the regions $r > 2\text{M}$ and $0 < r < 2\text{M}$ in the tr-half-plane $\boldsymbol{R}^1 \times \boldsymbol{R}^+$, each furnished with line element $-h\, dt^2 + h^{-1}\, dr^2$, where $h(r) = 1 - (2\text{M}/r)$. If S^2 is the unit sphere, then the warped product $N = P_{\text{I}} \times_r S^2$ is called *Schwarzschild exterior spacetime* and $B = P_{\text{II}} \times S^2$ the *Schwarzschild black hole*, both of *mass* M.

A star with properties as above is characterized by two numbers, its mass M and radius r^*. Since the star itself is not modeled, we are left with the region $r > r^*$ in $N \cup B$ as a model for the spacetime around it. In the usual case $r^* > 2\text{M}$, the surface of the star is outside the Schwarzschild radius so its exterior is given by the connected region $r > r^*$ in N. (For example, the sun has $r^* = 7 \times 10^5$ km $\gg \text{M} = 1.5$ km.)

The situation is more interesting if $r^* \leq 2\text{M}$. As we shall see, r^* can then only be 0: the star has disappeared, leaving a black hole B.

When the Schwarzschild radius r is sufficiently large, the metric on N is nearly Minkowskian and we can think of t as time and r as radial distance, but for smaller r these interpretations are suspect. Indeed, as r passes below 2M, the function h becomes negative so it is ∂_r that is timelike and ∂_t spacelike. Generally, if say $r < 10\text{M}$, the situation is already too relativistic for Newtonian analogies to be of much help.

Consider the warped product definitions above, indicated schematically in Figure 2.

(1) For each $(t, r) \in P_{\text{I}}$ the fiber $\pi^{-1}(t, r)$ is the sphere $S^2(r)$ in the rest-space of Schwarzschild time t. As always in a warped product, this sphere is totally umbilic in N and σ maps it homothetically onto S^2.

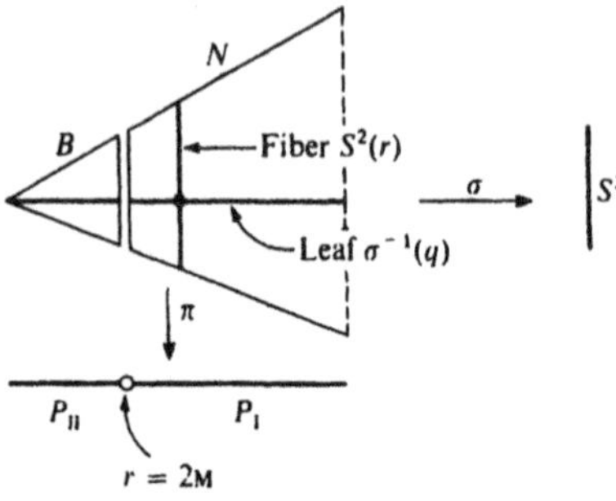

Figure 2. Schwarzschild warped product structure (schematic).

(2) For each $q \in S^2$ the leaf $\sigma^{-1}(q) = P_I \times q$ is, again by warped product generalities, totally geodesic in N and isometric under the projection π to the Schwarzschild half-plane P_I. Such a *radial plane* $\sigma^{-1}(q)$ is the two-dimensional spacetime consisting of *all events occurring directly over a single point of the star*. Its Newtonian analogue is a radial line from the center of the star together with Newtonian time. A particle in $\sigma^{-1}(q)$ is *moving radially*.

The definition $N = P_I \times S^2$ thus presents the Schwarzschild exterior as the product of the standard radial plane P_I with the unit sphere, the warping function turning the fibers into spheres of radius $r > 2\mathrm{M}$.

The geometric remarks above are valid also for the black hole B but some physical interpretations will differ.

GEOMETRY OF N AND B

Since the metric tensors of the Schwarzschild exterior N and black hole B are formally the same, the geometries can be computed simultaneously.

3. Lemma. On $P_I \cup P_{II}$, where $ds^2 = -h(r)\, dt^2 + h(r)^{-1}\, dr^2$,

(1) $D_{\partial_t} \partial_t = (\mathrm{M}h/r^2)\, \partial_r$,
$D_{\partial_t} \partial_r = D_{\partial_r} \partial_t = (\mathrm{M}/r^2 h)\, \partial_t$,
$D_{\partial_r} \partial_r = (-\mathrm{M}/r^2 h)\, \partial_r$.

(2) $\operatorname{grad} t = -\partial_t/h$; $\operatorname{grad} r = h\, \partial_r$.

(3) $H^r = (\mathrm{M}/r^2)g$.

(4) $K = 2\mathrm{M}/r^3$.

Proof. (1) and (4) follow from Proposition 3.44, and the gradients and Hessian are computed as usual. ■

Let ∂_t and ∂_r on $N \cup B$ be the lifts of the corresponding vector fields on $P_{\mathrm{I}} \cup P_{\mathrm{II}}$. $\mathfrak{L}(S^2)$ consists as usual of all lifts to $N \cup B$ of vector fields on S^2. Also, following warped product conventions, on each tangent space $T_p(N \cup B)$,

tan is orthogonal projection onto the tangent plane of the fiber $\pi^{-1}(t, r) = S^2(r)$ through p;

nor is orthogonal projection onto the tangent plane of the leaf $\sigma^{-1}(\sigma p)$ through p.

4. Proposition. On $N \cup B$, if $V, W \in \mathfrak{L}(S^2)$, then

(1) Same as (1) in the lemma above.

(2) $D_{\partial_t}(V) = D_V(\partial_t) = 0$; $D_{\partial_r}(V) = D_V(\partial_r) = (1/r)V$.

(3) $\mathrm{nor}(D_V W) = II(V, W) = -(\hbar/r)\langle V, W\rangle\, \partial_r$, where II is the shape tensor of each fiber.

(4) $\tan(D_V W) \in \mathfrak{L}(S^2)$ is the lift of $\nabla_V W$ from S^2.

This follows immediately from Proposition 7.35.

5. Proposition. Let V, W be vector fields on $N \cup B$ that are vertical, that is, tangent to all spheres $S^2(r)$. Then

(1) $R_{\partial_t \partial_r}(\partial_t) = (-2\mathrm{M}\hbar/r^3)\, \partial_r$; $R_{\partial_r \partial_t}(\partial_r) = (2\mathrm{M}/r^3 \hbar)\, \partial_t$.

(2) $R_{\partial_t V}(\partial_t) = (\mathrm{M}\hbar/r^3)V$;
$R_{\partial_r V}(\partial_r) = (-\mathrm{M}/\hbar r^3)V$;
$R_{\partial_t V}(\partial_r) = R_{\partial_r V}(\partial_t) = 0$.

(3) $R_{\partial_r \partial_t}(V) = R_{VW}(\partial_t) = R_{VW}(\partial_r) = 0$.

(4) $R_{XV} W = R_{XW} V = (\mathrm{M}/r^3)\langle V, W\rangle X$ for $X = \partial_t$ or ∂_r.

(5) $R_{VW} U = (2\mathrm{M}/r^3)[\langle U, V\rangle W - \langle U, W\rangle V]$, where U is also vertical.

Proof. These formulas follow from the correspondingly numbered ones in Proposition 7.42. In fact, letting X and Y denote any choices from ∂_t, ∂_r, and using Lemma 3:

(1) The curvature operator of $P_{\mathrm{I}} \cup P_{\mathrm{II}}$ is $R_{\partial_t \partial_r} X = (2\mathrm{M}/r^3)[\langle X, \partial_t\rangle \partial_r - \langle X, \partial_r\rangle \partial_t]$.

(2) $H^r = (\mathrm{M}/r^2)g$, hence $R_{XV} Y = -R_{VX} Y = -(\mathrm{M}/r^3)\langle X, Y\rangle V$.

(3) This is obvious.

(4) From Lemma 3 compute $D_X(\mathrm{grad}\ r) = (\mathrm{M}/r^2)X$.

(5) We saw in Chapter 7 that this is just the Gauss equation of each fiber $\pi^{-1}(t, r) = S^2(r)$, for which the lift ${}^F R$ gives the curvature tensor. Thus ${}^F R_{VW} U = r^{-2}[\langle U, V\rangle W - \langle U, W\rangle V]$. But $\langle \mathrm{grad}\ r, \mathrm{grad}\ r\rangle = \hbar = 1 - (2\mathrm{M}/r)$, since $\mathrm{grad}\ r = \hbar\, \partial_r$. ∎

Lemma 1 shows already that N and B are Ricci flat.

6. Definition. Let ϑ, φ be spherical coordinates on the unit sphere S^2 (Exercise 3.13). Let t, r be the usual Schwarzschild time and radius coordinates on $P_{\mathrm{I}} \cup P_{\mathrm{II}}$. The product coordinate system $(t, r, \vartheta, \varphi)$ in $N \cup B$ is called a *Schwarzschild-spherical coordinate system.*

The domain of these coordinates omits say $\sigma^{-1}(C)$, where C is a semicircle in S^2, but the coordinate vector fields $\partial_t, \partial_r, \partial_\varphi$ are well defined and smooth everywhere, and ∂_ϑ is singular only over the poles $\vartheta = 0, \pi$ of S^2. Note that ∂_t and ∂_r are tangent to radial planes (horizontal) while ∂_ϑ and ∂_φ are tangent to spheres (vertical).

For indexing purposes, let $x^0 = t$, $x^1 = r$, $x^2 = \vartheta$, $x^3 = \varphi$. Then the line element

$$-h\,dt^2 + h^{-1}\,dr^2 + r^2(d\vartheta^2 + \sin^2\vartheta\,d\varphi^2)$$

shows that the coordinates are orthogonal with

$$\langle \partial_t, \partial_t \rangle = g_{00} = -h, \qquad \langle \partial_r, \partial_r \rangle = g_{11} = h^{-1},$$
$$\langle \partial_\vartheta, \partial_\vartheta \rangle = g_{22} = r^2, \qquad \langle \partial_\varphi, \partial_\varphi \rangle = g_{33} = r^2 \sin^2 \vartheta.$$

Because of spherical symmetry, the coordinate system t, r, ϑ, φ can be rotated as follows. An isometry $A \in O(3)$ of S^2 determines an isometry $(t, r, q) \to (t, r, Aq)$ of $N \cup B$; thus $t, r, \tilde{\vartheta} = \vartheta \circ A$, $\tilde{\varphi} = \varphi \circ A$ is a coordinate system with line element given by the same formula as before.

By construction, ∂_t is a Killing vector field on $N \cup B$, and, as with any warped product, the lift of a Killing vector field on S^2 is Killing on $N \cup B$. These are essentially all.

7. Proposition. Every Killing vector field on N or B has the form $c\,\partial_t + V$, where V is a Killing vector field lifted from S^2.

Proof. Since radial planes are totally geodesic, the component of Y tangent to $\sigma^{-1}(q)$ is Killing (Exercise 9.7). The projection π is an isometry from $\sigma^{-1}(q)$ to P_{I} that preserves t and r; hence by Example 9.24 this component is a constant, say $f(q)$, times ∂_t. Thus $Y = \tilde{f}\,\partial_t + V$, where V is vertical and $\tilde{f}$ is the lift $f \circ \sigma$ of a function f on S^2. We must show that $V \in \mathfrak{L}(S^2)$ and $\tilde{f}$ is constant.

Since Y is Killing, DY is skew-adjoint and hence in particular $\langle D_{\partial_t} Y, Z \rangle + \langle D_Z Y, \partial_t \rangle = 0$ for $Z \in \mathfrak{L}(S^2)$. Using Proposition 4, we compute $\langle D_{\partial_t} Y, Z \rangle = (\partial/\partial t)\langle V, Z \rangle$ and $\langle D_Z Y, \partial_t \rangle = -hZ\tilde{f}$. Hence,

(1) $(\partial/\partial t)\langle V, Z \rangle = hZ\tilde{f}$.

Similarly, replacing ∂_t by ∂_r leads to

(2) $(\partial/\partial r)\langle V, Z \rangle = (2/r)\langle V, Z \rangle$.

Fix $q \in S^2$ and consider $\langle V, Z\rangle$ as a function of t and r on $\sigma^{-1}(q)$. Equation (2) implies $\langle V, Z\rangle = g(t)r^2$. Substituting this into (1) gives $g'(t)r^2 = h(r)zf$, where $z \in T_q(S^2)$ lifts to Z on $\sigma^{-1}(q)$. Since $h(r) = 1 - (2\text{M}/r)$, this equation is impossible unless both $g'(t)$ and zf are zero.

Then $g' = 0$ implies $\langle V, Z\rangle = kr^2$, where k is constant relative to t and r, but depends, of course, on z. Thus

$$k(z) = \langle V_{(t,r)}, Z_{(t,r)}\rangle / r^2 = (d\sigma(V_{(t,r)}), z).$$

Choosing z from a frame at q shows at once that $d\sigma(V_{(t,r)})$ is independent of t and r. Thus $V \in \mathfrak{L}(S^2)$.

Since $zf = 0$ for every tangent vector z to S^2, the functions f and hence $\tilde{f}$ are constant. Thus $Y = c\,\partial_t + V$ as required. ∎

8. Corollary. (1) On N, every timelike Killing vector field has the form $c\,\partial_t$. (2) On B, every Killing vector field is spacelike.

Proof. For a Killing vector field on N or B write $Y = c\,\partial_t + V$ as above; since V is vertical it is spacelike. (1) If Y is timelike on N, then

$$0 > \langle Y, Y\rangle = -c^2 h(r) + \langle V, V\rangle.$$

Since V is the lift of a vector field V_0 on S^2, $\langle V, V\rangle = (V_0, V_0)r^2 \geq 0$. In any radial plane $\sigma^{-1}(q)$, if $r \to \infty$, then $h(r) \to 1$. Hence the inequality above can be maintained only if $(V_0, V_0)(q) = 0$. Consequently V_0 and hence V are zero. (2) On B, ∂_t also is spacelike since $h < 0$ on B. ∎

SCHWARZSCHILD OBSERVERS

We consider some basic features of the Schwarzschild exterior N. The Killing vector field ∂_t is timelike on N, and *N is time-oriented by requiring that ∂_t be future-pointing*. By construction, N is static relative to the corresponding observer field

$$U = \partial_t/\sqrt{h}.$$

The integral curves α of U are called *Schwarzschild observers* (represented by vertical lines in Figure 1). As predicted by Lemma 12.37, U is synchronizable, and Schwarzschild time t is the average time, since, by Lemma 3, $U = -\sqrt{h}\,\text{grad}\, t$. If τ is the proper time of a Schwarzschild observer α, then

$$\frac{d(t \circ \alpha)}{d\tau} = \langle \alpha', \text{grad}\, t\rangle = \langle U, \text{grad}\, t\rangle = h(\alpha)^{-1/2},$$

and this time dilation is constant along α.

Since $0 < h < 1$ on N, Schwarzschild time is always faster than the proper time of a Schwarzschild observer. The two times are nearly the same if the observer is far from the star, where $h(\alpha) \sim 1$. But since $h(r) \to 0$ as $r \to 2M$, Schwarzschild time speeds up unboundedly for observers with $r(\alpha)$ ever closer to $2M$.

By construction, the restspaces of U are all naturally isometric to the standard *Schwarzschild restspace* $\mathbf{S}$: the region $r > 2M$ in $\mathbf{R}^+ \times S^2$, with line element $h^{-1}\, dr^2 + r^2\, d\sigma^2$. Again for r large, hence $h(r)$ near 1, $\mathbf{S}$ is close to Euclidean space (described spherically).

Thus Newtonian notions of both time and space are approximately valid far from the star.

If γ is a particle in N, deleting its t coordinate projects γ to a curve $\vec{\gamma}$ in the restspace $\mathbf{S}$. (Parametrizing $\vec{\gamma}$ by t would give the analogue of an associated Newtonian particle in special relativity.) For a Schwarzschild observer α, the curve $\vec{\alpha}$ is constant: these observers are *at rest*. Hovering over a fixed point of the star at constant height, they are definitely not freely falling:

9. Lemma. If α is a Schwarzschild observer, then $\alpha'' = D_U U = (M/r^2)\, \partial_r$.

Proof. Since $U = h^{-1/2}\, \partial_t$ and $\partial h/\partial t = 0$, Proposition 4 gives $D_U U = h^{-1/2}(h^{-1/2} D_{\partial_t}\, \partial_t) = (M/r^2)\partial_r$. ∎

Thus each observer must aim his rockets at the star and supply the acceleration required to remain at rest. In the analogous Newtonian case (Appendix C), the acceleration has magnitude M/r^2. Here, for r large, $|\partial_r| = 1/h(r)$ is close to 1, hence the lemma gives $|D_U U| \sim M/r^2$, supporting the designation of M as the mass of the star. (See also Exercises 1, 2, and 3.)

By Corollary 8, the black hole B is not static relative to any observer field; we consider its physical interpretation later on.

SCHWARZSCHILD GEODESICS

Schwarzschild-spherical coordinates $x^0 = t$, $x^1 = r$, $x^2 = \vartheta$, $x^3 = \varphi$ are orthogonal, hence by Exercise 3.15 the geodesic equations become

$$\frac{d}{ds}\left[g_{ii}\left(\frac{dx^i}{ds}\right)\right] = \frac{1}{2}\sum_{j=0}^{3}\frac{\partial g_{jj}}{\partial x^i}\left(\frac{dx^j}{ds}\right)^2 \qquad (0 \le i \le 3),$$

where $g_{00} = -h$, $g_{11} = h^{-1}$, $g_{22} = r^2$, $g_{33} = r^2 \sin^2 \vartheta$.

10. Lemma. If γ is a geodesic in $N \cup B$, then for constants E, L,

(1) $h\, dt/ds = E$,
(2) $r^2 \sin^2 \vartheta\, d\varphi/ds = L$,
(3) $(d/ds)(r^2\, d\vartheta/ds) = r^2 \sin \vartheta \cos \vartheta\, (d\varphi/ds)^2$.

Here (1) and (2) follow from the geodesic equations for $i = 1, 3$, since no g_{jj} involves either t or φ. Equation (3) will be simplified below. For the geodesic equation for r, see Exercise 5.

A curve in $N \cup B$ is *initially equatorial* relative to spherical coordinates ϑ, φ on S^2 provided it has $\vartheta(0) = \pi/2$ and $(d\vartheta/ds)(0) = 0$. Evidently a suitable rotation of coordinates will make any curve initially equatorial (assuming $0 \in I$).

11. Proposition. Let γ be a freely falling material particle in $N \cup B$ that is initially equatorial relative to Schwarzschild-spherical coordinates. Then

(G1) $h\, dt/d\tau = E$,
(G2) $r^2\, d\varphi/d\tau = L$,
(G3) $\vartheta = \pi/2$,

where E and L are constants. Furthermore the *energy equation* holds:

$$E^2 = (dr/d\tau)^2 + (1 + (L^2/r^2))h(r).$$

Proof. Equations (G1)–(G3) will follow from the corresponding formulas in Lemma 10, with the parameter s now proper time τ. (1) is unchanged, and evidently $\vartheta = \pi/2$ is the unique solution of (3) satisfying the equatorial initial conditions. Thus (2) implies (G2). Then $\gamma' = \sum (dx^i/d\tau)\, \partial_i$ becomes

$$\gamma' = (E/h)\, \partial_t + (dr/d\tau)\, \partial_r + (L/r^2)\, \partial_\varphi.$$

Hence

$$-1 = \langle \gamma', \gamma' \rangle = (E^2/h^2)(-h) + (dr/d\tau)^2 h^{-1} + (L^2/r^4)r^2.$$

The energy equation follows. ∎

These equations embody Galileo's principle that the motion of a freely falling body is independent of its mass, a result obtained in Newtonian theory only by equating inertial and gravitational mass (Appendix C.)

By (G3), the particle remains always over a great circle of S^2 (or the surface of the star).

On N at least, equation (G2) is formally identical with Kepler's second law (Appendix C); hence we call L the *angular momentum per unit mass* of the particle.

If γ has mass m and hence energy–momentum $m\gamma'$, then in N the Schwarzschild observers measure its energy as

$$-\langle m\gamma', U\rangle = -\langle m\gamma', h^{-1/2}\,\partial_t\rangle = m\sqrt{h}\,(dt/d\tau).$$

By (G1), $h\, dt/d\tau$ is the constant E. But $h \to 1$ as $r \to \infty$, hence E is called the *energy per unit mass at infinity* of the particle. Since γ' is future-pointing, $dt/d\tau$ and hence E are positive. (For brevity we sometimes call E and L merely *energy* and *angular momentum.*)

FREE FALL ORBITS

Let γ be a freely falling material particle in the Schwarzschild exterior N. The projection $\vec{\gamma}$ into the Schwarzschild restspace S lies in the *orbital plane* of γ, given in equatorial coordinates by $\vartheta = \pi/2$. There we can write $\vec{\gamma}(\tau) = (r(\tau), \varphi(\tau))$, inviting comparison with the polar coordinate description of Newtonian gravitation. The *orbit* of γ is the route followed by $\vec{\gamma}$ in the orbital plane. The particle is *ingoing* when $dr/d\tau < 0$, *outgoing* when $dr/d\tau > 0$.

In dealing with initial conditions, we write simply 0 instead of τ_0. Using Proposition 11 it is easy to check that if $r(0)$ and, of course, M are given, then E, L, and $\mathrm{sgn}(dr/d\tau)(0)$ uniquely determine the coordinates of the 4-velocity $\gamma'(0)$.

In order to relate the character of the orbit to the physical parameters E and L of the particle γ, write the energy equation as

$$E^2 = \left(\frac{dr}{d\tau}\right)^2 + V(r),$$

where

$$V(r) = \left(1 + \frac{L^2}{r^2}\right)h(r) = 1 - \frac{2M}{r} + \frac{L^2}{r^2} - \frac{2ML^2}{r^3}.$$

The energy equation can be regarded as expressing conservation of energy for a unit mass particle whose motion on the half-line $\mathbf{R}^+$ is given by $r = r \circ \gamma$. Ignoring a factor $\frac{1}{2}$, we call $V(r)$ the *effective potential energy* of γ. Energy diagrams can now be constructed as in Appendix C.

(1) Plot the graph of $V(r)$ and draw a horizontal line at height E^2.

(2) Since $(dr/d\tau)^2 \geq 0$, the energy equation restricts $r = r \circ \gamma$ to the component I of $\{r\colon V(r) \leq E^2\}$ containing its initial value $r(0)$.

(3) Since $2\, d^2r/d\tau^2 = -V'(r)$ (Exercise 5), a critical point of V represents a circular orbit $r = r_0$, stable for a minimum, unstable for a maximum. Furthermore, a noncritical point $V(r) = E^2$ is a *turning point*: r bounces off to retraverse I in the opposite direction.

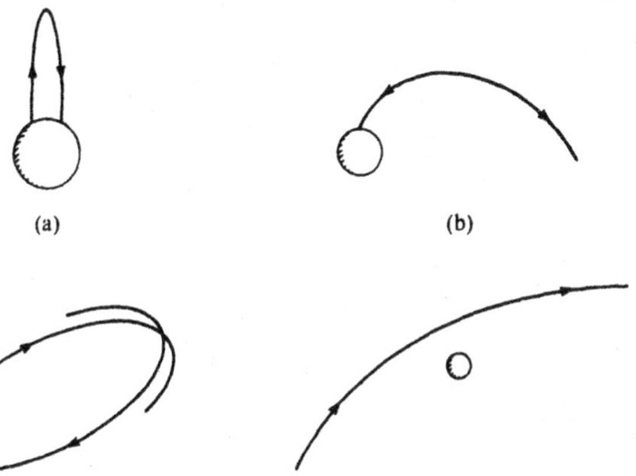

Figure 3. Ordinary relativistic orbits: (a) Crash orbit. (b) Crash/escape orbit. (c) Bound orbit. (d) Flyby orbit.

The effective potential $V(r)$ differs from that in Newtonian gravitation (Appendix C) by the relativistic correction term $-2\text{M}L^2/r^3$, constants being unimportant. Thus for r large we expect approximately Newtonian results. But in general the relativistic situation is more complex since the shape of the graph of $V(r)$ depends on the ratio L/M. In all cases, $\lim_{r\to 0} V(r) = -\infty$ and $\lim_{r\to\infty} V(r) = 1$. One major change in the shape of the graph is signaled by the appearance of critical points $r_1 \leq r_2$, computable from

$$V'(r) = (2/r^4)[\text{M}r^2 - L^2 r + 3\text{M}L^2].$$

When two critical points $r_1 < r_2$ exist, r_1 is a local maximum and r_2 is a local minimum.

An orbit is *exceptional* if $E^2 = V(r_1)$; otherwise it is *ordinary*. As we shall see, exceptional orbits are geometrically interesting, but they lack physical significance, since they can be made ordinary by arbitrarily small changes in E or L (while ordinary orbits remain ordinary). Considering at first only ordinary orbits, we now find the four types suggested in Figure 3.

Case I. Low Angular Momentum: $L^2 < 12\text{M}^2$. $V(r)$ has no critical points, hence is strictly increasing (see Figure 4). Thus there are two orbit types, depending on E:

(a) $E^2 < 1$: *crash orbit.* Ingoing particles crash directly into the star. (For $r^* > 2\text{M}$ this means that $r \circ \gamma$ approaches r^*, hence γ meets the star, thereby leaving our model. In the black hole case, $r \circ \gamma$ approaches 0, and the term "crash" is amply justified.) Initially outgoing particles move out to a turning point then back in to crash.

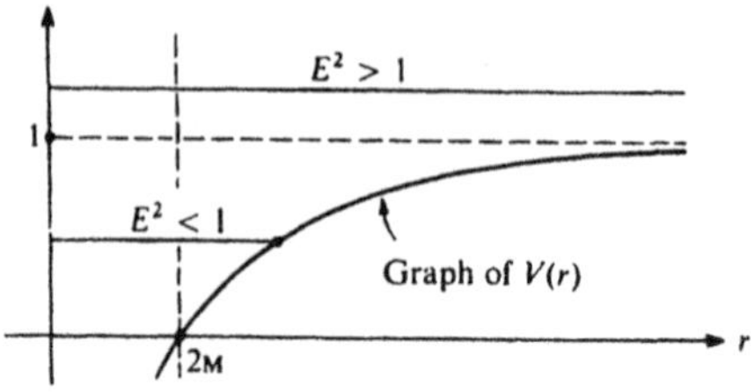

Figure 4. Low angular momentum.

(b) $E^2 \geq 1$: *crash/escape orbit.* Ingoing particles crash; outgoing particles escape to infinity.

When L^2 reaches 12M^2, a critical point appears at 6M with $V(6\text{M}) = \frac{8}{9}$.

Case II. Moderate Angular Momentum $12\text{M}^2 < L^2 < 16\text{M}^2$. There are now two critical points, $r_1 < 6\text{M} < r_2$, forming the crest and trough of a potential well (Figure 5). Three ordinary orbit types occur in four cases, depending on energy E and $r(0)$:

(a) $E^2 < V(r_1)$ and $r(0) < r_1$: crash orbit as in Case I(a).

(b) $V(r_2) \leq E^2 < V(r_1)$ and $r(0) > r_1$: *bound orbit.* The particle is in the potential well, its Schwarzschild radius oscillating between maximum and minimum, with a stable circular orbit at $r = r_2$. By (G2), $d\varphi/d\tau$ is thus bounded away from zero, so the particle travels perpetually around the star.

(c) $V(r_1) < E^2 < 1$: crash orbit.

(d) $E^2 \geq 1$: crash/escape orbit as in Case I(b).

At $L^2 = 16\text{M}^2$ the crest $V(r_1)$ of the potential well, rising with L/M, reaches $V(\infty) = 1$.

Case III. Large Angular Momentum: $L^2 > 16\text{M}^2$. The situation is now dominated by the rising potential barrier with crest $V(r_1)$. (See Figure 6.)

(a) $E^2 < V(r_1)$ and $r(0) < r_1$: crash orbit.

(b) $V(r_2) \leq E^2 < 1$ and $r(0) > r_1$: bound orbit as in Case II(b).

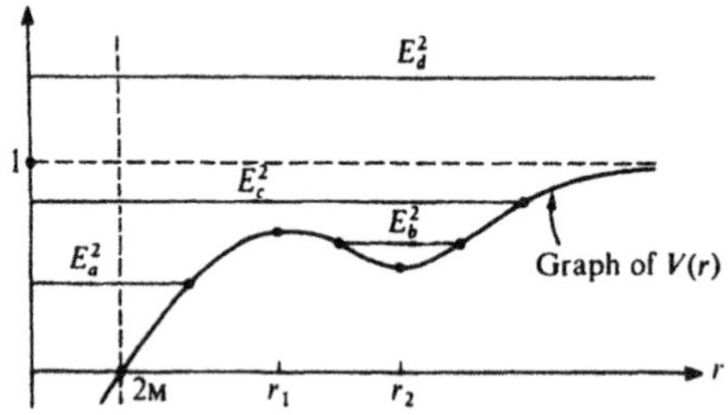

Figure 5. Moderate angular momentum.

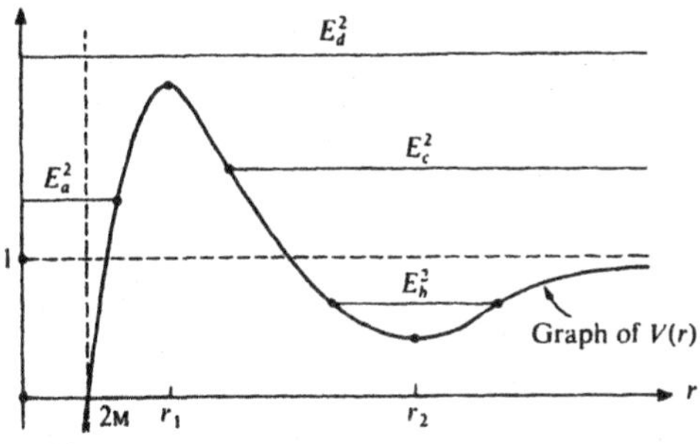

Figure 6. Large angular momentum.

(c) $1 \leq E^2 < V(r_1)$ and $r(0) > r_1$: *flyby orbit*. An outgoing particle has enough energy to escape. In the incoming case the potential barrier (derived from the large angular momentum) protects the particle from a crash and it turns back to escape to infinity.

(d) $E^2 > V(r_1)$: crash/escape orbit.

The simplest exceptional orbits are the unstable circular orbits $r = r_1$ at the local maxima of $V(r)$—or at $r_1 = r_2$ if $L^2 = 12\text{M}^2$. The other exceptional orbits are infinite spirals approaching ever closer to these circular orbits (see Example 35(1)). Nearby ordinary orbits are also very different from Keplerian ellipses and hyperbolas since, by continuity, they too must spiral to some extent while close to $r = r_1$.

A freely falling particle with $L = 0$ is moving radially, since equations (G2) and (G3) show that its φ and ϑ coordinates are constant. Its energy equation is

$$E^2 - 1 = (dr/d\tau)^2 - 2\text{M}/r,$$

which, but for notation, agrees with the equation for radial motion in Newtonian theory (Appendix C) and also with the Friedmann equation (Lemma 12.20). As for the latter, we can get parametric solutions. For example, consider the low energy case, in which the particle eventually falls back to the star.

12. Lemma. In N, let γ be a freely falling material particle with $L = 0$ and $E < 1$. The proper time τ and the Schwarzschild radius r of γ are cycloidally related by

$$\tau = \tfrac{1}{2}R\sqrt{R/2\text{M}}(\eta + \sin \eta), \qquad r = \tfrac{1}{2}R(1 + \cos \eta),$$

where $\tau = 0$ at the maximum radius R of γ.

Proof. At maximum radius, $dr/d\tau$ is zero, hence $E^2 = \hbar(R) = 1 - (2\text{M}/R)$. Thus the energy equation becomes

$$(dr/d\tau)^2 + 2\text{M}/R = 2\text{M}/r.$$

Direct substitution shows that the formulas provide the required solution. ■

PERIHELION ADVANCE

We want to obtain more precise information about the orbit of a freely falling material particle γ in the Schwarzschild exterior N. In the nontrivial case $L \neq 0$, equation (G2) implies that $d\varphi/d\tau$ is nonvanishing, hence the orbit can be described by expressing r as a function of φ.

13. Proposition. (The Orbit Equation). For a freely falling material particle γ in N with $L \neq 0$.

$$d^2u/d\varphi^2 + u = (\text{M}/L^2) + 3\text{M}u^2,$$

where $u = 1/r$, $r = r \circ \gamma$.

Proof. Using (G2),

$$\frac{dr}{d\tau} = \frac{dr/d\varphi}{d\tau/d\varphi} = \frac{L}{r^2}\frac{dr}{d\varphi};$$

hence the energy equation becomes

$$E^2 = \frac{L^2}{r^4}\left(\frac{dr}{d\varphi}\right)^2 + \left(1 + \frac{L^2}{r^2}\right)\left(1 - \frac{2\text{M}}{r}\right).$$

Setting $r = 1/u$ gives

$$E^2 = L^2(du/d\varphi)^2 + (1 + L^2u^2)(1 - 2\text{M}u).$$

Then taking another derivative gives the result. ■

This equation differs from its Newtonian analogue (Appendix C) by the relativistic correction term $3\text{M}u^2$. Obviously $3\text{M}u^2 \ll u$ for large r. Furthermore $3\text{M}u^2 \ll \text{M}/L^2$ if the tangential component $r\, d\varphi/d\tau$ of orbital velocity is small compared to the speed of light, since by (G2)

$$\frac{\text{M}u^2}{\text{M}/L^2} = \frac{L^2}{r^2} = \left(r\frac{d\varphi}{d\tau}\right)^2.$$

Thus the relativistic orbit will be nearly Keplerian when these conditions hold. (The comparison is valid, since, for $r \gg M$, the orbital plane is nearly Euclidean.) In the bound case each trip around the star does indeed have an approximately elliptical orbit, but the ellipses are not the same for successive trips. This *precession* can be measured by change of *perihelion* (point at which r is a minimum).

14. Corollary. Let γ be a freely falling particle in bound orbit around a Schwarzschild star of mass M. If $L \gg M$ and $r(d\varphi/d\tau) \ll 1$, then the orbit of γ is approximately elliptical with angular perihelion advance

$$\delta \sim \frac{6\pi M^2}{L^2} = \frac{6\pi M}{a(1-e^2)} \quad \text{radians per revolution,}$$

where a is the semimajor axis and e the eccentricity of the ellipse.

To see this, recall from Appendix C that the Newtonian orbit equation $(d^2u/d\varphi^2) + u = M/L^2$ has $\tilde{u} = (M/L^2)(1 + e\cos\varphi)$ as solution with perihelion at $\varphi = 0$. The hypotheses imply $r \gg L \gg M$ so this is close to the relativistic solution. To obtain a more refined approximation, we use $\tilde{u}$ in the relativistic correction term thus

$$(d^2u/d\varphi^2) + u = (M/L^2) + 3M\tilde{u}^2.$$

A routine computation gives

$$u = \frac{M}{L^2}(1 + e\cos\varphi) + \frac{3M^3}{L^4}\left(1 + \frac{e^2}{2} - \frac{e^2}{6}\cos 2\varphi + e\varphi\sin\varphi\right)$$

as the solution with perihelion at $\varphi = 0$. To find the next perihelion we need

$$\frac{du}{d\varphi} = -\frac{Me}{L^2}\sin\varphi + \frac{3M^3e}{L^4}\left[\frac{e}{3}\sin 2\varphi + \sin\varphi + \varphi\cos\varphi\right].$$

Since $L \gg M$, the dominant terms here are the first and the one involving $\varphi\cos\varphi$. Thus the next perihelion will occur near $\varphi = 2\pi$ at say $2\pi + \delta$, where

$$0 = \left.\frac{du}{d\varphi}\right|_{2\pi+\delta} \sim \frac{Me}{L^2}\left[-\sin\delta + \frac{3M^2}{L^2}(2\pi + \delta)\cos\delta\right].$$

Hence

$$\delta \sim \tan\delta \sim \frac{3M^2}{L^2}(2\pi + \delta).$$

Again since $L \gg M$, we obtain $\delta \sim 6\pi M^2/L^2$. The alternative formula follows since $a(1 - e^2) = L^2/M$ for a Keplerian ellipse (Appendix C).

Since perihelion advance is always positive, we can think of the particle as traversing an elliptical orbit in a plane that is itself slowly rotating in the same direction.

This early result of Einstein led to the first experimental test of general relativity. The orbit of the planet Mercury has eccentricity 0.206, largest of the well-charted planets, so its perihelion can be located with precision. All known perturbations of perfect elliptical orbit (largely due to the gravitational effects of the other planets) give by Newtonian theory a predicted perihelion advance for Mercury about 43 sec of arc per century *less* than the observed value. The relativistic advance accounts for this discrepancy. In fact the hypotheses of the corollary are fulfilled, and using

$$\text{M} = \text{mass of the sun} = 1.48 \times 10^5 \quad \text{cm},$$

$$a = \text{semimajor axis for Mercury} = 5.79 \times 10^{12} \quad \text{cm},$$

we compute $\delta \sim 5.02 \times 10^{-7}$ radians per revolution. There are about 2.06×10^5 sec of arc per radian, and with its period of 87.96 days, Mercury makes 415.2 revolutions per century. The relativistic perihelion advance is thus about

$$(5.02)(2.06)(415.2) \times 10^{-2} \sim 43 \quad \text{sec per century.}$$

LIGHTLIKE ORBITS

Now consider the behavior of a lightlike particle γ (photon, neutrino, graviton) in the Schwarzschild exterior. By definition, γ is a geodesic, so equations (G1)–(G3) in Proposition 11 are the same as for a material particle (with affine parameter s instead of proper time), but the energy equation is altered since $\langle \gamma', \gamma' \rangle$ is now 0 instead of -1.

15. Proposition. If γ is a lightlike particle in $N \cup B$, then relative to equatorial coordinates,

(G1) $\hbar\, dt/ds = E$,
(G2) $r^2\, d\varphi/ds = L$,
(G3) $\vartheta = \pi/2$,

and γ satisfies the *energy equation*

$$E^2 = (dr/ds)^2 + (L^2/r^2)\hbar(r).$$

As before, on N we interpret the constants E and L as the *energy at infinity* and *angular momentum* of γ.

A lightlike particle with $L = 0$ has both φ and ϑ constant, hence moves radially. Since each radial plane is naturally isometric to P_{I}, Example 5.41 gives explicit formulas for the coordinates t and r of the particle.

As with a material particle, the Schwarzschild view of a lightlike particle γ is as its projection $\vec{\gamma} = (r, \varphi)$ into its orbital plane in the Schwarzschild restspace. In Newtonian theory, light rays are straight lines, but relativistically they are influenced by gravitation: γ is geodesic, but $\vec{\gamma}$ is accelerating, producing the so-called "bending of light" by the star. By contrast with the material case, the orbit of γ does not depend on E and L separately, but only on the ratio $b = |L|/E$, called the *impact parameter* of γ. For $b \neq 0$, r becomes a function of φ. After rewriting the energy equation equation as

$$\left(\frac{1}{L}\frac{dr}{ds}\right)^2 + \frac{\hbar}{r^2} = \frac{1}{b^2},$$

we use (G2) to get

16. Corollary. For a lightlike particle in N with impact parameter $b \neq 0$,

$$\left(\frac{1}{r^2}\frac{dr}{d\varphi}\right)^2 + \frac{\hbar}{r^2} = \frac{1}{b^2}.$$

As usual we interpret $\hbar/r^2$ as an *effective potential*. This function rises from limit $-\infty$ at $r = 0$ and, crossing the r axis at $2\mathrm{M}$, reaches a maximum of $1/27\mathrm{M}^2$ at $3\mathrm{M}$—then declines toward zero. The qualitative character of lightlike orbits can then be read from Figure 7.

Case I. Small Impact Parameter: $b < 3\sqrt{3}\,\mathrm{M}$. Depending on initial conditions, γ either crashes into the star or escapes to infinity.

For impact parameter $3\sqrt{3}\mathrm{M}$ there is an unstable circular orbit $r = 3\mathrm{M}$. A Schwarzschild observer at this radius can see the back of his own head. Two exceptional orbit types spiral toward this circular orbit.

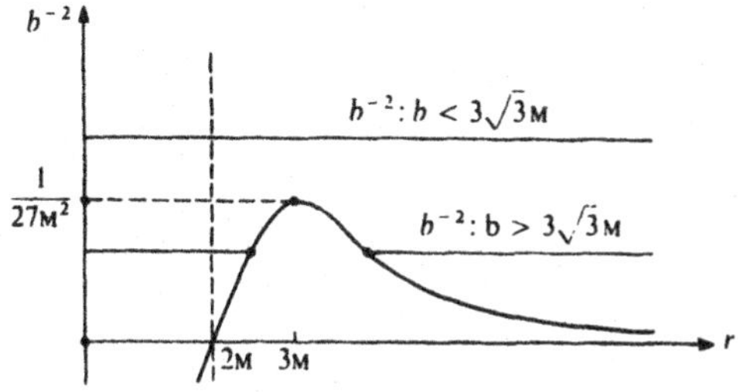

Figure 7. Graph of $\hbar/r^2$.

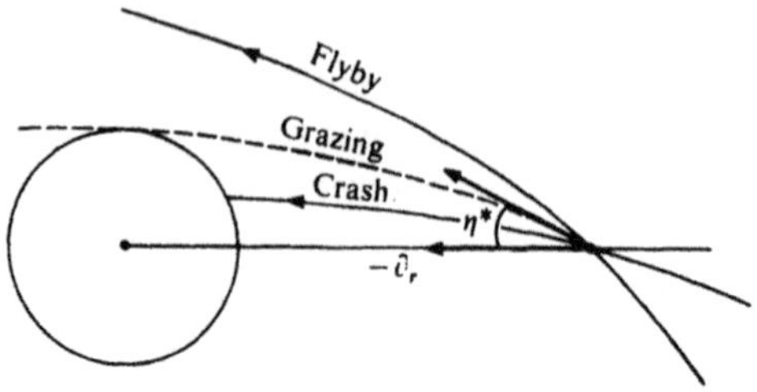

Figure 8. Photons in an orbital plane.

Case II. Large Impact Parameter: $b > 3\sqrt{3}$M. (a) $r(0) < 3$M: crash orbit; (b) $r(0) > 3$M: flyby orbit.

The only bound orbit is the unstable one at $r = 3$M; practically speaking, any photon in N either escapes to infinity or meets the star.

The impact parameter b of γ has a more geometric characterization in terms of the angle between orbital velocity $\vec{\gamma}'$ and the inward radial direction $-\partial_r$ (Figure 8).

17. Lemma. If γ is a lightlike particle in N with impact parameter b, then $r \sin \eta = b\sqrt{h(r)}$, where $\eta = \measuredangle(-\partial_r, \vec{\gamma}')$.

Proof. In terms of equatorial coordinates, adding $\vec{\gamma}' = (dr/ds)\partial_r + (d\varphi/ds)\partial_\varphi$ and $(dt/ds)\partial_t$ gives the null vector γ'. Hence by (G1),

$$|\vec{\gamma}'| = |(dt/ds)\,\partial_t| = (E/h)|\partial_t| = E/\sqrt{h}.$$

Then using (G2),

$$\sin \eta = \frac{|d\varphi/ds\;\partial_\varphi|}{|\vec{\gamma}'|} = \frac{|L|/r}{E/\sqrt{h}} = \frac{b\sqrt{h}}{r}. \qquad \blacksquare$$

We can now find what a Schwarzschild observer sees of a star of radius $r^* > 3$M. Drawing the line $r = r^*$ in Figure 7 shows that a photon sent from $r > r^*$ will escape to infinity if and only if its impact parameter b is greater than $r^*/\sqrt{h(r^*)}$. An ingoing photon with $b^* = r^*/\sqrt{h(r^*)}$ has a turning point $dr/d\tau = 0$ at radius r^*, so it just grazes the star (Figure 8). By the lemma, the initial angle η^* of such a grazing photon satisfies $r \sin \eta^* = b^*\sqrt{h(r)}$, with $\eta < \pi/2$ since it is ingoing. But the *orbits* of particles are reversible (Exercise 10), so the star will appear as a disk of angular radius η^*. Eliminating b^* from the two equations above thus gives the following result.

18. Corollary. For a Schwarzschild observer at radius r, a star of radius $r^* > 3\text{M}$ has angular radius $\eta^* \leq \pi/2$ such that

$$\sin \eta^* = \frac{r^*}{r}\left(\frac{h(r)}{h(r^*)}\right)^{1/2}.$$

The square root factor here is greater than 1, showing that as expected, the bending of light makes the star appear larger than the straight line Newtonian analogue $\sin \eta^* = r^*/r$.

As with material particles, substituting $r = 1/u$ in Corollary 16 and differentiating gives

19. Corollary (The Orbit Equation). For a lightlike particle γ in N with $L \neq 0$,

$$d^2u/d\varphi^2 + u = 3\text{M}u^2,$$

where $u = 1/r$, $r = r \circ \gamma$.

As the Schwarzschild radius approaches infinity, the relativistic correction term $3\text{M}u^2$ above becomes negligibly small and the orbital plane approaches flatness. Thus for a lightlike particle in flyby orbit, as its parameter s approaches $\pm\infty$, its orbit approaches limiting straight lines. If the entire orbit is far from the star, we can measure its total bending by means of the *deflection angle* Δ between the limiting lines (Figure 9).

20. Corollary. If γ is a lightlike particle in flyby orbit in N with perihelion $r_0 \gg \text{M}$, then its deflection angle is approximately $4\text{M}/r_0$.

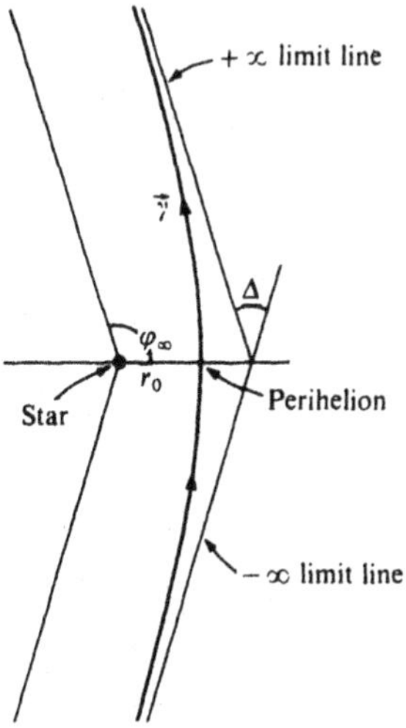

Figure 9

We use the same scheme as for Corollary 14. The (straight line) solution of $d^2u/d\varphi^2 + u = 0$ with perihelion at $\varphi = 0$ is $\tilde{u} = a \cos \varphi$, where $a = 1/r_0$. Replacing the right-hand side of the orbit equation by $3\text{M}\tilde{u}^2$, we compute

$$u = a(1 - \text{M}a) \cos \varphi + \text{M}a^2(2 - \cos^2 \varphi)$$

as the solution for which $u(0) = a$. Since $a \gg a^2$ and as $s \to \infty$ both u and $\cos \varphi$ approach 0, the dominant terms in the limit are

$$0 \sim a \cos \varphi_\infty + 2\text{M}a^2.$$

Thus $\cos \varphi_\infty \sim -2\text{M}a$. It is clear from Figure 9 that this limit angle is $\pi/2$ plus half the deflection angle Δ. Hence

$$\Delta \sim 2 \sin \tfrac{1}{2}\Delta \sim -2 \cos (\tfrac{1}{2}\pi + \tfrac{1}{2}\Delta) \sim 4\text{M}a = 4\text{M}/r_0.$$

During a solar eclipse, light from a distant star that just grazes the sun en route to earth is observed to be slightly displaced from its direction with the same earth–star position but the sun elsewhere. The earth–sun distance is sufficiently large compared to the radius r_0 of the sun that the deflection angle is essentially at its limiting value in the preceding corollary. For solar mass $\text{M} = 1.5$ km and $r_0 = 7 \times 10^5$ km, the deflection angle is

$$\Delta \sim 4\text{M}/r_0 \sim \tfrac{6}{7} \times 10^{-5} \quad \text{rad} \sim 1.7 \quad \text{sec},$$

which is close to observed values.

STELLAR COLLAPSE

The first role of the Schwarzschild model was to confirm the essential validity of general relativity by accounting for minor flaws in the Newtonian model of the solar system. Since the solar system is not very relativistic, such refinements, though profound theoretically, are qualitatively not too significant. More recently, however, stellar objects have been discovered for which the Newtonian approximation is crude and in the extreme case fails completely.

A star is formed when a cloud of gas is drawn together by gravity; as its density increases it gets hotter and nuclear burning begins. Then for a long period—say a billion years—near equilibrium prevails: energy loss by radiation is balanced by energy gain from the nuclear reaction; gravitational tendency toward contraction is balanced by pressures from the hot dense core. (Our sun is now in this equilibrium state.) Eventually the supply of nuclear fuels runs low and gravitational contraction begins. This phase is short-lived and ends violently; though the process is not fully understood, several possible endstates have been singled out, including the following:

(1) For masses less than about $\frac{5}{4}$ that of the sun, the collapsing star may stabilize as a *white dwarf*, at a size comparable with the earth. This produces enormous densities, but since r^* is on the order of 1000M, relativistic effects are not decisive. White dwarfs are fairly common throughout the universe.

(2) For somewhat larger masses the collapse may continue, stripping atomic nuclei of their electrons and packing them together as a *neutron star*. With radii of say 10–15 km, enormously high densities are attained—billions of tons per teaspoon. Since r^* is on the order of 8M, relativistic effects are crucial. *Pulsars* are believed to be rapidly rotating neutron stars.

(3) For masses more than half again that of the sun such relatively temperate endstates are ruled out, and if the star cannot somehow manage to eject sufficient mass, no known physical mechanism can prevent its radius from shrinking below 2M. A catastrophic collapse ensues, and it is predicted that in a fraction of a second an endstate is reached that can only be said to have radius 0 and density ∞: a *black hole* has formed.

That black holes actually exist in our universe is widely accepted. The simplest mathematical model will join the Schwarzschild black hole B with its exterior N. This raises the question: How bad is the singularity $r = 2\text{M}$ that separates them? Let us drop a pebble into a black hole. If it starts from rest, say at $r = 8\text{M}$, then the pebble falls radially and by Lemma 12 its Schwarzschild radius r and proper time τ are given by the cycloidal parametrizations

$$\tau = 8\text{M}(\eta + \sin\eta), \qquad r = 4\text{M}(1 + \cos\eta).$$

Thus $r \to 2\text{M}$ as $\eta \to 2\pi/3$, and the elapsed proper time is about 23.7M. To see what happens to the pebble as $r \to 2\text{M}$, we consider the tidal forces on it.

21. Remarks. Radial Schwarzschild Tidal Forces. Let u be a radial instantaneous observer, for example a 4-velocity vector of the pebble above. Thus u is a timelike unit vector of the form $a\,\partial_t + b\,\partial_r$. In the restspace $u^\perp$, u regards the vector $x = (b/h)\,\partial_t + ah\,\partial_r$ as pointing directly toward (or away from) the star. Vectors $w \in u^\perp$ that are orthogonal to x are simply vectors tangent to the fiber $S^2(r)$; u regards them as *transverse*. The tidal forces measured by u are readily computed from Proposition 5:

$$F_u(x) = (2\text{M}/r^3)x, \qquad F_u(w) = -(\text{M}/r^3)w \qquad \text{for all} \quad w.$$

For example,

$$\begin{aligned} F_u(w) = R_{wu}u &= a^2 R_{w\partial_t}\partial_t + b^2 R_{w\partial_r}\partial_r \\ &= (\text{M}/r^3)(-a^2 h + b^2/h)w = -(\text{M}/r^3)w. \end{aligned}$$

Thus in u's radial space direction there is *tension* of twice the magnitude of the transverse *compression*. In terms of acceleration, this is in agreement with the situation in Figure 12.1, where $\pm x$ would point from o to b and d, while suitable $\pm y$ point to a and c. (See also Exercise 3.)

Since the tidal forces on the falling pebble are of order M/r^3, there is no reason to expect any profound physical change at $r = 2\text{M}$. So let the pebble continue on into B. It reaches the central singularity $r = 0$ at $\eta = \pi$, taking proper time $8\pi\text{M}$ for the entire fall. This not very long: if M is k solar masses, about k ten-thousands of a second.

Schwarzschild observers tell a different story. The time dilation $dt/d\tau = E/\hbar$ approaches infinity as $r \to 2\text{M}$, and in fact for Schwarzschild observers *the pebble will never reach* $r = 2\text{M}$, though it eventually drifts in arbitrarily close (Exercise 4). (By Example 5.41, the same is true for a beam of light directed radially inward.) Thus the pebble seems to experience no difficulty in falling through $r = 2\text{M}$, but Schwarzschild observers cannot describe the trip. This suggests that the trouble at $r = 2\text{M}$ is due to the Schwarzschild time function t and not to the spacetimes N and B.

THE KRUSKAL PLANE

In the remainder of this chapter we define and study Kruskal spacetime K, half of which joins N and B to give a (connected) spacetime. For some history of its discovery see Box 31.1 of [MTW]. Like N and B, K will be a warped product with fiber a 2-sphere; thus the problem is reduced to joining the Schwarzschild half-plane P_{I} and strip P_{II}. Fix the notation

$$f(r) = (r - 2\text{M})e^{(r/2\text{M})-1} \qquad \text{for} \quad r \in \boldsymbol{R}^+,$$

with M a positive constant. Since $f' > 0$ on $\boldsymbol{R}^+$, f is a diffeomorphism onto the half-line $(-2\text{M}/e, \infty)$.

Now let Q be the region in the uv-plane given by $uv > -2\text{M}/e$. (See Figure 10.) If f^{-1} is the inverse diffeomorphism of f, then $r = f^{-1}(uv)$ defines a smooth positive function on Q that is characterized implicitly by the equation $f(r) = uv$

Thus the level curves r constant in Q are the hyperbolas $uv = \text{const}$, except that $r = 2\text{M}$ gives the coordinate axes. The function r has limit value 0 on the boundary hyperbola $uv = -2\text{M}/e$, which is not part of Q.

Removing the coordinate axes from Q leaves its four *open quadrants*, denoted as usual by $Q_{\text{I}}, \ldots, Q_{\text{IV}}$.

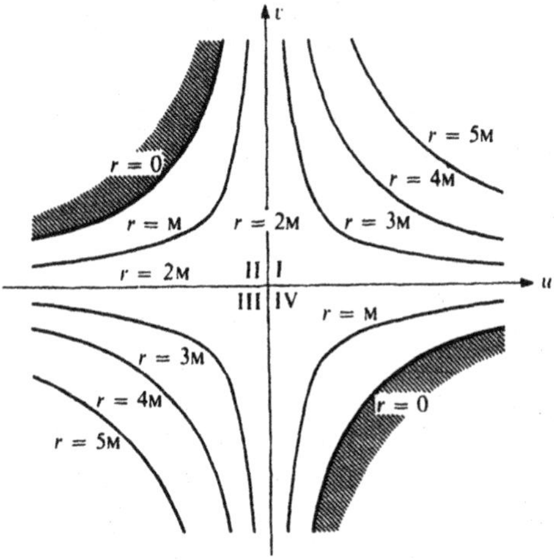

Figure 10. The Kruskal plane Q.

22. Definition. With notation as above, the region Q in the uv-plane, furnished with line element

$$ds^2 = 2F(r)\,du\,dv, \qquad \text{where} \quad F(r) = (8\text{M}^2/r)e^{1-(r/2\text{M})},$$

is called the *Kruskal plane* of mass M.

Since the natural coordinates u, v are a null coordinate system on Q (Definition 5.42), *the null geodesics of Q are parametrizations of the coordinate lines u constant and v constant.* The metric tensor of Q is

$$F(r)(du \otimes dv + dv \otimes du).$$

Thus $\langle \partial_u, \partial_v \rangle = F(r) > 0$. The mapping $(u, v) \to (-u, -v)$ of Q preserves uv, hence r, hence $F(r)$, hence the line element, and is thus an isometry of Q that reverses the quadrants Q_{I} and Q_{III}, and also reverses Q_{II} and Q_{IV}.

Using Exercise 5.8 we find that Q has sectional curvature $2\text{M}/r^3$.

On the open quadrants $uv \neq 0$ of Q, define $t = 2\text{M} \ln|v/u|$. The level curves t constant are thus rays from the origin in Q.

23. Lemma. On the Kruskal plane,

(1) $Ff = 8\text{M}^2 \hbar$, $Ff' = 4\text{M}$, and $f/f' = 2\text{M}\hbar$, where as usual $\hbar(r) = 1 - (2\text{M}/r)$.

(2) $dt = 2\text{M}((dv/v) - (du/u))$, $\qquad dr = 2\text{M}\hbar((du/u) + (dv/v))$ for $uv \neq 0$.

(3) $\text{grad}\, r = (1/4\text{M})(u\,\partial_u + v\,\partial_v)$.

Proof. (1) The first identity is immediate. The second follows from

$$f'(r) = (r/2M)e^{((r/2M)-1)}.$$

The third is a consequence of the first two.

(2) $dt = 2M(u/v)\,d(v/u) = 2M(dv/v - du/u)$. Differentiating $f(r) = uv$ gives $f'(r)\,dr = v\,du + u\,dv$. Since $2M\hslash = f(r)/f'(r) = uv/f'(r)$, the formula for dr follows.

(3) The vector fields metrically equivalent to du and dv are ∂_v/F and ∂_u/F, respectively. Hence by (2),

$$\text{grad } r = (2M\hslash/Ff)(u\,\partial_u + v\,\partial_v) = (1/4M)(u\,\partial_u + v\,\partial_v). \quad \blacksquare$$

We can now show that Q_I and hence Q_{III} are isometric to the Schwarzschild half-plane P_I, while Q_{II} and Q_{IV} are isometric to the Schwarzschild strip P_{II}.

24. Proposition. The function $\psi: Q_I \cup Q_{II} \to P_I \cup P_{II}$ sending (u, v) to $(t(u, v), r(u, v))$ is an isometry that preserves quadrants and the functions t and r.

Proof. On $P_I \cup P_{II}$, t and r are just the natural coordinate functions, and we write them temporarily as $\tilde{t}$ and $\tilde{r}$. Then by the definition of ψ, $\tilde{t} \circ \psi = t$ and $\tilde{r} \circ \psi = r$. Hence $\psi^*(d\tilde{t}) = d(\tilde{t} \circ \psi) = dt$ and similarly $\psi^*(d\tilde{r}) = dr$.

The line element of $P_I \cup P_{II}$ is $-\hslash\, d\tilde{t}^2 + \hslash^{-1}\, d\tilde{r}^2$. Since ψ preserves r it also preserves $\hslash$, so ψ^* applied to this line element is $-\hslash\, dt^2 + \hslash^{-1}\, dr^2$. Substituting the formulas for dt and dr gives

$$8M^2\hslash(du\,dv/uv) = (16M^2\hslash/f(r))\,du\,dv = 2F(r)\,du\,dv,$$

which is the line element of Q.

It remains only to check that ψ is a diffeomorphism on each quadrant. But we can solve the coordinate formula for ψ to show, for example, that $\psi: Q_I \to P_I$ has inverse function given by

$$u = \sqrt{f(r)}e^{-t/4M}, \qquad v = \sqrt{f(r)}e^{t/4M}. \quad \blacksquare$$

The essential problem in joining N and B is now solved, since $Q_I \approx P_I$ and $Q_{II} \approx P_{II}$ fit together naturally in Q along the positive v axis.

By definition, the mapping ψ preserves level curves of t and of r, hence ψ can be visualized in Figure 11 as (1) exploding the origin (0, 0) of Q into the whole vertical line $r = 2M$ and thus lifting the pairs of radial lines t constant into horizontal lines t constant in $P_I \cup P_{II}$, and (2) sending the u axis to a point at $-\infty$ so that the hyperbolas r constant are carried to the vertical lines r constant in $P_I \cup P_{II}$.

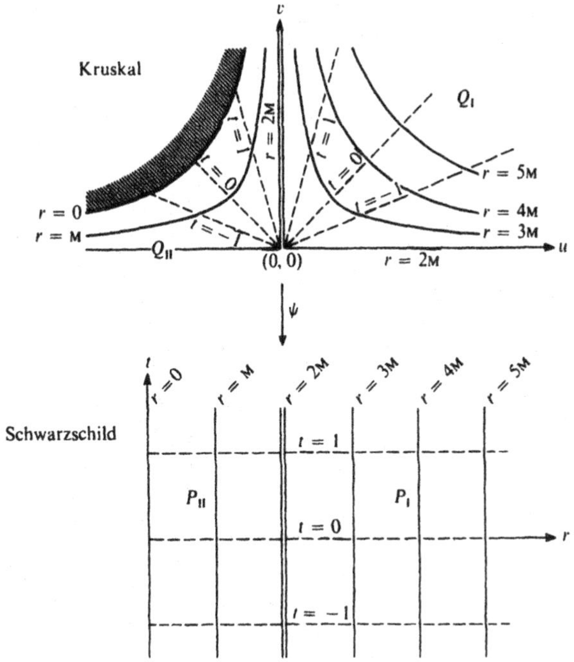

Figure 11. Mapping Kruskal to Schwarzschild.

KRUSKAL SPACETIME

25. Definition. Let Q be a Kruskal plane of mass M, and let S^2 be the unit 2-sphere. The *Kruskal spacetime* of mass M is the warped product $K = Q \times_r S^2$, where r is the function on Q characterized by $f(r) = uv$.

Explicitly, K is the smooth manifold $Q \times S^2$ furnished with line element $2F(r)\,du\,dv + r^2\,d\sigma^2$. The time-orientation of K is specified below.

Let π and σ be the projections of K onto Q and S^2. As always, each leaf $\sigma^{-1}(q)$ is totally geodesic and isometric to Q, and each fiber $\pi^{-1}(u, v)$ is a 2-sphere $S^2(r(u, v))$ of radius $r(u, v)$ that is totally umbilic in K.

We continue to denote by u, v, r the lifts (prefix by π) of the natural coordinate functions u, v and the radius function r on Q.

For each $\mathcal{N} = $ I, II, III, IV, let $K_{\mathcal{N}}$ be the open submanifold $\pi^{-1}(Q_{\mathcal{N}})$ over the quadrant $Q_{\mathcal{N}}$ of Q. Deleting these quadrants from K leaves the *horizon* H, consisting of all points over the coordinate axes of Q. Removing the *central sphere* $\pi^{-1}(0, 0)$ from H leaves four hypersurfaces, each diffeomorphic to $\boldsymbol{R}^+ \times S^2$.

The *time function* $t = 2\text{M} \ln|v/u|$ is defined only on the open quadrants of Q, hence its lift into K, given by the same formula, is defined only on $K - H$.

It is easy to show that the quadrants K_{I} and K_{III} are isometric to N and the quadrants K_{II} and K_{IV} to B. For example, let $\psi: Q_{\text{I}} \to P_{\text{I}}$ be the isometry given by Proposition 24. Since ψ preserves r, the mapping $\psi \times \text{id}$ is an isometry from $K_{\text{I}} = Q_{\text{I}} \times_r S^2$ onto $N = P_{\text{I}} \times_r S^2$. Similarly, since the isometry $(u, v) \to (-u, -v)$ of Q preserves r, it induces an isometry $\phi(u, v, p) = (-u, -v, p)$ of K called its *central symmetry*. Evidently ϕ reverses K_{I} and K_{III} and also K_{II} and K_{IV}. We record these results as

$$K_{\text{III}} \overset{\phi}{\approx} K_{\text{I}} \overset{\psi}{\approx} N, \qquad K_{\text{IV}} \overset{\phi}{\approx} K_{\text{II}} \overset{\psi}{\approx} B.$$

Hence *Kruskal spacetime also is Ricci flat but not flat.*

The isometries above obviously preserve the functions t and r; thus *on any quadrant of K these functions are geometrically the same as the usual Schwarzschild time and radius functions on N or B.* We emphasize that, unlike t, the function r is well defined on all of K.

26. Remark. Coordinate Systems on K. (1) The natural coordinate functions u, v on the Kruskal plane and spherical coordinates on S^2 give a product coordinate system u, v, ϑ, φ, that, as in the Schwarzschild case, is effective on all of K except over the poles of S^2. The line element for these *Kruskal-spherical coordinates* is

$$2F(r)\, du\, dv + r^2(d\vartheta^2 + \sin^2 \vartheta\, d\varphi^2).$$

(2) As noted above Schwarzschild-spherical coordinates t, r, φ, ϑ are valid on $K - H$ (but for poles), with formally the same geometric properties as on $N \cup B$.

Covariant derivatives and curvature on K can be expressed in Kruskal terms using Exercise 5.8 and warped product generalities as in the corresponding Schwarzschild case.

On $K_{\text{I}} \approx N$ and $K_{\text{II}} \approx B$ the coordinate vector field ∂_t is by construction a Killing vector field. Though the function t fails on the horizon, nevertheless ∂_t can be uniquely extended as a Killing field over all of K.

27. Lemma. The vector field $X = (v\, \partial_v - u\, \partial_u)/4\text{M}$ on K is a Killing vector field that equals ∂_t on $K - H$.

Proof. X is the lift of the vector field on Q given by the same formula. Thus using Schwarzschild-spherical coordinates we have $X = Xt\, \partial_t + Xr\, \partial_r$ on $K - H$. Since $d(t \circ \pi) = \pi^*(dt)$, the formula for dt remains valid when lifted to $K - H$. Thus

$$Xt = dt(X) = \frac{2\text{M}}{4\text{M}}\left(\frac{dv}{v} - \frac{du}{u}\right)(v\, \partial_v - u\, \partial_u) = 1.$$

Similarly $Xr = 0$. Thus $X = \partial_t$ on $K - H$.

X is Killing on $K_I \cup K_{II} \approx N \cup B$ since ∂_t is. The central symmetry $\phi(u, v, p) = (-u, -v, p)$ reverses the signs of u, v, ∂_u, and ∂_v; hence $d\phi X = X$. Since ϕ is an isometry, X is Killing on $K_{III} \cup K_{IV}$ and hence by continuity on all of K. ■

The proof shows that X is tangent to radial planes and on each is tangent to the hyperbolas r constant.

28. Corollary. Every Killing vector field Y on K has the form $cX + V$, for X as above and V a Killing field lifted from S^2.

Proof. Since $X = \partial_t$ on $K_I \approx N$, it follows from Proposition 7 that Y can be expressed in the given form $cX + V$ on K_I. The remark following Lemma 9.27 implies that $Y = cX + V$ everywhere on K. ■

Kruskal spacetime is time-orientable, since, for example, $\partial_v - \partial_u$ is a timelike vector field. In order that the natural isometry $\psi\colon K_I \to N$ preserve the physical meaning of these two spacetimes, it must preserve time-orientation. Thus we time-orient K by requiring that on K_I as on N the vector field ∂_t be future-pointing.

29. Lemma. On Kruskal spacetime the null vector fields $-\partial_u$ and ∂_v are future-pointing. On $K_{II} \approx B$, grad r is timelike future-pointing.

Proof. On K_I, ∂_t is timelike future-pointing, and by Lemma 27, $\partial_t = (v\,\partial_v - u\,\partial_u)/4\text{M}$. Both functions $-\langle \partial_u, \partial_t \rangle = -vF(r)/4\text{M}$ and $\langle \partial_v, \partial_t \rangle = -uF(r)/4\text{M}$ are negative on K_I; hence $-\partial_u$ and ∂_v are future-pointing on K_I. Since these are null vector fields, they remain future-pointing on all of K (see Figure 12).

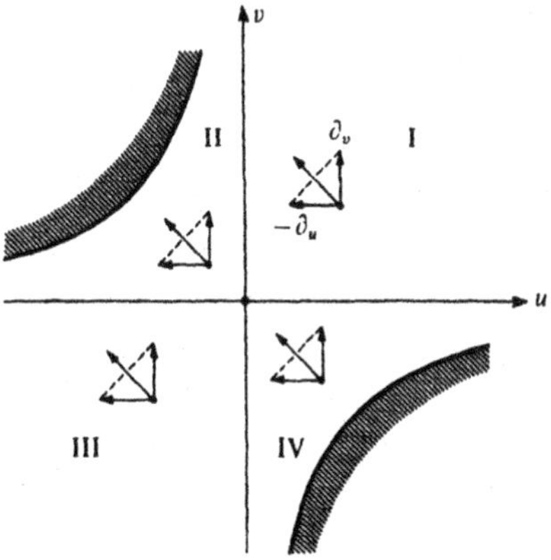

Figure 12. Kruskal future-cones.

By Lemma 7.34 the formula for grad r in Lemma 23 is valid on K. It shows that on K_{II}, grad r is a linear combination of $-\partial_u$ and ∂_v with positive coefficients $-u/4M$ and $v/4M$; hence grad r is timelike and future-pointing on K_{II}. ∎

BLACK HOLES

The region $v > 0$ in Kruskal spacetime K is called a *truncated Kruskal spacetime.* If Q' is the region $v > 0$ in Q, then

$$K' = \pi^{-1}(Q') = Q' \times_r S^2.$$

Thus K' joins the Schwarzschild exterior $K_I \approx N$ $(r > 2M)$ and the black hole $K_{II} \approx B$ $(r < 2M)$ along the horizon $H' = H \cap K'$ $(r = 2M)$. It is the sought-for spacetime containing a single Schwarzschild black hole and exterior.

When a particle α from K_I reaches the horizon H, it actually reaches H'. To see this, it suffices to show that the v coordinate of α is nondecreasing. Write α' as $(du/ds)\partial_u + (dv/ds)\partial_v + \tan \alpha'$. Since $-\partial_u$ is future-pointing, $\langle \alpha', -\partial_u \rangle \leq 0$; hence $dv/ds \geq 0$.

30. Proposition. No particle, whether material or lightlike, can escape from the black hole $K_{II} \approx B$. Furthermore any particle in K_{II} moves inward, ending (on a finite parameter interval) at the central singularity $r = 0$, if not before.

Proof. Let α be a particle such that $\alpha(0) \in K_{II}$. Since $r = 2M$ on the boundary of K_{II} in K' (or in K), if $dr/ds \leq 0$, then α remains in K_{II}. By Lemma 29, grad r is timelike and future-pointing on K_{II}. Since α' is nonspacelike and future-pointing,

$$dr/ds = \langle \alpha', \text{grad } r \rangle < 0.$$

A material particle can always be ended by overacceleration, and hence need not reach $r = 0$. What we must show is that α cannot be extended to $[0, \infty)$. For this it suffices to prove that dr/ds, which is negative, is bounded away from zero. If α is lightlike, then differentiating the energy equation in Proposition 15 shows that $d^2r/ds^2 < 0$ for $r < 3M$; hence $(dr/ds)(s) \leq (dr/ds)(0) < 0$ for all $s \geq 0$. If α is a material particle, then for initially equatorial coordinates,

$$-1 = \langle \alpha', \alpha' \rangle = -h(dt/d\tau)^2 + ((dr/d\tau)^2/h) + r^2(d\varphi/d\tau)^2.$$

Since $h < 0$ on K_{II}, this implies $-1 \geq (dr/d\tau)^2/h$, and thus $-h \leq (dr/d\tau)^2$. But $-h$ increases as r decreases, so again $(dr/d\tau)(\tau) \leq (dr/d\tau)(0) < 0$. ∎

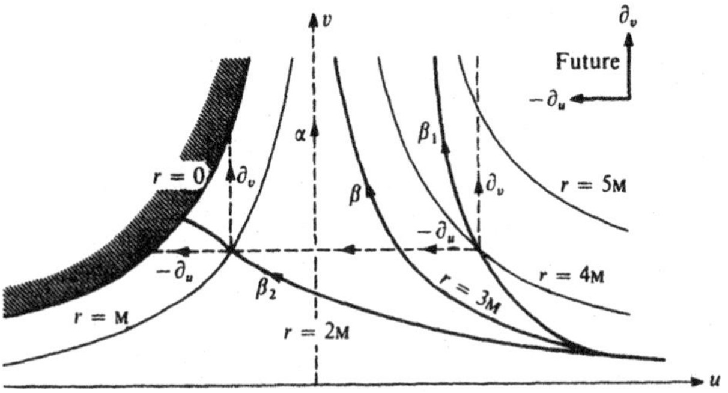

Figure 13. Observers: β at rest, β_1 outgoing, β_2 ingoing. Dashed lines indicate photons: α is at rest; β_1's message is received by β_2 but there is no answer.

This result is easiest to visualize for particles in a radial plane $\sigma^{-1}(p)$, identified as usual with Q' (Figure 13). Future timecones are marked off by the future-pointing null vectors $-\partial_u$ and ∂_v. In the exterior Q_{I} a material particle β can hover at rest (r constant) by supplying the radial acceleration specified by Lemma 9. With larger acceleration, β becomes outgoing and cuts through the r constant curves to move away from the black hole. But if the acceleration drops, β becomes ingoing and descends into the black hole quadrant Q_{II}. There, as the figure indicates, no acceleration can save β.

The extreme cases of radial motion are provided by lightlike particles, say photons. As null geodesics these parametrize the coordinate lines u constant and v constant, in the future-pointing directions ∂_v and $-\partial_u$, respectively. Evidently the $-\partial_u$ photons are always ingoing. Outside the black hole the ∂_v photons are outgoing, but inside (in Q_{II}) they too are ingoing. Thus observers in the black hole can receive messages or even material particles from outside, and in favorable cases can exchange both with each other—but can send nothing outside.

In the exceptional case of a ∂_v photon parametrizing the positive v axis, $r = 2\text{M}$, though it is racing "outward" at the speed of light the pull of the black hole holds it hovering at rest. Since the v axis is the intersection of the radial plane with the horizon $r = 2\text{M}$ of K', *the horizon H' is the union of the world-lines of all such rest photons.*

31. Corollary. Except for rest photons, every particle in K' that meets the horizon continues into the black hole.

Proof. Suppose that $\alpha(0)$ is in the horizon, given by $u = 0$ in the region K'. There by Lemma 23(3), grad $r = v\,\partial_v/4\text{M}$, which is null and future-pointing. Since α' is nonspacelike and future-pointing, it follows from Exercise 5.3(a) that

$$(dr/ds)(0) = \langle \alpha'(0), \text{grad } r \rangle \leq 0,$$

with equality holding if and only if $\alpha'(0)$ and grad r are collinear null vectors. Thus α enters K_{II} except in the latter case, in which α is lightlike. But then α is a geodesic, and since $\alpha'(0)$ and ∂_v are collinear, α parametrizes the v axis of $\sigma^{-1}(p) \approx Q'$ and is thus a rest photon. ∎

The attitudes of the future-cones on K' can now be visualized as follows. In the limit as $r \to \infty$ they are Minkowskian. As r decreases, they tip inward making escape to infinity more difficult, demonstrating the gravitation attraction of the black hole. At $r = 2\text{M}$, the inward tilting future-cones are tangent to the horizon along the worldline of a rest photon. For $r < 2\text{M}$, the entire future is inward, toward the central singularity.

The gravitational attraction of a black hole is no stronger than that of a normal star of the same mass, but because a black hole has radius $r^* = 0$ it offers no protection from the spacetime region $r < 2\text{M}$ where gravity dominates causality. Every material object entering this region is destroyed. In fact the tidal force computations in Remark 21 are valid in B just as in N; since these forces are of order M/r^3 they become infinite as $r \to 0$. A steel ball falling radially is stretched in the radial direction and compressed transversally until it shatters to the shape of a radial line. Acceleration only shortens life (see Exercise 7), while nonradial motion produces even larger tidal forces (Exercise 13).

Now we consider what a black hole looks like.

32. Corollary. For a Schwarzschild observer at radius r, the angular radius η^* of the black hole $K_{II} \approx B$ is given by

$$\sin \eta^* = 3\text{M}\sqrt{3h(r)}/r,$$

where η^* is in the first quadrant if $r \geq 3\text{M}$, second quadrant if $2\text{M} \leq r \leq 3\text{M}$.

Proof. As in Corollary 18, consider which photons γ starting at radius r fail to escape to infinity.

If $r > 3\text{M}$, then Figure 7 shows that γ fails to escape if and only if it is ingoing with $b \leq 3\sqrt{3}\text{M}$. If $b^* = 3\sqrt{3}\text{M}$, the photon asymptotically approaches the horizon and by Lemma 17, $r \sin \eta^* = b^*\sqrt{h(r)}$. Such photons are ingoing, hence $\eta^* = \measuredangle(-\partial_r, \bar{\gamma}'(0)) \leq \pi/2$.

If $r \leq 3\text{M}$, no ingoing photon escapes, and even outgoing ones with $b > 3\sqrt{3}\text{M}$ are drawn back to the black hole. Thus $\sin \eta^*$ is the same as before but $\eta^* \geq \pi/2$. ∎

For example, at radius $r = 10\text{M}$ the black hole appears against the stars of the background sky as a black disk of angular radius ~28°. For Schwarzschild observers closer to it, the black disk is larger, and at $r = 3\text{M}$ it fills half the sky. Closer still, the visible sky is concentrated in an ever smaller disk overhead that disappears at $r = 2\text{M}$.

Let K'' be the other half of Kruskal spacetime, namely the region over the lower half $v < 0$ of the Kruskal plane. The central symmetry ϕ provides an isometry from K' to K'' that carries ∂_v to $-\partial_v$, hence reverses time-orientation. Thus K'' is K' with past and future reversed. In particular, $K_{IV} \approx B$ is not a black hole but a *white hole*: every particle in K_{IV} must leave it; no particle in K_{III} can enter it. The horizon $H'' = H \cap K''$ consists of the worldlines of photon racing "inward" but only managing to hover at rest at $r = 2\text{M}$.

KRUSKAL GEODESICS

The Schwarzschild geodesic equations can be extended to Kruskal spacetime as follows.

33. Proposition. Let γ be a geodesic in K with $\langle \gamma', \gamma' \rangle = \varepsilon = -1, 0,$ or 1. Let r and t be the usual Schwarzschild functions and let ϑ, φ be initially equatorial spherical coordinates. Then

(G1) $h(r)t' = E$ on $K - H$,
(G2) $r^2\varphi' = L$,
(G3) $\vartheta = \pi/2$.

Furthermore, $E^2 = r'^2 + V(r)$, where $V(r) = (L^2/r^2 - \varepsilon)h(r)$.

Proof. (G3) is obvious since, as in any warped product, the projection of γ into S^2 is pregeodesic. By the conservation lemma (9.26), if X is the Killing extension of ∂_t, then $-\langle \gamma', X \rangle$ is a constant, equal to $-\langle \gamma', \partial_t \rangle = h(dt/ds) = E$ on $K - H$. For (G2), ∂_φ is Killing on S^2, hence so is its lift to K. Since $\vartheta = \pi/2$, the constant $\langle \gamma', \partial_\varphi \rangle$ is $r^2\varphi'$. The energy equation holds as usual on N and B, and since the central symmetry preserves r and t, it holds on the four open quadrants of K. The hypersurfaces $u = 0$ and $v = 0$ of K are totally geodesic (Exercise 9). Thus if γ meets H, then either (a) γ remains in H, where the energy equation is trivial since $r' = 0 = h$ or (b) γ cuts transversally through H; hence by continuity the energy equation holds (with the same constant on both sides of H). ∎

Conversely, if these equations hold for a curve γ, and r' is rarely zero, then γ is a geodesic (Exercise 5).

The energy equation can be used to find formulas for geodesics—and, in particular, to determine their maximal domains.

34. Remark. Integral Formulas for Geodesics. Let γ be a geodesic in $K_I \approx N$ or $K_{II} \approx B$. For definiteness, take γ to be a freely falling material particle with $r(\gamma(0)) = r_0$ and $(dr/d\tau)(0) \neq 0$. Motivated by the energy equation, define

$$\tau(r) = \pm \int_{r_0}^{r} (E^2 - V(r))^{-1/2}\, dr,$$

where the sign depends on whether γ is initially outgoing or ingoing. If $E^2 > V(r)$ on some r-interval J, then the function $\tau(r)$ is a diffeomorphism from J onto an interval I that may well be larger than the original domain of γ. On I the inverse function $r(\tau)$ has $dr/d\tau = \pm(E^2 - V(r))^{1/2}$, hence $r(\tau)$ satisfies the energy equation. For initially equatorial spherical coordinates, $\vartheta = \pi/2$, and we further define

$$t(\tau) = t_0 + E \int_0^{\tau} \frac{d\tau}{h(r(\tau))}, \qquad \varphi(\tau) = \varphi_0 + L \int_0^{\tau} \frac{d\tau}{r(\tau)^2}.$$

Then equations (G1)–(G3) also hold on I. By Exercise 5(b), these functions t, r, ϑ, φ are the Schwarzschild coordinates of a geodesic—one that agrees at first with γ. Hence they provide a geodesic extension of γ over the interval I.

35. Examples. (1) *Exceptional orbits in N.* Let γ be a freely falling material particle in N moving outward from $r_0 < r_1$ with angular momentum $L^2 > 12\text{M}^2$. Suppose that $E^2 = V(r_1)$, so the orbit of γ is exceptional. Using Remark 34, define

$$\tau(r) = \int_{r_0}^{r} (E^2 - V(r))^{-1/2} dr \qquad \text{on} \quad J = [r_0, r_1).$$

Since $V'(r_1) = 0$ and $V''(r_1) < 0$, we have $E^2 - V(r) = V(r_1) - V(r) \sim C(r_1 - r)^2 > 0$ for r near r_1. It follows that $\tau(r) \to \infty$ as $r \to r_1$, hence the inverse function $r(\tau)$ is defined on $I = [0, \infty)$. This shows that γ can be geodesically extended to $[0, \infty)$, and that $r \to r_1$ and $d\varphi/d\tau \to L/r_1^2$ as $\tau \to \infty$. Thus γ spirals in ever closer to the circular orbit at $r = r_1$.

(2) *Free fall in B.* If γ is a freely falling material particle in B, then $dr/d\tau < 0$, hence $E^2 - V(r) > 0$. Thus we can define

$$\tau(r) = \int_{r}^{r_0} (E^2 - V(r))^{-1/2}\, dr \qquad \text{on} \quad (0, r_0].$$

As $r \to 0$, $E^2 - V(r) \to \infty$, hence $\tau(r)$ approaches a finite limit $b > 0$. It follows as before that γ can be geodesically extended to $[0, b)$ and that $r(\tau) \to 0$ as $\tau \to b$.

36. Proposition. An inextendible timelike geodesic $\gamma: I \to K$ in Kruskal spacetime is incomplete if and only if $r\gamma(\tau) \to 0$ as τ approaches a finite endpoint of I.

Proof. We can suppose I has the form $[0, B)$, $B \leq \infty$. First we consider the possibilities for $\gamma(0)$. The central symmetry of K and the isometries $(u, v, p) \to (v, u, p)$ and $(u, v, p) \to (-v, -u, p)$ preserve geodesics, but only the latter preserves time orientation. Considering the effect of these on the quadrants and horizon of K leaves just three cases.

Case 1: $\gamma(0) \in K_{\mathrm{I}}$, *and* γ *is future-pointing.* If γ has an exceptional orbit, then it is complete (that is, $B = \infty$). In fact, if its orbit is circular, γ is periodic, hence complete; otherwise, by Example 35(1) or a variant, γ spirals and is complete. If γ escapes to infinity—more precisely, if $r \circ \gamma$ is outgoing with no points of graph $V(r)$ ahead on its E^2 line—then the method of Remark 34 shows easily that γ is complete and that $r(\gamma(\tau)) \to \infty$ as $\tau \to \infty$.

The crash case is more interesting. Suppose that $r \circ \gamma$ is ingoing with no points of graph $V(r)$ ahead on its E^2 line. The integral

$$\tau(r) = \int_r^{r_0} (E^2 - V(r))^{-1/2}\, dr$$

is well defined for all $r \leq r_0$, but Schwarzschild coordinates fail at $r = 2\text{M}$, so the method of Remark 34 extends γ only over $[0, \tau(2\text{M}))$. To reach $\tau(2\text{M})$ we use Kruskal coordinates. That γ is future-pointing implies $u' < 0$ and $v' > 0$, so u is decreasing and v is increasing. As $\tau \to \tau(2\text{M})$, $r \to 2\text{M}$, hence $u \to 0$, but $v \geq v(0) > 0$. The energy equation shows that $dr/d\tau \to -E < 0$, and then manipulation of the energy equation shows that $(E + r')/\hbar$ approaches a positive limit as $\tau \to \tau(2\text{M})$. Hence, by Exercise 16(b), v'/v approaches a positive limit. Thus $v'/v \leq A$ for some $A > 0$. But then $v(\tau) \leq v(0)e^{A\tau}$, so v approaches a positive limit as $\tau \to \tau(2\text{M})$. As in Remark 34, it follows that $\gamma(\tau)$ approaches a limit as $\tau \to \tau(2\text{M})$. Hence by Lemma 5.8, γ can be geodesically extended past $\tau(2\text{M})$. Since $dr/d\tau = -E < 0$ there, γ passes through H' into K_{II}, thus reducing the situation to Case 3.

Finally, we consider turning points. For definiteness, suppose that $r \circ \gamma$ is moving outward, approaching ρ such that $V(\rho) = E^2$ and $V'(\rho) > 0$. Defining $\tau(r)$ as usual, we see that the limit value $\tau(\rho)$ is finite. Now extend $r(\tau)$ on $[0, \tau(\rho)]$ symmetrically to $[0, 2\tau(\rho)]$. There $r(\tau)$ has a continuous second derivative, so by Remark 34, γ can also be extended over this interval. Thus γ passes the turning point in finite time, and the situation is reduced to one of the subcases above. (Another turning point, of course, would give a bound orbit.)

Case 2: $\gamma(0) \in H'$, *or* γ *is future-pointing and* $\gamma(0) \in \pi^{-1}(0, 0)$. It is clear from the warped product metric that $d\pi(\gamma'(0))$ is also timelike, hence not tangent to either coordinate axis in Q. It follows that for small $\tau > 0$, $\gamma(\tau)$ is in either K_{I} or K_{II}. The latter gives Case 3 and the former is reducible to Case 1.

Case 3: $\gamma(0) \in K_{II}$. If γ is future-pointing, Example 35(2) shows that it cannot be extended past some $b > 0$, and that $r(\gamma(\tau)) \to 0$ as $\tau \to b$. If γ is past-pointing, a simpler variant of the crash subcase in Case 1 shows that γ can be extended through the horizon H. Then the situation can be reduced (without circularity) to future-pointing geodesics in this case or Case 1. ∎

37. Corollary. Kruskal spacetime K is incomplete and inextendible.

Since tidal forces approach infinity as $r \circ \gamma \to 0$, inextendibility follows from Remark 5.45.

Exercises

1. Let γ be a freely falling material particle in N with $L = 0$. Prove (a) For τ small and $r \gg \mathrm{M} : r \sim r_0 + v_0\tau - g_0\tau^2/2$, where $v_0 = (dr/d\tau)(0)$ and $g_0 = \mathrm{M}/r_0^2$. (b) The smallest value of v_0 for which γ escapes to infinity is $(2\mathrm{M}/r_0)^{1/2}$, in which case $r^{3/2} = r_0^{3/2} + 3\sqrt{\mathrm{M}/2}\,\tau$.

2. In N consider a freely falling material particle in circular orbit of radius r. (a) If $\omega = d\varphi/dt$ is the (constant) angular velocity relative to Schwarzschild time, verify Kepler's formula $\omega^2 r^3 = \mathrm{M}$. (b) For $\bar{\omega} = d\varphi/d\tau$, relative to proper time, prove $\bar{\omega}^2 r^2(r - 3\mathrm{M}) = \mathrm{M}$.

3. Define the *tidal force tensor* of a Newtonian force field $Z \in \mathfrak{X}(\boldsymbol{R}^3)$ to be $DZ \in \mathfrak{T}_1^1(\boldsymbol{R}^3)$. For the gravitational force field due to a mass M at $0 \in \boldsymbol{R}^3$ (as in Appendix C) compute the tidal force tensor and compare with Remark 21.

4. The Schwarzschild time for any freely falling particle in N to reach $r = 2\mathrm{M}$ is infinite. (Hint: Express this time as an integral.)

5. Show that: (a) In Proposition 11 the geodesic differential equation for r can be written as $2r'' = -dV/dr$. (b) If the equations in Proposition 11 [Proposition 33] hold for some curve γ, and r' is zero on no interval, then γ is a geodesic.

6. Find expressions for the critical points $r_1 \leq r_2$ of the potential function $V(r)$ of a material particle in N. Show that as $L \to \infty : r_1 \to 3\mathrm{M}$ and $V(r_1) \to \infty$; $r_2 \to \infty$ and $V(r_2) \to 1$.

7. Let $U = -\sqrt{-h}\,\partial_r$ on the black hole B. Prove: (a) U is an observer field. (b) U is geodesic and proper time synchronizable. (c) Every U-observer goes from (limit values) $r = 2\mathrm{M}$ to $r = 0$ in proper time $\pi\mathrm{M}$; all other material particles take less time.

8. *Kruskal curvature.* Prove: (a) On $N \cup B$, the curvature tensor is zero on any three choices from mutually orthogonal vectors ∂_t, ∂_r, v, w. (b) On

K, the curvature invariant $\sum R^{ijkl}R_{ijkl}$ is $48\text{M}^2/r^6$. (Hint: For orthogonal coordinates, $I = \sum (R_{ijkl})^2/(g_{ii}\, g_{jj} g_{kk}\, g_{ll})$.)

9. (a) Prove the analogue of Proposition 4 with Kruskal coordinates u, v replacing Schwarzschild coordinates t, r. Show that (b) in K, the hypersurfaces $u = 0$ and $v = 0$ are totally geodesic but not semi-Riemannian (Exercise 4.9), and (c) the central sphere $\pi^{-1}(0, 0)$ is Riemannian and totally geodesic.

10. If γ is a freely falling particle (material or lightlike) in N, let $\gamma_-(s) = \psi(\gamma(-s))$, where $\psi(t, r, p) = (-t, r, p)$. Show that γ_- is a freely falling particle of the same causal character, with $E_- = E$ and $L_- = -L$. Since $\vec{\gamma}_-(s) = \vec{\gamma}(-s)$, the particles traverse the same orbit in opposite directions.

11. *Exceptional orbits.* (a) Consider the freely falling material particles in N with $E^2 = V(r_1)$. Interpreting *spiral* as in Example 34, show that there are three noncircular orbital types for such particles: crash/spiral, spiral/spiral, and spiral/escape. Find the conditions on L/M and $r(0)$ that characterize each type. (b) Show that for lightlike particles in N with impact parameter $b = 3\sqrt{3}\text{M}$ there are two noncircular orbital types: crash/spiral and spiral/escape.

12. *Turn angles.* If γ is a freely falling particle in N with $L \neq 0$, let $\Delta\varphi$ be the total (equatorial) angle of turn for $\tau \geq 0$. (By Exercise 10, corresponding results will hold for $\tau \leq 0$.) Show that $\Delta\varphi$ is infinite if γ is bound or spirals, finite if γ escapes to infinity or crashes. (Hint: Express $d\varphi/dr$ in terms of r.)

13. An arbitrary instantaneous observer on N can be written as $u = a\, \partial_t + b\, \partial_r + y$, where y is tangent to fibers. (a) If $w \perp y$ is also tangent to fibers and hence is in $u^\perp$, compute the tidal force

$$F_u(w) = -(M/r^3)(1 + 3\, \langle u,u\rangle)w.$$

(b) Find the other eigenvectors and eigenvalues of $F_u\colon u^\perp \to u^\perp$. (Hint: trace $F_u = -\text{Ric}(u, u)$.)

14. *Isometries of Q.* Prove: (a) The four mappings of Q sending (u, v) to (u, v), $(-u, -v)$, (v, u), and $(-v, -u)$, respectively, constitute a group G of isometries. (b) The flow isometries $\mathbb{R} = \{\psi_s\colon s \in \boldsymbol{R}\}$ of the Killing field X (Lemma 27) are given by $\psi_s(u, v) = (ue^{-s/4\text{M}}, ve^{s/4\text{M}})$. (c) $\phi(0, 0) = (0, 0)$ for all $\phi \in I(Q)$. (d) $I(Q) \approx O(1, 1) = O_1(2)$. (Hint: For (c), look at curvature; for (d), consider $d\phi$ at $(0, 0)$.)

15. $I(K) = I(Q) \times O(3)$. Prove: (a) If $\xi \in I(Q)$ and $\eta \in I(S^2) = O(3)$, then $\xi \times \eta \in I(K)$. (b) If $\psi \in I(K)$, there exist unique ξ, η such that $\psi = \xi \times \eta$. (Hint: Find η first by showing that ψ preserves r.)

16. Let γ be a geodesic in K. (a) If Kruskal coordinates are used in Proposition 33, then (G1) becomes $E = (F(r)/4\text{M})(uv' - vu')$, valid on all of K, and the energy equation becomes $2F(r)u'v' + L^2/r^2 = \varepsilon$. (b) Deduce from Lemma 23 that $E + r' = 4\text{M}\hbar v'/v$ if $v \neq 0$, and $-E + r' = 4\text{M}\hbar u'/u$ if $u \neq 0$.

17. *Null geodesics in Q.* Prove: (a) If $\gamma(s) = (u(s), v(s))$ has both v and $u'F(r)$ constant, then γ is a geodesic. (See Exercise 5.8.) (b) The u and v axes can be parametrized as complete geodesics. (c) Given $(u_o, v_o) \in Q$, with $v_o \neq 0$, there is a unique function $u(s)$ such that $r(u(s), v_o) = -s + r_o$. (d) Then the curve $\gamma(s) = (u(s), v_o)$, defined for $s < r_o$, is an inextendible geodesic.

18. *Null geodesics in K.* Prove Proposition 36 with *timelike* replaced by *null*. In particular, for an inextendible null geodesic $\gamma\colon [0, B) \to K$ show that as $s \to B$ there are five possibilities: $r \to 0$, $r = 2\text{M}$, $r = 3\text{M}$, $r \to 3\text{M}$, and $r \to \infty$. (Hint: Use the energy equation preceding Corollary 16 and the method of Remark 34.)

19. *Spacelike geodesics in K.* Prove Proposition 36 with *timelike* replaced by *spacelike*. (Hint: The case $L = 0$ is easy; for the many spacelike geodesics with $|L| = 2\text{M} = r$, use Proposition 7.38.)

14 CAUSALITY IN LORENTZ MANIFOLDS

By *causality* we refer to the general question of which points in a Lorentz manifold can be joined by causal curves: relativistically, which events can influence (be influenced by) a given event. In a particular manifold M, causality may be trivial, but under fairly mild conditions it is closely related to fundamental geometrical properties of M. For example the study of causality leads to sufficient conditions for points to be joinable by a (longest) causal geodesic and also for there to be a normal geodesic from a spacelike hypersurface to a point. In both cases a useful aid is a Lorentz analogue of Riemannian distance.

Essential parts of this chapter were developed by the physicists R. Penrose and S. W. Hawking in an effort to understand why the most fundamental relativistic models—of the universe or of individual stars—turn out to have "singularities." (We use this term in its broadest sense to mean *timelike* or *null geodesic incompleteness*: some freely falling observer or light ray is prematurely ended by a flaw in the manifold M.) In this context a *singularity theorem* is a general result of Lorentz geometry that asserts the existence of singularities. We prove two such theorems—one, due to Penrose, motivated by black hole singularities; the other, due to Hawking, motivated by cosmological (e.g., big bang) singularities. For more results of this type, see [HE], our basic reference.

Through the chapter, M will denote a connected time-oriented Lorentz manifold of dimension n.

CAUSALITY RELATIONS

The *causality relations* on M are defined as follows. If $p, q \in M$, then

(1) $p \ll q$ means there is a future-pointing *timelike* curve in M from p to q;

(2) $p < q$ means there is a future-pointing *causal* curve in M from p to q.

Evidently $p \ll q$ implies $p < q$. As usual, $p \leq q$ means that either $p < q$ or $p = q$. For a subset A of M, the subset

$$I^+(A) = \{q \in M : \text{there is a } p \in A \text{ with } p \ll q\}$$

is called the *chronological future* of A, and

$$J^+(A) = \{q \in M : \text{ there is a } p \in A \text{ with } p \leq q\}$$

is called the *causal future* of A.

Thus $J^+(A) \supset A \cup I^+(A)$. For a single point, $I^+(p) = \{q : p \ll q\}$, and for a subset, $I^+(A) = \bigcup \{I^+(p) : p \in A\}$; similarly for J^+.

Dual to the preceding definitions are corresponding *past* versions. Thus $I^-(A) = \{q \in M : \text{there is a } p \in A \text{ with } q \ll p\}$ is the *chronological past of A*. In general, *past* definitions and proofs follow from the *future* versions (and vice versa) merely by reversing time-orientation.

The standard for causality is Minkowski space $\boldsymbol{R}_1^4$. There $I^+(p)$ is just the future timecone of p, that is, $\{q : \overrightarrow{pq} \text{ is timelike future-pointing}\}$, and $J^+(p)$ is p together with $\{q : \overrightarrow{pq} \text{ is causal future-pointing}\}$. Thus $I^+(p)$ is an open set with closure $J^+(p)$, and the latter is the union of $I^+(p)$, the future nullcone at p, and p itself.

At the other extreme, the Lorentz cylinder $S_1^1 \times \boldsymbol{R}^1$ has trivial causality: even for a single point, $I^+(p) = J^+(p)$ is the entire manifold.

The relations defined above are transitive; furthermore, if $x \ll z$ there are infinitely many y such that $x \ll y \ll z$ (and similarly for $<$). Proposition 10.46 has this fundamental consequence:

1. Corollary. *If $x \ll y$ and $y \leq z$, or if $x \leq y$ and $y \ll z$, then $x \ll z$.*

Such results are all summarized as

$$I^+(A) = I^+(I^+A) = I^+(J^+A) = J^+(I^+A) \subset J^+(J^+A) = J^+(A),$$

where A is an arbitrary subset of M.

An open subset $\mathscr{U}$ of M is a time-oriented Lorentz manifold in its own right, and the *intrinsic* causality relations of $\mathscr{U}$ imply the corresponding ones in M. For example, let $I^+(A, \mathscr{U})$ denote the chronological future in the manifold $\mathscr{U}$ of the set $A \subset \mathscr{U}$. Then $I^+(A, \mathscr{U}) \subset I^+(A) \cap \mathscr{U}$. This remark is particularly useful in the case of a convex open set $\mathscr{C}$, since the intrinsic causality of $\mathscr{C}$ is as simple as that of Minkowski space.

2. Lemma. *If $\mathscr{C}$ is a convex open set in M, then*

(1) *For $p \neq q$ in $\mathscr{C}$, $q \in J^+(p, \mathscr{C}) \Leftrightarrow \overrightarrow{pq}$ is future-pointing causal (analogously for I^+).*

(2) *$I^+(p, \mathscr{C})$ is open in $\mathscr{C}$ (hence in M).*

(3) *$J^+(p, \mathscr{C})$ is the closure in $\mathscr{C}$ of $I^+(p, \mathscr{C})$.*

(4) *The relation $\leq$ is closed on $\mathscr{C}$*; that is, if $\{p_n\} \to p$ and $\{q_n\} \to q$, with all points in $\mathscr{C}$, then $q_n \in J^+(p_n, \mathscr{C})$ for all n implies $q \in J(p, \mathscr{C})$.

(5) A causal curve α contained in a compact subset K of $\mathscr{C}$ is (continuously) extendible.

Proof. The first three properties follow from causality in $T_p(M) \approx \boldsymbol{R}_1^n$ and properties of the exponential map—particularly Lemma 5.33.

Because $\overrightarrow{pq}$ is a continuous function of $(p, q) \in \mathscr{C} \times \mathscr{C}$, (1) implies (4). In (5) we can suppose the domain of α is $[0,B)$, $B \leq \infty$. Since K is compact, there is a sequence $\{s_i\} \to B$ such that $\{\alpha(s_i)\}$ converges to a point $p \in K$. We must show that every such sequence converges to the same point p. Assume $\{t_i\} \to B$ but $\{\alpha(t_i)\} \to q \neq p$. Roughly speaking, α races back and forth between p and q. By (4), $q \in J^+(p, \mathscr{C})$ and $p \in J^+(q, \mathscr{C})$. Hence (1) requires $\overrightarrow{pq}$ to be both future- and past-pointing. ■

Of these properties it turns out that only (2) holds for arbitrary M; in fact, a stronger result is true.

3. Lemma. The relation $\ll$ is open; that is, if $p \ll q$ in M, there are neighborhoods $\mathscr{U}$ and $\mathscr{V}$ of p and q, respectively, such that $p' \ll q'$ for all $p' \in \mathscr{U}$, $q' \in \mathscr{V}$.

Proof. Let σ be a timelike curve from p to q. If $\mathscr{C}$ is a convex neighborhood of q, let q^- be a point of $\mathscr{C}$ on σ before q. Dually, let p^+ be a point of σ between p and q^- and contained in a convex neighborhood $\mathscr{C}'$ of p. By the lemma, $I^+(q^-, \mathscr{C})$ and dually $I^-(p^+, \mathscr{C})$ are open in M. Hence they have the properties required of $\mathscr{V}$ and $\mathscr{U}$, respectively. ■

This lemma links causality firmly to the topology of M. It implies, in particular, that the *chronological future $I^+(A)$ of any set A is open.*

By contrast with the situation in Minkowski space, for arbitrary M the sets $J^+(p)$ need not be closed.

4. Example. Let M be $\boldsymbol{R}_1^2$ with a point, say $(1, 1)$, deleted. Taking p to the origin, $I^+(p)$ is the usual Minkowskian cone. But no causal curve from

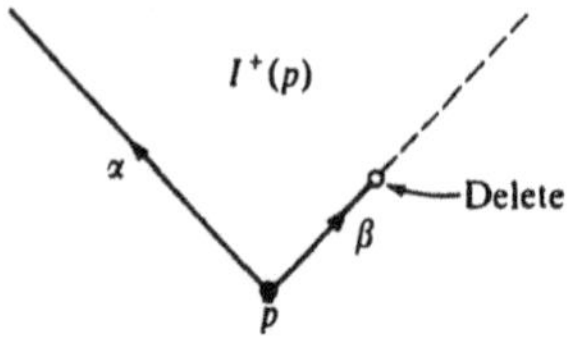

Figure 1

p can reach points indicated by the dashed line in Figure 1. Thus $J^+(p)$ consists of only $I^+(p)$ together with the null geodesics rays α and β in the figure. In particular, the closure of $I^+(p)$ is strictly larger than $J^+(p)$.

This example also illustrates the following results (taking the set A to be a single point p). Theorem 10.51 gives

5. Corollary. If α is a future-pointing causal curve from A to a point $q \in J^+(A) - I^+(A)$, then α is a null geodesic that has no conjugate points before q and does not meet $I^+(A)$.

Thus the causal future $J^+(A)$ of A is a union of A, $I^+(A)$, and (possibly) certain null geodesics from A.

6. Lemma. For a subset A, (1) int $J^+(A) = I^+(A)$, and (2) $J^+(A) \subset \overline{I^+(A)}$, with equality if and only if $J^+(A)$ is a closed set.

Proof. (1) $I^+(A)$ is open and contained in $J^+(A)$, hence is contained in int $J^+(A)$. If $q \in \text{int } J^+(A)$, then for a convex neighborhood $\mathscr{C}$ of q, $I^-(q, \mathscr{C})$ contains a point of $J^+(A)$. Hence $q \in I^+J^+(A) \subset I^+(A)$.

(2) The equality assertion is clear, since $I^+(A) \subset J^+(A)$. It will suffice to prove that $J^+(p) \subset \overline{I^+(p)}$ for a single point. Evidently $p \in \overline{I^+(p)}$, so suppose $p < q$. Let α be a future-pointing causal curve from p to q. If $\mathscr{C}$ is a convex neighborhood of q, let q^- be a point of α in $J^-(q, \mathscr{C})$. By Lemma 2, $q \in J^+(q^-, \mathscr{C}) \subset \overline{I^+(q^-, \mathscr{C})}$. But $I^+(q^-, \mathscr{C}) \subset I^+(J^+p) \subset I^+(p)$, hence $q \in \overline{I^+(p)}$. ∎

QUASI-LIMITS

The study of causality demands some notion of limit of a sequence of (piecewise smooth) causal curves. For any reasonable notion the limit curves need not be piecewise smooth; hence complications ensue. We shall ask merely for *quasi-limits*: broken geodesics that are only approximate limits, their accuracy measured by a convex covering of M. By this means, global causality questions can often be reduced to easy local ones.

7. Definition. Let $\{\alpha_n\}$ be an infinite sequence of future-pointing causal curves in M, and let $\mathfrak{K}$ be a convex covering of M. A *limit sequence for* $\{\alpha_n\}$ *relative to* $\mathfrak{K}$ is a (finite or infinite) sequence $p = p_0 < p_1 < \cdots$ in M such that

(L1) For each p_i there is a subsequence $\{\alpha_m\}$ and, for each m, numbers $s_{m0} < s_{m1} < \cdots < s_{mi}$ such that

(a) $\lim_{m\to\infty} \alpha_m(s_{mj}) = p_j$ for each $j \leq i$.

(b) For each $j < i$, the points p_j, p_{j+1} and the segments $\alpha_m | [s_{mj}, s_{m,j+1}]$ for all m are contained in a single set $\mathscr{C}_j \in \mathfrak{K}$.

(L2) If $\{p_i\}$ is infinite, it is nonconvergent. If $\{p_i\}$ is finite, it has more than one point and no strictly longer sequence satisfies (L1).

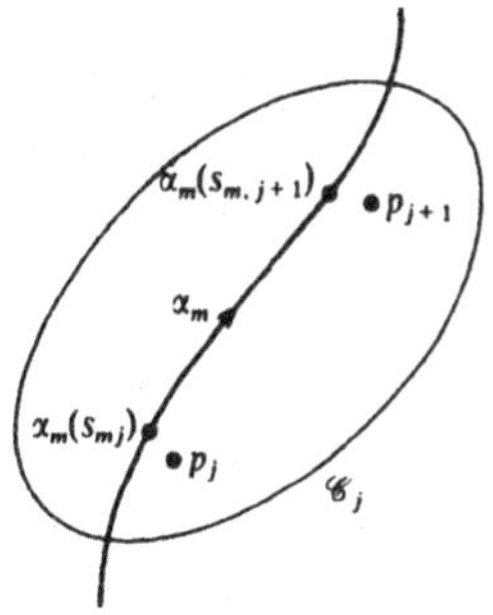

Figure 2

Here (L1a) is a natural limit requirement, (L2) is a technicality, and (L1b) accomplishes the reduction to individual convex sets (Figure 2). After proving existence of limit sequences we shall geodesically connect successive points $p_i < p_{i+1}$ to obtain quasi-limits. The existence will be assured by two mild initial conditions on $\{\alpha_n\}$:

(1) The sequence $\{\alpha_n(0)\}$ converges to a point p.

(2) There is a neighborhood of p that contains only finitely many of the curves α_n (notation: $\{\alpha_n\} \not\to p$).

8. Proposition. Let $\{\alpha_n\}$ be a sequence of future-pointing causal curves such that $\{\alpha_n(0)\} \to p$ but $\{\alpha_n\} \not\to p$. Then relative to any convex covering $\mathfrak{K}$, $\{\alpha_n\}$ has a limit sequence starting at p.

Proof. Since M is paracompact, it has a locally finite covering $\mathfrak{K}'$ by open sets $\mathscr{B}$ such that each $\overline{\mathscr{B}}$ is compact and contained in some member of $\mathfrak{K}$. By the hypotheses on $\{\alpha_n\}$, we can arrange for $\mathfrak{K}'$ to contain a $\mathscr{B}_0$ such that infinitely many α_ns start in $\mathscr{B}_0$ and leave $\overline{\mathscr{B}}_0$. Relabel these curves as $\{{}^1\alpha_n\}$, and for each let ${}^1\alpha_n(s_n)$ be the first point in bd $\mathscr{B}_0$. Passing to a further subsequence we can suppose that $\{{}^1\alpha_n(s_n)\}$ converges to a point p_1 of bd $\overline{\mathscr{B}}_0$.

Now choose $\mathscr{B}_1 \in \mathfrak{K}'$ containing p_1. *If* infinitely many ${}^1\alpha_n$s leave $\mathscr{B}_1$, we obtain as before a subsequence $\{{}^2\alpha_n\}$ whose first departure points from $\mathscr{B}_1$ converge to a point p_2 in bd $\mathscr{B}_1$. Repeat this step as many times as possible, with this proviso on subsequent choices of $\mathscr{B}_i$: if there is more than one candidate (element of $\mathfrak{K}'$ containing p_i), pick one that has been used fewest times before. Clearly condition (L1) of the preceding definition holds, with $\mathscr{C}_i$ any member of $\mathfrak{K}$ that contains $\overline{\mathscr{B}}_i$. Since the relation $\leq$ is closed on $\mathscr{C}_i$, it follows that $p_{i+1} \geq p_i$. By the construction, $p_{i+1} \neq p_i$, hence $p_{i+1} > p_i$.

If the sequence $\{p_i\}$ obtained above is infinite, we must show it is nonconvergent. Assume $\{p_i\} \to q$. Pick $\mathscr{B} \in \mathfrak{K}'$ containing q, then $p_i \in \mathscr{B}$ for all but a finite number of the integers $i \geq 0$. Since $\overline{\mathscr{B}}$ is compact and $\mathfrak{K}'$ is locally finite, only finitely many members of $\mathfrak{K}'$ meet $\mathscr{B}$. Hence some one must have been chosen as $\mathscr{B}_i$ for infinitely many i. But this violates the choice proviso, above, for $\mathscr{B}$ itself was a candidate infinitely many times, but was chosen at most finitely many times (because, since it contains almost all p_is, only finitely many can be in bd $\mathscr{B}$).

Finally suppose that the sequence $\{p_i\}$ obtained above is finite: $p = p_0 < p_1 < \cdots < p_k$. Since the construction cannot continue, it must be true that only a finite number of the curves ${}^k\alpha_n$ leave $\mathscr{B}_k$. Let $\{\alpha_m\}$ be those trapped in $\mathscr{B}_k$. By Lemma 2(5) they are extendible; by Exercise 5 we may as well assume that α_m is already defined on a closed interval $[0, b_m]$. Since $\overline{\mathscr{B}}_k$ is compact, for a further subsequence the endpoints $\alpha_m(b_m)$ converge to a point $q \in \overline{\mathscr{B}}_k$.

If $q = p_k$, then $p_0 < \cdots < p_k$ obviously cannot be extended still satisfying (L1); hence it is a limit sequence. If $q \neq p_k$, then both (L1) and (L2) hold for $p_0 < \cdots < p_k < p_{k+1} = q$. ∎

If $\{p_i\}$ is a limit sequence for $\{\alpha_n\}$ as above, let λ_i be the (future-pointing causal) geodesic from p_i to p_{i+1} in a convex set $\mathscr{C}_i$ as in (L1). Assembling these segments for all i gives a broken geodesic $\lambda = \sum \lambda_i$ called a *quasi-limit* of $\{\alpha_n\}$ *with vertices* p_i. Thus λ is a future-pointing causal broken geodesic that starts at p. *If* $\{p_i\}$ *is infinite, then by* (L2), λ *is future-inextendible*. In the finite case $p_0 < \cdots < p_k$, the curve λ runs from p_0 to p_k.

A sequence satisfying the initial conditions always has infinitely many limit sequences, though sometimes as in Example 9 they all give the same quasi-limit. (In this context we can ignore change of parametrization, since it has no effect on the causality properties of curves.)

A *quasi-limit* λ *of future-inextendible curves* $\{\alpha_n\}$ *is future-inextendible.* In fact, it is clear from the preceding proof that every limit sequence will be infinite—hence λ will be inextendible.

9. Example. In $\boldsymbol{R}_1^2$ let α_n be the timelike geodesic segment from the origin 0 to $(n + (1/n), n)$. In any limit sequence for $\{\alpha_n\}$ the vertices clearly must lie on the null geodesic ray $\lambda(s) = (s, s)$, $s \geq 0$. In fact, with 0 as initial point, λ is the unique quasi-limit. It is easy to test the construction from Proposition 8 (convex sets are the same as for $\boldsymbol{R}^2$).

Now, as in Example 4, let $M = \boldsymbol{R}_1^2 - (1, 1)$. Then $\{\alpha_n\}$ has the unique quasi-limit $\beta = \lambda | [0, 1)$, which is future-inextendible and incomplete in M. Note that only decreasingly small initial segments of the curves α_n are actually used to get β.

CAUSALITY CONDITIONS

If M contains no closed timelike curves, we say that the *chronology condition* holds on M. Physically this is a natural requirement since its absence leads to distressing paradoxes: an observer could take a trip from which he returns before his departure—and then decide not to go after all. More radically, by killing an ancestor he could prevent his own birth.

Compact spacetimes are of minor importance for relativity, since the chronology condition cannot hold for them:

10. Lemma. If M is compact, it contains a closed timelike curve.

Proof. By compactness the open covering $\{I^+(p): p \in M\}$ has a finite subcover $I^+(p_1), \ldots, I^+(p_k)$. We can assume that $I^+(p_1)$ is not contained in any later $I^+(p_i)$—otherwise discard $I^+(p_1)$. But then $p_1 \in I^+(p_1)$—that is, there is a closed timelike curve through p_1—for if p_1 is in some later $I^+(p_i)$, then $I^+(p_1) \subset I^+(p_i)$. ∎

The manifold M satisfies the *causality condition* provided there are no closed causal curves in M. Obviously this implies the chronology condition, but not conversely: For example, the translation $(t, x) \to (t + 1, x + 1)$ of $\boldsymbol{R}_1^2$ generates a properly discontinuous group of isometries, and the resulting Lorentz orbit manifold M *satisfies the chronology but not the causality condition.* In fact, the null geodesic $s \to (s, s)$ in $\boldsymbol{R}_1^2$ projects to a smoothly closed null geodesic in M, but a timelike loop σ in M would lift to a timelike curve in $\boldsymbol{R}_1^2$ from (t, x) to $(t + n, x + n)$—an impossibility.

The causality condition (and similarly for chronology) is said to hold *at a point p* if there are no closed causal curves through p, and *on a subset A* if it holds at each $p \in A$.

11. Definition. The *strong causality condition* holds at $p \in M$ provided that given any neighborhood $\mathscr{U}$ of p there is a neighborhood $\mathscr{V} \subset \mathscr{U}$ of p such that every causal curve segment with endpoints in $\mathscr{V}$ lies entirely in $\mathscr{U}$.

This says that causal curves that start arbitrarily close to p and leave some fixed neighborhood cannot return arbitrarily close to p. In short, there are no causal curves *almost closed* at p. Thus strong causality implies causality; but the converse fails. We give an example in the traditional pictorial style in which M is a region in $\boldsymbol{R}_1^2$ (sometimes with identifications), null geodesics run at $\pm 45°$, and the future is upward. If no other metric is mentioned, use the flat $\boldsymbol{R}_1^2$ metric.

12. Example. Following [HE], build M from $S_1^1 \times \boldsymbol{R}^1$ by deleting two spacelike half-lines whose endpoints were the endpoints of a short null geodesic (dashed line in Figure 3). The causality condition holds on M, but the strong causality condition fails at each point of the null geodesic.

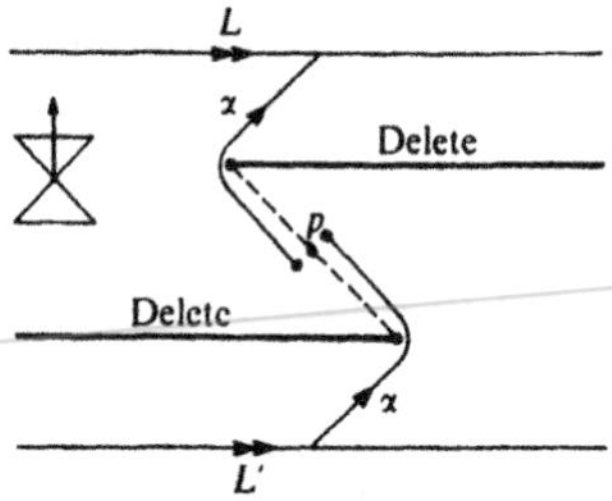

Figure 3. Lines L and L' are identified. There are no closed causal curves through p, but α is almost closed.

13. Lemma. Suppose the strong causality condition holds on a compact subset K of M. If α is a future-inextendible causal curve that starts in K, then α eventually leaves K never to return; that is, there is an $s > 0$ such that $\alpha(t) \notin K$ for all $t \geq s$.

Proof. Assume that the conclusion is false. Then α either remains in K or persistently returns. Thus, if the domain of α is $[0, B)$, $B \leq \infty$, there is a sequence $\{s_i\}$ in $[0, B)$ such that $\{s_i\} \to B$ and $\{\alpha(s_j)\}$ is contained in K. Then, for a subsequence, $\{\alpha(s_j)\}$ converges to a point $p \in K$. Since α has no future endpoint, there must be another sequence $\{t_j\}$ converging to B such that $\{\alpha(t_j)\}$ does not converge to p. By a further subsequence we can suppose that some neighborhood $\mathscr{U}$ of p contains no $\alpha(t_j)$. Since $\{s_j\}$ and $\{t_j\}$ both converge to B, they have subsequences that alternate: $s_1 < t_1 < s_2 < t_2 < \cdots$. Thus the curves $\alpha|[s_k, s_{k+1}]$ are "almost closed" at p, contradicting the strong causality of M at p. ∎

The following will be the main step in constructing geodesics joining points $p < q$.

14. Lemma. Suppose the strong causality condition holds on a compact subset K. Let $\{\alpha_n\}$ be a sequence of future-pointing causal curve segments in K such that $\{\alpha_n(0)\} \to p$ and $\{\alpha_n(1)\} \to q \neq p$. Then there is a future-pointing causal broken geodesic λ from p to q and a subsequence $\{\alpha_m\}$ of $\{\alpha_n\}$ such that $\lim_{m\to\infty} L(\alpha_m) \leq L(\lambda)$.

Proof. Proposition 8 implies that $\{\alpha_n\}$ has a limit sequence $\{p_i\}$ starting at p. If $\{p_i\}$ is infinite, the corresponding quasi-limit λ is a future-inextendible causal curve starting at p. Hence by the preceding lemma, λ must leave K and never return. In particular, some vertex p_i is not in K. But this implies that α_ns leave K: a contradiction.

Thus the limit sequence is finite, starts at p, and ends at $\lim \alpha_m(1) = q$. The quasi-limit λ with these vertices is a causal broken geodesic from p to q. In Definition 7, (L1b) lets us deal with one convex set $\mathscr{C}_i$ at a time. The length of the ith segment of α_m is at most the separation of its points in $\mathscr{C}_i$; that is,

$$L(\alpha_m | [s_{mi}, s_{m,i+1}]) \leq |\overrightarrow{p_{mi}p_{m,i+1}}|,$$

where $p_{mi} = \alpha_m(s_{mi})$. Hence

$$L(\alpha_m) \leq L_m = \sum_{i=0}^{k} |\overrightarrow{p_{mi}p_{m,i+1}}|.$$

By Lemma 5.9, the vector $\overrightarrow{pq}$ and hence its norm $|\overrightarrow{pq}|$ depend continuously on (p, q). Thus $\{L_m\}$ converges to $\sum p_i p_{i+1} = L(\lambda)$. Taking a further subsequence gives the result. ∎

TIME SEPARATION

There is a natural way to generalize the notion of the separation of points $p \leq q$ in $\mathbf{R}^n_1$ to the arbitrary time-oriented Lorentz manifold M.

15. Definition. If $p, q \in M$, the *time separation* $\tau(p, q)$ from p to q is

$$\sup\{L(\alpha)\colon \alpha \text{ is a future-pointing causal curve segment from } p \text{ to } q\}.$$

Here $\tau(p, q) = \infty$ if the set of lengths is unbounded, and $\tau(p, q) = 0$ if it is empty; that is, if $q \notin J^+(p)$.

When the supremum is taken on, we can think of $\tau(p, q)$ as the proper time of a slowest trip in M from p to q. In Minkowski spacetime, $\tau(p, q)$ is the separation $|\overrightarrow{pq}|$ if $p \leq q$ and otherwise is zero. Of course τ will behave badly if chronology conditions fail; for example, it is infinite for all p, q in $S^1_1 \times \mathbf{R}^1$.

The comparison between time separation and Riemannian distance is more dual than direct: τ maximizes, d minimizes. Because it involves time orientation, τ is symmetric only in trivial cases.

16. Lemma. (1) $\tau(p, q) > 0$ if and only if $p \ll q$.

(2) Reverse triangle inequality: If $p \leq q \leq r$, then $\tau(p, q) + \tau(q, r) \leq \tau(p, r)$.

Proof. (1) If $\tau(p, q) > 0$, there is a future-pointing causal curve α from p to q with $L(\alpha) > 0$. Thus α is not a null geodesic, so there is a fixed endpoint deformation of α to a timelike curve. The converse is obvious by definition of τ.

(2) If there are (future-pointing) causal curves α from p to q and β from q to r, then, given any number $\delta > 0$, we can choose α and β to have lengths within $\delta/2$ of $\tau(p, q)$ and $\tau(q, r)$, respectively. Hence

$$\tau(p, r) \geq L(\alpha + \beta) \geq \tau(p, q) + \tau(q, r) - \delta,$$

and the result follows.

If there is no (future-pointing) causal curve from say p to q, then $\tau(p, q) = 0$. Since $p \leq q$, we have $p = q$, so the result holds trivially. ∎

17. Lemma. The time-separation function $\tau: M \times M \to [0, \infty]$ is lower semicontinuous.

Proof. If $\tau(p, q) = 0$, there is nothing to prove. Suppose $q \in I^+(p)$ and $0 < \tau(p, q) < \infty$.

Given $\delta > 0$, we must find neighborhoods $\mathscr{U}$ and $\mathscr{V}$ of p and q, respectively, such that if $p' \in \mathscr{U}$ and $q' \in \mathscr{V}$, then $\tau(p', q') > \tau(p, q) - \delta$.

Let α be a timelike curve from p to q with $L(\alpha) > \tau(p, q) - \delta/3$. Let $\mathscr{C}$ be a convex neighborhood of q and let q_1 be a slightly earlier point of α in $\mathscr{C}$. Write $[r, r']$ for the geodesic segment in $\mathscr{C}$ from r to r'. Since the length of such segments depends continuously on endpoints, there is a neighborhood $\mathscr{V}$ of q such that if $q' \in \mathscr{V}$, then $[q_1, q']$ is causal and $L[q_1, q'] > L([q_1, q]) - \delta/3$. Since $[q_1, q]$ is geodesic, it is at least as long as the segment of α from q_1 to q.

The corresponding construction at the endpoint p then gives a neighborhood $\mathscr{U}$ of p such that points $p' \in \mathscr{U}$ and $q' \in \mathscr{V}$ can be joined (in an obvious way) by a causal curve of length $L > L(\alpha) - \delta/3 > \tau(p, q) - \delta$.

If $\tau(p, q) = \infty$, the same argument shows that for any $A > 0$ there are neighborhoods as above such that $\tau(p', q') > A$. ∎

The notation $J(p, q) = J^+(p) \cap J^-(q)$ is a convenient one; $J(p, q)$ is the smallest set containing all future-pointing causal curves from p to q (hence it is empty unless $p \leq q$).

18. Example. The time-separation function τ need not be continuous. Delete a spacelike segment from $\boldsymbol{R}^2_1$. Then for points p and q arranged as in Figure 4, τ is not continuous at (p, q). In fact, the shaded region is $J(p, q)$, so

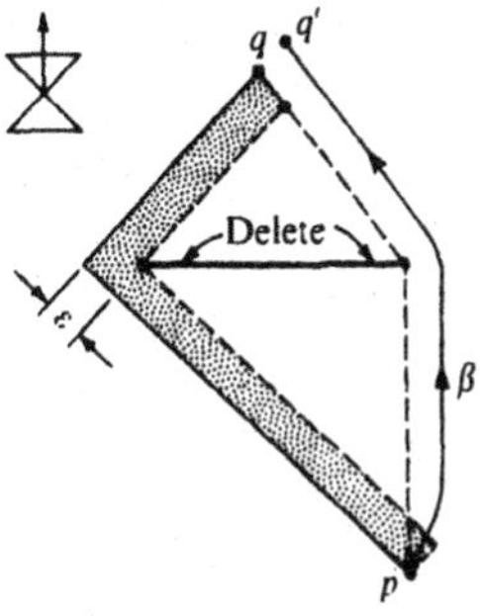

Figure 4

for $\varepsilon > 0$ small, every causal curve from p to q is nearly lightlike, hence short. However, for points q' (as in Figure 4) arbitrarily near q, new causal routes such as β appear with $L(\beta)$ large.

The *time separation* $\tau(A, B)$ of subsets A and B of M is $\sup\{\tau(a, b): a \in A, b \in B\}$. A variant of the preceding proof shows that the functions $x \to \tau(x, B)$ and $y \to \tau(A, y)$ are lower semicontinuous.

For the Lorentz manifold M we now establish reasonable sufficient conditions for the existence of a longest causal geodesic from p to q. An obvious necessary condition is $p < q$.

19. Proposition. For $p < q$, if the set $J(p, q) = J^+(p) \cap J^-(q)$ is compact and the strong causality condition holds on it, then there is a causal geodesic from p to q of length $\tau(p, q)$.

Proof. Let $\{\alpha_n\}$ be future-pointing causal curve segments from p to q whose lengths converge to $\tau(p, q)$. (A priori the latter could be ∞, but the proof shows it is finite.) These curves are all in $J(p, q)$, hence by Lemma 14 there is a causal broken geodesic λ from p to q with $L(\lambda) = \tau(p, q)$. If λ actually has breaks, there is a longer causal curve from p to q (10.3 and 10.46), hence λ is unbroken. ∎

This proposition motivates a fundamental definition: M is *globally hyperbolic* provided the strong causality condition holds, and for each $p < q$ the set $J(p, q)$ is compact. Then any pair of points that can be joined by a causal curve can be joined by a (longest) causal geodesic. This is the best possible conclusion, and opens the way for a variety of geodesic constructions that make globally hyperbolic manifolds a particularly convenient type. Evidently Minkowski space $\boldsymbol{R}_1^n$ is globally hyperbolic, and, as we shall see, so are Lorentz spheres S_1^n, Robertson–Walker spacetimes (with space S complete), and Kruskal spacetime. On the other hand, removing a single point from a globally hyperbolic M destroys the property, since there will be noncompact $J(p,q)$s.

Because global hyperbolicity of M is such a stringent condition, it is useful to weaken it as follows.

20. Definition. A subset $\mathscr{H}$ of M is *globally hyperbolic* provided (1) the strong causality condition holds on $\mathscr{H}$, and (2) if $p, q \in \mathscr{H}$ with $p < q$, then $J(p, q)$ is compact and contained in $\mathscr{H}$.

The definition relates the subset $\mathscr{H}$ to the causality of the manifold M; hence the property is definitely not intrinsic to $\mathscr{H}$.

By Proposition 19, in a globally hyperbolic set $\mathscr{H}$ there is a causal geodesic of M joining any points $p < q$.

21. Lemma. If $\mathscr{U}$ is a globally hyperbolic open set, the time-separation function τ of M is continuous on $\mathscr{U} \times \mathscr{U}$.

Proof. From before, τ is finite and lower semicontinuous. Assume it is not upper semicontinuous at $(p, q) \in \mathscr{U} \times \mathscr{U}$. Thus there is a number $\delta > 0$ and sequences $\{p_n\} \to p$ and $\{q_n\} \to q$ such that $\tau(p_n, q_n) \geq \tau(p, q) + \delta$ for all n.

Since $\tau(p_n, q_n) > 0$, there is a causal curve α_n from p_n to q_n such that $L(\alpha_n) > \tau(p_n, q_n) - 1/n$. Because $\mathscr{U}$ is open, it contains points $p^- \ll p$ and $q^+ \gg q$. We can suppose that the sequences $\{p_n\}$ and $\{q_n\}$ are contained in the open sets $I^+(p^-)$ and $I^-(q^+)$, respectively. It follows that the curves α_n are all in $J(p^-, q^+)$. Since $\mathscr{U}$ is globally hyperbolic, Lemma 14 applies (with $K = J(p^-, q^+)$) to show that there is a causal curve λ from p to q with $L(\lambda) \geq \tau(p, q) + \delta$. This is impossible in view of the definition of τ. ∎

22. Lemma. If $\mathscr{U}$ is a globally hyperbolic open set in M, the causality relation $\leq$ of M is closed on $\mathscr{U}$.

Proof. As in Lemma 2, the assertion is that if $\{p_n\} \to p$ and $\{q_n\} \to q$ with all these points in $\mathscr{U}$, then $p_n \leq q_n$ for all n implies $p \leq q$.

The proof is trivial if $p_n = q_n$ for infinitely many n, hence we can suppose $p_n < q_n$ for all n. Let α_n be a causal curve from p_n to q_n. Just as in the preceding proof, all α_n are in a suitable $J(p^-, q^+)$, and if $p \neq q$, the causal curve λ given by Lemma 14 shows that $p < q$. ∎

In particular, if M itself is globally hyperbolic, then all sets $J^+(p)$, $J^-(q)$, and $J(p, q)$ are closed.

ACHRONAL SETS

A subset A of M is *achronal* provided the relation $p \ll q$ never holds for $p, q \in A$; that is, provided no timelike curve meets A more than once. For example, in $\boldsymbol{R}_1^n$ a hyperplane t constant is achronal. Obviously any subset of an achronal set is achronal. By Lemma 3 the closure of an achronal set is achronal.

23. Definition. The *edge* of an achronal set A consists of all points $p \in \bar{A}$ such that every neighborhood $\mathscr{U}$ of p contains a timelike curve from $I^-(p, \mathscr{U})$ to $I^+(p, \mathscr{U})$ that does not meet A.

For example, in $\boldsymbol{R}_1^2$ the interval $A = \{(0, x): 0 \leq x < 1\}$ is achronal and has two edge points, (0, 0) and (0, 1). But if A is considered as a subset of $\boldsymbol{R}_1^3$, then edge $A = \bar{A}$.

In $\boldsymbol{R}_1^n$ a hyperplane t constant has empty edge. We want to show that every edgeless achronal set is a hypersurface—not necessarily a *smooth* hypersurface, since, for example, any nullcone $\Lambda^+(p)$ in $\boldsymbol{R}_1^n$ is achronal and edgeless.

An *n-dimensional topological manifold* T is a Hausdorff space such that each point has a neighborhood homeomorphic to an open set in $\boldsymbol{R}^n$. (Thus a smooth manifold is a topological manifold.) The classical Brouwer theorem on the invariance of domain [V] has the following useful consequence: *If $\phi: T \to T'$ is a one-to-one continuous mapping of n-dimensional topological manifolds, then ϕ is a homeomorphism onto an open set $\phi(T)$ of T'.*

24. Definition. A subspace S of T is a *topological hypersurface* provided that for each $p \in S$ there is a neighborhood $\mathscr{U}$ of p in T and a homeomorphism of $\mathscr{U}$ onto an open set in $\boldsymbol{R}^n$ such that $\phi(\mathscr{U} \cap S) = \phi(\mathscr{U}) \cap \Pi$, where Π is a hyperplane in $\boldsymbol{R}^n$.

Then S is an $(n-1)$-dimensional topological manifold. Evidently this definition is just a topological form of the slice criterion (1.31) for smooth submanifolds.

For example, the cone $\Lambda^+(0)$ in $\boldsymbol{R}_1^n$ is a topological hypersurface, since the homeomorphism $(t, x) \to (t - |x|, x)$ carries it to the hyperplane $t = 0$.

25. Proposition. An achronal set A is a topological hypersurface if and only if A contains no edge points (that is, A and edge A are disjoint).

Proof. First let A be a topological hypersurface. If $p \in A$, let $\mathscr{U}$ be an adapted topological coordinate neighborhood as in the preceding definition. We can suppose that $\mathscr{U}$ is connected and $\mathscr{U} - A$ has just two components.

Since A is achronal, the open sets $I^-(p, \mathcal{U})$ and $I^+(p, \mathcal{U})$ are open connected sets that are disjoint and do not meet A. Any timelike curve through p meets both sets, hence they are contained in different components of $\mathcal{U} - A$. Thus $p \notin \text{edge } A$.

Now suppose A and edge A are disjoint. If $p \in A$, then on a neighborhood $\mathcal{U}$ of p let ξ be a coordinate system with $\partial/\partial x^0$ timelike future-pointing. Using these coordinates we can get a smaller neighborhood $\mathcal{V}$ such that

(1) $\xi(\mathcal{V})$ has the form $(a - \delta, b + \delta) \times \mathcal{N} \subset \mathbf{R}^1 \times \mathbf{R}^{n-1} = \mathbf{R}^n$.

(2) The slice $x^0 = a$ of $\mathcal{V}$ is in $I^-(p, \mathcal{U})$; the slice $x^0 = b$ of $\mathcal{V}$ is in $I^+(p, \mathcal{U})$ (see Figure 5).

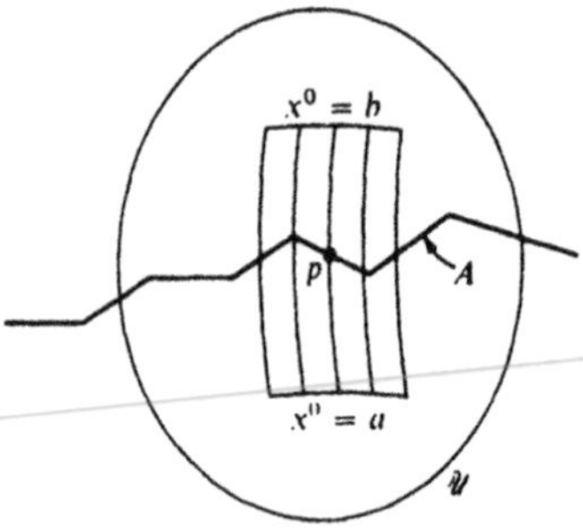

Figure 5

For $\mathcal{U}$ sufficiently small, if $y \in \mathcal{N} \subset \mathbf{R}^{n-1}$ the x^0 coordinate curve

$$s \to \xi^{-1}(s, y) \qquad (a \leq s \leq b)$$

must meet A, since p is not an edge point. Since A is achronal, the meeting point is unique; let $h(y)$ be its x^0 coordinate. It suffices to show that the function $h\colon \mathcal{N} \to (a, b)$ is continuous, for then $\phi = (x^0 - h(x^1, \ldots, x^{n-1}), x^1, \ldots, x^{n-1})$ is a homeomorphism that carries $A \cap \mathcal{V}$ to the slice $u^0 = 0$ of $\phi(\mathcal{V}) \subset \mathbf{R}^n$. Thus A is a topological hypersurface.

Let $\{y_n\}$ be a sequence that converges in $\mathcal{N}$ to y. Assume $\{h(y_n)\}$ does not converge to $h(y)$. Then some subsequence $\{h(y_m)\}$ converges to $r \neq h(y)$ (since h's values are bounded). But then $\xi^{-1}(y, r)$ is in the open set $I^-(q, \mathcal{V}) \cup I^+(q, \mathcal{V})$, where $q = \xi^{-1}(y, h(y)) \in A$. Hence the same is true for some $\xi^{-1}(y_n, h(y_n)) \in A$, contrary to the achronality of A. (In fact, the function h satisfies a Lipschitz condition). ∎

26. Corollary. An achronal set A is a closed topological hypersurface if and only if edge A is empty.

Proof. If A is a closed hypersurface, then by the proposition, A and edge A are disjoint. But edge $A \subset \bar{A} = A$, so edge A is empty.

Conversely, suppose edge A is empty. Then A is a topological hypersurface. That A is closed is shown by the achronal identity $\bar{A} - A \subset \text{edge } A$. To prove this inclusion, note that since $\bar{A}$ is also achronal, if $q \in \bar{A} - A$, then no timelike curve through q can ever meet A. It follows at once that $q \in \text{edge } A$. ∎

A subset F of M is a *future set* provided $I^+(F) \subset F$. For example, if B is any set, then $J^+(B)$ is a future set. Note that if F is a future set, its complement $M - F$ is a *past set* (closed under I^-).

27. Corollary. The (nonempty) boundary of a future set is a closed achronal topological hypersurface.

Proof. Let $p \in \text{bd } F$. If $q \in I^+(p)$, then $I^-(q)$ is a neighborhood of p and hence contains a point of F. Thus $q \in I^+(F) \subset F$. This proves $I^+(p) \subset F$; dually, $I^-(p) \subset M - F$. A first consequence is that $I^+(\text{bd } F)$ and $I^-(\text{bd } F)$ are disjoint, and hence bd F is achronal. A second is that the closed set bd F has no edge points, since, in fact, $I^+(p) \subset \text{int } F$ and $I^-(p) \subset \text{ext } F$ for $p \in \text{bd } F$. Thus the result follows from the preceding proposition. ∎

For example, in $\boldsymbol{R}^n_1$ this shows again that a nullcone $\Lambda^+(p) = \text{bd } J^+(p)$ is a closed achronal topological hypersurface.

A subset B of M is *acausal*, provided that the relation $p < q$ never holds for $p, q \in B$, that is, provided that no causal curve meets B more than once. This is a stronger requirement than achronality; for example, a null geodesic line in Minkowski space is achronal but not acausal.

CAUCHY HYPERSURFACES

28. Definition. A *Cauchy hypersurface* in M is a subset S that is met exactly once by every inextendible timelike curve in M.

In particular, S is achronal. In $\boldsymbol{R}^n_1$, the hyperplanes t constant are Cauchy hypersurfaces, but the achronal sets H^{n-1} and $\Lambda^+(p)$ are not. Examples will show that a given M need not contain a Cauchy hypersurface.

29. Lemma. A Cauchy hypersurface S is a closed achronal topological hypersurface and is met by every inextendible causal curve.

Proof. It is immediate from the definition that M is the disjoint union of the nonempty sets $I^-(S)$, S, $I^+(S)$. A timelike curve through any point of S instantly meets both $I^-(S)$ and $I^+(S)$. Thus S is the common boundary of the open sets $I^-(S)$ and $I^+(S)$. In view of Corollary 27 it remains to show that S is met, not just by every inextendible timelike curve, but by every inextendible causal curve α. Assume that α does not meet S. For definiteness, let

$\alpha(0) \in I^+(S)$. By Lemma 30, below, there is a past-inextendible timelike curve β starting in $I^+(S)$ that does not meet S. Any future-pointing timelike curve starting at $\beta(0)$ must remain in $I^+(S)$; thus adjoining it to β gives an inextendible timelike curve that avoids S. ■

We combine the "avoidance lemma" required above with a sharper version needed later.

30. Lemma. Let α be a past-inextendible causal curve starting at p that does not meet a closed set C.

(1) If $p_0 \in I^+(p, M - C)$, there is a past-inextendible timelike curve starting at p_0 that does not meet C.

(2) If α is not a conjugate-free null geodesic, there is a past-inextendible timelike curve starting at $\alpha(0)$ that does not meet C.

Proof. Since α is past-inextendible, we can suppose that it has domain $[0, \infty)$ and that the sequence $\{\alpha(n)\}$ does not converge.

(1) The scheme is to push α slightly to the future, the displacement dropping as one proceeds pastward on α. We work solely in the open submanifold $M - C$; all points are in $M - C$ and the relation $\ll$ is that of $M - C$ (implying that of M).

Since $p_0 \gg \alpha(0)$, also $p_0 \gg \alpha(1)$. There is a point p_1 such that $\alpha(1) \ll p_1 \ll p_0$. Continuing by induction we get a sequence $\{p_n\}$ with $\alpha(n) \ll p_n \ll p_{n-1}$ for all $n \geq 1$. Joining each p_{n-1} to p_n by a timelike segment gives a past-pointing timelike curve β in $M - C$ with $\beta(0) = p_0$. During the construction we can choose p_n so close to $\alpha(n)$ that $\{p_n\}$ does not converge (e.g., $d(p_n, \alpha(n)) < 1/n$ for some topological metric d on M). Thus β is inextendible.

(2) We can assume $\alpha|[0, 1]$ is not a conjugate-free null geodesic. Since $\alpha[0, 1]$ is compact and disjoint from C, it follows from Theorem 10.51 that this curve segment can be deformed to a timelike segment with the same endpoints, still avoiding C. Let α_1 be the causal curve gotten from α by replacing $\alpha|[0, 1]$ by this timelike segment.

Now $\alpha|[1, \infty)$ may be a conjugate-free null geodesic, but for $\delta_1 > 0$ small, $\alpha|[1 - \delta_1, 2]$ is not. As before, we get α_2 by replacing this segment by a timelike segment avoiding C. Iterating the last step gives a timelike past-pointing curve β from $\alpha(0)$ that avoids C.

Choosing the sequence $\{\delta_n\}$ to converge to 0 rapidly enough ensures, as in (1), that β is past-inextendible. ■

The exception in (2) is necessary; for example, let C be the lower imbedding of H^n in $\mathbf{R}_1^{n+1}$. Then every past-inextendible timelike curve through the origin 0 must meet C, but null geodesics through 0 do not.

31. Proposition. Let S be a Cauchy hypersurface in M, and let X be a timelike vector field on M. If $p \in M$, a maximal integral curve of X through p meets S at a unique point $\rho(p)$. Then $\rho: M \to S$ is a continuous open map onto S leaving S pointwise fixed. In particular, S is connected.

Proof. Lemma 1.56 and Exercise 1.16 show that maximal integral curves of X are inextendible. Let $\tilde{\psi}: \mathscr{D} \to M$ be the flow of X. $\mathscr{D}$ is an open set in $M \times \boldsymbol{R}$, and S is a topological hypersurface in M, hence $\mathscr{D}(S) = (S \times \boldsymbol{R}) \cap \mathscr{D}$ is a topological hypersurface in $\mathscr{D}$. The restriction $\psi: \mathscr{D}(S) \to M$ is continuous, and, since S is a Cauchy hypersurface, ψ is one-to-one onto. $\mathscr{D}(S)$ and M are topological manifolds of the same dimension, hence by the invariance of domain, ψ is a homeomorphism. The natural projection $\pi: S \times \boldsymbol{R} \to S$ is open, continuous, and onto. But $\rho = \pi \circ \psi^{-1}$, since $\rho\psi(p, t) = \rho\alpha_p(t) = \alpha_p(0) = p$ for $p \in S$. Thus ρ has the same properties as π. Clearly $\rho | S = \text{id}$. Since M is connected, so is S. ∎

32. Corollary. Any two Cauchy hypersurfaces in M are homeomorphic.

Proof. Let S and T be Cauchy hypersurfaces in M. For a fixed timelike vector field let ρ_S and ρ_T be the resulting retractions of M onto S and T. Then clearly $\rho_T | S$ and $\rho_S | T$ are inverse maps. ∎

WARPED PRODUCTS

We consider some causality relations on Lorentz warped products $M = B \times_f F$ that, like Schwarzschild spacetime, have B Lorentz and F Riemannian. Recall that for each $q \in F$ the leaf $\sigma^{-1}(q)$ is totally geodesic in M and isometric to B under the projection $\pi: M \to B$.

(W1) $\langle d\pi(v), d\pi(v) \rangle \leq \langle v, v \rangle$ *for each* $v \in TM$, *with equality if and only if* v *is horizontal.*

This follows immediately from Definition 7.33 since $f > 0$ and F is Riemannian. Hence

(W2) $d\pi$ *carries causal vectors onto causal vectors.*

(*Onto* because each causal vector in $T_{\pi p}(B)$ has a horizontal lift to $T_p(M)$.) In fact, timelike and nonhorizontal null vectors project to timelike vectors, while horizontal null vectors project to null vectors.

Thus causal [timelike] curves in M project to causal [timelike] curves in B. It follows also that M is time-orientable if and only if B is. As always in this chapter, M is assumed to be connected and time-oriented; then we time-orient B so that $d\pi$ carries future-pointing vectors to future-pointing vectors.

(W3) *A subset A of B is achronal [acausal] if and only if $\pi^{-1}(A)$ is achronal [acausal] in M.*

Proof. If α is timelike with endpoints in $\pi^{-1}(A)$, then $\pi \circ \alpha$ is timelike with endpoints in A. If γ is timelike with endpoints in A, then any horizontal lift $\tilde{\gamma}$ is timelike with endpoints in $\pi^{-1}(A)$. Similarly,

(W4) *M satisfies the chronology, causality, or strong causality conditions, respectively, if and only if B does.*

(W5) *Suppose F is complete and the strong causality condition holds on B, hence on M. If a causal curve α in M is inextendible, then $\pi \circ \alpha$ is inextendible in B* (and conversely).

Proof. Assume that $\pi \circ \alpha: [0, b) \to B$ is extendible to $\beta: [0, b] \to B$. By Exercise 5.14, $\pi \circ \alpha$ has finite length. Since $f > 0$ on the compact set $\beta[0, b]$, there is a $C > 0$ such that $f \circ \pi \circ \alpha \geq C$. The formula for the warped product metric then gives $|d\sigma(\alpha')| \leq |d\pi(\alpha')|/C$. Consequently, $\sigma \circ \alpha$ also has finite length. Since F is complete Riemannian, this means that $\sigma \circ \alpha$ stays in some compact set K. Thus α remains in the compact set $\beta[0, b] \times K$, contradicting Lemma 13. ∎

33. Lemma. With notation as above, if F is complete, then $M = B \times_f F$ has a Cauchy hypersurface if and only if B does.

Proof. If S is a Cauchy hypersurface in M, then $S \cap \sigma^{-1}(q)$ is a Cauchy hypersurface in a leaf $\sigma^{-1}(q)$, hence $\pi(S \cap \sigma^{-1}(q))$ is a Cauchy hypersurface in B.

Let Σ be a Cauchy hypersurface in B. Anticipating Corollary 39 (or, in special cases, Lemma 34), we assert that the strong causality condition holds on B. Thus (W5) applies and it follows immediately that $\pi^{-1}(\Sigma)$ is a Cauchy hypersurface. ∎

Some of the most important spacetimes in relativity theory are (spherically symmetric) warped products $M = B \times_f S^2$ with B a Lorentz surface. Then the preceding lemma is particularly effective, since a hypersurface in B is only a curve. Using null geodesic criteria such as Corollary 54, one can often tell by inspection whether B (hence M) has a Cauchy hypersurface. For example, in the Kruskal plane (Figure 13.10), the diagonal line $\Delta: u = v$ is a Cauchy hypersurface, hence the cylinder $\pi^{-1}(\Delta) = \Delta \times S^2$ is a

Cauchy hypersurface in the Kruskal spacetime $Q \times_f S^2$. In applications, B is often simply connected; then (W4) and the following lemma show that the strong causality condition holds on $B \times_f S^2$. (These remarks remain valid if S^2 is replaced by any complete Riemannian manifold.)

34. Lemma. The strong causality condition holds on any simply connected Lorentz surface B.

Proof. (See, for example, [V] for the topological results used.) Since B is simply connected, it is time-orientable and there are smooth future-pointing null vector fields U, V on B that are linearly independent. Hence $V - U$ is a nonvanishing spacelike vector field.

(1) The causality condition holds. Assume α is a closed causal curve. By taking its first loop we can suppose α is simply closed. Since it is a simply connected surface, B is homeomorphic to $\boldsymbol{R}^2$ or S^2 (the latter is ruled out by Proposition 5.37). Thus the Jordan curve theorem applies: α parametrizes the boundary of a 2-cell E in B. Since $V - U$ is never tangent to α, it (or $U - V$) always points *into* E: integral curves starting α are initially in E. For all sufficiently small t, the flow map ψ_t is defined on (compact) E and carries E into itself. By the Brouwer fixed point theorem, each $\psi_{1/n}$ has a fixed point in E. But at a limit point of such fixed points, $V - U$ must be zero: a contradiction.

(2) The strong causality condition holds. If not, we can readily find a causal curve α whose endpoints $\alpha(0)$ and $\alpha(1)$ are joined by a segment σ of an integral curve of $V - U$. Then the previous argument still applies to $\sigma + \alpha$. ∎

The result fails in higher dimensions (see Exercise 8).

CAUCHY DEVELOPMENTS

35. Definition. If A is an achronal subset of M, the *future Cauchy development* of A is the set $D^+(A)$ of all points p of M such that every past-inextendible causal curve through p meets A. (In particular, $A \subset D^+(A)$.)

Relativistically, $D^+(A)$ is that part of the causal future of A that is predictable from A: no past-inextendible particle or light ray can reach an event q in $D^+(A)$ without earlier having gone through A.

With the *past Cauchy development* $D^-(A)$ defined dually, $D(A) = D^-(A) \cup D^+(A)$ is the *Cauchy development* of A.

36. Examples of Cauchy Developments. (1) If A is a spacelike hyperplane $t = c$ in $\mathbf{R}^n_1$, then $D^+(A) = J^+(A) = \{(t, x): t \geq c\}$, and similarly for $D^-(A)$. Hence $D(A) = \mathbf{R}^n_1$.

(2) For the lower imbedding of H^n in Minkowski $(n + 1)$-space, $D^+(H^n) = J^+(H^n) \cap I^-(0)$, the union of H^n and the open region between H^n and the past nullcone of the origin—only a small part of $J^+(H^n)$. But $D^-(H^n) = J^-(H^n)$.

(3) Let M be $\mathbf{R}^1_1 \times S^1$ with a point deleted. For a spacelike circle S as in Figure 6, $D^+(S)$ is the union of S and the open region between S and the null geodesics α and β. Again $D^-(S) = J^-(S)$.

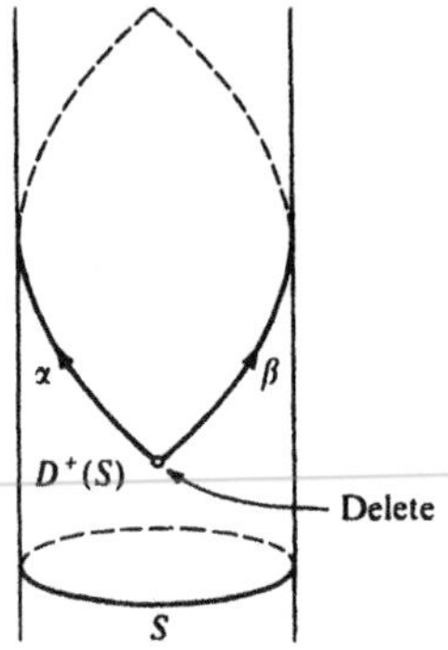

Figure 6

In view of Lemma 29, *an achronal set A in M is a Cauchy hypersurface if and only if $D(A) = M$*. Thus in the general case, we can think of $D(A)$ as the largest subset for which A plays the role of Cauchy hypersurface.

The definition of Cauchy development makes sense for any set, and $D^+(A) \subset A \cup I^+(A) \subset J^+(A)$. Since we assume A is achronal, $D^+(A)$ and $I^-(A)$ are disjoint, hence $D^+(A) \cap D^-(A) = A$ and $D^+(A) - A = D(A) \cap I^+(A)$.

For future reference we call the following fact *regression*: *a past-pointing causal curve α starting in $D^+(A)$ cannot leave $D^+(A)$ without first meeting A*. (*Proof*. If $\alpha(s) \notin D^+(A)$, there is a past-inextendible causal curve β starting at $\alpha(s)$ that does not meet A—but $\alpha|[0, s] + \beta$ must meet A, hence $\alpha[0, s)$ meets A.)

37. Lemma. If A is achronal and $p \in \operatorname{int} D(A)$, then every inextendible causal curve through p meets both $I^-(A)$ and $I^+(A)$.

Proof. We can suppose also that $p \in A \cup I^+(A)$. Let α be a past-inextendible causal curve starting at p. The proof of Lemma 30(1)—with C

empty—shows that there is past-inextendible causal curve β starting in $D(A) \cap I^+(A) \subset D^+(A)$, such that each $\beta(s)$ has a point of α in $I^-(\beta(s))$. Since β meets A, it follows that α meets $I^-(A)$.

Let γ be a future-inextendible causal curve starting at p. If $p \in A$, then the dual of the past-inextendible case asserts that γ meets $I^+(A)$, and there is nothing to prove if $p \in I^+(A)$. ■

38. Theorem. If A is an achronal set, then int $D(A)$ (if nonempty) is globally hyperbolic.

Proof. (1) *The causality condition holds on* $D(A)$. Assume there is a causal loop γ at $p \in D(A)$. Traversing γ repeatedly gives an inextendible causal curve $\tilde{\gamma}$ which must then meet A. But $\tilde{\gamma}$ meets A repeatedly, contrary to achronality.

(2) *Strong causality holds at* $p \in$ int $D(A)$. Assume not. Then there exist future-pointing causal curve segments α_n defined on $[0, 1]$ such that $\{\alpha_n(0)\}$ and $\{\alpha_n(1)\}$ both converge to p, but every α_n leaves some fixed neighborhood of p. Thus $\{\alpha_n\}$ has a future-directed limit sequence $\{p_i\}$ starting at p. If $\{p_i\}$ is finite, it ends at $\lim \alpha_n(1) = p$. But then $p < p$ contrary to (1). Thus $\{p_i\}$ is infinite, hence the corresponding quasi-limit λ is future-inextendible. By Lemma 37, λ enters $I^+(A)$ and hence remains there, so some vertex p_i is in $I^+(A)$. Thus there is a subsequence $\{\alpha_m\}$ and (by reparametrization) a number $s \in [0, 1]$ such that $\lim \alpha_m(s) = p_i$. Evidently we can suppose every $\alpha_m(s)$ is in $I^+(A)$.

Since $p_i \neq p$, we can apply Proposition 8 dually to $\{\alpha_m | [s, 1]\}$ to obtain a *past-directed* limit sequence $\{\bar{p}_i\}$ starting at p. If $\{\bar{p}_i\}$ is finite, it must end at $\lim \alpha_m(s) = p_i$. But then $p_i < p$. Since $p_i > p$, this gives the contradiction $p < p$ again.

Thus $\{\bar{p}_i\}$ is infinite. The resulting quasi-limit $\bar{\lambda}$ is a past-inextendible causal curve starting at p. By the lemma it meets $I^-(A)$. Hence as usual some $\alpha_m | [s, 1]$ must meet $I^-(A)$. Since α_m is future-pointing and has $\alpha_m(s) \in I^+(A)$, this is contrary to the achronality of A.

(3) *If* $p \leq q$ *in* int(A), then $J(p, q)$ *is compact.* If $p = q$, then by (1), $J(p, q) = \{p\}$. So suppose $p < q$. Let $\{x_n\}$ be a sequence in $J(p, q)$; we must show that some subsequence converges to a point of $J(p, q)$. Let α_n be a future-pointing causal curve segment from p through x_n to q. Let $\mathfrak{R}$ be a covering of M by convex open sets $\mathscr{C}$ such that $\overline{\mathscr{C}}$ is a compact and contained in a convex open set; all limit sequences will be relative to $\mathfrak{R}$. There is certainly such a sequence starting at p. Suppose it is finite, hence ends at $p_k = q$. Let $\{\alpha_m\}$ be a subsequence as in condition (L1) of Definition 7. By the pigeonhole argument, there is an $i < k$ such that, for infinitely many m, the point x_m lies on the ith segment $\alpha_m | [s_{mi}, s_{m,i+1}]$ of α_m. Pass to this subsequence. By

(L1) the segments, hence the points x_m, all lie in a single member $\mathscr{C}$ of $\mathfrak{R}$. The properties of $\mathscr{C}$ imply that $\{x_m\}$ converges to a point x and by Lemma 2 that $p_i \leq x \leq p_{i+1}$. Hence $p \leq x \leq q$, that is, $x \in J(p, q)$.

It remains to derive a contradiction to the assumption: *every limit sequence for* $\{\alpha_n\}$, *relative to* $\mathfrak{R}$ *and starting at* p, *is infinite*. Let λ be a quasi-limit. Since λ is a future-inextendible causal curve starting at p, we can find, as before, a subsequence $\{\alpha_m\}$ and (reparametrizing) a single s such that $\{\alpha_m(s)\}$ converges to a vertex p_i in $I^+(A)$.

The proof is increasingly like that for (2). Since $p_i \neq q$, we can apply Proposition 8 dually to $\{\alpha_m | [s, 1]\}$, getting a past-directed limit sequence $\{q_i\}$ starting at q. If $\{q_i\}$ is finite, it must end at $\lim \alpha_m(s) = p_i$. This means that

$$p < p_1 < \cdots < p_i < \cdots < q_1 < q$$

is a finite limit sequence for $\{\alpha_n\}$ starting at p—a contradiction to our current assumption.

Thus $\{q_i\}$ is infinite. The corresponding quasi-limit μ is a past-inextendible causal curve starting at q. As before, μ reaches $I^-(A)$, hence some $\alpha_m | [s, 1]$ does. Since $\alpha_m(s) \in I^+(A)$, this contradicts achronality.

(4) *If* $p \leq q$ *in* int $D(A)$, *then* $J(p, q) \subset \text{int } D(A)$. As before, we can assume $p < q$. By duality, only two cases need to be considered.

Case 1. $p, q \in I^+(A)$. Pick $q^+ \in I^+(q) \cap D(A) \subset D^+(A)$. Then $\mathscr{N} = I^+(A) \cap I^-(q^+)$ is an open set containing $J(p, q)$, so it suffices to prove $\mathscr{N} \subset D^+(A)$. Let σ be past-pointing timelike from q^+ to $y \in \mathscr{N}$. Since A is achronal and $y \in I^+(A)$, σ does not meet A. Hence, by regression, $y \in D^+(A)$.

Case 2. $p \in J^-(A)$ and $q \in J^+(A)$. The argument is similar; since $p, q \in \text{int } D(A)$, there exist points $p^- \in I^-(p) \cap D^-(A)$ and $q^+ \in I^+(q) \cap D^+(A)$. We assert that the neighborhood $\mathscr{N} = I^+(p^-) \cap I^-(q^+)$ of $J(p, q)$ is in $D(A)$. If $x \in \mathscr{N}$, let σ and τ be past-pointing timelike curve segments from q^+ to x and from x to p^-, respectively. Since $A \subset D(A)$ we can suppose $x \notin A$. By achronality at least one of the curves does not meet A. If σ, then, by regression, $x \in D^+(A)$; if τ, then $x \in D^-(A)$. ■

In the theorem, the restriction to the interior of $D(A)$ is unavoidable; for example, in the manifold of Figure 3, take A to be a spacelike closed segment with one endpoint on the null geodesic (dashed line).

In the next section, we find reasonable conditions for $D(A)$ to be open, hence globally hyperbolic. But the theorem already has an important consequence:

39. Corollary. If M has a Cauchy hypersurface, then M is globally hyperbolic.

Proof. If S is a Cauchy hypersurface, then $D(S) = M$, hence int $D(S) = M$. ∎

For example, a Robertson–Walker spacetime with space S complete is globally hyperbolic. In fact, any spacelike slice $t \times S$ is a closed acausal hypersurface and by Remark 12.27 every inextendible null geodesic meets $t \times S$. Corollary 54 will show that this makes $t \times S$ a Cauchy hypersurface. Since Kruskal spacetime has a Cauchy hypersurface, it too is globally hyperbolic. (For the converse of Corollary 39, due to R. P. Geroch, see [G] or [HE].)

40. Lemma. Let A be an achronal set. If $p \in \text{int}\, D(A) - I^-(A)$, then $J^-(p) \cap D^+(A)$ is compact.

Proof. If $p \in A$, then the set in question consists of p alone, so suppose $p \in I^+(A) \cap \text{int}\, D(A)$.

Let $\{x_n\}$ be an infinite sequence in $J^-(p) \cap D^+(A)$ and let α_n be a past-pointing causal curve segment from p to x_n.

There is nothing to prove if any subsequence of $\{x_n\}$ converges to p. Otherwise there is a past-directed limit sequence $\{p_i\}$ for $\{\alpha_n\}$ starting at p. If $\{p_i\}$ is infinite, then, as usual, it follows that some x_n is in $I^-(A)$: a contradiction. If $\{p_i\}$ is finite, some subsequence $\{x_m\}$ converges to a point $x \in J^-(p)$. Let σ be a timelike curve from $p^+ \in D^+(A) \cap I^+(p)$ to x. If σ meets A, then either $x \in A \subset D^+(A)$ or $x \in I^-(A)$, the latter implying some $x_n \in I^-(A)$; a contradiction. If σ avoids A, then, by regression, $x \in D^+(A)$. ∎

As Figure 7 indicates, this lemma and Lemma 37 both fail if the point p is not in the interior of $D(A)$.

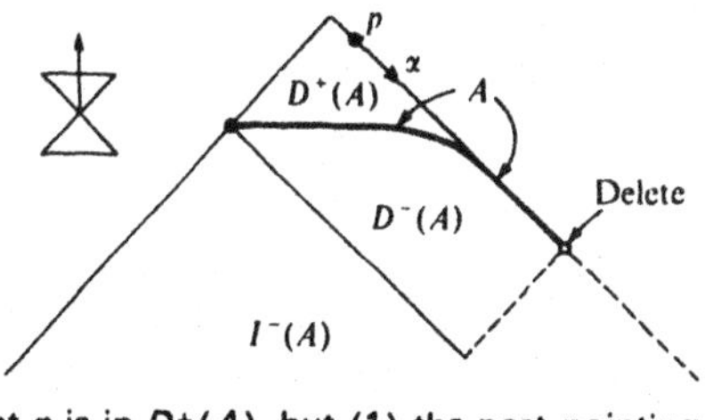

Figure 7. The point p is in $D^+(A)$, but (1) the past-pointing null geodesic α starting at p fails to reach $I^-(A)$, and (2) $J^-(p) \cap D^+(A)$ is not compact.

41. Example. Causality in $\tilde{H}^n_1$. (See definition, page 228.) Although $\tilde{H}^n_1$ is complete, diffeomorphic to $\boldsymbol{R}^n$, and has constant curvature $K = -1$, its causality is not trivial. The following facts are readily found using the model of $\tilde{H}^n_1$ discussed in Example 8.27. There we saw that the slices t constant (see

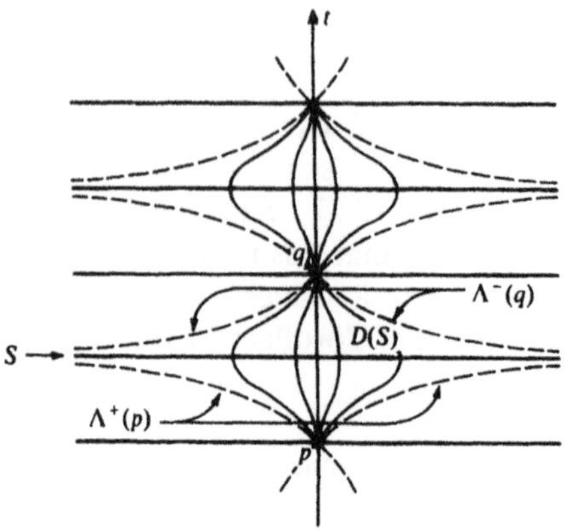

Figure 8. Causality in $\tilde{H}^n_1$. Dashed lines are null geodesics; horizontal lines are spacelike hypersurfaces; all other lines are timelike geodesics. (For a three-dimensional representation, rotate the figure about the t axis.)

Figure 8) are totally geodesic spacelike hypersurfaces, each isometric to hyperbolic $(n-1)$-space (a line for $n = 2$). The t axis is a timelike geodesic and ∂_t is everywhere timelike future-pointing. The points p and q, symmetrically placed relative to the slice $S: t = 0$, project to antipodal points in H^n_1. Maps $(t, x) \to (\pm t + c, x)$ are isometries. (Note that since $\tilde{H}^n_1$ is frame homogeneous, the t axis represents an *arbitrary* timelike geodesic.)

(1) The future-pointing null geodesics starting at p form a curving cone $\Lambda^+(p)$ that approaches but does not reach S. (Consider the corresponding null geodesics in H^n_1.)

(2) $J^+(p)$ is the entire closed half-space on the future side of $\Lambda^+(p) = \text{bd}\, J^+(p)$. (Vertical lines in the figure are timelike, though only the t axis is geodesic.)

(3) The strong causality condition holds. (Causal curves starting near p can never return to near p.)

(4) $J(p, q)$ is the closed region bounded by $\Lambda^-(q) \cup \Lambda^+(p)$. Since this set is not compact, $\tilde{H}^n_1$ is not globally hyperbolic.

(5) Timelike geodesics in $\tilde{H}^n_1$ project to closed geodesics in H^n_1; thus the future-pointing timelike geodesics starting at p all meet, after distance π, at the conjugate point q (and again periodically along the t axis). These geodesics are the normal geodesics to S, so p and q are its nearest focal points. Evidently, many points of $J^+(p)$ cannot be reached by geodesics from p.

(6) $D^+(S) = S \cup (\text{open region between } S \text{ and } \Lambda^-(g)) = J^+(S) \cap I^-(q)$. Thus $D(S) = I^+(p) \cap I^-(q)$.

SPACELIKE HYPERSURFACES

Causality can be expressed in topological terms, but in Lorentz geometry the initial data will often be smooth. Furthermore, the most obvious examples of achronal sets appearing earlier have been (smooth) spacelike hypersurfaces. For these we establish some small but significant advantages.

42. Lemma. An achronal spacelike hypersurface S is acausal.

Proof. Assume there exists a future-pointing causal curve segment α with endpoints $\alpha(0)$ and $\alpha(1)$ in S. If α is not a null geodesic, it admits a fixed endpoint deformation to a timelike curve—contradicting the achronality of S. If α is a null geodesic, then since $\alpha'(0)$ cannot be normal to S, Lemma 10.50 applies, giving again a contradiction to achronality. ■

43. Lemma. If S is a acausal topological hypersurface in M, then $D(S)$ is open (hence globally hyperbolic).

Proof. (1) First we show that $S \subset \text{int } D(S)$. Assume $p \in S - \text{int } D(S)$. Then there is a sequence $\{\alpha_n\}$ of inextendible causal curves such that $\{\alpha_n(0)\} \to p$ and each α_n avoids S. Since S contains no edge points, we can show that $I(S) = I^-(S) \cup S \cup I^+(S)$ is open. Let $\mathscr{N}$ be a neighborhood of p such that $\bar{\mathscr{N}}$ is compact, contained in a convex set, and contained in $I(S)$.

Without loss of generality, suppose all $\alpha_n(0)$ are in $\mathscr{N} \cap I^+(S)$. As in the construction of a limit sequence, let e_n be the first point after $\alpha_n(0)$ at which (past-pointing) α_n meets bd $\mathscr{N}$. For a convergent sequence $\{e_m\} \to e$, we thus have $e \in J^-(p)$. Hence $e \notin I^+(S)$. S is acausal and $e \neq p$, so $e \notin S$. Finally, e cannot be in $I^-(S)$ since, while in $I(S)$, no α_n can reach $I^-(S)$ without meeting S. This contradicts $e \in bd\, \mathscr{N} \subset I(S)$.

(2) Assume for the moment that S is closed in M. By duality it suffices to show that the set

$$D^+(S) - S = I^+(S) \cap D(S)$$

is open in M. Assume that $p \in D^+(S) - S$ is not an interior point. Thus there is a sequence $\{\alpha_n\}$ of past-inextendible causal curves such that $\{\alpha_n(0)\} \to p$ and every α_n avoids S. By (1) and the fact that S is closed, there is a convex covering $\mathfrak{R}$ whose members are either contained in int $D(S)$ or disjoint from S. Let λ be a quasi-limit of $\{\alpha_n\}$ relative to $\mathfrak{R}$. Since λ starts at p and is a past-inextendible causal curve, it meets S at a (unique) point $\lambda(s)$. Let p_i be the vertex of λ such that $p_i > \lambda(s) \geq p_{i+1}$. The member of $\mathfrak{R}$ containing this segment of λ meets S, hence by construction is contained in int $D(S)$. Since S is acausal, $p_i \notin S$, and by regression $p_i \in D^+(S)$. Thus $p_i \in I^+(S)$. Consequently

$I^+(S) \cap \text{int } D(S) \subset D^+(S)$ is a neighborhood of p_i. Some α_n must meet it, and hence meet S; a contradiction.

(3) For arbitrary S, we can assume S connected, since the Cauchy developments of different components would be disjoint. Clearly S is closed in $I(S)$, hence the preceding arguments apply with M replaced by its connected open submanifold $I(S)$. The Cauchy development of S is the same in M as in $I(S)$, and being open in $I(S)$, it is open in M. ■

This lemma fails if the hypothesis *acausal* is weakened to *achronal* (in Figure 7, extend A to a closed achronal hypersurface).

44. Theorem. Let S be a closed achronal spacelike hypersurface in M. If $q \in D^+(S)$, there is a geodesic from S to q of length $\tau(S, q)$. Hence γ is normal to S and has no focal points of S before q. (γ is timelike except in the trivial case $q \in S$.)

Proof. By the preceding lemma, $D(S)$ is a globally hyperbolic open set. By Lemma 40, $J^-(q) \cap D^+(S)$ is compact; hence its intersection with S—namely, $J^-(q) \cap S$—is compact. By Lemma 21, the function $x \to \tau(x, q)$ is continuous on $J^-(q) \cap S$, hence takes on a maximum at say p. Evidently this maximum is $\tau(p, q) = \tau(S, q)$. By Proposition 19, there is a geodesic segment γ from p to q of length $\tau(p, q) = \tau(S, q)$. Assuming $q \notin S$ gives $p \ll q$, hence $\tau(p, q) > 0$; so γ is timelike. Then Corollary 10.26 implies that γ is normal to S, and Theorem 10.37 forbids focal points. ■

The result extends by duality to the entire Cauchy development. Note that $D(S)$ is not necessarily a normal neighborhood of S, since the geodesic γ in the theorem is not unique. (For example, put a ripple in the x axis of $\boldsymbol{R}^2_1$.)

The failure of a desirable property in M is often not too serious if the property holds in some covering manifold of M. Our goal now is to show that any closed spacelike hypersurface of M can be made achronal (hence acausal) by lifting it into a suitable covering manifold.

The proof uses a result from *intersection theory* [GP]. A homotopy of closed curves (loops) in which the endpoint is allowed to move is called a *free homotopy*. A curve α is *transversal* to a submanifold S at $\alpha(t) \in S$ provided $\alpha'(t)$ not tangent to S. Thus α must cut cleanly through S at $\alpha(t)$. For example, causal curves are always transversal to spacelike submanifolds. The result we need is this: *a closed curve that meets a closed hypersurface S exactly once, and there transversally, is not freely homotopic to a closed curve that does not meet S.*

Since M is time-oriented, a spacelike hypersurface S in M is *two-sided*: If $\mathscr{N}$ is a normal neighborhood of S, then $\mathscr{N} - S$ is the union of the disjoint open sets $\mathscr{N}^- = \mathscr{N} \cap I^-(S)$ and $\mathscr{N}^+ = \mathscr{N} \cap I^+(S)$.

45. Lemma. Let S be a closed connected spacelike hypersurface in M.

(1) If the homomorphism $j_{\#}: \pi_1(S) \to \pi_1(M)$ induced by the inclusion map $j: S \subset M$ is *onto*, then S separates M (that is, $M - S$ is not connected).
(2) If S separates M, then S is achronal.

Proof. (1) The hypothesis simply means that—picking base point $p \in S$—each loop in M at p is fixed-endpoint homotopic to a loop in S.

With notation as above, let $\sigma: [-1, 1] \to \mathcal{N}$ be a timelike curve from $\mathcal{N}^-$ to $\mathcal{N}^+$ that meets S only at $\sigma(0) = p$. Now assume S does not separate M. Since $M - S$ is connected, there is a curve α from $\sigma(1)$ to $\sigma(-1)$ that does not meet S. Then $\gamma = \sigma|[0, 1] + \alpha + \sigma|[-1, 0]$ is a closed curve meeting S only at $\sigma(0)$ and there transversally. By hypothesis, γ is fixed endpoint homotopic, hence freely homotopic, to a closed curve in S. Evidently a small deformation of this curve in the future direction moves it into $\mathcal{N}^+$. So γ is freely homotopic to a closed curve disjoint from S. This contradicts the intersection theory fact mentioned above.

(2) Assume S is not achronal. Thus there is a future-pointing timelike curve segment α with endpoints $\alpha(0)$, $\alpha(1)$ in S. It is clear that there are numbers $0 < a < b < 1$ such that $\alpha(a) \in \mathcal{N}^+$ and $\alpha(b) \in \mathcal{N}^-$. $\mathcal{N}^+$ and $\mathcal{N}^-$ are connected, since S is; hence they are contained in the same component of $M - S$.

Since M is connected, each point of $M - S$ can be joined to $\mathcal{N} - S = \mathcal{N}^- \cup \mathcal{N}^+$ by a curve in $M - S$. Hence $M - S$ is connected, contrary to the hypothesis that S separates M. ∎

46. Corollary. If M is simply connected, then every closed spacelike hypersurface in M is achronal (hence acausal).

Given a spacelike hypersurface S of an arbitrary M, it will not suffice merely to pass to the simply connected covering of M since it may not be possible to find a copy of S there.

47. Example. In the Lorentz torus $M = S^1_1 \times S^1$, the circle $S = p \times S^1$ is spacelike and compact but not achronal (Figure 9). The simply connected covering manifold of M is $\boldsymbol{R}^2_1$, which contains no spacelike circles. However, the Lorentz cylinder $\tilde{M} = \boldsymbol{R}^1_1 \times S^1$ also covers M, with map $k = \exp \times \text{id}$. Each component of $k^{-1}(S)$ is a closed spacelike hypersurface in $\tilde{M}$ that is achronal and isometric to S.

The result in this example can be achieved in general.

48. Proposition. Let S be a closed, connected, spacelike hypersurface in M. Then there is a Lorentz covering $k: \tilde{M} \to M$ and a closed spacelike hypersurface $\tilde{S}$ in $\tilde{M}$ that is achronal and isometric under k to S.

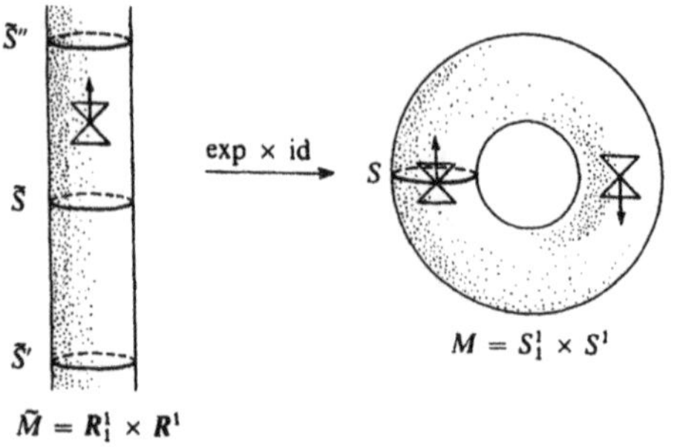

Figure 9

Proof. For an arbitrary Lorentz covering $k: \tilde{M} \to M$, $k^{-1}(S)$ is a closed spacelike hypersurface in $\tilde{M}$ and any connected component $\tilde{S}$ of $k^{-1}(S)$ shares these properties. Furthermore, $k|\tilde{S}: \tilde{S} \to S$ is a covering map and a local isometry, that is, $k|\tilde{S}$ is a Riemannian covering map.

By Corollary A.13 there is a connected covering $k: \tilde{M} \to M$ such that Image $k_{\#}$ = Image $j_{\#}$, where $j: S \subset M$. Assign $\tilde{M}$ the induced Lorentz metric and time-orientation.

(1) $k|\tilde{S}$ *is an isometry*. It remains only to show that $k|\tilde{S}$ is one-to-one. We can assume that $\tilde{S}$ contains the base point $\tilde{p}$ of the fundamental group of $\tilde{M}$. Then since $k|\tilde{S}$ is a covering map, it suffices to show that $\tilde{p}$ is the only point of $\tilde{S}$ sent to $p = k(\tilde{p}) \in S$.

If $q \in \tilde{S}$ and $k(q) = p$, let α be a curve in $\tilde{S}$ from $\tilde{p}$ to q. Then $k \circ \alpha$ is a loop in S at p. Thus its homotopy class $[k \circ \alpha] \in \pi_1(M, p)$ is in Image $k_{\#}$, so there is a loop β in $\tilde{M}$ at $\tilde{p}$ such that $k \circ \alpha \simeq k \circ \beta$. But by Corollary A.10, α and β end at the same point; that is, $q = \tilde{p}$.

(2) $\tilde{S}$ *is achronal*. By the preceding lemma it suffices to show that the homomorphism $i_{\#}$ induced by $i: \tilde{S} \subset \tilde{M}$ is *onto*. Let $x \in \pi_1(\tilde{M}, \tilde{p})$. By construction there is a $y \in \pi_1(S, p)$ such that $k_{\#}(x) = j_{\#}(y)$. By (1), $k|\tilde{S}$ is a homeomorphism, so there is a $z \in \pi_1(\tilde{S}, \tilde{p})$ such that $(k|\tilde{S})_{\#} z = y$. But $k \circ i = j \circ k|\tilde{S}$, hence $k_{\#} i_{\#}(z) = j_{\#}(y) = k_{\#}(x)$. Since $k_{\#}$ is always one-to-one (Appendix A), $i_{\#}(z) = x$.

CAUCHY HORIZONS

The future part of the boundary of the Cauchy development $D^+(A)$ is defined in causal terms as follows.

49. Definition. If A is an achronal set, its *future Cauchy horizon* $H^+(A)$ is

$$\overline{D^+}(A) - I^-(D^+A) = \{p \in \overline{D^+}(A) : I^+(p) \text{ does not meet } D^+(A)\}.$$

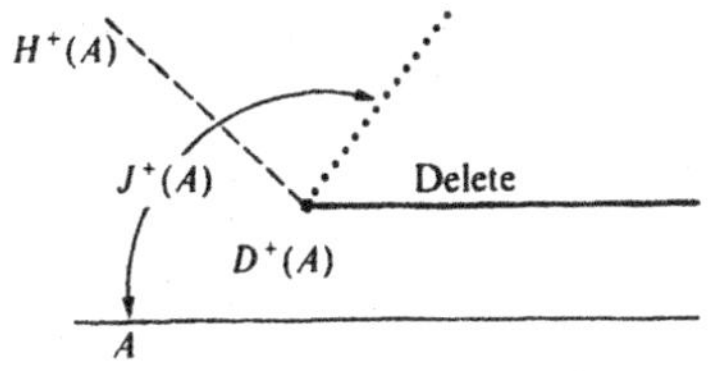

Figure 10. $H^+(A)$ separates $D^+(A)$ from the rest of $J^+(A)$.

Relativistically $H^+(A)$ marks the limit of the spacetime region controlled by A; if $H^+(A)$ is nonempty, the entire future of A cannot be predicted from A (see Figure 10).

With the *past Cauchy horizon* $H^-(A)$ defined dually, the *Cauchy horizon* of A is $H(A) = H^-(A) \cup H^+(A)$.

50. Examples of Cauchy Horizons. (See Example 36.) (1) For a restspace t constant in $\boldsymbol{R}^n_1$, both H^+ and H^- are empty.

(2) For the lower imbedding of hyperbolic space in $\boldsymbol{R}^{n+1}$, H^+ is the null cone $\Lambda^-(0)$ and H^- is empty.

(3) Referring to Figure 6, the future Cauchy horizon $H^+(S)$ is given by the null geodesics α and β, while $H^-(S)$ is empty.

(4) In Figure 8, $D^+(S) = \Lambda^-(q)$ and $D^-(S) = \Lambda^+(p)$.

It is clear from the definition that a future Cauchy horizon $H^+(A)$ is *closed*. Also $H^+(A)$ is *achronal*, since the open set $I^+(H^+A)$ is disjoint from $D^+(A)$, hence from $\overline{D^+}(A)$, hence from $H^+(A)$.

If A is not closed, $H^+(A)$ may not even be contained in $J^+(A)$ (consider an open interval in the x axis of $\boldsymbol{R}^2_1$). For A closed, $H^+(A)$ can meet A along null geodesics or at edge points (see Figure 7). However, we shall see that—as in the examples above—if A is a closed achronal spacelike hypersurface, then $H^+(A)$ is contained in $I^+(A)$, hence does not meet A.

51. Lemma. For a closed achronal set A, $\overline{D^+}(A)$ is the set T of points p such that every past-inextendible timelike curve through p meets A.

Proof. (1) $\overline{D^+}(A) \subset T$. Assume $p \in \overline{D^+}(A) - T$. Then there is a past-inextendible timelike curve α starting at p that does not meet A. Thus $p \notin A$, so p has a convex neighborhood $\mathscr{C}$ disjoint from A. Move from p in the past direction on α to a point r still in $\mathscr{C}$. Then $I^+(r, \mathscr{C})$ contains p and hence a point $q \in D^+(A)$. The geodesic segment from q to r in $\mathscr{C}$ followed by the part of α past r then constitutes a past-inextendible timelike curve that does not meet A, contradicting $q \in D^+(A)$.

(2) $\overline{D^+}(A) \supset T$. If $q \notin \overline{D^+}(A)$, pick $r \in I^-(q, M - \overline{D^+}A)$. There is a past-inextendible causal curve α starting at r that misses A. By Lemma 30, there is a past-inextendible timelike curve through q that misses A. Thus $q \notin T$. ∎

52. Lemma. If A is a closed achronal set, then $\text{bd}\, D^+(A) = A \cup H^+(A)$.

Proof. The definition of $H^+(A)$ and the fact that A is achronal give $A \cup H^+(A) \subset \text{bd}\, D^+(A)$. To prove the reverse inclusion, assume $p \in \text{bd}\, D^+(A) - A - H^+(A)$. Then $p \in \overline{D^+}(A) - A$; hence by the preceding lemma, $p \in I^+(A)$. Also $p \in \overline{D^+}(A) - H^+(A)$, so there is a point $q \in I^+(p) \cap D^+(A)$. Then $I^+(A) \cap I^-(q)$ is a neighborhood of p that, by regression, is contained in $D^+(A)$. This contradicts $p \in \text{bd}\, D^+(A)$. ■

53. Proposition. Let S be a closed acausal topological hypersurface. Then

(1) $H^+(S) = I^+(S) \cap \text{bd}\, D^+(S) = \overline{D^+}(S) - D^+(S)$. In particular, $H^+(S)$ and S are disjoint.

(2) $H^+(S)$, if nonempty, is a closed achronal topological hypersurface.

(3) Starting at each point of $H^+(S)$ there is a past-inextendible null geodesic without conjugate points that is entirely contained in $H^+(S)$. (Future-extended as far as possible in $H^+(S)$, such null geodesics are called *generators* of $H^+(S)$.)

Proof. (1) (a) Using Lemma 51 gives $H^+(S) \subset \overline{D^+}(S) \subset S \cup I^+(S)$. (b) $D^+(S)$ *does not meet* $H^+(S)$. If fact, by Lemma 43, $D(S)$ is open, so if $p \in D^+(S) \subset D(S)$, then $I^+(p)$ meets $D(S)$—but not $D^-(S)$. Thus $I^+(p)$ meets $D^+(S)$; so $p \notin H^+(S)$. (c) Since $S \subset D(S)$, (a) and (b) imply $H^+(S) \subset I^+(S)$. (d) Intersecting $I^+(S)$ with the sets equated in Lemma 52 then gives $I^+(S) \cap \text{bd}D^+(S) = H^+(S)$. (e) Then by (b), $H^+(S) \subset \overline{D^+}(S) - D^+(S)$. Suppose $p \in \overline{D^+}(S) - D^+(S)$. Then if $q \in I^+(p)$, there is a past-pointing timelike curve from q to p, and it avoids S since $p \notin S \cup I^-(S)$. Also there is a past-inextendible causal curve starting at p that avoids S. Hence $q \notin D^+(S)$, and consequently $p \in H^+(S)$.

(2) $P = D^+(S) \cup I^-(S)$ is a past set, by regression. Thus Corollary 27 asserts that $\text{bd}\, P$ is a topological hypersurface. By (1), $H^+(S) = I^+(S) \cap \text{bd}\, D^+(S)$. Since $I^-(S)$ is an open set disjoint from $I^+(S)$, it follows that $H^+(S) = I^+(S) \cap \text{bd}\, P$. Clearly this implies that $H^+(S)$ is also a topological hypersurface, and we know it is closed and achronal.

(3) If $p \in H^+(S)$, then by (1) there is a past-inextendible causal curve γ starting at p that does not meet S. By Lemma 51, γ cannot be timelike. Thus it follows from Lemma 30(2) that γ is a conjugate-free null geodesic. It remains to show that γ never leaves $H^+(S)$. Evidently γ cannot meet $D^+(S)$ or it would meet S. If $\gamma(s) \notin \overline{D^+}(S)$ for some $s > 0$, there is a past-pointing past-inextendible timelike curve β starting at $\gamma(s)$ that misses S. But then applying Lemma 30(2) to $\gamma | [0, s] + \beta$ gives a contradiction to $\gamma(0) = p \in \overline{D^+}(S)$. ■

54. Corollary. Let S be a closed acausal topological hypersurface. If every inextendible null geodesic meets S, then S is a Cauchy hypersurface.

Proof. First we prove that *S is a Cauchy hypersurface if and only if $H(S)$ is empty*. Since the boundary of a union is contained in the union of the boundaries, Lemma 52 gives bd $D(S) \subset S \cup H(S)$. But $D(S)$ is open and contains S, so bd $D(S) \subset H(S)$. The reverse inclusion is always true, hence bd $D(S) = H(S)$. Since M is connected, $H(S)$ is empty if and only if $D(S) = M$. Definition 28 and Lemma 29 show that the latter is equivalent to S being a Cauchy hypersurface.

Now we show $H(S)$ is empty. Assume there is a point p in say $H^+(S)$. By the proposition, the (past-inextendible, null geodesic) generator γ of $H^+(S)$ through p does not meet S. The inextendible extension of γ cannot meet S in the future since S is achronal and $p \in I^+(S)$: a contradiction. ∎

In view of Lemma 42 the two preceding results apply to a closed achronal spacelike hypersurface S.

HAWKING'S SINGULARITY THEOREM

The idea of the theorem is to extract the geometrical essence of the cosmological argument for the existence of a past singularity in a Robertson–Walker model of our universe: On the spacelike slice of present galactic time the galaxies are diverging (shape tensor); since gravity attracts (Ricci tensor), they have been diverging no less rapidly in the past. Thus trouble can be expected in the sufficiently distant past.

As usual we state the result in future terms: convergence producing a future singularity.

For a spacelike hypersurface S in a time-oriented n-dimensional Lorentz manifold, the convergence $\boldsymbol{k}$ (Definition 10.36) reduces to a real-valued function on S:

$$\boldsymbol{k} = \langle U, H \rangle = \frac{1}{n-1} \operatorname{trace} S_U,$$

where U is a future-pointing unit normal on S, and H is its mean normal curvature vector field. We call $\boldsymbol{k}$ here the *future convergence* of S.

There are two versions of Hawking's theorem; the first assumes more and proves more.

55A. Theorem. Suppose $\operatorname{Ric}(v, v) \geq 0$ for every timelike tangent vector to M. Let S be a spacelike future Cauchy hypersurface with future convergence $\boldsymbol{k} \geq b > 0$. Then every future-pointing timelike curve starting in S has length at most $1/b$.

Proof. If $q \in D^+(S) - S$, by Theorem 44 there is a (timelike) normal geodesic γ from S to q with $L(\gamma) = \tau(S, q)$ and no focal points before q. But Proposition 10.37 asserts that there will be a focal point along γ before q if $L(\gamma) > 1/b$. Consequently,

$$D^+(S) \subset \{p \in M : \tau(S, q) \leq 1/b\}.$$

That S is a future Cauchy hypersurface means $H^+(S)$ is empty (see Exercise 9). If a future-pointing timelike curve starting in S leaves $D^+(S)$, it must meet bd $D^+(S)$, but by Proposition 53 this would imply that $H^+(S)$ is nonempty. Hence $I^+(S) \subset D^+(S)$. In view of the inclusion above, the result follows from the definition of τ. ∎

55B. Theorem. Suppose $\text{Ric}(v, v) \geq 0$ for every timelike tangent vector to M. Let S be a compact spacelike hypersurface with future convergence $\mathbf{k} > 0$. Then M is future timelike incomplete.

Proof. We can suppose that S is connected. Let $b > 0$ be the minimum of $\mathbf{k}$ on S. Actually we show

(a) *There is an inextendible future-pointing normal geodesic starting in S that has length* $\leq 1/b$.

By Proposition 48 there is no loss of generality in assuming that S is achronal (hence acausal). Thus, as in the preceding proof, $D^+(S) \subset \{p \in M : \tau(S, q) \leq 1/b\}$. If $H^+(S)$ is empty, then Theorem 55A applies and proves the present theorem. Hence we can suppose that $H^+(S)$ is nonempty.

Assuming (a) *is false*, we shall derive a contradiction. Two preliminary facts are needed.

(b) *If* $q \in H^+(S)$, *there is a normal geodesic from S to q of length* $\tau(S, q) \leq 1/b$.

In the normal bundle of S, let B consist of all zero vectors and all future-pointing vectors v with $|v| \leq 1/b$. Evidently B is compact, since S is. There is a sequence $\{q_n\}$ in $D^+(S)$ that converges to q. For each q_n there is a geodesic with the asserted properties. Hence for each q_n there is a vector $v_n \in B$ such that $\exp(v_n) = q_n$. Since B is compact, $\{v_n\}$ converges to some $v \in B$. By continuity, $\{q_n\} \to \exp(v)$, hence the latter is q. Now $|v_n|$ converges to $|v| \leq 1/b$ and, by construction, $|v_n| = \tau(S, q_n)$. Since the function $p \to \tau(S, p)$ is lower semicontinuous, $|v| \geq \tau(S, q)$. Since (a) is assumed false, the geodesic γ_v is defined on $[0, 1]$; thus it runs from S to q and has length $|v|$. Hence $\tau(S, q) = |v|$.

(c) *The function* $p \to \tau(S, p)$ *is strictly decreasing on past-pointing generators of* $H^+(S)$ (see Proposition 53).

Let α be such a generator; in its domain suppose $s < t$. By (b) there is a past-pointing timelike geodesic σ from $\alpha(t)$ to S of length $\tau(S, \alpha(t))$. Since α is null, the causal curve $\alpha|[s, t] + \sigma$ is broken and hence can be lengthened by a small fixed-endpoint deformation. Thus

$$\tau(S, \alpha(s)) > L(\alpha|[s, t] + \sigma) = L(\sigma) = \tau(S, \alpha(t)).$$

The contradiction: Since (a) is false, the normal exponential map is defined on all of B. It follows that $H^+(S)$ is compact, since it is closed and by (b) is contained in the continuous image of the compact set B. But $p \to \tau(S, p)$ is lower semicontinuous, so its restriction to $H^+(S)$ takes on a finite minimum at some point. This contradicts (c) since there is a generator extending pastward from each point of $H^+(S)$. ∎

If S is neither Cauchy nor compact, the theorem fails utterly. For example, the lower imbedding of H^n in $\mathbf{R}_1^{n+1}$ is a closed achronal spacelike hypersurface with $\boldsymbol{k} = 1$, and $\mathbf{R}_1^{n+1}$ is flat—but complete. The "galaxies" from H^n all collide at the origin, but without harm to the containing manifold.

When time-orientation is reversed in Hawking's theorem, past convergence (future expansion) implies past singularities. Obviously the result can be refined by only requiring $\text{Ric}(\gamma', \gamma') \geq 0$ on geodesics normal to S. Then it provides a powerful augmentation of the Robertson–Walker result in Proposition 12.15.

Let us verify that Theorem 55A generalizes this proposition when the space S in the latter is complete. Let $M = I \times_f S$ and let the hypersurface in Theorem 55A be the spacelike slice $t_0 \times S$. The geodesics normal to the slice are the galaxies; hence Corollaries 12.10 and 12.12 show that $\text{Ric}(\gamma', \gamma') \geq 0$ is equivalent to $\rho + 3\not{p} \geq 0$. By Corollary 12.8(3) the totally umbilic slice has mean curvature $f'(t_0)/f(t_0)U$, hence constant convergence $\boldsymbol{k} = -f'(t_0)/f(t_0)$. Thus Hubble expansion at galactic time t_0 is equivalent to $t_0 \times S$ having past convergence $\boldsymbol{k} \leq a < 0$. If the space S is complete, the slice is a Cauchy hypersurface; hence Theorem 55A applies and gives the conclusion of Proposition 12.15.

Since our universe seems to be at least approximately Robertson–Walker, the hypotheses of Theorem 55A are not unreasonable, and this result strongly suggests that our universe is catastrophically singular in the past. This conclusion is thus freed from the specific Robertson–Walker model; in particular, the global hypothesis of exact spatial isotropy is no longer needed.

Though Theorem 55B proves less, its hypotheses are strikingly weaker. M is required to satisfy only the timelike convergence condition, which is natural both mathematically and physically (see page 340). The replacement of global hyperbolicity of M by compactness of S means, for example, that when versions 55A and 55B both apply, deleting a closed set from $I^+(S)$ destroys 55A but not 55B.

56. Example. *The conclusion of 55A need not hold under the hypothesis of 55B.* Delete from $\boldsymbol{R}_1^2$ the points $(\frac{1}{2}, x)$ with $|x| \geq 1$ and the points $(1, x)$ with $|x| \leq 1.1$. Alter the Minkowski metric on the resulting open set M as follows: Suppose $0 \leq f(x) \leq 1$, with $f(x) = 1$ if $|x| \leq 1$ and 0 if $|x| \geq 1.1$. Then let

$$g(t, x) = \begin{cases} 1 - t & \text{if } t \leq \frac{1}{2}, \\ 1 - tf(x) & \text{if } \frac{1}{2} \leq t \leq 1, \\ 1 & \text{if } 1 \leq t. \end{cases}$$

With line element $-dt^2 + g^2\,dx^2$, M is Minkowskian above the deleted sets (Figure 11) and "conical" below. Then ∂_t is a future-pointing unit normal on the (spacelike) x axis. There $S_{\partial_t}(\partial_x) = -D_{\partial_t}\partial_x = \partial_x$, so the x axis has constant future convergence $\boldsymbol{k} = 1$. The metric is independent of x for $|x| \geq 1.1$. Identifying the lines $x = \pm 2$ turns M into a topological cylinder and the x axis into a circle S. Since M is flat, the hypotheses of Theorem 55B are satisfied. The future-pointing geodesics normal to S are t-parameter curves; none is longer than 1. But, for example, the timelike curve $(t, x) = (2s, \frac{3}{4} + s)$ has infinite length for $s \geq 0$. Thus all galaxies end, but a well-directed spaceship can go on forever.

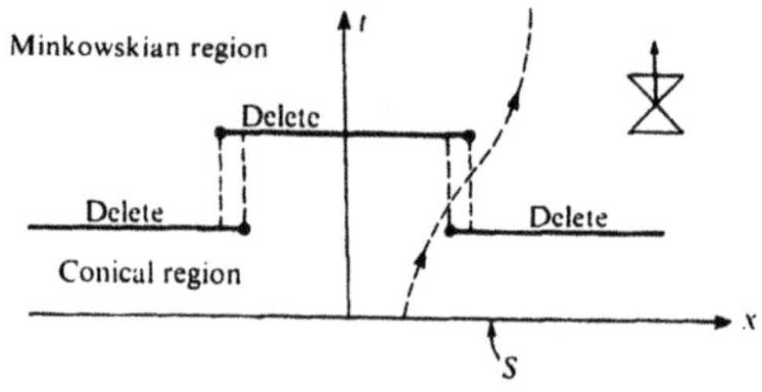

Figure 11

PENROSE'S SINGULARITY THEOREM

This result aims at recognizing in the geometry of a collapsing star, general conditions that imply the existence of future "black hole" singularities.

In a Schwarzschild black hole B, the most remarkable feature is that from a t-constant sphere $S^2(r)$ even the "outgoing" light rays are actually going inward: $dr/ds < 0$. These null geodesics have a better chance to escape than any particle starting inside $S^2(r)$, hence their failure to do so looks like a good warning of the ensuing singularity.

In the general case no such function r is available, but Example 58 will show that the initial condition $dr/ds < 0$ on these null geodesics to $S^2(r)$ is equivalent to positive convergence. This condition makes sense for any spacelike submanifold P of an arbitrary M, and under the appropriate Ricci curvature condition, the initial collapse will continue. The problem then (as analogously in the preceding section) is to show that this concatenation of null geodesics actually involves a singularity in M.

Let P be a spacelike submanifold of M with codimension ≥ 2. If H is the mean curvature vector field of P, then linear algebra in each (Lorentz) normal space $T_p(P)^\perp$ shows that the following are equivalent:

(1) $k(v) = \langle H, v\rangle > 0$ for all future-pointing null vectors v normal to P.

(2) $k(w) = \langle H, w\rangle > 0$ for all future-pointing causal vectors w normal to P.

(3) H is past-pointing timelike.

57. Definition. A spacelike submanifold of M is *future-converging* provided its mean curvature vector field H is past-pointing timelike.

The definition makes sense even for hypersurfaces. For example, the hypersurface S in Theorems 55A and 55B is future-converging. In relativistic contexts, a future-converging surface is called a *trapped surface.*

58. Example. *Trapped Surfaces in a Schwarzschild Black Hole B.* By Lemma 13.3 and Proposition 13.4(3) the sphere $S^2(r)$ in restspace t constant has mean curvature vector field $H = -(1/r)\text{grad } r$. Lemma 13.29 shows that grad r is timelike future-pointing on B, hence H is timelike past-pointing. If v is a future-pointing causal vector normal to $S^2(r)$, then then $k(v) = \langle H, v\rangle = -v(r)/r$, so decrease of r in the v direction is equivalent to positive convergence.

The term "trapped" has another—purely causal—meaning, which we now discuss.

For a subset A of M let $E^+(A) = J^+(A) - I^+(A)$. It is easy to see that $E^+(A)$ is achronal and that $A \subset E^+(A)$ if and only if A is achronal. By Corollary 5, $E^+(A)$ is generated by conjugate-free null geodesics: if $q \in E^+(A)$, there is a null geodesic in $E^+(P)$ from A to q that has no conjugate points before q.

59. Definition. A closed achonal subset A of M is *future-trapped* provided $E^+(A)$ is compact.

Dually, *past-trapped* means $E^-(A)$ is compact. In either case A itself must be compact. For example, in the cylinder $R^1_1 \times S^1$, each individual point is both future- and past-trapped (see Figure 6).

A main step in Penrose's proof of Theorem 61 is to show that under reasonable hypotheses a trapped surface (differential geometric concept) is a trapped subset (causal concept).

60. Proposition. Suppose

(1) $\mathrm{Ric}(v, v) \geq 0$ for all null tangent vectors v to M.
(2) M is future null complete.

If P is a compact achronal spacelike $(n-2)$-submanifold of M that is future-converging, then P is future-trapped.

Proof. Globally, P need not admit a (continuous) normal null vector field, but we can choose at each $p \in P$ a pair of future-pointing null vectors that locally fall into two independent (smooth) normal vector fields. The subspace $\tilde{P} \subset N(P)$ consisting of these null normals is a double covering of P, hence in particular is compact.

By Proposition 10.43, for each $v \in \tilde{P}$ there is a focal point of P along the geodesic $\gamma_v | [0, 1/\boldsymbol{k}(v)]$, where $\boldsymbol{k}(v) = \langle H, v \rangle > 0$. Since $\boldsymbol{k}$ is continuous on compact $\tilde{P}$, there is a number $b > 0$ such that for every $v \in P$ there is a focal point of P along $\gamma_v | [0, b)$.

If $q \in E^+(P)$, then as noted above there is a null geodesic γ from P to q. Since $q \notin I^+(P)$, Theorem 10.51 shows that γ is normal to P and has no focal points of P before q. Thus γ is a reparametrization of some γ_v, $v \in P$, and we conclude that

$$E^+(P) \subset \exp(K), \qquad \text{where} \quad K = \{sv \in N(P) : v \in \tilde{P} \text{ and } 0 \leq s \leq b\}.$$

Since $\tilde{P}$ is compact, K is compact, hence $\exp(K)$ is compact. To see that $E^+(P)$ is compact, let $\{q_n\}$ be a sequence in $E^+(P)$. Some subsequence $\{q_m\}$ converges to a point $q \in \exp(K) \subset J^+(P)$. But q cannot be in $I^-(P)$ since no q_n is. Thus $q \in E^+(P)$. ∎

We state the main result in positive terms, followed by two versions as singularity theorems.

61. Theorem (Penrose). Suppose

(1) $\mathrm{Ric}(v, v) \geq 0$ for all null tangent vectors to M.
(2) M has a Cauchy hypersurface S.
(3) P is a compact achronal spacelike $(n-2)$-submanifold in M that is future-converging.
(4) M is future null complete.

Then $E^+(P)$ is a Cauchy hypersurface in M.

The theorem is almost illustrated by taking P to be a point of $\boldsymbol{R}_1^1 \times S^1$ (compare Figure 6). The only hypothesis to fail is future-convergence, but P is nevertheless future-trapped and $E^+(P)$ is indeed a Cauchy hypersurface.

Proof. Since M has a Cauchy hypersurface, it is globally hyperbolic. Then Lemma 22 implies that the sets $J^-(p)$ and $J^+(p)$ are closed for all $p \in M$. Since P is compact, Exercise 4 shows that $J^+(P)$ is a closed set. By Lemma 6, int $J^+(P) = I^+(P)$, hence $E^+(P) = \text{bd}\, J^+(P)$. As the boundary of a future set, $E^+(P)$ is a topological manifold, and by the preceding lemma it is compact.

For the Cauchy hypersurface S let $\rho: E^+(P) \to S$ be the restriction to $E^+(S)$ of a retraction as in Proposition 31. Thus ρ is continuous, and since $E^+(P)$ is achronal, the uniqueness of integral curves implies that ρ is one-to-one. By the invariance of domain, ρ is a homeomorphism of $E^+(P)$ onto an open subset of S. Since $E^+(P)$ is compact, $\rho E^+(P)$ is compact, hence closed in S. Since S is connected, $\rho E^+(P) = S$; that is, $\rho: E^+(P) \to S$ is a homeomorphism. Hence

Corollary A. *If* (1), (2), and (3) *hold, and the Cauchy hypersurface is noncompact, then M is future null incomplete.*

Continuing the proof of the theorem: Since $E^+(P)$ is achronal, it remains to show that every inextendible timelike curve β meets $E^+(P)$. The strong causality condition holds, hence by Exercise 11, β is an integral curve of some timelike vector field X. We saw above that the retraction produced by X gives a homeomorphism $E^+(P) \approx S$. Thus since β meets S, it also meets $E^+(P)$. ∎

Corollary B. *If* (1), (2), and (3) *hold, and there is an inextendible causal curve in M that does not meet $E^+(P)$, then M is future null incomplete.*

In fact, by Lemma 29 the curve hypothesis means that $E^+(P)$ is not a Cauchy hypersurface.

Penrose's result, the first singularity theorem (1965), improved theoretical prospects for the existence of black holes, since the "trapped surface" sufficient conditions for singularities are free of the symmetry of particular models such as Schwarzschild's.

Exercises

M denotes a connected time-oriented Lorentz manifold.

1. Let λ be a quasi-limit of $\{\alpha_n\}$. If either λ is null or every α_n is geodesic, show that every neighborhood of a point on λ meets infinitely many α_ns.

2. Give an example of a sequence $\{\alpha_n\}$ of causal curves in $\boldsymbol{R}_1^2$ with infinite limit sequence $\{p_i\}$ such that there does not exist one single subsequence $\{\alpha_m\}$ such that every p_i is a limit of points on the curves α_m.

3. Prove that the set of points in M at which the chronology [causality] condition fails is a (possibly empty) disjoint union of sets of the form $I^+(p) \cap I^-(p)$ $[J^+(p) \cap J^-(p)]$.

4. Suppose that for every $p \in M$ the sets $J^-(p)$ and $J^+(p)$ are closed. Prove that if K is compact, then $J^+(K)$ is closed.

5. Let $\alpha\colon [0, b) \to M$ be a future-pointing causal curve starting at p. If α is future-extendible, with endpoint q, show that: (a) The continuous extension $\tilde{\alpha}$ of α to $[0, b]$ need not be smooth. (b) If α is merely piecewise smooth, then $\tilde{\alpha}$ need not be piecewise smooth. (c) $p \leq q$. (d) Given any neighborhood $\mathscr{U}$ of $\tilde{\alpha}[0, b]$ there exists in $\mathscr{U}$ a (smooth) causal curve segment from p to q. (Hint: For (a) and (b), zigzag; for (c), use Lemma 2.)

6. The Schwarzschild exterior N and black hole B are globally hyperbolic.

7. If V and W are linearly independent vector fields on a smooth manifold P, show that there is a Lorentz metric on P such that (a) V is timelike and W is null; (b) both V and W are null.

8. Give examples of (a) a Lorentz metric on $\boldsymbol{R}^3$ for which there are many closed timelike curves and many closed null curves (compare Lemma 34); (b) a Lorentz metric on $S^1 \times \boldsymbol{R}^1$ for which some $J^+(p)$ is an open set with compact closure. (Hint: Use the preceding exercise.)

9. Let A be a closed achronal set. Prove : (a) $H^+(A)$ is empty if and only if $J^+(A) \subset D^+(A)$—and in this case A is a closed topological hypersurface (called a *future Cauchy hypersurface*). (b) A is a Cauchy hypersurface if and only if it is both a future and past Cauchy hypersurface.

10. *Causality neighborhoods.* Prove: (a) The strong causality condition holds at $p \in M$ if and only if every neighborhood of p contains a neighborhood $\mathscr{N} = I^+(x) \cap I^-(y)$. (b) If $\mathscr{N}$ is such a neighborhood, and $\bar{\mathscr{N}}$ is compact and contained in a convex open set, then any future-inextendible causal curve starting in $\mathscr{N}$ leaves it and after its first departure never returns.

11. Let $\alpha\colon I \to M$ be an inextendible timelike curve. If the strong causality condition holds on M, prove: (a) $\alpha(I)$ is a closed submanifold of M. (b) α is an integral curve of a timelike vector field on M. (Hint: Use Exercise 3.12, Lemma 5.32, and a partition of unity.)

12. Give examples showing that neither converse in Lemma 45 is true.

13. Let A be a subset of M. Prove: (a) $E^+(A)$ has the properties stated before Definition 59. (b) The *event horizon* $\mathscr{E} = \operatorname{bd} J^+(A)$ is a closed topological hypersurface. (c) $E^+(A) = \mathscr{E} \cap J^+(A)$. (d) Describe $\mathscr{E} = \operatorname{bd} J^+(\alpha)$ if α

is (i) a timelike geodesic in $\boldsymbol{R}_1^n$, (ii) the accelerating observer given on page 181, (iii) a timelike geodesic in S_1^n.

14. Prove: (a) If N is a compact simply connected four-dimensional manifold, its Euler number $\chi(N)$ is at least 2. (Hint: Since N is orientable, Poincaré duality [V] can be used.) (b) The simply connected covering manifold of a four-dimensional Lorentz manifold is noncompact.

15. In view of its relation to Theorem 55A (see page 433), prove the analogue of Proposition 12.15 with $\rho + 3p > 0$ replaced by $\rho + 3p \geq 0$. (There can be a static future.)

Appendixes

A FUNDAMENTAL GROUPS AND COVERING MANIFOLDS†

Both topics are topological, however we describe them in terms of manifolds M. This is sufficient for our purposes and simplifies matters somewhat.

The fundamental group of M is a group whose elements are certain classes of continuous curves in M. Let I be the closed unit interval $[0, 1]$ in $\boldsymbol{R}^1$. A *path from p to q* in M is a continuous map $\alpha: I \to M$ such that $\alpha(0) = p$ and $\alpha(1) = q$. Let $P(p, q)$ be the set of all such paths.

1. Definition. If $\alpha, \beta \in P(p, q)$, a *fixed-endpoint homotopy from* α *to* β is a continuous map $H: I \times I \to M$ such that for all $s, t \in I$

$$H(t, 0) = \alpha(t), \qquad H(0, s) = p,$$

$$H(t, 1) = \beta(t), \qquad H(1, s) = q.$$

Defining $\alpha_s(t) = H(t, s)$ shows that in effect H is a one-parameter family of paths $\alpha_s \in P(p, q)$, varying continuously from $\alpha_0 = \alpha$ to $\alpha_1 = \beta$. If such a homotopy exists, α and β are *fixed-endpoint homotopic*, written $\alpha \simeq \beta$.

2. Lemma. Fixed-endpoint homotopy $\simeq$ is an equivalence relation on $P(p, q)$.

The equivalence class containing $\alpha \in P(p, q)$ is denoted by $[\alpha]$ and called the *fixed-endpoint homotopy class of* α.

In favorable cases two paths combine to give a single path.

† See [Ma] and [ST].

3. Definition. If $\alpha \in P(p, q)$ and $\beta \in P(q, r)$, let

$$(\alpha * \beta)(t) = \begin{cases} \alpha(2t) & \text{for } 0 \leq t \leq \frac{1}{2}, \\ \beta(2t - 1) & \text{for } \frac{1}{2} \leq t \leq 1. \end{cases}$$

Then $\alpha * \beta$ is a path from p to r called the *path product* of α and β.

If $\alpha \in P(p, q)$, the reverse path $\bar{\alpha} \in P(q, p)$ is defined by $(\bar{\alpha})(t) = \alpha(1 - t)$. These two operations on paths respect fixed-endpoint homotopy, hence yield well-defined operations on homotopy classes: $[\alpha][\beta] = [\alpha * \beta]$ and $[\bar{\alpha}]$.

The set of all such homotopy classes in M fails to be a group solely because path products are defined only when the paths of one class end at the start of those of the other class. This difficulty is eliminated by considering only loops at some point $p \in M$. A *loop* is an element of $P(p, p)$, that is, a path starting and ending at p.

4. Proposition (Poincaré). If $p \in M$, let $\pi_1(M, p)$ be the set of all fixed-endpoint homotopy classes in $P(p, p)$. The multiplication $[\alpha][\beta] = [\alpha * \beta]$ makes $\pi_1(M, p)$ a group, called *the fundamental group of M at p.*

Proof. Only associativity, $\alpha * (\beta * \gamma) \simeq (\alpha * \beta) * \gamma$, demands care in constructing the required homotopy.

It is easy to see that the identity element of $\pi_1(M, p)$ is the homotopy class consisting of all loops at p that are fixed-endpoint homotopic to the constant loop e_p at p. Then the group inverse of $[\alpha]$ is $[\bar{\alpha}]$. ∎

The general idea is that if there are "holes" in M, the loops surrounding a hole cannot be shrunk back to the base point p. Homotopic loops surround the same holes, and thus $\pi_1(M, p)$ gives an algebraic description of them.

We list some easily verified properties of fundamental groups.

(a) *If M is connected, then the groups $\pi_1(M, p)$ for all p are isomorphic.* In fact, if γ is a path from p to q, then $[\beta] \to [\gamma * \beta * \bar{\gamma}]$ is a well-defined isomorphism $\pi_1(M, q) \approx \pi_1(M, p)$.

Thus we can speak of *the* fundamental group of a connected M, writing merely $\pi_1(M)$ when the particular base point p is unimportant.

5. Definition. M is *simply connected* provided M is connected and its fundamental group is trivial, that is, reduces to the identity element.

(Thus every loop in M is homotopic to a constant.) For example, $\boldsymbol{R}^n$ is simply connected, since any loop α at 0 is fixed-endpoint homotopic to e_0 under the homotopy $H(t, s) = s\alpha(t)$. Also S^n is simply connected if $n \geq 2$, but Chapter 7 shows that $\pi_1(S^1)$ is infinite cyclic.

(b) *For a connected product manifold*, $\pi_1(M \times N) \approx \pi_1(M) \times \pi_1(N)$. In fact, any loop in $M \times N$ can be written as $t \to (\alpha(t), \beta(t))$. Then $[(\alpha, \beta)] \leftrightarrow ([\alpha], [\beta])$ is the required isomorphism.

(c) *A continuous map* $\phi: M \to N$ *induces a homomorphism* $\phi_{\#}: \pi_1(M, p) \to \pi_1(N, \phi p)$ *given by* $\phi_{\#}[\alpha] = [\phi \circ \alpha]$. Thus $\phi_{\#}$ provides an algebraic description of the map ϕ.

In dealing with a manifold we prefer to use paths that are at least piecewise smooth so that manifold machinery can be applied to them.

6. Lemma. If α is a path in M from p to q, there is a piecewise smooth path β that is fixed-endpoint homotopic to β.

This can be proved by a direct argument in coordinate neighborhoods. If M is a semi-Riemannian manifold, we can take the neighborhoods to be convex and construct β as a broken geodesic.

Now we turn to a notion that seems quite different but turns out to be closely related to fundamental groups.

7. Definition. A smooth map $k: \tilde{M} \to M$ onto M is a *covering map* provided each point $p \in M$ has a connected neighborhood $\mathscr{U}$ that is *evenly covered* by k; that is, k maps each component of $k^{-1}(\mathscr{U})$ diffeomorphically onto $\mathscr{U}$.

For example, the exponential map $t \to (\cos t, \sin t)$ from R^1 to S^1 is a covering map, but its restriction to an interval $J \neq R^1$ is not.

8. Lemma. If $k: \tilde{M} \to M$ is a covering and M is connected, then the number of points in $k^{-1}(p)$—an integer or ∞—is the same for all $p \in M$. (This number is called the *multiplicity* of the covering.)

Proof. For each $m \leq \infty$, let $\mathcal{O}_m$ consist of all $p \in M$ such that $k^{-1}(p)$ has exactly m points. The even covering condition shows each $\mathcal{O}_m$ is open. Hence $M = \mathcal{O}_m$ for some one m. ∎

For example, considering S^1 as the unit circle in the complex plane, for each $n > 0$ the map $z \to z^n$ is a covering map $S^1 \to S^1$ of multiplicity n.

In general, given maps $\pi: E \to M$ and $\phi: P \to M$, a *lift of ϕ through π* is a map $\tilde{\phi}: P \to E$ such that $\pi \circ \tilde{\phi} = \phi$.

9. Lemma. Let $k: \tilde{M} \to M$ be a covering. Let $\alpha: J \to M$ be a continuous [smooth] curve, and let q be a point of $\tilde{M}$ such that $k(q) = \alpha(0)$. Then there is a unique continuous [smooth] lift $\tilde{\alpha}: J \to \tilde{M}$ of α through k such that $\tilde{\alpha}(0) = q$.

Proof. Decompose J into subintervals so that each subcurve α_i lies in a (connected) evenly covered open set $\mathscr{U}_i$ of M. There is no choice as to the lift of α_1: If $\mathscr{V}$ is the component of $k^{-1}(\mathscr{U})$ containing q, then $\tilde{\alpha}_1$ must be $(k|\mathscr{V})^{-1} \circ \alpha_1$. Continue by induction, replacing q by the appropriate endpoint of $\tilde{\alpha}_i$. ∎

In short, *paths can be lifted uniquely to any level.* A similar proof establishes the two-dimensional analogue: If H is a homotopy in M and $k(q) = (0, 0)$, then H has a unique lift $\tilde{H}$ through k such that $\tilde{H}(0, 0) = q$. The following consequence links covering maps to fixed-endpoint homotopy:

10. Corollary. Let $k: \tilde{M} \to M$ be a covering, and let α and β be fixed-endpoint homotopic paths in M. If $\tilde{\alpha}$ and $\tilde{\beta}$ are lifts of α and β through k such that $\tilde{\alpha}(0) = \tilde{\beta}(0)$, then $\tilde{\alpha}$ and $\tilde{\beta}$ are fixed-endpoint homotopic. In particular, $\tilde{\alpha}(1) = \tilde{\beta}(1)$.

It follows, for example, that the homomorphism $k_{\#}$ induced by k is one-to-one.

11. Proposition. Let $k: \tilde{M} \to M$ be a covering map and $\phi: P \to M$ a smooth map. Let $p_0 \in P$ and $q_0 \in \tilde{M}$ be such that $\phi(p_0) = k(q_0)$. Then (1) if P is connected, there is at most one lift $\tilde{\phi}$ of ϕ through k such that $\tilde{\phi}(p_0) = q_0$; (2) if P is simply connected, such a lift exists.

Assertion (1) follows from the uniqueness in Lemma 9. In (2), if $p \in P$, let α be a path from p_0 to p. Let β be the lift of $\phi \circ \alpha$ starting at q_0. Then $\tilde{\phi}(p) = \beta(1)$ can be shown to provide the required map $\tilde{\phi}$.

Topological or manifold properties attributed to a covering $k: \tilde{M} \to M$ refer to the *covering manifold* $\tilde{M}$. Thus a simply connected covering is one for which $\tilde{M}$ is simply connected (hence M is connected).

12. Theorem. Every connected manifold has a simply connected covering.

Using Proposition 11, it is easy to show that any two simply connected coverings $k_i: M_i \to M$ of the same manifold are *equivalent*; that is, there is a diffeomorphism $\psi: M_1 \to M_2$ such that $k_2 \circ \psi = k_1$. Thus we can speak of *the* simply connected covering of M (also called the *universal covering* of M).

13. Corollary. If M is connected, then, given any subgroup H of $\pi_1(M, p)$, there is a connected covering $k: \tilde{M} \to M$ and a point $\tilde{p} \in \tilde{M}$ such that $k_{\#}(\pi_1(\tilde{M}, \tilde{p})) = H$.

A covering $k: \tilde{M} \to M$ is *trivial* if each component of M is evenly covered by k. Thus if M is connected, k is a diffeomorphism of each component C of $\tilde{M}$ onto M, so $\lambda = (k|C)^{-1}$ is a global cross section of k.

14. Corollary. Every covering of a simply connected manifold is trivial.

The proof is a straightforward application of Proposition 11.

B LIE GROUPS†

A *Lie group* G is a smooth manifold that is also a group with smooth group operations; that is, the maps

$$\mu: G \times G \to G \quad \text{sending } (a, b) \text{ to } ab$$

and

$$\zeta: G \to G \quad \text{sending } a \text{ to } a^{-1}$$

are both smooth. We always assume G is second countable. The identity element of G is denoted by e.

1. Example. The Full Linear Group $GL(n, \boldsymbol{R})$. The set $\mathrm{gl}(n, \boldsymbol{R})$ of all $n \times n$ real matrices is in a natural way a real vector space, hence a manifold. Stringing out the entries of each $x \in \mathrm{gl}(n, \boldsymbol{R})$ in some fixed order would give a linear isomorphism (hence diffeomorphism, hence coordinate system) from $\mathrm{gl}(n, \boldsymbol{R})$ to $\boldsymbol{R}^{n^2}$. The set $GL(n, \boldsymbol{R}) = \{g : \det g \neq 0\}$ of all invertible matrices in $\mathrm{gl}(n, \boldsymbol{R})$ is evidently a group under matrix multiplication. The formula for the determinant of a matrix shows that the determinant function $\det: \mathrm{gl}(n, \boldsymbol{R}) \to \boldsymbol{R}$ is smooth; thus $GL(n, \boldsymbol{R})$ is an open submanifold of $\mathrm{gl}(n, \boldsymbol{R})$. The formulas for matrix multiplication and inverses then show that for $GL(n, \boldsymbol{R})$, the maps μ and ζ above are smooth. In this way $GL(n, \boldsymbol{R})$ becomes a Lie group.

A Lie group H is a *Lie subgroup* of a Lie group G provided H is both an abstract subgroup and an immersed submanifold of G. This notion is subtle

† See [Ch], [H], [W].

because H need not have the induced topology. For our purposes something much simpler suffices.

2. Definition. A *closed subgroup* H of a Lie group G is an abstract subgroup that is a closed set of G.

3. Theorem. If H is a closed subgroup of a Lie group G, then H is a submanifold of G and hence a Lie subgroup of G.

In particular, a closed subgroup has the induced topology.

4. Examples of Closed Subgroups. (1) The kernel

$$K = \{a \in G: \phi(a) = e\}$$

of a smooth homomorphism $\phi: G \to H$ is a closed subgroup of G.

(2) The determinant function $\det: GL(n, \boldsymbol{R}) \to \boldsymbol{R} - 0$ is a smooth homomorphism. Its kernel $SL(n, \boldsymbol{R}) = \{a: \det a = 1\}$ is thus a closed subgroup of $GL(n, \boldsymbol{R})$, called the *special linear group.*

(3) For a Lie group G, its component G_0 containing the identity element e is a closed (and open) subgroup.

In all three cases above, the subgroup is *normal*, that is, invariant under $a \to gag^{-1}$ for all $g \in G$.

LIE ALGEBRAS

5. Definition. A *Lie algebra* over $\boldsymbol{R}$ is a real vector space $\mathfrak{g}$ furnished with bilinear function $[\ ,\ ]: \mathfrak{g} \times \mathfrak{g} \to \mathfrak{g}$, called its *bracket operation*, such that for all $X, Y, Z \in \mathfrak{g}$,

(1) $[X, Y] = -[Y, X]$ (skew-symmetry),

(2) $[[X, Y], Z] + [[Y, Z], X] + [[Z, X], Y] = 0$ (Jacobi identity).

For example, $\mathfrak{gl}(n, \boldsymbol{R})$ will always be made a Lie algebra by defining $[x, y] = xy - yx$, where xy is matrix multiplication. Lie algebras are assumed to be finite-dimensional unless the contrary is mentioned.

We shall now see that there is a Lie algebra canonically associated with each Lie group. The essence of Lie theory is to study the groups in terms of their algebras.

If a is an element of a Lie group G, define $L_a(g) = ag$ and $R_a(g) = ga$ for all $g \in G$. Then $L_a: G \to G$ is a smooth map, in fact a diffeomorphism since $L_{a^{-1}}$ is its inverse (similarly for R_a).

By convention, *left-multiplication* L_a is the standard way to get around in a Lie group. In particular, any $a \in G$ can be moved to e by $L_{a^{-1}}$.

6. Definition. A vector field X on a Lie group G is *left-invariant* provided $dL_a(X) = X$ for all $a \in G$.

Explicitly, $dL_a(X_g) = X_{ag}$ for all $a, g \in G$. Thus left-multiplication merely permutes the tangent vectors constituting X. It is not hard to show that *a left-invariant vector field is smooth.*

Let $\mathfrak{g}$ be the set of all left-invariant vector fields on a Lie group G. The usual addition of vector fields and scalar multiplication by real numbers make $\mathfrak{g}$ a vector space. By Lemma 1.22, $\mathfrak{g}$ is closed under the bracket operation, hence the bracket properties in Lemma 1.18 hold on $\mathfrak{g}$. Thus $\mathfrak{g}$ is a Lie algebra, called *the Lie algebra of* G. Furthermore, $\mathfrak{g}$ has (finite) dimension $n = \dim G$, since

7. Lemma. The function $\mathfrak{g} \to T_e(G)$ sending each $X \in \mathfrak{g}$ to its value $X_e \in T_e(G)$ is a linear isomorphism.

Proof. The function is obviously linear, and it is one-to-one since, if $X_e = 0$, then $X_a = dL_a(X_e) = 0$ for all a. To prove *onto*: If $x \in T_e(G)$, define $X_a = dL_a(x)$ for all $a \in G$. Then X is left-invariant and $X_e = x$. ∎

This isomorphism is so natural that, where convenient, we can neglect it and think of the Lie algebra of G as $T_e(G)$ with induced bracket operation.

A *Lie algebra homomorphism* is a linear transformation of Lie algebras that preserves brackets. A *subalgebra* of a Lie algebra $\mathfrak{g}$ is a vector subspace $\mathfrak{h}$ that is closed under brackets (hence $\mathfrak{h}$ is also a Lie algebra).

We show now that the Lie algebra of $GL(n, \boldsymbol{R})$ is (canonically isomorphic to) the matrix Lie algebra $\mathrm{gl}(n, \boldsymbol{R})$. Let u_{ij} be the real-valued function on $\mathrm{gl}(n, \boldsymbol{R})$ such that $u_{ij}(a)$ is the ij entry a_{ij} of each matrix a. Then

$$\{u_{ij}: 1 \le i, j \le n\}$$

is a coordinate system on $\mathrm{gl}(n, \boldsymbol{R})$, hence on its open submanifold $GL(n, \boldsymbol{R})$.

8. Lemma. The map $X \to (X_e(u_{ij}))$ is a Lie algebra isomorphism from the Lie algebra of $GL(n, \boldsymbol{R})$ to $\mathrm{gl}(n, \boldsymbol{R})$.

Proof. The map is a composition of canonical isomorphisms:

$$\text{Lie algebra of } GL(n, \boldsymbol{R}) \overset{1}{\leftrightarrow} T_e(GL(n, \boldsymbol{R})) \overset{2}{\leftrightarrow} T_e(\mathrm{gl}(n, \boldsymbol{R}) \overset{3}{\leftrightarrow} \mathrm{gl}(n, \boldsymbol{R}),$$

where (1) is from Lemma 7, (2) is the usual open submanifold identification, and (3) is the canonical isomorphism for vector spaces as manifolds. That brackets are preserved is a straightforward coordinate computation. ∎

The preceding results extend to complex matrices as follows. Let $\mathfrak{gl}(n, \boldsymbol{C})$ be the set of all $n \times n$ matrices with complex entries. Although $\mathfrak{gl}(n, \boldsymbol{C})$ is in a natural way an n-dimensional vector space over $\boldsymbol{C}$, we always restrict to real scalars. Thus $\mathfrak{gl}(n, \boldsymbol{C})$ is a real vector space of dimension $2n^2$. The matrix bracket $[x, y] = xy - yx$ makes it a real Lie algebra. For $a \in \mathfrak{gl}(n, \boldsymbol{C})$ let

$$u_{ij}(a) = \operatorname{Re} a_{ij}, \qquad v_{ij}(a) = \operatorname{Im} a_{ij}.$$

Then $\{u_{ij}, v_{ij}: 1 \leq i, j \leq n\}$ is a natural coordinate system on $\mathfrak{gl}(n, \boldsymbol{C})$.

The set $GL(n, \boldsymbol{C}) = \{a: \det a \neq 0\}$ of invertible matrices in $\mathfrak{gl}(n, \boldsymbol{C})$ is, as before, a group under matrix multiplication and an open submanifold of $\mathfrak{gl}(n, \boldsymbol{C}) \approx \boldsymbol{R}^{2n^2}$. By the usual formulas, these structures make $GL(n, \boldsymbol{C})$ a Lie group, the *complex full linear group.* Computations as for Lemma 8 show that $\mathfrak{gl}(n, \boldsymbol{C})$ is (canonically isomorphic to) the Lie algebra of $GL(n, \boldsymbol{C})$.

THE LIE EXPONENTIAL MAP

9. Definition. A *one-parameter subgroup* in a Lie group G is a smooth homomorphism α from $\boldsymbol{R}$ (under addition) to G.

Thus $\alpha: \boldsymbol{R} \to G$ is a curve such that $\alpha(s + t) = \alpha(s)\alpha(t)$ for all s, t. Hence $\alpha(0) = e$, $\alpha(-t) = \alpha(t)^{-1}$, and $\alpha(s)\alpha(t) = \alpha(t)\alpha(s)$ for all s, t. For example,

$$\alpha(t) = \begin{pmatrix} \cos t & \sin t & 0 \\ -\sin t & \cos t & 0 \\ 0 & 0 & e^t \end{pmatrix}$$

is a one-parameter subgroup of $GL(3, \boldsymbol{R})$.

10. Proposition. The one-parameter subgroups of G are exactly the maximal integral curves, starting at e, of the elements of its Lie algebra $\mathfrak{g}$.

The proof is not difficult.

11. Definition. Let $\mathfrak{g}$ be the Lie algebra of G. The *Lie exponential map* $\exp: \mathfrak{g} \to G$ sends X to $\alpha_X(1)$, where α_X is the one-parameter subgroup of $X \in \mathfrak{g}$ (as above).

Like its geometric analogue, exp carries lines through the origin 0 of $\mathfrak{g}$ to one-parameter subgroups (geodesics), and its differential map at 0 is the canonical isomorphism $T_0(\mathfrak{g}) \approx \mathfrak{g} \approx T_e(G)$. Hence, by the inverse function theorem,

12. Lemma. Some neighborhood of 0 in $\mathfrak{g}$ is mapped diffeomorphically by exp onto a neighborhood of e in G.

Its interpretation as $T_e(G)$ shows that the Lie algebra $\mathfrak{g}$ is determined by any arbitrarily small neighborhood of e in G. Conversely, $\mathfrak{g}$ has direct influence on the set $\exp(\mathfrak{g})$, which necessarily lies in the identity component G_0 of G (but need not fill it). The preceding lemma and the one that follows show that the influence of $\mathfrak{g}$ extends to all of G_0.

13. Lemma. Given any neighborhood $\mathscr{U}$ of e, every element of G_0 can be expressed as a finite product of elements of $\mathscr{U}$.

Let H be a Lie subgroup of G, and let $\mathfrak{h}$ be the Lie algebra of H. Each $X \in \mathfrak{h}$ has a unique extension to $\tilde{X} \in \mathfrak{g}$, namely $\tilde{X}_a = dL_a(X_e)$ for all $a \in G$. The function $X \to \tilde{X}$ is linear, one-to-one, and bracket-preserving. It is customary to ignore this natural map and treat $\mathfrak{h}$ as a subalgebra of $\mathfrak{g}$.

14. Lemma. Let H be a Lie subgroup of G. Then $\mathfrak{h}$ is the set of all $X \in \mathfrak{g}$ such that $\exp(tX) \in H$ for $|t|$ sufficiently small.

The reason for the term "exponential map" is clear in the case of the full linear group, as follows.

15. Example. The Lie exponential map $\exp : \mathfrak{gl}(n, \boldsymbol{C}) \to GL(n, \boldsymbol{C})$ sends x to e^x, where the latter is given by the convergent series

$$e^x = \mathrm{id} + x + \cdots + x^n/n! + \cdots.$$

The proof consists in showing that the one-parameter subgroup of $x \in \mathfrak{gl}(n, \boldsymbol{C})$ is $t \to e^{tx}$.

For instance, the one-parameter subgroup α given after Definition 9 is $t \to e^{tx}$, where

$$x = \begin{pmatrix} 0 & 1 & 0 \\ -1 & 0 & 0 \\ 0 & 0 & 1 \end{pmatrix}.$$

THE CLASSICAL GROUPS

If G is a Lie subgroup of $GL(n, \boldsymbol{C})$, then its Lie algebra is a subalgebra of $\mathfrak{gl}(n, \boldsymbol{C})$. We shall compute the Lie algebras for some important closed subgroups of $GL(n, \boldsymbol{C})$ that are defined using the following linear functions on $\mathfrak{gl}(n, \boldsymbol{C})$:

(1) The *trace* of $x \in \mathrm{gl}(n, \boldsymbol{C})$ is $\sum x_{ii}$, hence $\mathrm{trace}(xy) = \mathrm{trace}(yx)$.
(2) The *transpose* ${}^t x$ of x has $({}^t x)_{ij} = x_{ji}$, hence ${}^t(xy) = {}^t y {}^t x$.
(3) The *complex conjugate* $\bar{x}$ of x has $(\bar{x})_{ij} = \overline{(x_{ij})}$, hence $\overline{xy} = \bar{x}\bar{y}$.

Furthermore, if $a \in GL(n, \boldsymbol{C})$, then ${}^t(a^{-1}) = ({}^t a)^{-1}$ and similarly for the other two operations. Here are some related properties of the matrix exponential map.

16. Lemma. If $x, y \in \mathrm{gl}(n, \boldsymbol{C})$ and $a \in GL(n, \boldsymbol{C})$, then

(1) $e^{x+y} = e^x e^y$ if $xy = yx$, (4) $\overline{e^x} = e^{\bar{x}}$,
(2) $e^{-x} = (e^x)^{-1}$, (5) $e^{axa^{-1}} = ae^x a^{-1}$,
(3) $e^{{}^t x} = {}^t(e^x)$, (6) $\det e^x = e^{\mathrm{trace}\, x}$.

Each of the great number systems $\boldsymbol{R}$ (reals), $\boldsymbol{C}$ (complexes), and $\boldsymbol{H}$ (quaternions) has for each n a natural n-dimensional geometry whose group of linear isometries is a Lie group.

(1) *The orthogonal group* $O(n)$ is the group of linear isometries of $\boldsymbol{R}^n$ with its natural inner product $v \cdot w = \sum v_i w_i$. A simple computation shows that $gv \cdot gw = v \cdot w$ for all v, w is equivalent to ${}^t g = g^{-1}$. Hence $O(n) = \{g \in GL(n, \boldsymbol{R}) : {}^t g = g^{-1}\}$.

(2) *The unitary group* $U(n)$ is the group of $\boldsymbol{C}$-linear isometries of $\boldsymbol{C}^n$ with its natural Hermitian product $(v, w) = \sum v_i \bar{w}_i$. It follows that $U(n) = \{g \in GL(n, \boldsymbol{C}) : {}^t \bar{g} = g^{-1}\}$.

(3) *The symplectic group* $Sp(n)$ is the group of $\boldsymbol{H}$-linear isometries of $\boldsymbol{H}^n$ with natural symplectic product [Ch]. Since quaternion multiplication is not commutative, the equivalent definition $Sp(n) = \{g \in U(2n) : {}^t gJ = Jg^{-1}\}$ may be preferred, where $J = \left(\begin{smallmatrix} 0 & -I \\ I & 0 \end{smallmatrix}\right)$.

Because the operations defining them are continuous, these are closed subgroups of $GL(n, \boldsymbol{C})$ and hence are Lie groups.

17. Lemma. The Lie groups above are all compact, and all except $O(n)$ are connected.

The compactness can be shown directly by verifying that each G is a closed bounded subset of $\mathrm{gl}(n, \boldsymbol{C}) \approx \boldsymbol{R}^{2n^2}$ for suitable n. Connectedness is shown in Chapter 11.

Using Lemmas 14 and 16, it is easy to compute the (matrix) Lie algebras of these groups.

(1) The Lie algebra $\mathfrak{o}(n)$ of the orthogonal group $O(n)$ consists of all $n \times n$ (real) skew-symmetric matrices:

$$\mathfrak{o}(n) = \{x \in \mathrm{gl}(n, \boldsymbol{R}) : {}^t x = -x\}.$$

(2) The Lie algebra $\mathfrak{u}(n)$ of the unitary group $U(n)$ consists of all (complex) skew-Hermitian matrices:

$$\mathfrak{u}(n) = \{x \in \mathfrak{gl}(n, \boldsymbol{C}) : {}^t\bar{x} = -x\}.$$

(3) The Lie algebra $\mathfrak{sp}(n)$ of the symplectic group $Sp(n)$ consists of matrices in $\mathfrak{gl}(2n, \boldsymbol{C})$ of the form $\left(\begin{smallmatrix} x & y \\ -\bar{y} & \bar{x} \end{smallmatrix}\right)$, where $x \in u(n)$ and ${}^t y = y$.

The dimensions of these Lie algebras are readily counted, and $\dim O(n) = \dim \mathfrak{o}(n) = n(n-1)/2$; $\dim U(n) = \dim \mathfrak{u}(n) = n^2$; $\dim Sp(n) = \dim \mathfrak{sp}(n) = 2n^2 + \text{n}$.

The three types respond differently when we take their special subgroups as in Example 4(2).

(1) *The special orthogonal group* $SO(n) = \{a \in O(n) : \det a = 1\}$. If $a \in O(n)$, then ${}^t a = a^{-1}$ implies $\det a = \pm 1$. Lemma 9.6 shows that $SO(n)$ is connected, hence it is the identity component of $O(n)$, and $\det a = -1$ gives the only other component. Since $O(n)$ and $SO(n)$ have common neighborhoods of e, remarks above show that they have the same Lie algebra: $\mathfrak{so}(n) = \mathfrak{o}(n)$.

(2) *The special unitary group* $SU(n) = \{a \in U(n) : \det a = 1\}$. The Lie algebra is $\mathfrak{su}(n) = \{x \in \mathfrak{u}(n) : \text{trace } x = 0\}$. Thus $\dim SU(n) = \dim \mathfrak{su}(n) = n^2 - 1$.

(3) Every element of Sp(n) has determinant 1.

Note that $U(1)$ is the unit circle in the complex plane, and $SO(2) \approx U(1)$.

The matrix descriptions above are convenient, but invariant descriptions are sometimes more natural in applications. For example, if V is an n-dimensional real vector space, then the vector space $\mathfrak{gl}(V)$ of all linear operators on V becomes a Lie algebra under $[A, B] = AB - BA$. The invertible elements of $\mathfrak{gl}(V)$ form a group $GL(V)$ under composition of functions. Assigning to each operator its matrix relative to some fixed basis for V gives a Lie algebra isomorphism $\mathfrak{gl}(V) \approx \mathfrak{gl}(n, \boldsymbol{R})$ and a group isomorphism $GL(V) \approx GL(n, \boldsymbol{R})$, the latter making $GL(V)$ a Lie group, with $\mathfrak{gl}(V)$ its Lie algebra.

C NEWTONIAN GRAVITATION

For the sake of comparison with the relativistic version in Chapter 13, we outline briefly the Newtonian description of planetary motion. As in Chapter 13, geometric units are used (Remark 6.7).

A mass M is located at the origin of Euclidean space $\boldsymbol{R}^3$, and α is a particle of mass $m \ll \mathrm{M}$ in $\boldsymbol{R}^3$. By Newton's law of gravitation, the central mass exerts a force $F = -(\mathrm{M}m/r^2)U$ on α where $r = |\alpha|$ and U is the outward radial unit vector. By canonical identification, $U = \alpha/r$. By Newton's second law of motion, $F = m\alpha''$, where primes indicate derivatives relative to Newtonian time. Actually, two different notions of the mass of α appear above; it is assumed that they have the same value m. Less subtly, since $m \ll \mathrm{M}$, we ignore the motion of the mass M. Then

(1) $\alpha'' = -(\mathrm{M}/r^3)\alpha.$

The vector field $\vec{L} = \alpha \times \alpha'$ is the *angular momentum vector of α per unit mass.*

(2) $\vec{L}$ is parallel, hence can be identified with a point of $\boldsymbol{R}^3$. If $\vec{L} \neq 0$, then α lies in a plane through the origin and does not pass through the origin.

Proof. $\vec{L}' = (\alpha \times \alpha')' = \alpha' \times \alpha' + \alpha \times \alpha'' = 0$, since α and α'' are collinear. Also $\alpha \cdot L = \alpha \cdot (\alpha \times \alpha') = 0$, and obviously $\alpha(t) = 0$ for some t implies $\vec{L} = 0$. ∎

If $\vec{L} = 0$, then α lies in a line through the origin, hence we can always assume that α lies in the xy-plane of $\boldsymbol{R}^3$. Then the *angular momentum of α per unit mass* is the number L such that $\vec{L} = L\,\partial_z$.

Shifting to polar coordinates replaces (1) by

(3) $r'' - r\varphi'^2 = -\mathrm{M}/r^2, \qquad r\varphi'' + 2r'\varphi' = 0.$

The second equation here shows that $r^2\varphi'$ is constant. In fact a polar computation of L gives

(4) $r^2\varphi' = L$ (*Kepler's second law*).

Assume $L \neq 0$, so r and φ' are never 0. Then the substitution $u = 1/r$ transforms the first equation in (3) to

(5) $d^2u/d\varphi^2 + u = \mathrm{M}/L^2$ (*the orbit equation*).

This has general solution $u = \mathrm{M}/L^2 + A\cos(\varphi - \varphi_0)$. By rotation of coordinates, we can arrange that $\varphi_0 = 0$ and $A \geq 0$. Then resubstituting $u = 1/r$ gives

(6) $$r = \frac{L^2/\mathrm{M}}{1 + e\cos\varphi}, \quad \text{where } e = AL^2/\mathrm{M} \geq 0.$$

Consulting an analytic geometry book we conclude that

(7) If $L \neq 0$, the particle α parametrizes a conic section of eccentricity e with a focus at the origin (*Kepler's first law*).

This conic section is the *orbit* of α: an ellipse ($0 \leq e < 1$), parabola ($e = 1$), or hyperbola ($e > 1$).

Recall that if $X \in \mathfrak{X}(\mathscr{U})$ is a force field on a connected open set $\mathscr{U}$ in $\boldsymbol{R}^3$, and $V \in \mathfrak{F}(\mathscr{U})$ is a function such that $X = -\text{grad } V$, then X is said to be *conservative* and V is a *potential function* for X. In this case, if α lies in $\mathscr{U}$, its *total mechanical energy* $\mathscr{E}$ is the sum of its *kinetic energy* $m\alpha' \cdot \alpha'$ and its *potential energy* $mV(\alpha)$. A simple computation verifies that $\mathscr{E}$ is constant. Henceforth we deal with $E = \mathscr{E}/m$, the *total energy per unit mass* of α.

(8) On $\boldsymbol{R}^3 - 0$ the gravitational force field $F = -(\mathrm{M}/r^2)\,\partial_r$ is conservative, and its potential function such that $V(\infty) = 0$ is $V = -\mathrm{M}/r$.

Then in polar terms:

(9) $2E = r'^2 + L^2/r^2 - 2\mathrm{M}/r$ (*the energy equation*).

Proof. We can suppose $m = 1$. In polar coordinates the kinetic energy $\alpha' \cdot \alpha'/2$ is $(r'^2 + r^2\varphi'^2)/2$. By (4), this becomes $\frac{1}{2}(r'^2 + (L^2/r^2))$. Since $V(\alpha) = -\mathrm{M}/r$, the result follows. ∎

The energy equation can also be interpreted as follows. If $r = r(t)$ is regarded as the position of a unit mass particle moving on the half-line $\boldsymbol{R}^+$, then its kinetic energy is $\frac{1}{2}r'^2$. Equation (9) will express conservation of its energy if we introduce the artificial potential energy $V(r)/2$, where now $V(r) = L^2/r^2 - 2\mathrm{M}/r$. In fact, (9) gives $E = \frac{1}{2}r'^2 + \frac{1}{2}V(r)$, and since $r'^2 \geq 0$, $E \geq V(r)/2$. Plot the graph of $V(r)/2$; then drawing horizontal lines at various

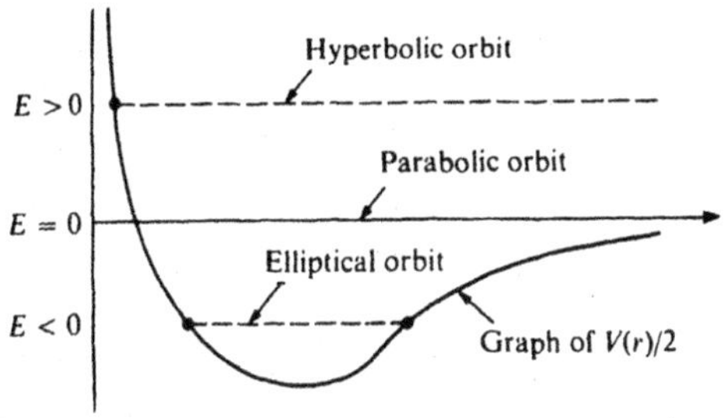

Figure C1

heights E, as in the figure, shows how E determines the range of r. Evidently $E < 0$ for elliptical orbits since then r oscillates between finite endpoints, and $E > 0$ for hyperbolic orbits where incoming r bounces off the graph to return to infinity.

(10) If a is the semimajor axis of an elliptical orbit, then $a(1 - e^2) = L^2/\text{M}$.

Proof. By (6), $r_{\min} = L^2/\text{M}(1 + e)$, but by the geometry of ellipses, $r_{\min} = a(1 - e)$. ∎

(11) If T is the time for one elliptical orbit, then $\text{M}T^2 = 4\pi^2 a^3$ (*Kepler's third law*).

Proof. By (4), the rate at which α sweeps out area is $L/2$. Hence the area enclosed is $TL/2$. By the geometry of ellipses, this area is $\pi a^2(1 - e^2)^{1/2}$. Thus, using (10),

$$T^2L^2/4 = \pi^2 a^4(1 - e^2) = \pi^2 a^3 L^2/\text{M}. \qquad ∎$$

REFERENCES

Following the works specifically referred to in the text, a few general references are listed.

[A] Ambrose, W. Parallel translation of Riemannian curvature, *Ann. of Math.* **64** (1956), 337–363.

[BE] Beem, J. K., and P. E. Ehrlich. *Global Lorentzian Geometry*, Dekker, New York, 1981.

[BG] Bishop, R. L., and S. I. Goldberg. *Tensor Analysis on Manifolds*, Dover, New York, 1980.

[BO] Bishop, R. L., and B. O'Neill. Manifolds of negative curvature, *Trans. Amer. Math. Soc.* **145** (1969), 1–49.

[CE] Cheeger, J., and D. G. Ebin. *Comparison Theorems in Riemannian Geometry*, North-Holland Publ., Amsterdam, 1975.

[Ch] Chevalley, C. *Theory of Lie Groups*, Princeton Univ. Press, Princeton, New Jersey, 1946.

[E] Einstein, A., *et al. The Principle of Relativity*, Dover, New York, 1952.

[G] Geroch, R. P. Domain of dependence, *J. Math. Phys.* **11** (1970), 437–449.

[GKM] Gromoll, D., W. Klingenberg, and W. Meyer. *Riemannsche Geometrie im Grossen*, Springer-Verlag, Berlin and New York, 1968.

[GP] Guillemin, V., and A. Pollack. *Differential Topology*, Prentice-Hall, Englewood Cliffs, New Jersey, 1974.

[H] Helgason, S. *Differential Geometry, Lie Groups, and Symmetric Spaces*, Academic Press, New York, 1978.

[HE] Hawking, S. W., and G. F. R. Ellis. *The Large Scale Structure of Space–time*, Cambridge Univ. Press, London and New York, 1973.

[KN] Kobayashi, S., and K. Nomizu. *Foundations of Differential Geometry*, Wiley (Interscience), New York, Vol. I, 1963; Vol. II, 1969.

[L] Lang, S. *Introduction to Differentiable Manifolds*, Wiley (Interscience), New York, 1962.

[M] Marcus, L. *Cosmological Models in Differential Geometry* (mimeographed notes), University of Minnesota, Minneapolis, 1963.

[Ma] Massey, W. S. *Algebraic Topology: An Introduction*, Springer-Verlag, New York and Berlin, 1977.

[Mat] Matsushima, Y. *Differentiable Manifolds*, Dekker, New York, 1972.

[Mi] Milnor, J. *Morse Theory*, Princeton Univ. Press, Princeton, New Jersey, 1963.

[MTW] Misner, C. W., K. S. Thorne, and J. A. Wheeler. *Gravitation*, Freeman, San Francisco, 1973.

[O1] O'Neill, B. *Elementary Differential Geometry*, Academic Press, New York, 1966.

[O2] O'Neill, B. The fundamental equations of a submersion, *Michigan Math. J.* **13** (1966), 459–469.

[P] Palais, R. S. A global formulation of the Lie theory of transformation groups, *Mem. Amer. Math. Soc.* **22** (1957).

[Sp] Spivak, M. *A Comprehensive Introduction to Differential Geometry*, Vols. I–V, Publish or Perish, Berkeley, California, 1970, 1975.

[ST] Singer, I. M., and J. A. Thorpe. *Lecture Notes on Elementary Topology and Geometry*, Springer-Verlag, New York and Berlin, 1976.

[St] Steenrod, N. *Topology of Fibre Bundles*, Princeton Univ. Press, Princeton, New Jersey, 1951.

[SW] Sachs, R. K., and H. Wu. *General Relativity for Mathematicians*, Springer-Verlag, New York and Berlin, 1977.

[TW] Taylor, E. F., and J. A. Wheeler. *Spacetime Physics*, Freeman, San Francisco, 1966.

[V] Vick, J. W. *Homology Theory*, Academic Press, New York, 1973.

[W] Warner, F. W. *Foundations of Differential Manifolds and Lie Groups*, Scott, Foresman, Glenview, Illinois, 1971.

[Wo] Wolf, J. A. *Spaces of Constant Curvature*, Publish or Perish, Berkeley, California, 1977.

Bishop, R. L., and R. J. Crittenden. *Geometry of Manifolds*, Academic Press, New York, 1964.

Boothby, W. *An Introduction to Differentiable Manifolds and Riemannian Geometry*, Academic Press, New York, 1975.

Chern, S. S. Pseudo-Riemannian geometry and Gauss–Bonnet formula, *Ann. Acad. Brasil Ciênc.* **35** (1963), 17–26.

Frankel, T. *Gravitational Curvature*, Freeman, San Francisco, 1979.

Geroch, R. P. Spacetime structure from a global viewpoint. In *General Relativity and Cosmology* (R. K. Sachs, ed.), Academic Press, New York, 1971.

Penrose, R. *Techniques of Differential Topology in Relativity*, Regional Conference Series in Applied Mathematics, Vol. 7, SIAM Publications, Philadelphia, 1972.

Smith, J. W. Lorentz structures on the plane. *Trans. Amer. Math. Soc.* **95** (1960), 226–237.

Thorpe, J. A. *Elementary Topics in Differential Geometry*, Springer-Verlag, New York and Berlin, 1979.

Weinberg, S. *Gravitation and Cosmology*, Wiley, New York, 1972.

Wu, H. Holonomy groups of indefinite metrics, *Pacific J. Math.* **20** (1967), 351–392.

INDEX

Printed and bound by CPI Group (UK) Ltd, Croydon, CR0 4YY
03/10/2024
01040422-0018